MIKE JORDAN
287-7651

INTRODUCTORY CALCULUS

INTRODUCTORY
CALCULUS

Second Edition, with Analytic Geometry and Linear Algebra

A. WAYNE ROBERTS

Macalester College

ACADEMIC PRESS   New York and London

ACADEMIC PRESS, INC.
111 Fifth Avenue, New York, New York 10003

United Kingdom Edition published by
ACADEMIC PRESS, INC. (LONDON) LTD.
24/28 Oval Road, London NW1 7DD

LIBRARY OF CONGRESS CATALOG CARD NUMBER: 72-182601

AMS(MOS) 1970 Subject classifications: 26-01, 26A06

PRINTED IN THE UNITED STATES OF AMERICA

To all my D's

CONTENTS

PREFACE TO THE SECOND EDITION

A chemist must be familiar with the periodic chart of the elements, but we doubt if very many chemists chose their field because they found the chart so interesting. In the same way, certain essentials, such as limits and the properties of real numbers, must be mastered if one is to understand calculus. We doubt, however, that a study of these rather sophisticated notions will be initially responsible for attracting very many people to the study of calculus. In our view, calculus should be presented in a way that appeals to the student's intuitive feeling of what ought to be. This book represents such an effort.

We have wished to sacrifice neither accuracy nor honesty in our attempt to be interesting. In fact, the definitions we give for function, graph, linear, derivative, and other terms used in advanced mathematics differ from those ordinarily given in elementary texts in that they do not need revision later on. Where a proof is not given, its omission is clearly pointed out. Often an example is given to show why careful attention to a proof is warranted in a future course.

We said in the preface to the first edition that a conscious effort had been made to reverse the trend of calculus texts to become encyclopedic. One concession has been made by including in this edition an introductory chapter on elementary analytic geometry (and a concession of another type was made to the editor when we agreed not to title this chapter *The Plane Facts*). We have also included a separate section to discuss limits more fully. Nevertheless, we have adhered to the principle that conciseness should take precedence over the temptation to say everything one can think of about the subject at hand.

The first edition stressed the role of the derivative as a tool of approximation by defining the derivative to be a *linear* transformation. Though the spirit of this idea is retained in the second edition, formal terminology (and notation toward this end) has been left out of the treatment of one variable except for one optional section. Thus, for a traditional course in one variable calculus, this notion can be ignored by those who choose to do so.

Keeping in mind the needs of increasing numbers of social science students appearing in the mathematics classroom, we have continued the emphasis on functions of several variables. Here there are clear advantages to thinking of the derivative as a *linear* transformation, and the presentation is set in this context. The background in linear algebra necessary to this approach, somewhat scattered through the first edition, has in this edition been collected into a separate chapter. It is to be stressed that we have included the linear algebra here for the same reason that it was included in the first edition; not because it is the *modern* thing to do so, but because it plays an essential role in what is to follow.

The opening chapter on analytic geometry begins with a discussion of vectors in the plane. By using vectors to develop some of the results of analytic geometry, we hope to avoid the deadening effect that this material might otherwise have cn those for whom it is essentially a review. We also prepare the way for using the same definitions and methods in the later chapter on linear algebra.

We distinguish between problems and exercises in this text. Problems are scattered throughout the text and are to be attempted by every reader. Their purpose is to clinch ar idea of a previous paragraph or to prepare the way for what is to come. They should be attempted immediately when they are met in reading. Because they form an important part of the text, complete solutions are given in the rear of the book. The reader should not be discouraged if he finds the problems very challenging; some are given more in the spirit of raising questions than in the expectation of getting correct answers.

Exercises occur at the end of most sections. In the nature of drill problems, they are arranged in what seems to the author to be ascending order of difficulty. Some sections include starred exercises at the end which are more difficult and serve to extend the abilities of the more able students. Exercises are arranged so that the same skills are required by the even-numbered problems (which have no answers given at the rear) and by the odd-numbered ones (for which the answers are given). Some sections and their corresponding exercise lists from the first edition have been broken into two parts so as to correspond more directly with the amount of material that might normally be covered in one lecture hour.

This text is intended for a two- or three-semester course in introductory calculus. The student is expected to have proficiency at high school algebra and trigonometry, though specific topics are reviewed in the text where experience suggests this to be helpful. Chapter 1 includes the analytic geometry so many missed in the first edition. It may be used with an emphasis suited to the preparation of the class. The heart of a traditional one variable calculus course is contained in Chapters 2–8 and 14 with Chapters 9 and 15 being optional. A third-semester course focusing on functions of several variables would include Chapters 10–13 with Chapters 9 and 15 again optional. In that Chapter 15 has been rewritten to take advantage of the linear algebra terminology, it would be understood more readily by a student having the background of Chapter 10.

Once again I wish to take the opportunity afforded by a Preface to thank the students and faculty at Macalester College who have now made substantial contributions to two editions of this text. Special thanks are due to Carol Svoboda who helped with many of the mechanical details of the second edition, checked all the answers in the back of the book, and read the galley proof of the entire book. Finally, it is a pleasure to acknowledge again the helpfulness and good spirit of the staff of Academic Press.

ACKNOWLEDGMENTS

The quotations appearing in this book are printed with the permission of the following publishers.

Chapter 2, page 44: Phillip E. B. Jourdain, "The Nature of Mathematics," *The World of Mathematics* (James R. Newman, ed.). New York: Simon and Schuster, 1956, p. 39.

Chapter 5, page 136: Giorgio de Santillana, *The Crime of Galileo*. New York: Time, Inc., Book Division, 1962, p. 18. Reprinted courtesy University of Chicago Press.

Chapter 12, page 408: J. A. Dieudonné, *Foundations of Modern Analysis*. New York: Academic Press, 1960.

Chapter 15, page 508: Albert Einstein, *Ideas and Opinions*. New York: Crown Publishers, Inc., p. 233.

SOME ANALYTIC GEOMETRY

Everyone has problems, but not everyone is happy about having problems. A mathematician is always happy to have a good problem. A good problem is one that stimulates the imagination, relates to other problems that are topics of lively investigation, and seems to be possible to solve by mental as opposed to physical effort.

We distinguish between problems and exercises in this text. Problems are scattered throughout the text and are to be attempted by every reader. Their purpose is to clinch an idea of a previous paragraph or to prepare the way for what is to come. They should be attempted immediately when they are met in reading. Because they form an important part of the text, complete solutions are given in the rear of the book. The reader should not be discouraged if he finds the problems very challenging; some are given more in the spirit of raising questions than in the expectation of getting correct answers.

1·1   What Is Analytic Geometry?

Points may be located in a plane with respect to two mutually perpendicular lines in the following way: Draw a vertical and a horizontal line, calling their intersection the *origin*. Mark off equal divisions on both of these lines. It is common to call the vertical axis the y axis, the horizontal axis the x axis. Number the divisions on each axis, using positive numbers to the right and above the origin, negative numbers to the left and below the origin. In this way it is possible to specify a certain point in the plane by means of an ordered pair (x, y) of numbers. It is usual to call the first number given (always the distance from the vertical axis) the *abscissa*. The second number, representing the distance from the horizontal axis, is called the *ordinate*.

In Fig. 1.1, $(-2, 3)$ locates the indicated point. The abscissa is -2; the ordinate is 3. Note that $(3.1416, 7)$ and $(\pi, 7)$ are different points even though

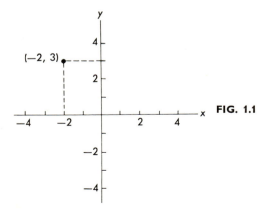

FIG. 1.1

the distinction is one that we will have to make in our minds, since the scale chosen on the axes is usually too small to permit us to see any difference in the plotted points.

This correspondence between ordered pairs of numbers and points in a plane affords us two ways to describe a certain set. We may give a geometric description, or we may give an analytic description by specifying a relation between the values of x and y. For instance, we know from the theorem of Pythagoras that the set described geometrically as a circle of radius 2 centered at the origin may also be described by the relationship $x^2 + y^2 = 4$ (Fig. 1.2). Geometric pictures of several other sets, together with analytic descriptions of the sets, are given in Fig. 1.3 and 1.4.

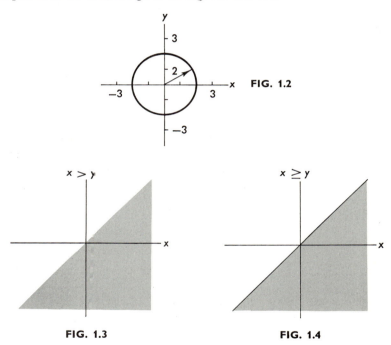

FIG. 1.2

FIG. 1.3   FIG. 1.4

The study that relates analytic descriptions of a set to the corresponding geometric figure in the plane is called *analytic geometry*. It addresses itself to two principal questions:

(1) Given an equation or an inequality involving two variables x and y, can we give a description of the corresponding set in the plane?

(2) Given a geometric description of a set in the plane, can we find a corresponding equation or inequality?

In the spirit of our comments above, we illustrate these general questions with two problems.

PROBLEMS

A · Sketch the set of all points (x, y) satisfying $x^2 - y^2 = 0$.

B · Let S be the set of all points (x, y) that are equidistant from the y axis and the point $(2, 0)$. Find an equation relating x to y for all (x, y) in S.

1·2 Vectors in the Plane

We wish to discuss a problem involving two tractors being used to pull a tree stump from the ground. One tractor is larger, exerting twice as much force as the other. Both tractors are secured to the stump by cables, and the angle between the cables is 40°. Our problem is to determine the actual force being exerted on the stump, and the direction of this force. (Alternatively, we are to determine how to replace the two tractors with one tractor. With what force must the new tractor pull, and in what direction?)

The situation is summarized in Fig. 1.5. Note that the arrow representing

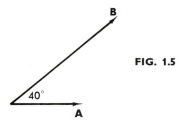

FIG. 1.5

tractor B is twice as long as that of A, corresponding to the information that one tractor exerts twice the force of the other. The picture also suggests a geometric solution often taught to students in an elementary physics course. The procedure is to draw a parallelogram using the given arrows as two adjacent sides. Then draw, starting at the common initial point of the given arrows, the diagonal of the parallelogram (Fig. 1.6). This diagonal, called the

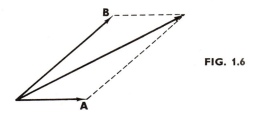

FIG. 1.6

resultant, indicates the actual force being exerted upon the stump, and the direction in which this force acts.

Apparently it was Newton who first noticed that the results obtained in constructing the resultant correspond to what actually happens. Thus, the use of arrows in this and a variety of other problems is a valuable tool in applied mathematics. These arrows, useful because they represent both direction and magnitude (by their length), are called *vectors*. Before proceeding with a formal presentation of vectors, we pause to state concisely what we have learned about the use of vectors in applications.

(1) Addition Given two vectors having the same initial point, their sum vector (also called the resultant vector) is obtained by drawing from this common initial point the diagonal of the parallelogram having the two given vectors as adjacent sides.

(2) Multiplication by a Real Number Given a vector v and a real number r,

> for $r > 1$, rv is a vector longer than v;
> for $r \in (0, 1)$, rv is a vector shorter than v;
> for $r = 0$, $0v = 0$, the zero vector;
> for $r < 0$, rv is a vector with the same length as $|r|v$ but pointing in the opposite direction. (We say in this case that the vectors have opposite sense.)

(3) Equality of Vectors In our problem concerning the removal of a tree stump, the effect on the stump should be the same whether the tractors pull or push. In fact it should not matter if one pushes while the other pulls. Such considerations make it useful in many applications to regard as equal any two vectors that:

> (a) are parallel;
> (b) have the same sense (point in the same direction);
> (c) have the same length.

Geometrically we think of two vectors v and w as equal if we can slide v, always keeping it parallel to its original position, so that it will coincide with w. The idea of being able to "slide vectors through parallel displacements" without changing them gives us alternative methods to represent $v + w$ and $v - w$. Draw v anywhere. Beginning at the terminal point of v, draw w. Then the vector drawn from the initial point of v to the terminal point of w represents $v + w$ (Fig. 1.7). If v and w are drawn from the same initial point, then the

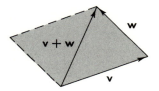

FIG. 1.7

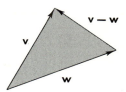

FIG. 1.8

vector drawn from the terminal point of **w** to the terminal point of **v** must be **v** − **w** (Fig. 1.8). (We see this clearly when we note that according to the addition pictured in Fig. 1.7, the vector we labeled **v** − **w** when added to **w** must give **v**.)

Guided by these considerations, we now set about representing vectors in a plane with respect to a set of coordinate axes. We begin by giving names to two special vectors.

> **i** is the horizontal vector of length 1 pointing in the direction of the positive x axis.
> **j** is the vertical vector of length 1 pointing in the direction of the positive y axis.

Since parallel vectors having the same sense and length are to be regarded as equal, the vectors **i** and **j** are represented by any of the arrows indicated in Fig. 1.9. In Fig. 1.10 we have used the rule for addition of vectors to indicate the vectors $\mathbf{v} = 2\mathbf{i} - \frac{3}{2}\mathbf{j}$, and $\mathbf{w} = -2\mathbf{i} + \frac{3}{2}\mathbf{j}$. Notice that $\mathbf{v} = 2\mathbf{i} - (\frac{3}{2})\mathbf{j} = 2\mathbf{i} + (-\frac{3}{2})\mathbf{j}$.

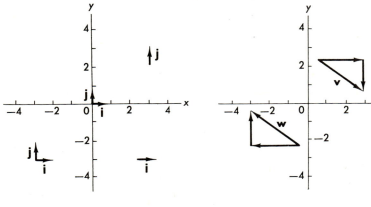

FIG. 1.9   **FIG. 1.10**

It is now clear that any vector $\mathbf{v}$ may be represented in the form $\mathbf{v} = a\mathbf{i} + b\mathbf{j}$. In particular, suppose the vector $\mathbf{v}$ has $P_1(x_1, y_1)$ as its initial point, $P_2(x_2, y_2)$ as its terminal point. Then the vector $\mathbf{v}$, also designated by $\mathbf{P_1P_2}$ (or $\overrightarrow{P_1P_2}$ when boldface is unavailable), is (Fig. 1.11)

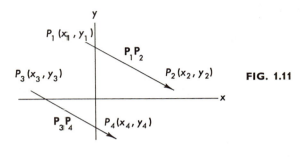

FIG. 1.11

$$\mathbf{v} = \mathbf{P_1P_2} = (x_2 - x_1)\mathbf{i} + (y_2 - y_1)\mathbf{j}.$$

If $\mathbf{w}$ is a second vector, this one having initial point $P_3(x_3, y_3)$ and terminal point $P_4(x_4, y_4)$. Then

$$\mathbf{w} = \mathbf{P_3P_4} = (x_4 - x_3)\mathbf{i} + (y_4 - y_3)\mathbf{j},$$

and it is clear from what we know about congruent triangles that $\mathbf{v} = \mathbf{w}$ (that is, $\mathbf{v}$ and $\mathbf{w}$ are parallel, having the same sense and length) if and only if

$$(x_2 - x_1) = (x_4 - x_3),$$
$$(y_2 - y_1) = (y_4 - y_3).$$

This suggests that all the concepts about which we have been talking might be defined analytically.

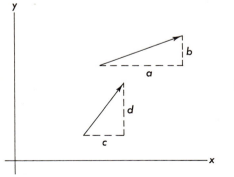

FIG. 1.12

Let **v** and **w** be two vectors in the plane. We have seen that they may be written in the form (see Figure 1.12)

$$\mathbf{v} = a\mathbf{i} + b\mathbf{j},$$
$$\mathbf{w} = c\mathbf{i} + d\mathbf{j}.$$

DEFINITION A

(Equality) $\mathbf{v} = \mathbf{w}$ if and only if $a = c$, $b = d$.
(Scalar Multiplication) $r\mathbf{v} = ra\mathbf{i} + rb\mathbf{j}$ (r real).
(Addition) $\mathbf{v} + \mathbf{w} = (a + c)\mathbf{i} + (b + d)\mathbf{j}$.

With **i** and **j** representing the vectors of unit length along the coordinate axes, we find the following definition helpful.

DEFINITION B   (Length)

The length of vector $\mathbf{v} = a\mathbf{i} + b\mathbf{j}$ is $|\mathbf{v}| = \sqrt{a^2 + b^2}$.

The length of the vector $\mathbf{P_1 P_2}$ according to this definition corresponds to the usual definition for the distance between the points P_1 and P_2. A *unit vector* is a vector of length one. The vector **0** having the property that $\mathbf{0} + \mathbf{v} = \mathbf{v}$ for all **v** is called the *zero* vector. Note that $0\mathbf{v} = \mathbf{0}$.

If an arbitrary nonzero vector **v** is multiplied by the scalar $1/|\mathbf{v}|$, the resulting vector will be a unit vector parallel to **v**. Since two parallel nonzero vectors **v** and **w** must be scalar multiples of the same unit vector, it is clear that **v** and **w** will be parallel if and only if there is a scalar r such that $\mathbf{v} = r\mathbf{w}$. Such vectors are said to be *linearly dependent*.

Note that if $\mathbf{v} = a\mathbf{i} + b\mathbf{j}$, and $\mathbf{w} = c\mathbf{i} + d\mathbf{j}$ are linearly dependent, then $a = rc$ and $b = rd$. If a and c are nonzero, then $b/a = d/c$. This ratio is called the *slope* of the vector. Thus, vector $\mathbf{v} = a\mathbf{i} + b\mathbf{j}$ has slope b/a. We have seen that vectors are parallel if and only if they have the same slope. Vectors for which the slope is undefined (that is, where the coefficient of **i** is 0) are said to be vertical.

PROBLEMS

A · Find a unit vector pointing in the direction of $\mathbf{v} = 3\mathbf{i} + 5\mathbf{j}$.

B · Show that $(1, -9)$, $(3, -4)$, and $(7, 6)$ are collinear.

There is a third definition that will be most useful to us. We shall motivate it by an example, but we need to pause to recall from trigonometry the law of cosines. This is the formula that enables one to determine the angles of a triangle if the sides are given. Specifically, with notation chosen as in Fig. 1.13,

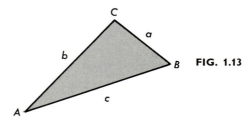

FIG. 1.13

the law of cosines says

$$a^2 = b^2 + c^2 - 2bc \cos A$$

EXAMPLE A

Given the two vectors $\mathbf{v} = a\mathbf{i} + b\mathbf{j}$ and $\mathbf{w} = c\mathbf{i} + d\mathbf{j}$, find the angle between them (Fig. 1.14).

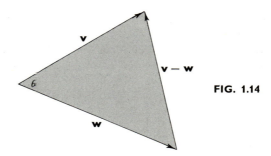

FIG. 1.14

From the law of cosines

$$|\mathbf{v} - \mathbf{w}|^2 = |\mathbf{v}|^2 + |\mathbf{w}|^2 - 2|\mathbf{v}||\mathbf{w}| \cos \theta,$$
$$(a - c)^2 + (b - d)^2 = a^2 + b^2 + c^2 + d^2 - 2|\mathbf{v}||\mathbf{w}| \cos \theta,$$
$$- 2ac - 2bd = - 2|\mathbf{v}||\mathbf{w}| \cos \theta,$$
$$\cos \theta = \frac{ac + bd}{|\mathbf{v}||\mathbf{w}|}. \quad \blacksquare \tag{1}$$

The expression in the numerator of (1) turns out to be very important in mathematics. For this reason it is given a special name.

DEFINITION C   (Dot Product)

Given two vectors $\mathbf{v} = a\mathbf{i} + b\mathbf{j}$ and $\mathbf{w} = c\mathbf{i} + d\mathbf{j}$, we define the dot product to be

$$\mathbf{v} \cdot \mathbf{w} = ac + bd.$$

With this definition, we may write (1) in the form

$$\cos \theta = \frac{\mathbf{v} \cdot \mathbf{w}}{|\mathbf{v}| |\mathbf{w}|} \tag{2}$$

where θ is the angle between the vectors $\mathbf{v}$ and $\mathbf{w}$.

PROBLEMS

C · A triangle has vertices $A(\frac{3}{2}, \frac{5}{2})$, $B(\frac{17}{2}, \frac{3}{2})$, and $C(\frac{9}{2}, -\frac{3}{2})$. Find $\angle ABC$.

D · Prove that for any vector $\mathbf{v} = a\mathbf{i} + b\mathbf{j}$, $\mathbf{v} \cdot \mathbf{v} = |\mathbf{v}|^2$.

We see immediately (2) that two nonzero vectors are perpendicular if and only if their dot product is zero (so the cosine of their included angle is zero). Note that $\mathbf{i} \cdot \mathbf{j} = 0$.

Given a vector to represent graphically, we may place its initial point anywhere. Since we often wish to specify that the initial point is to be taken at the origin, such a vector is given the name *radius vector*. Thus, the radius vector $a\mathbf{i} + b\mathbf{j}$ has (a, b) as its terminal point. This fact, together with the other properties of vectors, can be used to solve certain kinds of problems.

EXAMPLE B

Find the point of intersection of the medians of the triangle having as its vertices the points $A(-1, 4)$, $B(3, -3)$, and $C(4, 2)$.

A theorem from plane geometry informs us that the medians intersect in a common point two-thirds of the way along any median from the corresponding vertex. We begin by finding the coordinates of M. From Fig. 1.15, since M is the midpoint between A and C,

$$\mathbf{OM} = \mathbf{OA} + \tfrac{1}{2}\mathbf{AC}.$$

$\mathbf{OA}$ is a radius vector, so $\mathbf{OA} = -\mathbf{i} + 4\mathbf{j}$, $\mathbf{AC} = 5\mathbf{i} - 2\mathbf{j}$. (Caution: A common mistake is to get the negative result here. Remember that you subtract the coordinates of A from those of C.)

$$\mathbf{OM} = -\mathbf{i} + 4\mathbf{j} + \tfrac{1}{2}(5\mathbf{i} - 2\mathbf{j}) = \tfrac{3}{2}\mathbf{i} + 3\mathbf{j}.$$

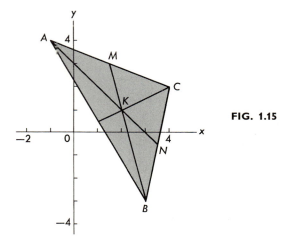

FIG. 1.15

Since **OM** is a radius vector, the coordinates of M are $(\frac{3}{2}, 3)$. This looks reasonable from our figure, so we proceed. (If we make the common mistake mentioned above, our answer will not even look reasonable. This is a good place to check on ourselves.)

We now use the theorem from plane geometry and the same ideas already employed to write

$$\mathbf{OK} = \mathbf{OB} + \tfrac{2}{3}\mathbf{BM},$$
$$\mathbf{OK} = 3\mathbf{i} - 3\mathbf{j} + \tfrac{2}{3}(-\tfrac{3}{2}\mathbf{i} + 6\mathbf{j}) = 2\mathbf{i} + \mathbf{j}.$$

Therefore K is at $(2, 1)$. ▌

The reader should verify his understanding of the preceding example by finding K from the relation

$$\mathbf{OK} = \mathbf{ON} + \tfrac{1}{3}\mathbf{NA}.$$

We may use our knowledge of vectors to prove some of the theorems of elementary geometry.

EXAMPLE C

Prove that the diagonals of a rhombus are perpendicular.

We take the base of the rhombus to be the radius vector $a\mathbf{i}$. Then another of the sides must be a radius vector, say $b\mathbf{i} + c\mathbf{j}$ (Fig. 1.16). Let C designate the vertex opposite the origin.

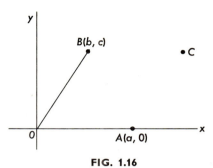

$$\mathbf{OC} = \mathbf{OB} + \mathbf{OA} = (a + b)\mathbf{i} + c\mathbf{j},$$
$$\mathbf{AB} = (b - a)\mathbf{i} + c\mathbf{j}.$$

Therefore,

$$\mathbf{OC} \cdot \mathbf{AB} = b^2 - a^2 + c^2,$$
$$= (b^2 + c^2) - a^2,$$
$$= |\mathbf{OB}|^2 - |\mathbf{OA}|^2.$$

FIG. 1.16

Since the sides of a rhombus have the same length, this says that the dot product of **OC** and **AB** is 0. The diagonals are perpendicular. ▌

1·2   **EXERCISES**

Exercises 1–14 all refer to the points $A(-3, 4)$, $B(3, 2)$, $C(-2, 2)$.

1· Write vector **AC** in the form $a\mathbf{i} + b\mathbf{j}$.

2· Write vector **AB** in the form $a\mathbf{i} + b\mathbf{j}$.

3· Find a unit vector pointing in the direction of **CA**.

4· Find a unit vector pointing in the direction of **BA**.

5· Find a point D so that $ABCD$ are the consecutive vertices of a parallelogram.

6· Find a point D so that **AC** and **BD** have the same sense and are opposite sides of a parallelogram.

7· Find the coordinates of a point two-thirds of the way from A to C.

8· Find the coordinates of a point three-quarters of the way from B to C.

9· The vector **AC** is placed so that its initial point is at $(1, 2)$. Where is the terminal point?

10· The vector **BC** is placed so that its initial point is at $(-3, 1)$. Where is the terminal point?

11· Find the acute angle of $\triangle ABC$ at A.

12· Find the acute angle of $\triangle ABC$ at B.

13· Find a unit vector that is perpendicular to **AC**.

14· Find a unit vector that is perpendicular to **AB**.

15· Show that $(2, 2)$, $(4, 1)$, and $(3, 4)$ are vertices of a right triangle.

16· Show that $(2, 2)$, $(0, 3)$, and $(-1, -4)$ are vertices of a right triangle.

Use vectors to prove the following theorems of plane geometry.

17. The midpoint of the hypotenuse of a right triangle is equidistant from the vertices.

18. The sum of the lengths of the bases of a trapezoid is twice the distance between the midpoints of the nonparallel sides.

19. The line segments joining the midpoints of the adjacent sides of an arbitrary quadrilateral form a parallelogram.

20. Any triangle inscribed in a semicircle must be a right triangle.

21. If the diagonals of a parallelogram are perpendicular, then the parallelogram must be a rhombus.

22. An isosceles trapezoid has equal diagonals.

23. The perpendicular bisectors of the sides of a triangle are concurrent.

24. In any triangle the sum of the squares of the medians is equal to three-fourths the sum of the squares of the three sides.

1·3   The Straight Line

Two points $P_1(x_1, y_1)$ and $P_2(x_2, y_2)$ determine a straight line. The *slope* m of such a line is defined to be the slope of the vector $\mathbf{P_1 P_2}$;

$$m = \frac{y_2 - y_1}{x_2 - x_1}.$$

Note that the slope is undefined if and only if the line is vertical. An analytic description of the line through P_1 and P_2 is easily determined. Choose any point $P(x, y)$ on the line (Fig. 1.17). Vector $\mathbf{P_1 P}$, being linearly dependent upon vector $\mathbf{P_1 P_2}$, has the same slope. Hence,

$$m = \frac{y - y_1}{x - x_1}.$$

This may be rewritten in the form

$$y - y_1 = m(x - x_1). \tag{1}$$

Using this form, we can immediately write the equation of a line if we know its slope and a point through which the line passes.

The point at which a straight line cuts the y axis, called the *y intercept*, is usually designated by $(0, b)$. If in (1) we set $(x_1, y_1) = (0, b)$, then (1) may be written in the form

$$y = mx + b.$$

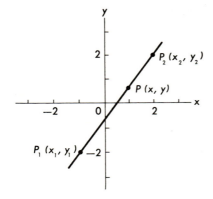

FIG. 1.17

PROBLEM

A · Write the equation of a straight line having a slope of $\frac{1}{2}$ and passing through $(-4, 1)$. Sketch a graph of the line. From the graph, determine the y intercept of the line. Knowing the slope and the y intercept, write the equation of the line in the form $y = mx + b$. You now have the equation of the same line written in two forms. Verify that they are equivalent.

Any equation of the form $Ax + By + C = 0$ can be put into the form $y = mx + b$ unless $B = 0$ (in which case $x = -C/A$, and the graph is a vertical line). Since this tells us the slope m and the place b where the line cuts the y axis, it is easy to graph any set of points (x, y) satisfying $Ax + By + C = 0$.

EXAMPLE A
Plot those points (x, y) that satisfy $3x + 5y = 12$.

It is a simple mental exercise to write (or even to think without writing)

$$y = -\tfrac{3}{5}x + \tfrac{12}{5}$$

This line intersects the y axis at $\frac{12}{5}$. The slope is $-\frac{3}{5}$. Count 5 units (of any convenient length) from the right of $(0, \frac{12}{5})$. Then go down 3 units (of the same length, of course). Alternately, go to the left 5 units and up 3 units. Sketch in the line (see Fig. 1.18). ∎

By using the method of the preceding example, we immediately find the y intercept. A closely related method for graphing a straight line is to rewrite the equation in the form

$$\frac{x}{a} + \frac{y}{b} = 1.$$

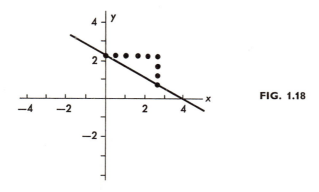

FIG. 1.18

It is easily seen now that the intercept on the x axis (where $y = 0$) is a, and the y intercept is b. The line is easily drawn, then, through these intercepts. The line of Example A, written in this form, is

$$\frac{x}{4} + \frac{y}{\frac{12}{5}} = 1$$

so the x intercept is 4 and the y intercept is (still) $\frac{12}{5}$.

We may summarize what we have learned about the straight line as follows.

Standard Form

$Ax + By + C = 0$ is the equation of a straight line.

Slope–Intercept Form

$y = mx + b$ is the equation of a straight line in a form where we may recognize the slope $m = s/r$ and the y intercept b (Fig. 1.19a).

Point–Slope Form

$y - y_0 = m(x - x_0)$ is the equation of the straight line that passes through (x_0, y_0) with a slope of $m = s/r$ (Fig. 1.19b).

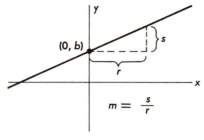

FIG. 1.19a   FIG. 19b

Intercept Form

$(x/a) + (y/b) = 1$ is the equation of the straight line that cuts the x axis at $x = a$, the y axis at $y = b$ (Fig. 1.19a).

Given a nonvertical line L with the standard equation $Ax + By + C = 0$, it is easily seen that its slope is $m = -A/B$. Thus, it is parallel to the vector $\mathbf{v} = B\mathbf{i} - A\mathbf{j}$ having the same slope.

A second line $A_1 x + B_1 y + C_1 = 0$ is similarly parallel to vector $B_1\mathbf{i} - A_1\mathbf{j}$. Then the two lines are parallel if and only if the two vectors are linearly dependent; that is, the lines are parallel if and only if their respective coefficients of x and y are proportional.

Returning to our line L with slope m, let us consider a line perpendicular to L meeting it at point $P_1(x_1, y_1)$. Note that vector $\mathbf{v} = \mathbf{i} + m\mathbf{j}$ is parallel to L (Fig. 1.20). Choosing a point P on the perpendicular line, we see that

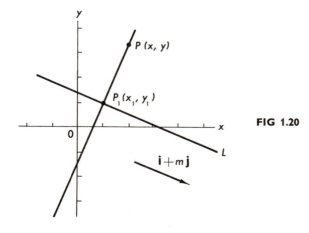

FIG 1.20

$\mathbf{P_1P} = (x - x_1)\mathbf{i} + (y - y_1)\mathbf{j}$ and $\mathbf{v}$ must satisfy $\mathbf{v} \cdot \mathbf{P_1P} = 0$. Hence, $(x - x_1)$ $+ m(y - y_1) = 0$, and the slope of vector $\mathbf{P_1P}$ is given by

$$\frac{y - y_1}{x - x_1} = -\frac{1}{m}.$$

In words, this says the slope of the perpendicular line is the negative reciprocal of the slope of line L.

EXAMPLE B

Find the equation of the perpendicular bisector of the line segment from $P_1(-3, -2)$ to $P_2(1, -4)$.

The line passes through the midpoint $M(-1, -3)$ of the given segment (Fig. 1.21). Since $\mathbf{P_1P_2} = 4\mathbf{i} - 2\mathbf{j}$ has a slope of $-2/4$, the slope of the desired line is 2. Using the point–slope form of the straight line,

$$y + 3 = 2(x + 1).$$

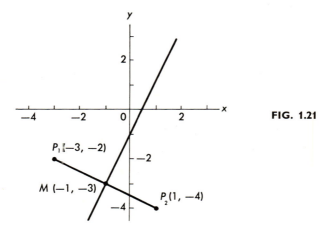

FIG. 1.21

It is instructive to note that we could also solve this problem directly with vector methods. Choose $P(x, y)$ on the desired line. Vector $\mathbf{P_1P} = (x + 1)\mathbf{i} + (y + 3)\mathbf{j}$ is perpendicular to $\mathbf{P_1P_2}$, so

$$4(x + 1) - 2(y + 3) = 0. \quad \blacksquare$$

EXAMPLE C

Find the acute angle formed by the intersection of the lines $y + 2x + 2 = 0$ and $3y + x - 9 = 0$.

The lines are respectively parallel to the vectors $\mathbf{v} = \mathbf{i} - 2\mathbf{j}$ and $\mathbf{w} = 3\mathbf{i} - \mathbf{j}$. The angle θ between the lines is determined from

$$\cos \theta = \frac{\mathbf{v} \cdot \mathbf{w}}{|\mathbf{v}||\mathbf{w}|} = \frac{3 + 2}{\sqrt{5}\sqrt{10}} = \frac{1}{\sqrt{2}}.$$

Thus, $\theta = \pi/4$. ∎

We wish now to find the distance from a point to a line. We begin with the special case in which the point is the origin. Since distance is measured along a perpendicular, we draw a perpendicular from the origin meeting the given line at $P_1(x_1, y_1)$ and making an angle of α with the positive x axis (Fig. 1.22).

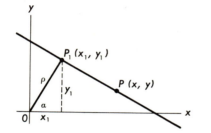

FIG. 1.22

Let ρ designate the distance from the origin to P_1. Elementary trigonometry tells us that

$$\begin{aligned} x_1 &= \rho \cos \alpha, \\ y_1 &= \rho \sin \alpha. \end{aligned} \tag{2}$$

Choosing any $P(x, y)$ on the given line, we see that the vectors

$$\mathbf{P_1 P} = (x - x_1)\mathbf{i} + (y - y_1)\mathbf{j},$$

and

$$\mathbf{u} = (\cos \alpha)\mathbf{i} + (\sin \alpha)\mathbf{j},$$

are perpendicular, so their dot product is zero:

$$(x - x_1)\cos \alpha + (y - y_1)\sin \alpha = 0,$$
$$(\cos \alpha)x + (\sin \alpha)y = x_1 \cos \alpha + y_1 \sin \alpha.$$

Using the expressions (2) for x_1 and y_1 together with the trigonometric identity $\cos^2 \alpha + \sin^2 \alpha = 1$,

$$(\cos \alpha)x + (\sin \alpha)y = \rho.$$

This is called the *normal form* of a straight line.

To put $Ax + By + C = 0$ into normal form, we need to find a unit vector **u** that is perpendicular to the line. Since vector $\mathbf{v} = A\mathbf{i} + B\mathbf{j}$ is perpendicular to the line, it is clear that **u** may be chosen as either

$$u = \pm\,\frac{\mathbf{v}}{|\mathbf{v}|} = \frac{A}{\pm\sqrt{A^2 + B^2}}\,\mathbf{i} + \frac{B}{\pm\sqrt{A^2 + B^2}}\,\mathbf{j}.$$

Thus, to obtain the desired form of the equation, we divide by $\pm\sqrt{A^2 + B^2}$, obtaining

$$\frac{A}{\pm\sqrt{A^2 + B^2}}\,x + \frac{B}{\pm\sqrt{A^2 + B^2}}\,y = \frac{-C}{\pm\sqrt{A^2 + B^2}}.$$

Ambiguity is removed by agreeing to choose the sign so as to make the coefficient of y (the sine of α) positive. This limits $\alpha \in (0, \pi)$ for nonvertical lines, and gives $\rho > 0$ for lines above the origin, $\rho < 0$ for lines below the origin.

It is now easy, given a line, to determine its distance from the origin. But we began with the more ambitious goal of finding the distance between any point and any line. This is now possible, however, as an easy consequence of our work.

EXAMPLE D
Find the distance from the point $(-3, 1)$ to the line $3x - 4y - 10 = 0$.

In this case, we divide the equation of the given line by $-\sqrt{A^2 + B^2} = -\sqrt{9 + 16} = -5$ to make the coefficient of y positive. This gives $-\frac{3}{5}x + \frac{4}{5}y = -2$. The graph is indicated in Fig. 1.23.

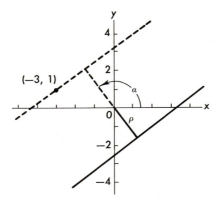

(-3, 1)

FIG. 1.23

Now let us pass a line through $(-3, 1)$ that is parallel to the given line. It has an equation of the form $3x - 4y = D$, and we easily see that $D = -13$ by noting that $(-3, 1)$ is to be on the line. Normal form for this line is $-\frac{3}{5}x + \frac{4}{5}y = \frac{13}{5}$. We now see that the distance from the given point to the given line is $\frac{13}{5} - (-2) = \frac{23}{5}$. ∎

The method of the previous example may be used to obtain a general formula for the distance d from a point (x_1, y_1) to a line $Ax + By + C = 0$. The line parallel to the given line and through (x_1, y_1) is $Ax + By = Ax_1 + By_1$. Putting both equations into normal form gives

$$\frac{A}{\pm\sqrt{A^2 + B^2}}x + \frac{B}{\pm\sqrt{A^2 + B^2}}y = \frac{Ax_1 + By_1}{\pm\sqrt{A^2 + B^2}},$$

$$\frac{A}{\pm\sqrt{A^2 + B^2}}x + \frac{B}{\pm\sqrt{A^2 + B^2}}y = \frac{-C}{\pm\sqrt{A^2 + B^2}},$$

where the ambiguous sign is chosen the same way in both equations. Subtracting as in Example D,

$$d = \frac{Ax_1 + By_1 + C}{\pm\sqrt{A^2 + B^2}}.$$

Reflection (or perhaps just faith in the printed word) will convince the reader that d is positive if and only if the given point lies above the given line.

PROBLEM

B · Use the formula for d to find the distance of the line $3x - 4y - 10 = 0$ from the point $(0, 0)$. Compare with ρ as obtained from the normal form of this same line (Example D). Explain.

==   1·3   EXERCISES

In Exercises 1–8, the equation of a line is given in standard form. Write each equation in (a) slope–intercept form, (b) intercept form, (c) normal form. Then graph the line.

1· $-4x + 3y - 12 = 0$.
2· $-5x + 12y - 60 = 0$.
3· $2x + y = 14$.
4· $3x + y = 5$.
5· $3x + 2y = 4$.

6 · $2x + 3y = 8$.

7 · $2x - 4y = 5$.

8 · $3x - 6y = 10$.

9 · Find a vector **v** parallel to and a vector **w** perpendicular to the line of Exercise 1.

10 · Find a vector **v** parallel to and a vector **w** perpendicular to the line of Exercise 2.

11 · Find the acute angle formed by the intersection of the lines of Exercises 1 and 3.

12 · Find the acute angle formed by the intersection of the lines of Exercises 2 and 4.

In Exercises 13–30, write the equation of the line.

13 · Through $(4, 2)$ and $(-6, 3)$.

14 · Through $(-2, 1)$ and $(5, 3)$.

15 · Through $(1, 3)$, parallel to $3x - y = 5$.

16 · Through $(4, 2)$, perpendicular to $x + 2y = 1$.

17 · y intercept is 2, perpendicular to $3x + y = 0$.

18 · y intercept is $\frac{2}{3}$, x intercept is $\frac{3}{4}$.

19 · Perpendicular to $3x + y = 4$, 3 units above the origin.

20 · Parallel to $2x = 5y$, tangent to a circle of radius 4 centered at the origin.

21 · The altitude through A of the triangle with vertices $A(5, 3)$, $B(-1, 1)$, and $C(3, -2)$.

22 · The altitude through B of the triangle of Exercise 21.

23 · Bisects the acute angle formed by the lines of Exercises 1 and 3.

24 · Bisects the acute angle formed by the lines of Exercises 2 and 4.

25 · Through $(5, 0)$, tangent to a circle of radius $\sqrt{2}$ centered at $(1, 2)$.

26 · Through $(7, 0)$, tangent to a circle of radius 5 centered at $(0, 1)$.

27 · Through $(2, 3)$, having an x intercept twice as large as the y intercept.

28 · Through $(3, 2)$, forming with the positive axes a triangle of area 16.

29 · Through $(6, -5)$, the segment intercepted by the axes having length $\sqrt{34}$.

30 · Through $(-3, 10)$, the product of the intercepts on the axes being 15.

1 · 4 The Conic Sections

We said that one of the principal questions of analytic geometry is, "Given an equation, can we give a geometric description of the corresponding set in the plane?" In the last section we answered this question for any equation of the form $Ax + By + C = 0$. In this section we shall answer it for the equation

$$Ax^2 + Bxy + Cy^2 + Dx + Ey + F = 0. \tag{1}$$

For reasons we give at the end of this section, the corresponding curves in the plane are called conic sections.

We have already studied the special case of (1) in which $A = B = C = 0$; the desired graph is a straight line. Another special case of (1) is obtained by deriving the equation of a circle of radius r centered at the origin. Directly from the theorem of Pythagoras (Fig. 1.24) we have $x^2 + y^2 = r^2$. This is in

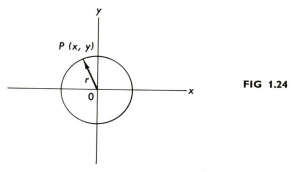

FIG 1.24

the form of (1), where $A = C = 1$, $F = -r^2$, and $B = D = E = 0$. Moreover, it is clear that whenever $A = C > 0$ and $B = D = E = 0$, then (1) may be written

$$Ax^2 + Ay^2 = -F.$$

If $F > 0$, there is no graph; and if $F \leq 0$, the graph is a circle of radius r where $r^2 = -F/A$. When $F = 0$, the graph reduces to a point which we view in this context as a degenerate circle.

One final special case is instructive. Consider a circle of radius r centered at (h, k) (Fig. 1.25). Impose a second coordinate system on the plane, denoting

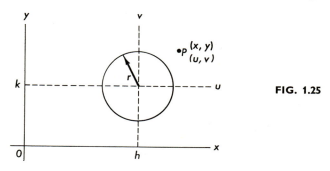

FIG. 1.25

the line $y = k$ as the u axis and the line $x = h$ as the v axis. Then any point P in the plane has two "addresses," (x, y) and (u, v) related by

$$x = u + h, \qquad y = v + k.$$

Now we know that the equation of the circle with respect to the (u, v) coordinates is $u^2 + v^2 = r^2$. Hence, by substitution, the equation with respect to the (x, y) coordinates is

$$(x - h)^2 + (y - k)^2 = r^2. \tag{2}$$

And, of course, any equation that can be written in this form describes a circle of radius r centered at (h, k). This means we are able to describe completely the locus of any equation (1) in which $A = C > 0$ and $B = 0$.

EXAMPLE A
Graph $4x^2 - 12x + 4y^2 + 4y + 6 = 0$.

Since the coefficients of x^2 and y^2 are equal, we know that the graph must be a circle (perhaps degenerate or nonexistent in the case of a negative radius). To write this in the form (2), we use the technique of *completing the square*. (To do this, manipulate as in the first step below to get a positive 1 as the coefficient of the squared term; then add the square of half the coefficient of the first degree term.)

$$4(x^2 - 3x) + 4(y^2 + y) = -6,$$
$$4(x^2 - 3x + \tfrac{9}{4}) + 4(y^2 + y + \tfrac{1}{4}) = -6 + 9 + 1 = 4,$$
$$(x - \tfrac{3}{2})^2 + (y + \tfrac{1}{2})^2 = 1.$$

The circle is centered at $(\tfrac{3}{2}, -\tfrac{1}{2})$ with a radius of 1 (Fig. 1.26). ∎

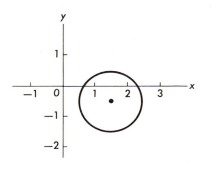

FIG. 1.26

Put the following equations in the form of (2). Graph.

1. $x^2 + 4x + y^2 = 0$.
2. $x^2 + y^2 + 6y = 0$.
3. $3x^2 + 3y^2 + 6y = 1$.

4. $2x^2 + 8x + 2y^2 = 1$.

5. $x^2 + 2x + y^2 - 6y + 10 = 0$.

6. $x^2 + y^2 - 3x + 5y + 7 = 0$.

7. $2x^2 + 2y^2 = 5y - 4x - 2$.

8. $2x^2 - 6x + 2y^2 + 10y + 18 = 0$.

In each of the following cases, find the equations of all circles that satisfy the given conditions.

9. Passes through (4, 0) and (8, 0) and is tangent to the y axis.

10. Passes through (4, 7) and (−3, 0) and has its center on the line $x = 1$.

11. Has its center on the line $y = -\frac{1}{2}x + 1$ and is tangent to both axes.

12. Has its center on the line $y = -x + 4$ and is tangent to both axes.

13. Passes through (6, 10), (0, 2), and (2, 8).

14. Passes through (0, 6), (2, 2), and (3, 5).

15. Touching the line $x - 2y = 3$ at (−1, −2), and having radius $\sqrt{5}$.

16. Touching the y axis and passing through the points (4, −1) and (−3, −2).

A Locus Problem

A *locus* is a collection of all those points and only those points which satisfy a given condition. Thus, a circle is the locus of all points in a plane that are some prescribed distance from a fixed point called the center.

Consider now a fixed vertical line L and a fixed point F not on the line. We shall be interested in the following locus problem (see Fig. 1.27):

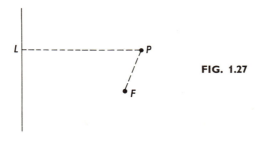

FIG. 1.27

Find all points P in the plane of the line L and the point F for which the ratio of the distance from F to P and from P to L remains constant. In symbols,

$$\frac{FP}{LP} = e \qquad (e \text{ some constant}).$$

The fixed line L is called the *directrix*. The fixed point F is called the *focus*, and the constant e is called the *eccentricity*. The line perpendicular to the directrix passing through the focus is called the *major axis*, though this term is also applied to just a segment of this line at times.

The solution to the problem is most conveniently obtained if we consider three separate cases: $e < 1$, $e > 1$, and $e = 1$. In each case it is easy to find several points that must be on the locus. The general idea is illustrated by the three choices of e indicated in Fig. 1.28. In each case, the points V_i are on the desired locus.

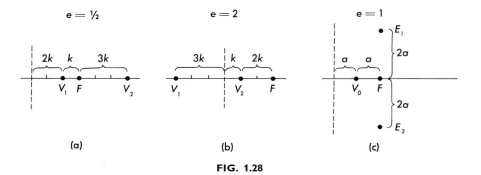

FIG. 1.28

Case I; $e < 1$

As we did in Fig. 1.28a, we may always locate two points V_1 and V_2 on the major axis of the locus. Denote the midpoint between V_1 and V_2 by 0, and label the distances from 0 to V_1, V_2, F, and the line L as indicated in Fig. 1.29. Since V_1 and V_2 are members of the locus,

$$\frac{FV_1}{LV_1} = e, \quad \text{or} \quad \frac{a-c}{d-a} = e \quad \text{so} \quad a - c = e(d-a),$$

$$\frac{FV_2}{LV_2} = e, \quad \text{or} \quad \frac{a+c}{d+a} + e \quad \text{so} \quad a + c = e(d+a).$$

Adding,

$$a = de, \quad \text{so} \quad d = a/e.$$

Subtracting,

$$c = ae.$$

Thus, if we impose a coordinate system on Fig. 1.29, we see that the coordinates of F are $(-ae, 0)$, and L is described by $x = -a/e$ (Fig. 1.30).

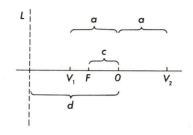

FIG. 1.29

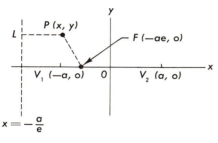

FIG. 1.30

Now for an arbitrary point $P(x, y)$ on the locus,

$$\frac{FP}{LP} = \frac{\sqrt{(x + ae)^2 + y^2}}{x + (a/e)} = e.$$

Algebraic simplification gives

$$(x + ae)^2 + y^2 = (xe + a)^2,$$
$$x^2 + 2aex + a^2e^2 + y^2 = x^2e^2 + 2aex + a^2,$$
$$x^2(1 - e^2) + y^2 = a^2(1 - e^2),$$
$$\frac{x^2}{a^2} + \frac{y^2}{a^2(1 - e^2)} = 1.$$

Since $e < 1, 1 - e^2 > 0$, and we may set $b^2 = a^2(1 - e^2)$. The resulting equation is

$$\frac{x^2}{a^2} + \frac{y^2}{b^2} = 1.$$

It is obvious from the equation that the desired locus of points is symmetric with respect to the y axis (since replacing x by $-x$ does not alter the equation), and is similarly symmetric with respect to the x axis. The locus clearly passes through $(\pm a, 0)$ and $(0, \pm b)$. Plotting a few points reveals the shape of the locus (Fig. 1.31).

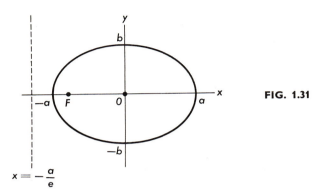

FIG. 1.31

This curve is called an *ellipse*. It is obvious from the symmetry of the curve that we would have obtained the same locus if we had started with a focus at $(ae, 0)$ and a directrix $x = a/e$. For this reason, it is customary to speak of the foci and directrices of an ellipse.

Related to the two foci of an ellipse is another important geometrical fact. Let P be a point on the ellipse (Fig. 1.32). Using F_1 and L_1, we have, by

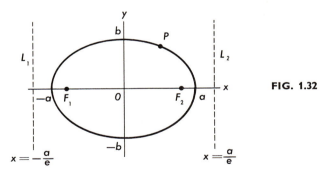

FIG. 1.32

definition of the ellipse,

$$F_1 P = eL_1 P = e\left(x + \frac{a}{e}\right) = ex + a.$$

Similarly, using F_2 and L_2,

$$F_2 P = eL_2 P = e\left(\frac{a}{e} - x\right) = a - ex.$$

Adding, $F_1P + F_2P = 2a$. Thus, the ellipse may be characterized as the locus of all points P for which the sum of the distances to two fixed points (the foci) is a constant.

Note that the major axis of the ellipse (the line segment from $(-a, 0)$ to $(a, 0)$) is longer than the line (called the *minor axis*) from $(0, b)$ to $(0, -b)$. This follows from the definition of b, since $b^2 = a^2(1 - e^2) < a^2$. If we had started out with a horizontal directrix, then of course the major axis would have been along the y axis. In a particular example we determine whether the directrix is vertical or horizontal by noting the direction of the major (longer) axis.

EXAMPLE B

Graph $\dfrac{x^2}{4} + \dfrac{y^2}{9} = 1$. Locate the foci and directrices.

The graph is most easily sketched (Fig. 1.33). Since $4 = 9(1 - e^2)$, and since $e > 0$, $e = \sqrt{5}/3$. The distance of the foci from the center is $ae = 3\sqrt{5}/3 = \sqrt{5}$ and the distance from the center to the directrices is

$$\frac{a}{e} = \frac{3}{\sqrt{5/3}} = \frac{9}{\sqrt{5}}.$$

See Fig. 1.34 for location of the foci and directrices.   ∎

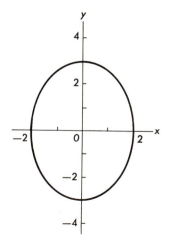

FIG. 1.33

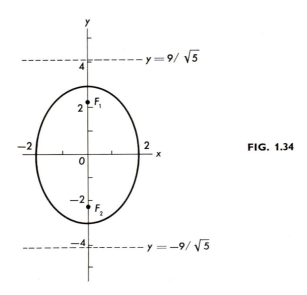

FIG. 1.34

The principal facts about ellipses may now be summarized:

(1)   The major axis, length $2a$, is the longest axis of the ellipse;
(2)   The length of the minor axis, $2b$, is determined from $b^2 = a^2(1 - e^2)$;
(3)   The distance from the center to the directrices is a/e;
(4)   The distance from the center to the foci is ae;
(5)   Two geometrical descriptions of an ellipse may be given:
    i.   The locus of all points P for which $FP/LP = e < 1$.
    ii. The locus of all points P so that $F_1P + F_2P$ is constant (and in fact equal to the length of the major axis).

Something more has now been learned about the curves described by (1). We know what the curve looks like if $B = D = E = 0$, while A and C are both positive. Moreover, if we profit from what we learned about the circle, it is clear that we may drop the restriction on D and E since if they are nonzero we can complete the square in x and y, obtaining either of the standard forms indicated in Fig. 1.35.

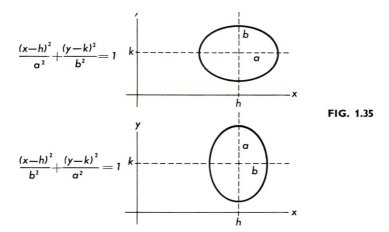

$$\frac{(x-h)^2}{a^2}+\frac{(y-k)^2}{b^2}=1$$

$$\frac{(x-h)^2}{b^2}+\frac{(y-k)^2}{a^2}=1$$

FIG. 1.35

EXAMPLE C

Graph $x^2 + 6x + 4y^2 - 8y + 9 = 0$. Locate the foci and directrices.

We first perform the algebra necessary to put the given equation in standard form:

$$(x^2 + 6x + 9) + 4(y^2 - 2y + 1) = 4,$$

$$\frac{(x + 3)^2}{4} + \frac{(y - 1)^2}{1} = 1.$$

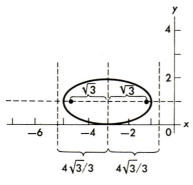

FIG. 1.36

In this case, $a^2 = 4$ and $b^2 = 1$. Hence,

$$1 = 4(1 - e^2),$$

$$e^2 = \tfrac{3}{4} \quad \text{so} \quad e = \sqrt{3}/2,$$

$$ae = 2\sqrt{3}/2 = \sqrt{3},$$

$$\frac{a}{e} = \frac{2}{\sqrt{3/2}} = \tfrac{4}{3}\sqrt{3},$$

and the graph is as indicated in Fig. 1.36. ∎

Graph the following. Locate the foci and directrices in each case.

1 · $12x^2 + 4y^2 = 9$.

2 · $27x^2 + 12y^2 = 36$.

3 · $4x^2 - 8x + 9y^2 - 54y + 49 = 0$.

4 · $x^2 - 4x + 4y^2 - 8y - 8 = 0$.

5 · $4x^2 + 8x + y^2 - 4y + 4 = 0$.

6 · $9x^2 + 36x + 4y^2 + 8y + 4 = 0$.

7 · $12x^2 - 36x + 4y^2 + 4y + 25 = 0$.

8 · $4x^2 - 4x + 20y^2 + 60y + 41 = 0$.

Write the equation of each of the following ellipses:

9 · Foci $(\pm 2, 0)$; directrices $x = \pm 3$.

10 · Foci $(0, \pm 3)$; directrices $y = \pm 4$.

11 · Foci $(\pm 2, 0)$; vertices $(\pm 3, 0)$.

12 · Directrices $y = \pm 4$; vertices $(0, \pm 3)$.

13 · Directrices $y = -2$ and $y = 6$; vertices $(-2, -1)$ and $(-2, 5)$.

14 · Foci $(1, 3)$ and $(5, 3)$; vertices $(0, 3)$ and $(6, 3)$.

Case II; $e > 1$

This case parallels in many ways the one just considered. In particular, notation may be chosen to minimize the number of new formulas to be memorized.

As we did in Fig. 1.28b, we may always locate two points V_1 and V_2 of the locus on the major axis. Denote the midpoint between V_1 and V_2 by 0, and label the distances from 0 to V_1, V_2, F, and the line L as indicated in Fig. 1.37.

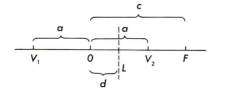

FIG. 1.37

Since V_1 and V_2 are members of the locus,

$$\frac{FV_1}{LV_1} = e \quad \text{or} \quad \frac{c+a}{d+a} = e \quad \text{so} \quad c + a = e(d + a),$$

$$\frac{FV_2}{LV_2} = e \quad \text{or} \quad \frac{c-a}{a-d} = e \quad \text{so} \quad c - a = e(a - d).$$

Adding,

$$c = ae.$$

Subtracting,

$$a = ed \quad \text{so} \quad d = a/e.$$

Thus, if we impose a coordinate system on Fig. 1.37, we see that the coordinates of F are $(ae, 0)$ and L is described by $x = a/e$. For an arbitrary P (Fig. 1.38) on the locus,

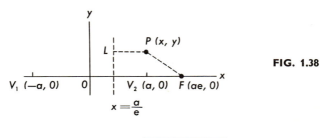

FIG. 1.38

$$\frac{FP}{LP} = \frac{\sqrt{(x - ae)^2 + y^2}}{x - (a/e)} = e.$$

Simplifying as in Case I, we again obtain

$$\frac{x^2}{a^2} + \frac{y^2}{a^2(1 - e^2)} = 1.$$

Guided by the results obtained for an ellipse, we are tempted to set the denominator of the second term equal to b^2. But $1 - e^2 < 0$ here, so we must set $-b^2 = a^2(1 - e^2)$. The resulting equation is

$$\frac{x^2}{a^2} - \frac{y^2}{b^2} = 1.$$

It is obvious that the desired locus is again symmetric with respect to both the x and y axes. The locus clearly passes through $(\pm a, 0)$, the vertices. It does not pass through $(0, \pm b)$ or any other point for which the x coordinate is between $-a$ and a. Plotting a few points reveals the shape of the locus (Fig. 1.39).

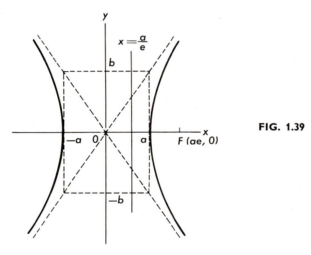

FIG. 1.39

This curve is called a *hyperbola*. It is obvious from the symmetry of the curve that we would have obtained the same curve if we had started with a focus at $(-ae, 0)$ and a directrix $x = -a/e$. We again, therefore, speak of the foci and directrices.

A remark about the dotted lines in Fig. 1.39 is in order. They are best understood in light of the following question. Among all lines $y = mx$ through the origin, which ones intersect the hyperbola? The answer is obtained by solving simultaneously the equations

$$\frac{x^2}{a^2} - \frac{y^2}{b^2} = 1 \qquad \text{and} \qquad y = mx.$$

Substitution gives

$$\frac{x^2}{a^2} - \frac{m^2 x^2}{b^2} = 1,$$

from which we obtain

$$x^2 = \frac{a^2 b^2}{b^2 - a^2 m^2}.$$

There will be a solution if and only if the right side is positive; that is, if and only if $b^2 - a^2 m^2 > 0$. This means $m^2 < b^2/a^2$. The line $y = bx/a$ is thus critical in the sense that the hyperbola never touches it, but it intersects any line having a smaller (positive) slope. Similar remarks apply to the line $y = -bx/a$. These lines are called *asymptotes*. They are often sketched in as an aid to drawing the hyperbola itself.

We saw that an ellipse may be defined geometrically in a second way as the locus of points, the sum of whose distances from two fixed points (the foci) is a constant $(2a)$. The reader should have no trouble verifying, that a hyperbola may be similarly characterized as the locus of points, the difference of whose distances from two fixed points (the foci) is a constant $(2a$ or $-2a)$.

If we had started out with a horizontal directrix, then of course the major axis would have been along the y axis, and the two portions of the hyperbola would have opened up and down instead of to the left and right. Notice that in the case of the hyperbola of Fig. 1.39, the major axis joining $(-a, 0)$ and $(a, 0)$ is not necessarily longer than the segment joining $(0, b)$ and $(0, -b)$. In a particular example we determine the major axis by remembering that when the equation is in standard form (so that the right-hand side is a positive 1), the a^2 factor appears as the denominator of the positive term on the left side.

The principal facts about hyperbolas may now be summarized:

(1) The curve cuts the major axis at distances of a on either side of the center. The constant a^2 is the denominator of the positive term on the left side of the equation when in standard form.

(2) The curve has asymptotes which are diagonals of a rectangle centered at the center of the hyperbola. Dimensions of the rectangle are $2b$ by $2a$, where $b^2 = a^2(e^2 - 1)$.

(3) The distance from the center to the directrices is a/e.

(4) The distance from the center to the foci is ae.

(5) Two geometrical descriptions of a hyperbola may be given:
    i.   The locus of all points P for which $FP/LP = e > 1$.
    ii. The locus of all points P so that $|F_1 P - F_2 P|$ is constant.

We now know something else about curves described by (1). Using the usual technique of completing the square, we can now describe completely the curves obtained when $B = 0$ and A and C are of opposite sign.

EXAMPLE D

Graph $4x^2 - 32x - y^2 - 6y + 59 = 0$. Find the directrices, the foci, and the equations of the asymptotes.

Completing the square in x and y,

$$4(x^2 - 8x + 16) - (y^2 + 6y + 9) = -59 + 64 - 9.$$

To obtain a positive 1 on the right, we divide by -4:

$$\frac{(y + 3)^2}{4} - \frac{(x - 4)^2}{1} = 1.$$

Since $a^2 = 4$ and $b^2 = 1$, $1 = 4(e^2 - 1)$ and so $e = \sqrt{5/2}$:

$$ae = 2\sqrt{5/2} = \sqrt{5}, \qquad \frac{a}{e} = \frac{2}{\sqrt{5/2}} = \frac{4}{5}\sqrt{5}.$$

The easiest way to determine the slope of the asymptotes is to look at Fig. 1.40. Their equations are

$$y + 3 = \pm 2(x - 4). \quad \blacksquare$$

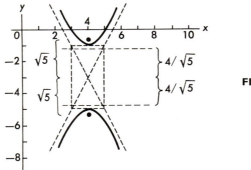

FIG. 1.40

Our concluding observation has to do with the degenerate case of a hyperbola illustrated by $x^2 - y^2 = 0$. The graph in this case "collapses" into the asymptotes $y = \pm x$, and this will always be the case when the procedure illustrated in Example D results in a 0 on the right side.

Graph the following. Locate the foci and directrices.

1. $x^2 - 2x - y^2 - 4y - 4 = 0.$

2. $y^2 + 2y - x^2 + 4x - 4 = 0.$

3. $4y^2 - x^2 - 4x - 5 = 0.$

4. $9x^2 - y^2 + 4y - 5 = 0.$

5. $4y^2 - 4y - 9x^2 - 12x - 4 = 0.$

6. $9x^2 - 6x - 4y^2 - 4y - 1 = 0.$

7. $x^2 - 2x = y^2 + 2y.$

8. $x^2 + 2x = y^2 - 2y.$

Write the equations of the following hyperbolas:

9. Asymptotes $y - 3 = \pm 2(x + 1)$, through $(0, 3)$.

10. Asymptotes $y + 2 = \pm 3(x - 1)$, through $(1, 1)$.

11. Asymptotes $y - 3 = \pm 2(x + 1)$, through $(-1, 5)$.

12. Asymptotes $y + 2 = \pm 3(x - 1)$, through $(2, -2)$.

13. Center at the origin with directrix $x = 4$ and $e = 2$.

14. Center at the origin with directrix $x = -4$ and $e = 2$.

15. Directrix $y = 6$, focus at $(2, 2)$ and $e = 3$.

16. Directrix $y = 4$, focus at $(3, 2)$ and $e = 2$.

Case III; $e = 1$

We impose a coordinate system on Fig. 1.28c as indicated in Fig. 1.41. As usual, P represents an arbitrary point on the desired locus. Since $e = 1$,

$$1 = \frac{FP}{LP} = \frac{\sqrt{(x - a)^2 + y^2}}{x + a}.$$

Algebraic simplification gives

$$y^2 = 4ax.$$

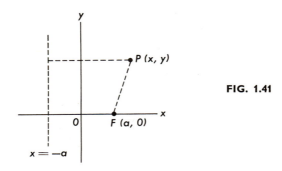

FIG. 1.41

As one simple check on our work, we note that the points E_1 and E_2 of Fig. 1.41 (which have coordinates $(a, 2a)$ and $(a, -2a)$, respectively) both satisfy our equation. The line segment E_1E_2 is called the *latus rectum*. The graph, indicated in Fig. 1.42, is called a *parabola*.

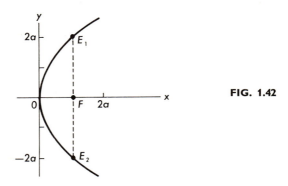

FIG. 1.42

Parabolas have a number of interesting properties. For example, the rays of light from a bulb placed at the focus will all be reflected by the parabola in lines parallel to the major axis. (See Example 15.3G.)

Nothing is sacred, of course, about the original positions of the directrix and focus. The directrix might have been horizontal; the focus might have been to the left of a vertical directrix, and either above or below a horizontal directrix. From the symmetry of the various possibilities, it is easy to see that the four graphs and their associated equations are as indicated in Fig. 1.43.

We are now able to graph equations of the form (1) if $B = 0$ and either A or C is also 0. In so saying, we again anticipate the possibility of a translation of axes.

The principal facts about the parabola are summarized as follows:

(1)   The major axis is perpendicular to the coordinate axis corresponding to the squared term.

(2)   When the equation is written in standard form, the coefficient of the first degree term is $4a$.

(3)   The focus, located on the major axis in the direction in which the parabola opens, is at a distance a from the vertex.

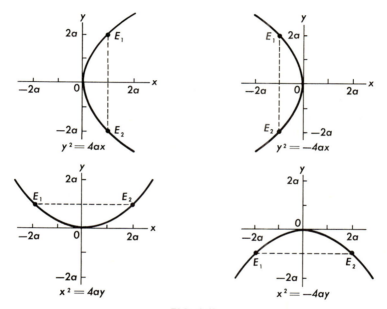

FIG. 1.43

(4)   The curve opens "away" from the directrix which is located at a distance a from the vertex.

(5)   A parabola is described geometrically as the locus of all points P such that $FP/LP = e = 1$.

EXAMPLE E
Graph $12x^2 - 12x + 32y + 35 = 0$.

The usual algebraic manipulation gives

$$12(x^2 - x + \tfrac{1}{4}) = -32y - 35 + 3,$$
$$(x - \tfrac{1}{2})^2 = -\tfrac{32}{12}(y + 1) = -4(\tfrac{2}{3})(y + 1).$$

and the graph is indicated in Fig. 1.44. ∎

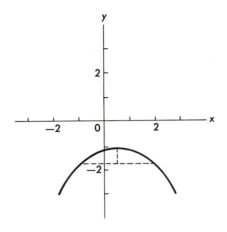

FIG. 1.44

1·4D EXERCISES

Graph the following parabolas. Locate the focus in each case.

1· $2x^2 = 3y$.

2· $3x^2 = 10y$.

3· $4x^2 - 4x - 24y - 23 = 0$.

4· $4y^2 + 4y - 24x + 25 = 0$.

5· $x^2 + 2x + 5y + 6 = 0$.

6· $y^2 - 2y - 6x - 2 = 0$.

7· $4y^2 - 4y + 10x - 19 = 0$.

8· $4x^2 + 12x + 8y + 13 = 0$.

Write the equation of each of the parabolas described.

9· Focus at (4, 2) and directrix $y = 6$.

10· Focus at $(-2, 1)$ and directrix $y = -5$.

11· Focus at (3, 1), directrix parallel to the y axis, latus rectum of length 3, opening to the left.

12· Focus at $(-4, 2)$, directrix parallel to the y axis, latus rectum of length 5, opening to the left.

By drawing together what we have learned about ellipses, hyperbolas, and parabolas, the graphs of the following equations may be rapidly sketched. Do so.

13· $x^2 - 2x - y^2 - 4y = 0$.

14· $x^2 - 2x - 2y = 0$.

15. $4x^2 + 4x + y^2 - 4y = 0.$

16. $3y^2 + 4y - x + 2 = 0.$

17. $9x^2 - 6x - 4y^2 - 4y - 1 = 0.$

18. $4x^2 - 20x + 4y^2 + 4y - 4 = 0.$

19. $6y^2 - 6y - x + 2 = 0.$

20. $9x^2 - 12x + y^2 = 0.$

21. $9x^2 - 6x + 1 + 9y^2 + 12y = 0.$

22. $x^2 + 2x - y^2 + 6y = 0.$

23. $x^2 - 6x - 4y = 0.$

24. $4x^2 + 4x - 9y^2 + 12y - 4 = 0.$

Rotation

We have now classified all curves described by (1) provided that $B = 0$. The happy truth is that no new curves are encountered when $B \neq 0$. This is illustrated by the following example.

EXAMPLE F

Graph $4xy - 3y^2 = 8$.

In Fig. 1.25 we imposed a second coordinate system on the xy plane. An arbitrary point P then had two "addresses," (x, y) and (u, v). Suppose we again impose a second coordinate system, this time as indicated in Fig. 1.45. Again a point has two addresses, and they are related by appeal to trigonometry.

$x = OR' - RR',$

$x = OR' - QQ',$

$x = u \cos \theta - v \sin \theta,$

$y = RQ + QP,$

$y = R'Q' + QP,$

$y = u \sin \theta + v \cos \theta,$

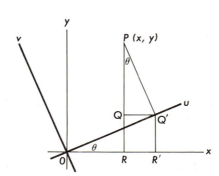

FIG. 1.45

The equation of the desired curve with respect to the uv axes is obtained by substitution.

$$4(u \cos \theta - v \sin \theta)(u \sin \theta + v \cos \theta) - 3(u \sin \theta + v \cos \theta)^2 = 8.$$

After multiplying and collecting terms we have

$$u^2(4 \cos \theta \sin \theta - 3 \sin^2 \theta) + uv(4 \cos^2 \theta - 4 \sin^2 \theta - 6 \cos \theta \sin \theta)$$
$$+ v^2(-4 \cos \theta \sin \theta - 3 \cos^2 \theta) = 8. \quad (3)$$

Now we would know how to graph this with respect to the uv axes if the coefficient of the uv term was zero. Hence, we seek a value of θ for which

$$4(\cos^2 \theta - \sin^2 \theta) - 6 \cos \theta \sin = 0.$$

Calling upon a little trigonometry,

$$4 \cos 2\theta - 3 \sin 2\theta = 0,$$
$$\tan 2\theta = \tfrac{4}{3}.$$

This means that $\cos 2\theta = \tfrac{3}{5}$, and the half-angle formulas then give

$$\cos \theta = \pm \sqrt{\frac{1 + \tfrac{3}{5}}{2}} = \frac{\pm 2}{\sqrt{5}},$$

$$\sin \theta = \pm \sqrt{\frac{1 - \tfrac{3}{5}}{2}} = \frac{\pm 1}{\sqrt{5}}.$$

If we agree to choose a value of θ in the first quadrant, we may use the $+$ sign. Substitution in (3) gives

$$u^2(\tfrac{8}{5} - \tfrac{3}{5}) + v^2(-\tfrac{8}{5} - \tfrac{12}{5}) = 8,$$

$$\frac{u^2}{8} - \frac{v^2}{2} = 1.$$

The graph is now easily drawn as in Fig. 1.46. Note that the uv axes are drawn so that $\tan \theta = \tfrac{1}{2}$. ∎

General formulas can be worked out (we shall return to this problem in Sections 10.6 and 10.7) to reduce the amount of computation necessary for a

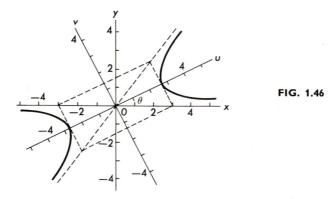

FIG. 1.46

particular example. But the principle is the same: The product term is elimi-
nated by a suitable rotation of the axes, and the resulting equation is one of
the forms we have already studied.

The curves we have now studied, that is, the curves described by equations
of the form (1) are often grouped together under the general heading of conic
sections. This is because these curves may be obtained as the intersection of
a right circular cone (both nappes) and a plane (Fig. 1.47).

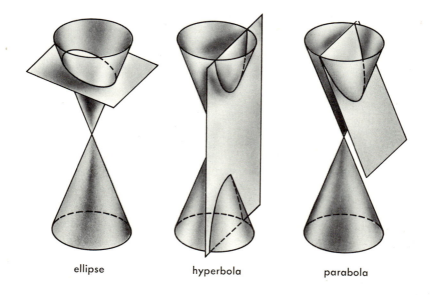

ellipse   hyperbola   parabola

Fig. 1.47

Remove the product term by an appropriate rotation and graph.

1· $xy = 4.$

2· $xy = -4.$

3· $x^2 + 2\sqrt{3}xy - y^2 = 2.$

4· $3x^2 - 2\sqrt{3}xy - 3y^2 = 4.$

5· $x^2 + 3xy + 5y^2 = 11.$

6· $4xy - 3y^2 = 4.$

FUNCTIONS

LAWS OF NATURE EXPRESS THE DEPENDENCE
UPON ONE ANOTHER OF TWO OR MORE VARI-
ABLES. THIS IDEA OF DEPENDENCE OF VARI-
ABLES IS FUNDAMENTAL IN ALL SCIENTIFIC
THOUGHT, AND REACHES ITS MOST THOROUGH
EXAMINATION IN MATHEMATICS AND LOGIC
UNDER THE NAME OF FUNCTIONALITY.

—*Philip E. B. Jourdain*
The Nature of Mathematics

Most words have a variety of meanings in ordinary language. Some words have additional technical meanings when used in the context of mathematics. It is customary, when such a word is to be used in a technical sense, to introduce it with a precise definition.

An elegant definition, however, does not instill an intuitive feeling for the way in which the word is to be used. For this reason we introduce the word "function" in Section 2.1 by comparing a function to various machines: meat grinders, snow blowers, and others.

The formal definition of a function appears in Section 2.2. We also introduce the notion of a graph of a function. The remainder of the chapter is devoted to collecting some of the basic facts about functions that we will need to know in our work.

2·1   Number Machines

Home meat grinders are simple machines. You put food to be ground up into the top, turn a crank, and get your product in a bowl at the side. The exact form of the product depends on what is put in at the top and on the internal construction of the machine; that is, on the type of cutting blades being used. Besides meat, such machines will grind up vegetables, certain kinds of fruit, and other foods. One can also think of a host of things (bones, glass bottles) that cannot be successfully fed through the machine.

Now imagine such a machine that takes numbers as its raw material. You put in a number, turn the crank, and the machine produces a second number. We may imagine machines that can accept any number as raw material; other machines might accept only certain kinds of numbers, such as integers or fractions, or any number between 0 and 1. Examples of the kinds of machines we have in mind follow. Such machines are commonly called *functions*.

EXAMPLE A

This machine takes as raw material any number. It produces the square of that number. Feed in a 3. Turn the crank. Out comes 9. Feed in x. Out comes x^2. Call the machine S. Write $S(x) = x^2$ (and read this, "the value of S at x equals x^2").

EXAMPLE B

This machine takes any nonnegative number as raw material. It produces a nonpositive number that, when squared, gives the original number. Feed in a 4. Turn the mental crank. Out comes a -2. Feed in any positive x. Out comes $-\sqrt{x}$. Call the machine N. Write $N(x) = -\sqrt{x}$. ∎

We pause here to observe that when we write $\sqrt{x}$, we always mean the nonnegative number that, when squared, gives x. Thus $N(x)$ is always negative or zero. Observe that $\sqrt{y^2}$ will be y if any only if y is positive or zero. If y is negative, then in order to get a positive number for $\sqrt{y^2}$, we must write $\sqrt{y^2} = -y$. Summarizing this whole paragraph,

$$\sqrt{y^2} = \begin{cases} y, & y \ge 0, \\ -y, & y < 0. \end{cases} \tag{1}$$

EXAMPLE C

This machine takes any nonnegative number as raw material and produces a positive number or zero that, when squared, gives the original number. Feed in a 4. Out comes 2. Feed in x. Out comes $\sqrt{x}$. Call the machine P. Write $P(x) = \sqrt{x}$. ∎

EXAMPLE D

This machine takes any nonnegative number as raw material. Its behavior is somewhat complicated to describe. Put in 0; 0 comes out. Put in any number greater than 0 but less than or equal to 1. Out comes 8. Put in any number greater than 1, less than or equal to 2. Out comes 16. Put in any number greater than the integer n but less than or equal to $n + 1$. Out comes $8(n + 1)$. This machine would be useful in a post office. Call it M. ∎

As in the preceding example, we often wish to talk about all the numbers x greater than a, less than or equal to b. We shall express this by writing, "x is a member of the interval $(a, b]$," or more concisely yet, $x \in (a, b]$. The left-hand parenthesis means a is not included; the right-hand square bracket means b is included. There are obvious variations of this notation. This entire paragraph may be summarized as follows.

$$\begin{array}{lll} x \in (a, b) & \text{means} & a < x < b; \\ x \in (a, b] & \text{means} & a < x \le b; \\ x \in [a, b) & \text{means} & a \le x < b; \\ x \in [a, b] & \text{means} & a \le x \le b. \end{array} \tag{2}$$

The interval (a, b) in which neither of the end points is included is called an *open interval*. The interval $[a, b]$ is called a *closed interval*.

A meat grinder, or any other decent machine, has the following property. Put in the same raw material at two different times. The output is always the same. It is perfectly all right if different inputs produce the same output. Thus, either -2 or 2 put into machine S (Example A) produces 4. But machines N and P cannot be somehow combined into one machine producing for an input of 4 either a -2 or a 2. This ambiguous behavior would drive us crazy.

Suppose we imagine the numbers that can be fed into a particular machine as being plotted on a horizontal line that we designate as the x axis. Its output, again numbers, will be plotted on a vertical line, the y axis. Pictorially we have in mind a machine not unlike those used in big cities to remove snow from the streets. The intake is at the bottom; the output is blown into a truck.

It will be our habit to place the x and y axes so that their 0 points coincide. It is clear that if our number machine is capable of taking very many numbers as raw material, the picture (Fig. 2.1) with an arrow showing the destination of each input number will not be practical. A better way to indicate the destination of x_1 and x_2 would be as illustrated in Fig. 2.2.

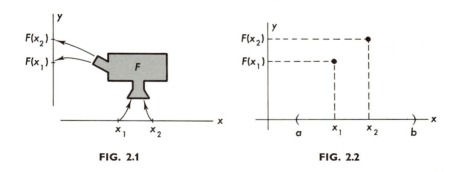

FIG. 2.1　　　　　　　**FIG. 2.2**

For each input number, a point in the xy plane indicates the destination. The aggregate of all such points (x, y) is called the graph of F. If F is defined for all numbers between a and b, then its graph, when plotted in the xy plane may (for well-behaved F) look like a curve in the plane.

Note the open circle corresponding to a, the closed circle corresponding to b in Fig. 2.3. We shall consistently use this device to indicate that F is not defined at a, but is defined at b. An obvious variation of this convention is used in Fig. 2.4d. The number machine functions described in Examples A–D have the graphs indicated in Fig. 2.4a–d, respectively.

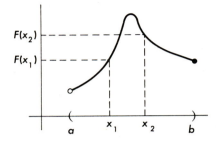

FIG. 2.3

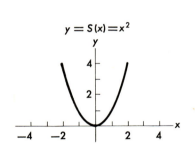

$y = S(x) = x^2$

FIG. 2.4a

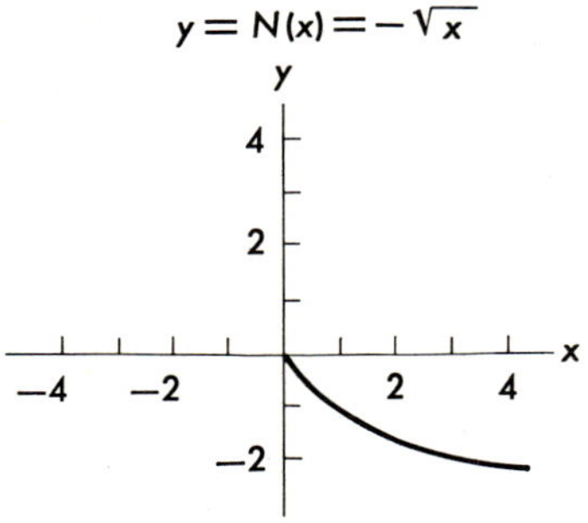

$y = N(x) = -\sqrt{x}$

FIG. 2.4b

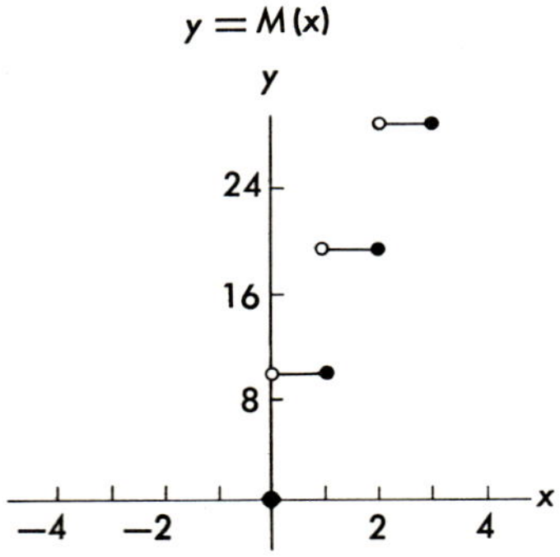

$y = M(x)$

FIG. 2.4c

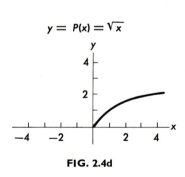

$y = P(x) = \sqrt{x}$

FIG. 2.4d

2·1 EXERCISES

1. Find $\sqrt{(x+1)^2}$ for $x < -3$.

2. Find $\sqrt{(x-3)^2}$ for $x < 1$.

3. Simplify $\dfrac{x}{\sqrt{x^2 + 3x^3}}$.

4. Simplify $\dfrac{2x}{\sqrt{x^3 - 2x^4}}$.

5. Find $\sqrt{(x^2 + 1)^2}$.

6. Find $\sqrt{(2 + \sin x)^2}$.

Using the functions S, N, P, and M as defined in Examples A–D, find the following.

7. $N((-3)^2)$.

8. $P((-4)^2)$.

9. $-N((-3)^2)$.

10. $-P((-4)^2)$.

11. $M(\pi)$.

12. $M(\pi/2)$.

13. $N(P(16))$.

14. $N(P(81))$.

Draw x and y axes as indicated in Section 2.1; then do the following.

15. Sketch the graph of the function G, only defined for positive integers, and taking at n the value $3/n$.

16. Sketch the graph of the function H, only defined for positive integers, and taking at n the value $4/n^2$.

17. Sketch the graph of $G(x) = \sqrt{x^2}$.

18. Sketch the graph of $H(x) = \sqrt{(x+1)^2}$.

19. The function R is defined as

$$R(x) = \begin{cases} -x+2, & x < 1, \\ \sqrt{x^2}, & x \in [1, 2), \\ 3, & x \geq 2. \end{cases}$$

(a) Find $R(-4)$.   (b) Find $R(0)$.
(c) Find $R(1)$.   (d) Find $R(2)$.
(e) Find $R(\pi/2)$.   (f) Find $R(6)$.
(g) Graph the function R.

20. The function S is defined as

$$S(x) = \begin{cases} 4, & x \leq -1, \\ \sqrt{x^2}, & x \in (-1, 0], \\ \tfrac{1}{2}x + 1, & x > 0. \end{cases}$$

(a) Find $S(-4)$.   (b) Find $S(0)$.
(c) Find $S(-\pi/4)$.   (d) Find $S(-1)$.
(e) Find $S(4)$.   (f) Graph the function S.

2·2 Functions

The number machine described in the preceding section is called *a real-valued function of a real variable*. We now describe a function more formally.

DEFINITION A (Function)

A function is a rule of correspondence F that assigns to each member x of one set (called the *domain* of the function) a unique element $F(x)$ (often called simply F of x) in a second set. The collection of all elements $F(x)$ is called the *range* of the function. See Fig. 2.5.

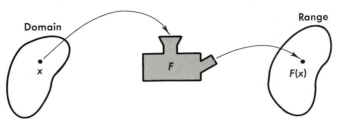

FIG. 2.5

Stated in this generality, the domain could be the set of all cars in a certain parking lot. The range could be a set of colors. The function would be a rule assigning a color to each car. This rule must be absolutely clear. It must cover two-tone cars, cars in some stage of being repainted, cars of different shades, and so on. Two people sent into this lot to classify the cars according to color by using this rule must be able to classify each car in exactly the same way without consulting one another at any time. This is what we mean by assigning to each member of the first set a unique member of the second. Also note that the range consists only of the colors for which a car is found. Orange is a color, but would very likely not be in the range of this function.

In the examples of Section 2.1, the domain was always some subset of the real numbers. (We did not allow imaginary or complex numbers to enter the discussion.) Similarly, the range was some subset of the real numbers. Hence the name, real-valued function of a real variable. This is the type of function with which we shall be concerned in the first eight chapters. When the domain is not specifically indicated, the reader should assume it to be as large as is possible for the given function. For instance,

$$\text{the function } F \text{ described by } F(t) = \frac{1}{t^2 - 1} \tag{1}$$

should be understood to have as its domain all real numbers not equal to ± 1.

The function itself is designated by F. If x is a member of the domain, the corresponding member of the range is designated by $F(x)$. The reader should be warned, however, that although it is correct to refer to "the function F," he will often find references to "the function $F(x)$." This designation is an abuse of language, but it turns out to be a useful one that our efforts are not likely to change. There are several places in mathematics where consistent use of language or notation is abused in deference to both custom and utility.

Note line (1). As stated, it is correct. The common phrasing of this same idea, not precisely correct, but understandable and certainly less cumbersome, is

$$\text{the function } F(t) = \frac{1}{t^2 - 1}. \tag{2}$$

We shall speak of "the function F" whenever this is either convenient or essential to clarity. But when no confusion seems likely, we shall also allow ourselves the luxury of such abuses as are illustrated in (2) and the problems that follow.

We have already introduced the notion of a graph. This, too, can be formalized.

DEFINITION B (Graph of a Function)

The graph of a function F is the set of all pairs $(p, F(p))$ where p is a point of the domain of F.

Our definition of the graph is very similar to a definition some authors use to describe the function itself. Much has been written about how best to define "function." (See C. P. Nicholas, "A dilemma in definition," *Amer. Math. Monthly* **73** (1966), pp. 762–768.) With no desire to do battle in this arena, we have merely chosen a definition that lends itself to the spirit of our presentation.

PROBLEMS

A · Let H be the function $H(x) = 3 + \sqrt{4 - x^2}$. (Note the abuse of language here.)

(a)   What is the domain of H?

(b)   What is the range of H?

B · Consider the set S of all pairs (x, y) for which $(y - 2)^2 = 4(x + 3)$.

(a)   Explain why no function G can be found so that the set S above will be the graph of $(x, G(x))$.

(b)   Define two functions G_1 and G_2 so that the graphs $(x, G_1(x))$ and $(x, G_2(x))$ include all the points and only those points of the set S.

C • Let $R(x) = (2x^2 + 3)/(x - 3)^2$. (The abuse of language gets worse.)
   (a) What is the domain of R?
   (b) What is the range of R? If you cannot determine the range exactly,
can you give any information about it?

Pairs of real numbers may be used to locate points in a plane with respect
to some fixed set of axes. When F is a real-valued function of a real variable,
the graph of F is a set of pairs of real numbers. Thus, the graph may be
identified with a certain subset of points in the plane. This is the formal
reasoning that leads to the pictures we drew in Section 2.1. In the next
section we aim at developing our skills in sketching graphs.

2·2 EXERCISES

In Exercises 1–6, give the domain (the largest possible) and the range of the indicated
function. Express your answers by using the notation of (2) of Section 2.1.

1 • $F(x) = \sqrt{9 - x^2}$.
                    2 • $F(x) = \sqrt{16 - x^2}$.

3 • $F(x) = \sqrt{9 - (x + 2)^2}$.
                4 • $F(x) = \sqrt{16 - (x + 5)^2}$.

5 • $F(x) = \sqrt{9 - (4x + 3)^2}$.
               6 • $F(x) = \sqrt{16 - (3x + 2)^2}$.

In Exercises 7–12, give the domain of F.

7 • $F(x) = \dfrac{x}{x^2 + 5x + 6}$.
            8 • $F(x) = \dfrac{2x - 3}{x^2 - 1}$.

9 • $F(x) = |x + 3|$.
                 10 • $F(x) = |2x + 5|$.

11 • $F(x) = \dfrac{\sqrt{3 - x^2}}{x}$.
         12 • $F(x) = \dfrac{\sqrt{5 - x^2}}{x^2 - 1}$.

Note that Exercises 5 and 6 could have been written as follows.

5′ • $F(x) = \sqrt{-16x^2 - 24x}$.
           6′ • $F(x) = \sqrt{12 - 9x^2 - 12x}$.

In Exercises 13–16, give the domain and range of the indicated function.

13 • $F(x) = \sqrt{-5 - 6x - x^2}$.
        14 • $F(x) = \sqrt{-3 - 4x - x^2}$.

15 • $F(x) = \sqrt{8 - 4x^2 - 4x}$.
         16 • $F(x) = \sqrt{15 + 6x - 9x^2}$.

2·3 Graphs; Getting the Picture

Graphs of real-valued functions of a real variable describe certain subsets of the plane. Given F, set $y = F(x)$. Then the points $(x, y) = (x, F(x))$ give a picture of the graph. It is often useful when studying a function to have its graph before you. Suppose, for example, that you knew that the functions H and R of Problems 2.2A and 2.2C had the graphs drawn in Figs. 2.6 and

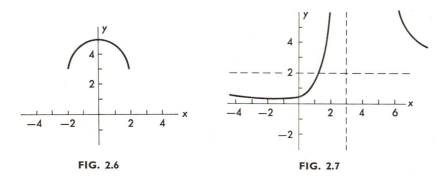

FIG. 2.6   FIG. 2.7

2.7. By looking at the graphs along with the algebraic descriptions of the functions, you would have been able to answer the questions about these functions more easily.

It is to be emphasized, however, that not every subset of the plane is the graph of some function. For a given x, a function can produce only one $y = F(x)$.

PROBLEM

A· Describe a geometrical test that can be used to show that none of the subsets pictured in Figs. 1.2, 1.3, or 1.4 (Chapter 1) can be the graph of a single function.

Our primary goal in this section is to develop facility in sketching graphs of functions that are well behaved wherever they are defined. We shall have more to say later about what we mean by well-behaved functions. For the time being, we simply say that if F is well behaved on an interval, then the graph of F appears as an unbroken line. Thus, the function F graphed in Fig. 2.7 appears by this test to be well behaved on the interval [0, 2] but not on the interval [2, 4]. One procedure for drawing the graph of a well-behaved function is to plot a few points and then sketch a smooth, unbroken line through them.

There are, of course, more sophisticated methods for drawing graphs. We saw in our study of analytic geometry that it is possible to learn to recognize certain expressions, write them in a standard form, and graph them very quickly and accurately. Any skill the reader has in such techniques is an obvious advantage in drawing graphs.

We wish to emphasize, however, that no matter how much one knows (or thinks he knows) about elaborate methods of drawing curves, the primitive method of plotting a few points should not be forgotten. Some students go limp and helpless when faced with the task of graphing a simple function because it does not fit into known standard forms or because the required standard form has been forgotten. Nothing wilts the spirit of a teacher faster; he is tempted to advise the student immobilized by such a problem, "Don't think. Plot!"

Let us illustrate the previous paragraph by graphing those points satisfying $4x^2 - 24x - y^2 - 4y + 16 = 0$. It is naturally understood that any reader of this book immediately sees that this equation describes a hyperbola, that the equation can be put into a standard form, and a graph drawn presto. But to make our point, let us suppose that among our readers there is someone who has forgotten some detail about conic sections. We maintain that such an unfortunate can still obtain a graph in reasonable time.

EXAMPLE A
Plot those points (x, y) that satisfy

$$4x^2 - 24x - y^2 - 4y + 16 = 0.$$

We note immediately that this cannot be a function of the form $y = F(x)$, for if we choose a value for x, we get a quadratic equation in y, which in general has either two or no real solutions; hence there is not one value of y determined by one value of x.

The foregoing remarks tell us that a vertical line corresponding to a fixed value of x cuts the desired curve in two places or not at all, if we allow ourselves to count a possible point of tangency as a point of double contact.

We could begin by choosing an x at random and solving the resulting equation for y, but we prefer a little more subtlety. We have learned that where squared terms and terms to the first degree in the same variable appear, it is almost always helpful to complete the square:

$$4(x^2 - 6x) - (y^2 + 4y) = -16,$$
$$4(x - 3)^2 - (y + 2)^2 = 16.$$

Solving for y,

$$(y + 2)^2 = 4(x - 3)^2 - 16,$$
$$y + 2 = \pm\sqrt{4[(x - 3)^2 - 4]},$$
$$y = -2 \pm 2\sqrt{(x - 3)^2 - 4}.$$

From this we see that there can be no points where $(x - 3)^2 - 4 < 0$. Thus, $x \geq 5$ or $x \leq 1$. A table of values gives

x	y
0	$-2 \pm 2\sqrt{5}$
1	-2 ± 0
5	-2 ± 0
6	$-2 \pm 2\sqrt{5}$

It is clear that there is symmetry about the line $y = -2$. Plotting our points and sketching a curve through them gives the desired graph (Fig. 2.8). ∎

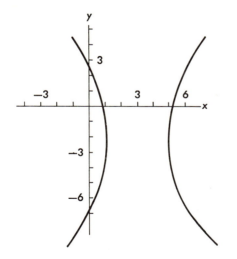

FIG. 2.8

Now let us turn our attention to graphing a function where no standard form exists to enable us to draw the graph immediately.

EXAMPLE B

Draw the graph for $y = G(x) = \dfrac{(x - 1)^2}{(x + 3)}$.

Whenever the numerator is zero, the graph has a point in common with the x axis. In this case, the so-called double root means the curve will be tangent, but will not cross the axis at $x = 1$. In any expression of this type, it is helpful to ask what happens to the values of y as x gets large. In order to find out, multiply the numerator and denominator by $1/x$.

$$y = \frac{(x^2 - 2x + 1)(1/x)}{(x + 3)(1/x)} = \frac{x - 2 + 1/x}{1 + 3/x}.$$

As x gets large, it is clear that the numerator increases while the denominator approaches 1. The quotient therefore increases. When x is large and negative, similar reasoning shows that the quotient becomes large and negative. Whenever the denominator is zero, we are in trouble. There is no corresponding value for y. Let us sketch what we know so far, erecting an invisible wall at $x = -3$, where we know no graph exists. The curve is well behaved except where $x = -3$, so the graph is a connected curve away from $x = -3$. The picture we have so far is indicated in Fig. 2.9.

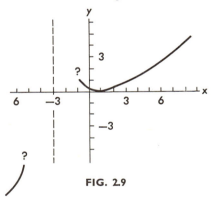

FIG. 2.9

The trouble spot at $x = -3$ now needs our attention. Let us indulge human nature. Aware that there is trouble at $x = -3$, we naturally do our best to get as close to it as possible. We thus plot values for x close to -3.

x	y
-2	9
$-\frac{5}{2}$	$\dfrac{\frac{49}{4}}{\frac{1}{2}} = \frac{49}{2}$
-4	-25
$-\frac{7}{2}$	$\dfrac{\frac{81}{4}}{-\frac{1}{2}} = -\frac{81}{2}$

We can now complete our graph (Fig. 2.10).

The line $x = -3$ that the curve approaches but never reaches is called an asymptote. ▮

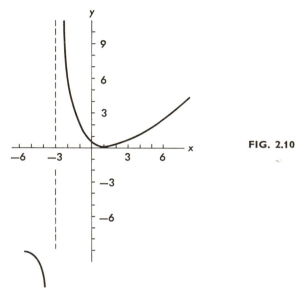

FIG. 2.10

PROBLEMS

Plot those points (x, y) that satisfy the following equations.

B · $y = \dfrac{x^2 - 1}{x + 1}$.

C · $y = \dfrac{x^2 + x - 6}{x^2 - 1}$.

D · $y = \dfrac{x^2}{(x - 3)^2}$.

Example B together with Problems B, C, and D have directed attention to the problem of graphing functions defined by the quotient of two polynomials. These are called *rational functions*. Problem B suggests how to proceed with such functions when the numerator and denominator have a common factor. Recognize that the function is not defined where this factor is zero, but that for all other x the common factor(s) may be divided out. When the numerator and denominator are free of common factors, we then observe the general rule that if a factor of $(x - a)^n$ occurs in the numerator, the graph either cuts or is tangent to the x axis at $x = a$ according as n is odd or even. If $(x - b)^m$ occurs as a factor in the denominator, there is a vertical asymptote at b, and the graph disappears in opposite directions (as in Problem C) or in the same direction (as in Problem D) on the two sides of the asymptote according as m is odd or even.

2·3 EXERCISES

Plot those points (x, y) that satisfy the following equations.

1. $y = 3x - 2$.
2. $y = x^2 - 3x - 4$.
3. $y = x^2 - 5x + 6$.
4. $y = 2x + 3$.
5. $y = (x + 2)(x - 1)(x - 4)$.
6. $y = (x + 3)(x - 1)(x - 5)$.
7. $x^2 - 2x - y^2 - 4y = 0$.
8. $6y^2 - 6y - x + 2 = 0$.
9. $x^2 - 2x - 2y = 0$.
10. $x^2 - 4x - y^2 - 6y - 5 = 0$.
11. $4x^2 + 4x + y^2 - 4y = 0$.
12. $9x^2 - 6x + 1 + 9y^2 + 12y = 0$.
13. $4x^2 - 20x + 4y^2 + 4y - 4 = 0$.
14. $9x^2 - 12x + y^2 = 0$.

15. $y = \dfrac{x - 3}{x^2 + x}$.
16. $y = \dfrac{x + 2}{x^2 - 3x}$.

17. $y = \dfrac{x^2 - 5x + 6}{x(x - 1)^2}$.
18. $y = \dfrac{x}{x^2 - 5x + 6}$.

19. $y = \sqrt{(x - 3)^2}$.
20. $y = \sqrt{(x + 1)^2}$.

2·4 This Is the Limit

Before proceeding with more graphs, we pause to introduce some terminology that is helpful in describing the behavior of a function. A function F having a graph like the one indicated in Fig. 2.11 is commonly said to have a " hole " at $x = a$ (as in Problem 2.3B where the " hole " occurs at $x = -1$). It also has a "jump" at $x = b$ (as does the postage function M of Example 2.1D). Finally, as x continues to increase, the values of $F(x)$ get closer and closer to q. Describing the behavior of F with a little more detail:

(1) As x approaches a (from either left or right), the values $F(x)$ get closer and closer to r.

(2) As x approaches b from the left, the values $F(x)$ get closer and closer to t.

(3) As x approaches b from the right, the values $F(x)$ get closer and closer to s.

(4)   As x increases without bound, the values $F(x)$ get closer
and closer to q.

The concept of a set of values "approaching" or "getting closer and
closer" to a fixed number is usually embodied by mathematicians in the
word *limit*. When we speak of an independent variable approaching a as a
limiting value, we write $x \to a$. When we think of a function $F(x)$ having r
as a limiting value when x is approaching a, we write

$$\lim_{x \to a} F(x) = r, \tag{1'}$$

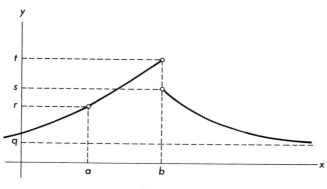

FIG. 2.11

which summarizes statement (1). It is important to notice that nothing at
all is said about the value of F at $x = a$. $F(a)$ may not even be defined. This
was the case in Example 2.3B where $G(-3)$ was undefined. We were still able
to ask what happened to the values of $G(x)$ as x got close to -3.

A second kind of behavior is illustrated by our function F graphed in
Fig. 2.11. As x gets close to b, it makes a difference as to the direction from
which it comes. To distinguish the two possibilities, we write $x \to b^-$ when
x comes toward b from the left side (from the negative side of the x axis)
and $x \to b^+$ when x comes from the right (the positive side of the x axis).
Statements (2) and (3) can then be written

$$\lim_{x \to b^-} F(x) = t, \tag{2'}$$

$$\lim_{x \to b^+} F(x) = s. \tag{3'}$$

We again emphasize that statements like (2') and (3') say nothing at all
about the value of $F(b)$. Figure 2.11 indicates that $F(b) = t$, but there is no
way to get this information from (2') or (3').

Statements (2') and (3') do tell us that it is not meaningful to talk about the limit of $F(x)$ as x approaches b. We must specify the direction of approach. When the direction of approach does not matter, then we may talk about the limit as in (1). One way to indicate that we do not care whether x is to the left or to the right of a is to introduce the notion of *absolute value*.

DEFINITION A

$$|x - a| = \begin{cases} x - a & \text{if} & x \geq a, \\ a - x & \text{if} & x \leq a. \end{cases}$$

PROBLEM

A · Show that absolute value has the following properties:

(a) $|x - a| = |a - x|$.

(b) $|x| = \sqrt{x^2}$.

(c) $|t + b| \leq |t| + |b|$ for any two numbers t and b.

(d) $|t - b| \leq |t| + |b|$ for any two numbers t and b.

(e) The graph of $F(x) = |x|$ has a sharp corner.

It is now easy to require that x be close to a without saying whether x is to the left or to the right of a. We merely require that $|x - a|$ be small. Similarly, to say that $F(x)$ is close to r without committing ourselves as to which of $F(x)$ or r is bigger, we may say that $|F(x) - r|$ is small.

DEFINITION B

If $|F(x) - r|$ can be made arbitrarily small by requiring that $|x - a|$ be small, $x \neq a$, then we say that the limit of $F(x)$ as x approaches a is r; that is, $\lim_{x \to a} F(x) = r$.

It should be clear that $\lim_{x \to a} F(x) = r$ if and only if $\lim_{x \to a^-} F(x) = r$ and $\lim_{x \to a^+} F(x) = r$. We have also been careful in our definition to say that $F(x)$ does not have to be close to r at $x = a$. It may not even be defined.

PROBLEM

B · Suppose

$$F(x) = \begin{cases} -x + 3 & \text{for} & x < 2, \\ -\tfrac{1}{4}x^2 + 2 & \text{for} & x \in (2, 4), \\ x - 4 & \text{for} & x \geq 4. \end{cases}$$

(a) Find $\lim_{x \to 0^-} F(x)$ and $\lim_{x \to 0^+} F(x)$.
(b) Find $\lim_{x \to 2^-} F(x)$ and $\lim_{x \to 2^+} F(x)$.
(c) Find $\lim_{x \to 4^-} F(x)$ and $\lim_{x \to 4^+} F(x)$.
(d) Does $\lim_{x \to x_0} F(x)$ exist for $x_0 = 0$? For $x_0 = 2$? For $x_0 = 4$?

We have delayed a concise statement of (4) because this involves the notion of x increasing without bound, or as it is sometimes put, x going to infinity. The common symbol here is $x \to \infty$. We wish to emphasize that we are learning shorthand here. We should not think of ∞ as some mystical point "at the end of the line." Neither should we attempt some connection with our instruction in theology. We simply mean by $x \to \infty$ that x gets large without bound. Similarly, $x \to -\infty$ indicates that x is negative and getting numerically large without bound. The symbol ∞ also find its way into the description of intervals: $x \in [a, \infty)$ means $a \le x$; $x \in (-\infty, a)$ means $x < a$. We can now write

$$\lim_{x \to \infty} F(x) = q. \tag{4'}$$

With our notation, let us return to Example 2.3B. We can say

$$\lim_{x \to -\infty} G(x) = -\infty,$$

$$\lim_{x \to \infty} G(x) = \infty,$$

$$\lim_{x \to -3^-} G(x) = -\infty,$$

$$\lim_{x \to -3^+} G(x) = \infty.$$

It turns out that limits play a very important role in the study of calculus. Unfortunately, it also turns out that limits can be very tricky.

EXAMPLE A
Consider the function defined by $F(h) = (1 + h)^{1/h}$. What do you think happens to the values $F(h)$ as h gets close to 0? Most people, unless they have been forewarned, guess that $F(h)$ gets closer and closer to 1. Would you believe that the right answer is about 2.7? Give a reason—any kind of reason. (You might try a few small values of h, for example.) Call the answer e.

You are now supposed to have less confidence in your ability to "see" what the answer should be. What shall we do in these circumstances? A thorough (well, almost) study of limits has been suggested as a prelude to the study of calculus. Our own view is that this hard work discourages much of the potential audience before we ever get to the excitement of the course.

We shall proceed as follows. We will list the principal facts that we need about limits. We have, as a matter of fact, already used these facts a number of times. All of them seem intuitively reasonable, but then so did the answer 1 in Example A. The difference is that these facts, which we state as theorems, can be proved to be so.

Facts about Limits

THEOREM A

If the function F is a polynomial function, then $\lim_{x \to x_0} F(x) = F(x_0)$. (After discussing the meaning of continuity, we shall see that for any function F continuous at x_0, $\lim_{x \to x_0} F(x) = F(x_0)$.)

THEOREM B

If $\lim_{h \to a} F(h)$ and $\lim_{h \to a} G(h)$ both exist, at least one being finite, then

$$\lim_{h \to a} [F(h) \pm G(h)] = \lim_{h \to a} F(h) \pm \lim_{h \to a} G(h).$$

THEOREM C

If $\lim_{h \to a} F(h)$ and $\lim_{h \to a} G(h)$ both exist and are finite, then

$$\lim_{h \to a} F(h)G(h) = \lim_{h \to a} F(h) \lim_{h \to a} G(h).$$

THEOREM D

If $\lim_{h \to a} F(h)$ and $\lim_{h \to a} G(h)$ both exist, at least one being finite, then

$$\lim_{h \to a} \frac{F(h)}{G(h)} = \frac{\lim_{h \to a} F(h)}{\lim_{h \to a} G(h)} \qquad \text{whenever} \qquad \lim_{h \to a} G(h) \ne 0.$$

Let us now see how these facts can be used to aid us in the problem started in the last section. We had learned that in graphing rational functions, that is, functions of the form

$$R(x) = \frac{a_n x^n + \cdots + a_1 x + a_0}{b_m x^m + \cdots + b_1 x + b_0}, \tag{5}$$

it was helpful to know the location of the zeros and the vertical asymptotes. It is also useful to know and easy to find out what happens for large values of x, that is, to determine $\lim_{x \to \infty} R(x)$.

EXAMPLE B

Find $\lim_{x\to\infty} \dfrac{3x^2 + x}{2x - 7}$.

The trick is to think of multiplying by $1/x$. From then on, one only needs to apply the theorems:

$$\frac{(3x^2 + x)(1/x)}{(2x - 7)(1/x)} = \frac{3x + 1}{2 - (7/x)}.\tag{6}$$

Thus,

$$\lim_{x\to\infty} \frac{3x^2 + x}{2x - 7} = \lim_{x\to\infty} \frac{3x + 1}{2 - (7/x)}$$

$$= \frac{\lim_{x\to\infty} (3x + 1)}{\lim_{x\to\infty} (2 - (7/x))}, \qquad \text{by Theorem D}$$

$$= \frac{\lim_{x\to\infty} (3x + 1)}{\lim_{x\to\infty} 2 - \lim_{x\to\infty} (7/x)} \qquad \text{by Theorem B}$$

To justify fully our use of Theorem D above, we need to be certain that the limit in the denominator is not 0. This is clear, however, because the constant 2 is independent of x, and $7/x$ goes to zero with increasing x, so the denominator approaches 2. We cannot use Theorem B to justify writing $\lim_{x\to\infty} (3x + 1) = \lim_{x\to\infty} 3x + \lim_{x\to\infty} 1$ in the numerator, but it is easy to see directly that $3x + 1$ increases as x increases, so the numerator and hence the entire expression goes to infinity. ∎

Of course no one (except authors of calculus texts) ever writes out all the details included in Example B. The answer is apparent from (6). On the other hand, because of nasty illustrations like Example A, it is of comfort to know that in the case of rational functions, all the "obvious" steps can be justified.

PROBLEMS

C · Find $\lim_{x\to\infty} \dfrac{3x^2 + 6x}{2x^3 - 7x^2 + 11}$ and $\lim_{x\to-\infty} \dfrac{3x^2 + 6x}{2x^3 - 7x^2 + 11}$.

D · Find $\lim_{x\to\infty} \dfrac{x^2 + 4x - 3}{x^2 - 4x + 3}$. Graph the corresponding rational function.

E · Consider the rational function $R(x)$ described in (5). Discuss $\lim_{x\to\infty} R(x)$ for the three cases $n > m$, $n = m$, and $n < m$.

F • Draw a graph of $G(x) = \dfrac{(x^2 + 3x)(x - 1)}{x - 1}$.

G • Graph $S(x) = 4(x - 2)^{2/3}/x^2$. Caution: Is this a rational function? Read the paragraphs following Problem 2.3A.

The next example shows our ability to graph rational functions can be utilized in graphing functions of the form $y = \sqrt{R(x)}$.

EXAMPLE C
Draw the graph for

$$y = H(x) = \sqrt{\frac{4(x + 2)(x - 3)}{(x - 1)^2}}.$$

Our method is to sketch very lightly a graph of

$$R(x) = [4(x + 2)(x - 3)]/(x - 1)^2.$$

The graph of R cuts (not tangent this time) the x axis at -2 and 3. To investigate the behavior of R for large values of x, write

$$R(x) = \frac{(4x^2 - 4x - 24)(1/x^2)}{(x^2 - 2x + 1)(1/x^2)} = \frac{4 - 4/x - 24/x^2}{1 - 2/x + 1/x^2};$$

$$\lim_{x \to \infty} R(x) = 4 \qquad \text{and} \qquad \lim_{x \to -\infty} R(x) = 4.$$

The trouble spot this time is at $x = 1$. We need to observe values of $R(x)$ as x approaches 1.

x	$R(x)$
0	-24
$\dfrac{1}{2}$	$\dfrac{-25}{1/4} = -100$
$\dfrac{3}{2}$	$\dfrac{-21}{1/4} = -84$

We thus get the graph indicated in Fig. 2.12. Clearly

$$\lim_{x \to 1^-} R(x) = -\infty \qquad \text{and} \qquad \lim_{x \to 1^+} R(x) = -\infty.$$

To get the graph of H from that of R, note that we are seeking the square root of the values of $R(x)$. Where this is impossible (where $R(x)$ is negative),

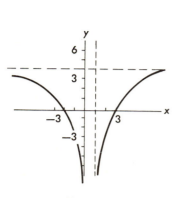

FIG. 2.12

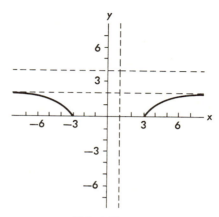

FIG. 2.13

there will be no graph for H. Elsewhere, we simply guess at the square root of our lightly sketched curve (Fig. 2.13). ∎

Suppose we had originally been required to sketch the points satisfying

$$y^2 = \frac{4(x + 2)(x - 3)}{(x - 1)^2}.$$

This is most easily accomplished by sketching the two functions

$$H_1(x) = \sqrt{\frac{4(x + 2)(x - 3)}{(x - 1)^2}},$$

$$H_2(x) = -\sqrt{\frac{4(x + 2)(x - 3)}{(x - 1)^2}}.$$

We have already sketched H_1; H_2 is the reflection of H_1 in the x axis (Fig. 2.14).

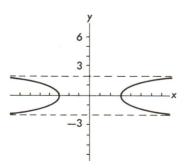

FIG. 2.14

We have already indicated that limits will play a central role in our future work. Example A alerted us to the fact that the limit of a function for specified behavior of the variable is not always easy to determine. Sometimes (Example C) a limit is more apparent after some algebraic manipulation, a technique we use again in Example D below. We hasten to add, however, that the principal reason for including this example is to have the result available when we need it in Chapter 3. This aims to assure the reader that the strenuous demands here made on our algebra are not typical of what is to come.

EXAMPLE D

$$\text{Find} \lim_{h \to 0} \frac{\sqrt[3]{8 + h} - \sqrt[3]{8}}{h} \tag{7}$$

Since both numerator and denominator approach zero as h approaches zero, it is not easy to say what happens to the quotient. We are on the right track when we notice that the numerator is the difference of two cube roots. This reminds us of the algebraic identity

$$a^3 - b^3 = (a - b)(a^2 + ab + b^2).$$

The reader may protest that the numerator of (7) does not remind him of this identity; that in fact he still does not get the connection. We defend our introduction of this identity on several counts:

(a)   The fact remains that in the numerator of (7) we do have the difference of two cube roots, and this, coupled with some experience, does suggest the indicated identity.

(b)   The reader was warned at the outset that we would make some real demands from his algebraic background in this section. (By way of comfort, we also promised that we will find some methods later on that offer some relief from this strain on our algebra.)

We see how our identity is to be used by setting $a = \sqrt[3]{8 + h}$; $b = \sqrt[3]{8} = 2$. Then $a^3 - b^3 = h$, and the identity in question gives

$$h = [\sqrt[3]{8 + h} - \sqrt[3]{8}][(\sqrt[3]{8 + h})^2 + 2\sqrt[3]{8 + h} + 4].$$

Our problem is equivalent to

$$\lim_{h \to 0} \frac{[\sqrt[3]{8 + h} - \sqrt[3]{8}]}{h} \cdot \frac{[(\sqrt[3]{8 + h})^2 + 2\sqrt[3]{8 + h} + 4]}{[(\sqrt[3]{8 + h})^2 + 2\sqrt[3]{8 + h} + 4]}.$$

In this form, however, the numerator is just h, so after obvious simplification, we get

$$\lim_{h \to 0} \frac{\sqrt[3]{8+h} - \sqrt[3]{8}}{h} = \lim_{h \to 0} \frac{1}{[(\sqrt[3]{8+h})^2 + 2\sqrt[3]{8+h} + 4]}$$

$$= \frac{1}{(\sqrt[3]{8})^2 + 2\sqrt[3]{8} + 4} = \frac{1}{12}. \quad \blacksquare$$

PROBLEM

H · Find $\lim\limits_{h \to 0} \dfrac{\sqrt{4+h} - \sqrt{4}}{h}$.

Compute the following limits.

1 · $\lim\limits_{h \to \infty} \dfrac{3h+1}{2h-3}$.

2 · $\lim\limits_{h \to \infty} \dfrac{4h - h^2}{2h+1}$.

3 · $\lim\limits_{h \to 0} \dfrac{3h+1}{2h-3}$.

4 · $\lim\limits_{h \to 0} \dfrac{4h - h^2}{2h+1}$.

5 · $\lim\limits_{h \to 0} h(1+h)^{1/h}$.

6 · $\lim\limits_{h \to 0} \dfrac{(1+h)^{1/h}}{h}$.

7 · $\lim\limits_{h \to 0} (1+h)^{2/h}$.

8 · $\lim\limits_{h \to 0} (1+h)^{1/(2h)}$.

9 · $\lim\limits_{h \to 0-} \dfrac{1}{h}$.

10 · $\lim\limits_{h \to 0+} \dfrac{1}{h}$.

11 · $\lim\limits_{h \to 0+} \dfrac{1}{1 + 2^{1/h}}$.

12 · $\lim\limits_{h \to 1+} \dfrac{1}{1 + 3^{1/(h-1)}}$.

13 · $\lim\limits_{h \to 0-} \dfrac{1}{1 + 2^{1/h}}$.

14 · $\lim\limits_{h \to 1-} \dfrac{1}{1 + 3^{1/(h-1)}}$.

15 ·* $\lim\limits_{h \to 0} (1 + 2h)^{1/h}$.

16 ·* $\lim\limits_{h \to \infty} \left(1 + \dfrac{2}{h}\right)^h$.

Graph the following.

17. $y = \dfrac{(x+3)^2(x-2)}{x^2+x}.$

18. $y = \dfrac{4x^2 - 8x - 12}{x^2 - x - 2}.$

19. $y^2 = \dfrac{(x+3)^2(x-2)}{x^2+x}.$

20. $y^2 = \dfrac{4x^2 - 8x - 12}{x^2 - x - 2}.$

21. $y = \dfrac{4x^2 - 4x - 8}{x^2 - 2x - 3}.$

22. $y = \dfrac{x^2 + x}{(x+3)(x-2)^2}.$

23. $y^2 = \dfrac{4x^2 - 4x - 8}{x^2 - 2x - 3}.$

24. $y^2 = \dfrac{x^2 + x}{(x+3)(x-2)^2}.$

25*. $y = \dfrac{\sqrt{x+4} - 2}{x}.$

26*. $y = \dfrac{2 - \sqrt{x+4}}{2x\sqrt{x+4}}.$

2·5   Composition of Functions; the Inverse

We are about to describe in words the situation pictured in Fig. 2.15. Continual attention to the picture should be helpful.

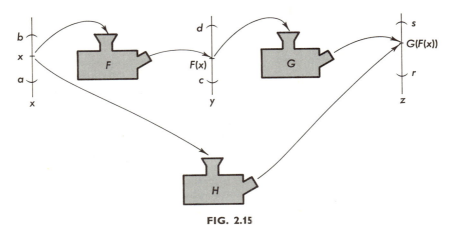

FIG. 2.15

Suppose F is a function defined for points $x \in (a, b)$. We will further suppose that the range of F is contained in (c, d), and that G is a second function, being defined for all $y \in (c, d)$. The range of G will be assumed to fall among the points $z \in (r, s)$. We have so arranged things as to make it possible to feed the output of F right into G. Thus, if x is put into F, $G(F(x))$ comes out of G.

Making use of the two functions just described, we now define a third function H. The domain of H is (a, b). We define $H(x) = G(F(x))$. H is called the composite of G with F. It is sometimes written $H = G \circ F$.

EXAMPLE A
Suppose $(a, b) = (0, \infty)$, $(c, d) = (-1, 1)$,

$$F(x) = \frac{x}{x + 1}, \quad \text{and} \quad G(x) = \frac{1}{x + 3}.$$

Find the formula for $H = G \circ F$; also, find one for $R = F \circ G$.

$$H(x) = G(F(x)) = \frac{1}{F(x) + 3} = \frac{1}{x/(x + 1) + 3}.$$

The easiest way to simplify this expression is to multiply the numerator and denominator by $(x + 1)$. This gives

$$H(x) = \frac{x + 1}{x + 3(x + 1)} = \frac{x + 1}{4x + 3}. \tag{1}$$

To demonstrate the use of this expression for H, consider "dropping" a 1 into F: $F(1) = \frac{1}{2}$. Drop this result in G: $G(\frac{1}{2}) = \frac{2}{7}$. Since $H(1) = G(F(1))$, we should get $\frac{2}{7}$ by setting $x = 1$ in our expression (1). We do.

$$R(x) = F(G(x)) = \frac{G(x)}{G(x) + 1} = \frac{1/(x + 3)}{1/(x + 3) + 1}.$$

The easiest way to simplify this expression is to multiply the numerator and denominator by $(x + 3)$. This gives

$$R(x) = \frac{1}{1 + 1(x + 3)} = \frac{1}{x + 4}. \ \blacksquare$$

It is instructive here to comment upon an error made in the first edition of this book. Example A appeared without specifying the intervals (a, b) and (c, d). This error violates in a fundamental way the very notion we are trying to illustrate.

Let us see why the error was so serious. With no restriction on (a, b), there seems to be no reason why we could not set $x = -1$; indeed $H(-1) = 0$. But if one notes that F, hence $G \circ F = H$ is not defined at $x = -1$, it is clear that something is wrong. We see immediately that (a, b) must be chosen to avoid $x = -1$.

There is still another problem. Suppose we chose $(a, b) = (-1, \infty)$. Then we could choose $x = -\frac{3}{4}$ and obtain $F(-\frac{3}{4}) = -3$. But according to the theory above, the range of F must be contained in a domain (c, d) on which G is defined. Thus, the range of F cannot contain -3; and this in turn means that the domain of F cannot contain $-\frac{3}{4}$.

The reader can convince himself that if the domain of F is $(0, \infty)$, the range is $(0, 1)$. Hence, (c, d) must be chosen to include $(0, 1)$ but to exclude -3. The knowledge that authors of mathematics texts also make mistakes or overlook something may comfort students. Authors themselves take little comfort in this.

There is an important special case of a composite function that deserves our attention. Suppose that it happens that (a, b) is contained in (r, s), and that every point x is sent back into itself. This means $H(x) = x$. In this situation, Fig. 2.15 could be drawn differently (Fig. 2.16) to emphasize the unusual behavior.

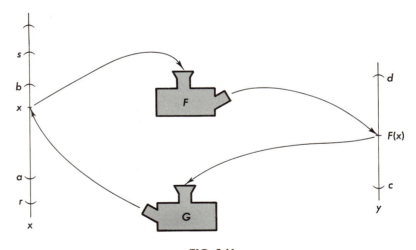

FIG. 2.16

When such a G can be found, it is called the *inverse of F*. The inverse of F is sometimes denoted by F^{-1} instead of introducing another letter. Since F stands for the function, $1/F$ is a meaningless symbol, so F^{-1} is not likely to cause any confusion. However, the notation $F^{-1}(y)$, meaning the *value of F^{-1} at y*, is different, in most cases, from $[F(y)]^{-1}$, which stands for the *reciprocal of the value of F evaluated at y*.

EXAMPLE B

Suppose $F(x) = \dfrac{x}{x+1}$.

  (a)   Find an expression for $[F(y)]^{-1}$.
  (b)   Show that the formula found in part (a) is not the formula for $F^{-1}(y)$.
  (c)   Find the formula for $F^{-1}(y)$.

(a)   By the formula for $F(x)$, it is immediately clear that

$$F(y) = \frac{y}{y+1},$$

$$[F(y)]^{-1} = \left[\frac{y}{y+1}\right]^{-1} = \frac{y+1}{y}.$$

(b)   Drop a 2 into F. Out comes $F(2) = \frac{2}{3}$. Drop this into $[F(y)]^{-1}$; the result is

$$[F(\tfrac{2}{3})]^{-1} = \frac{\frac{2}{3}+1}{\frac{2}{3}} = \frac{2+3}{2} = \frac{5}{2}.$$

Since we did not get our 2 back again, the formula for $[F(y)]^{-1}$ is not the formula for $F^{-1}(y)$. (There is nothing special about $x = 2$ here except that it is one of many numbers that illustrate our point. A point of logic: Suppose we had picked $x = a$ and found that $[F(F(a))]^{-1}$ was again a. This would not prove anything. Why ?)

(c)   To find the inverse, set $y = F(x)$ and solve the resulting equation for x.

$$y = \frac{x}{x+1},$$

$$xy - x = -y,$$

$$x = \frac{-y}{y-1}.$$

This now tells us how to get back to x if we know y.

$$F^{-1}(y) = \frac{-y}{y-1}.$$

We noted earlier that $F(2) = \frac{2}{3}$. From the formula just obtained,

$$F^{-1}(\tfrac{2}{3}) = \frac{-\frac{2}{3}}{\frac{2}{3} - 1} = \frac{-2}{2 - 3} = 2. \quad \blacksquare$$

Not all functions have an inverse. We recall that we might have $F(x_1) = F(x_2)$ for $x_1 \neq x_2$. Then G, being a function, cannot possibly send the point $y = F(x_1) = F(x_2)$ back to both x_1 and x_2. (See Fig. 2.17.) It is clear that F can have an inverse if and only if $F(x_1) = F(x_2)$ implies $x_1 = x_2$. Such a function is said to be one to one.

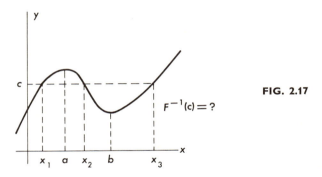

FIG. 2.17

$F^{-1}(c) = ?$

EXAMPLE C
Show that the function $S(x) = x^2$ defined in Example 2.1A has no inverse.

This follows from the fact that S is not one to one. For example, $S(2) = S(-2)$. $\quad \blacksquare$

We have emphasized that not every function has an inverse. We might also point out that even when the inverse exists, it is not always possible to find it; $y = F(x) = x + 10^x$ illustrates this unhappy truth.

Suppose that G is the inverse of F, so that if $y = F(x)$; then $G(y) = x$. It follows (see Fig. 2.18) that

$$F(G(y)) = F(x) = y. \tag{2}$$

This means that F is the inverse of G, providing that we restrict G to that subset of its domain that may be obtained as the range of F. The latter qualification is necessary. G may be defined on more than the range of F, but we could hardly expect (2) to hold if the y involved in this equation were not obtainable as the image of some x under F. This is illustrated in Example D.

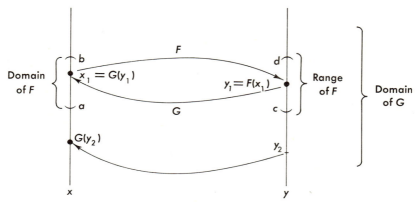

FIG. 2.18

EXAMPLE D

Show that the function $P(x) = \sqrt{x}$, $x \geq 0$ (see Fig. 2.19), defined in Example 2.1C has the function S (Example 2.1A) as its inverse.

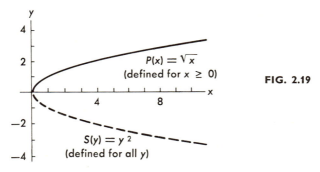

FIG. 2.19

We set $y = \sqrt{x}$. Solving for x, $x = y^2$. Therefore, $P^{-1}(y) = y^2 = S(y)$. We may also verify this fact from the computation

$$S \circ P(x) = S(P(x)) = S(\sqrt{x}) = (\sqrt{x})^2 = x.$$

Quoting from the paragraph that precedes Example D, with appropriate substitutions for the functions involved, "this means that P is the inverse of S, providing that we restrict S to that subset of its domain which may be obtained as the range of P." In this case, the range of P consists of all nonnegative numbers. Hence, if we restrict the domain of S to the nonnegative numbers, then P will be an inverse.

We saw in Example C that S, defined on all real numbers, has no inverse. We now see that S, if restricted to the nonnegative real numbers, has P for its inverse.

$$P \circ S(x) = P(S(x)) = P(x^2) = \sqrt{x^2} = x.$$

(The last step makes clear use of the fact that x is nonnegative, for it appeals to (1) of Section 2.1.)

This technique is very important. It will be used again. Let us review it. We begin with a function F that sends several members of its domain into the same member of the range. F thus has no inverse. We then restrict the domain of F in some way (that is, we imagine it to be defined for only a portion of the original domain) so that the correspondence between the restricted domain and the range is one to one; that is, there is one and only one member in the domain for each member of the range. Then we can define an inverse. The function F pictured in Fig. 2.16 has no inverse. But if we restrict F to the interval $[a, b]$, then an inverse could be defined.

The choice of the restricted domain is not unique. This means that if different people select the restricted domain, the function may be defined on different points, and the inverses will therefore be different.

EXAMPLE E

Suppose we restrict the domain of $S(x) = x^2$ to the nonpositive real numbers. Show that S^{-1} then exists, and that it is the function $N(x) = -\sqrt{x}$ defined in Example 2.1B.

Set

$$y = x^2, \qquad x \le 0;$$

then

$$x = -\sqrt{y}$$

where we choose the minus sign because $\sqrt{y}$ is always positive or zero while x is always negative or zero.

Thus

$$S^{-1}(y) = -\sqrt{y} = N(y).$$

We may also verify this fact from the computation

$$N \circ S(x) = N(S(x)) = N(x^2) = -\sqrt{x^2} = -(-x) = x.$$

Note our use here of the fact that since $x < 0$, $\sqrt{x^2} = -x$. ∎

Our last example illustrates the way in which a graph is sometimes useful when we seek an inverse.

EXAMPLE F

Let

$$F(x) = \frac{x^2 - 2x}{(x - 1)^2}.$$

Suppose we wish to find an expression for F^{-1} (if it exists) near $x = 3$.

From the graph (Fig. 2.20) we see that the function is not one to one. It also appears, however, that it will be one to one if we restrict the domain to $x > 1$. (Why not choose $x < 1$?)

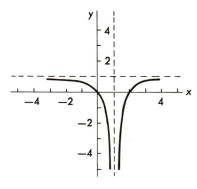

FIG. 2.20

Set

$$y = \frac{x^2 - 2x}{(x - 1)^2},$$

$$y(x^2 - 2x + 1) = x^2 - 2x.$$

We wish to solve this expression for x; we therefore treat it as a quadratic equation in x.

$$(y - 1)x^2 + (2 - 2y)x + y = 0,$$

$$x = \frac{2y - 2 \pm \sqrt{4y^2 - 8y + 4 - 4y(y - 1)}}{2(y - 1)},$$

$$x = 1 \pm \frac{\sqrt{1 - y}}{y - 1}. \tag{3}$$

It is clear from the graph that $y - 1$ is always negative, since $y < 1$. Of course, $\sqrt{1 - y}$ is positive, so if x is to be greater than 1, it is clear that we must use the minus sign in (3):

$$F^{-1}(y) = 1 - \frac{\sqrt{1 - y}}{y - 1} = 1 + \frac{\sqrt{1 - y}}{1 - y} = 1 + \frac{1}{\sqrt{1 - y}}.$$

We were interested in an inverse that worked at $x = 3$.

$$F(3) = \frac{9 - 6}{4} = \tfrac{3}{4},$$

$$F^{-1}(\tfrac{3}{4}) = 1 + \frac{1}{\sqrt{1 - \tfrac{3}{4}}} = 1 + 2 = 3. \quad \blacksquare$$

PROBLEMS

A • Find an inverse for $G(x) = (2x + 1)/(x - 3)$ that works for values of x near 5.

B • Find an inverse for $H(x) = x/(x^2 + 1)$ that works for values of x near 2.

2·5 EXERCISES

For each of the functions that follow, find an inverse valid for values of x near $x_0 = 2$.

1 • $F(x) = 2x + 3$.

2 • $G(x) = 3x - 2$.

3 • $F(x) = \dfrac{1}{x + 2}$.

4 • $G(x) = \dfrac{2}{3x + 1}$.

5 • $F(x) = 4 + \sqrt{2x}$.

6 • $G(x) = 3 - \sqrt{2x}$.

7· $F(x) = \dfrac{2x-1}{x+1}$.

8· $G(x) = \dfrac{3x+2}{x+2}$.

9· $F(x) = x^2 - 3x$.

10· $G(x) = 2x^2 - 10x$.

11· $F(x) = \dfrac{4x^2}{(x+1)^2}$

12· $G(x) = \dfrac{4(x-3)^2}{(x-1)^2}$.

13· For the functions F and G as defined in Exercises 3 and 4, find a formula for $F \circ G$.

14· For the functions F and G as defined in Exercises 7 and 8, find a formula for $F \circ G$.

15· For the functions F and G as defined in Exercises 3 and 4, find a formula for $G \circ F$.

16· For the functions F and G as defined in Exercises 7 and 8, find a formula for $G \circ F$.

2·6   Well-Behaved Functions

Although we do not want to be identified with the military establishment, we would like to exploit an analogy between the function pictured in Fig. 2.21 and a cannon. We have already insisted that our function F should

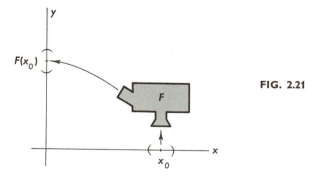

FIG. 2.21

always hit exactly the same point $F(x)$ whenever the same x is fed in. Now in the manufacture of ammunition, it is not possible to make every shell exactly the same size. The practice is to specify the desired size and tolerance limits within which the shells may deviate from the exact size. The assumption is that a (well-behaved) cannon will fire shells of approximately the same size into approximately the same spot. The tolerance permitted for the shells depends on the characteristics of the cannon itself, and perhaps on the deviation from the exact target point that will be acceptable.

So it is with a well-behaved function. If $y_0 = F(x_0)$, then points close to x_0 should land close to y_0. In this respect, the function M of Example 2.1D is not well behaved. Numbers close to $x_0 = 1$ but just a little bigger do not land close to $M(x_0) = 8$. It is unfortunate that out of the four examples presented in Section 2.1, the one for which we noted some practical use (in a post office) is the one that is not well behaved. The reader may be happy to know that many functions that arise in practical applications of mathematics are well behaved.

The particular kind of good behavior in which we have been interested here is formally called continuity. A function F defined in some interval about x_0 is called continuous at x_0 if the points close to x_0 are "fired" into points close to $y_0 = F(x_0)$.

We now work toward a more formal statement of what we mean by saying F is continuous at x_0. The first requirement is that F must be defined in some open interval containing x_0. Then we suppose that someone has specified a target area for us around $F(x_0)$. This is most easily done by giving an interval I about $F(x_0)$, say

$$I = (F(x_0) - \varepsilon, \ F(x_0) + \varepsilon),$$

where ε (the Greek letter epsilon) represents some positive number. If our cannon (function) is well behaved (continuous), we should then be able to specify an interval J of tolerance around x_0 from which we may choose shells (points) guaranteed to land in the target area. Again, the tolerance interval J is most easily described by

$$J = (x_0 - \delta, \ x_0 + \delta),$$

where δ (the Greek letter delta) represents some positive number.

> Continuity of F at x_0 merely means that given any target area I as described in the foregoing, we can specify a tolerance interval J so that if x is chosen from J, then F fires it into I.   (1)

The reader who understands the preceding statement understands what we mean by continuity. Only with reservation do we attempt any further definition of continuity in this section. A formal definition avoids such picturesque (and informative) notions as targets and shells being fired. It substitutes precise (and to beginners, frightening) symbols. Yet, there is something to be said in favor of at least an exposure to this language. So to provide such an exposure we shall digress briefly in order to introduce notation.

In describing an interval about a point b, we wish to indicate all the points t within a certain distance of b. If the distance is r, the difference between t and b is to be less than r. We do not care whether t is larger or smaller than b, just so the difference between the larger and the smaller is less than r. We therefore find it convenient to appeal again to the notion of absolute value, using it this time to define a *neighborhood* of b.

DEFINITION A (Neighborhood)

Let $r > 0$. The r neighborhood of b is the collection of all t for which $|t - b| < r$. We write

$$N_r(b) = \{t : |t - b| < r\},$$

which is read, "$N_r(b)$ is the set of all numbers t for which $|t - b| < r$."

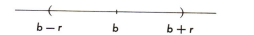

$$b - r \qquad b \qquad b + r$$

FIG. 2.22

The intervals I and J are thus

$$I = N_\varepsilon(F(x_0)) = \{y : |y - F(x_0)| < \varepsilon\},$$

$$J = N_\delta(x_0) = \{x : |x - x_0| < \delta\}.$$

We now give the formal definition of continuity that we promised (threatened). The reader should compare it carefully with (1), since the formal definition is a rewording of this statement.

DEFINITION B (Continuity at a Point)

The function F, defined in some open interval containing x_0, is said to be continuous at x_0 if and only if it satisfies the following. Given any neighborhood $N_\varepsilon(F(x_0))$ of $F(x_0)$, we can find a neighborhood $N_\delta(x_0)$ of x_0 so that if $x \in N_\delta(x_0)$, then $F(x) \in N_\varepsilon(F(x_0))$.

The condition for continuity can also be phrased this way. Given any $\varepsilon > 0$, we can find a $\delta > 0$ so that if $|x - x_0| < \delta$, then

$$|F(x) - F(x_0)| < \varepsilon.$$

Establishing Continuity of a Function F at x_0

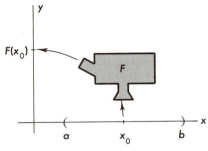

(a) Imagine yourself challenged by someone who doubts the continuity of the function F, known to be defined on an interval containing x_0.

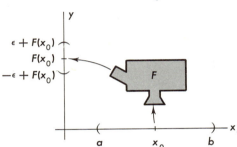

(b) Your opponent selects a target area T about $F(x_0)$. You may expect him to make it rather small, since his object is to make it hard for you to land inside the target area he specifies. But once the target area has been chosen, it cannot be changed while you are taking aim.

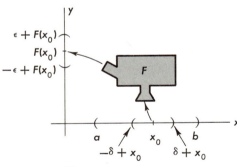

(c) After consideration of the function and the target area given, you must announce an interval of tolerance about x_0.

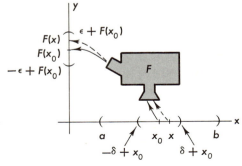

(d) Anyone may now choose a point x in your interval. The function must fire this point into the target interval, or else you have not been able to do your work correctly.

FIG. 2.23

A function F is said to be continuous on an interval (a, b) if it is continuous at each point of the interval. Suppose F is continuous on (a, b), and suppose $x_1 \in (a, b)$ and $x_2 \in (a, b)$. Given an $\varepsilon > 0$, we must be able to find δ_1 and δ_2 so that

$$\text{if} \quad |x - x_1| < \delta_1, \qquad \text{then} \qquad |F(x) - F(x_1)| < \varepsilon;$$
$$\text{if} \quad |x - x_2| < \delta_2, \qquad \text{then} \qquad |F(x) - F(x_2)| < \varepsilon.$$

Examination of Fig. 2.24 should help the reader understand that the value δ_1 which works near x_1 may not work near x_2. That is, we may be able to choose a larger interval of tolerance at one point than at another to get the same accuracy around the target area.

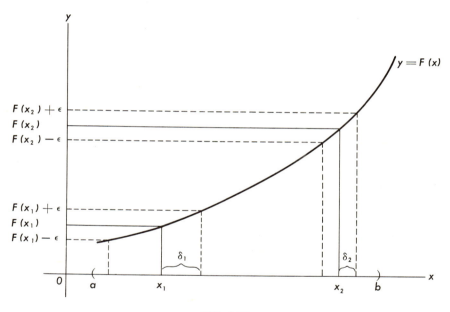

FIG. 2.24

The reader may already have seen some connection between the continuity of a function F at x_0 and the study of

$$\lim_{x \to x_0} F(x)$$

as discussed in Section 2.4. Since that discussion was kept informal, we will not attempt to prove that anything is continuous by reference to limits. But we can state the facts.

F is continuous at a point x_0 if and only if

(a) $\lim_{x \to x_0^-} F(x) = \lim_{x \to x_0^+} F(x) = b$;
(b) $F(x_0)$ is defined;
(c) $F(x_0) = b$.

Consider the function of Problem 2.4B:

$$F(x) = \begin{cases} -x + 3 & \text{for} & x < 2, \\ -\tfrac{1}{4}x^2 + 2 & \text{for} & x \in (2, 4), \\ x - 4 & \text{for} & x \geq 4. \end{cases}$$

At $x_0 = 0$, all three conditions are fulfilled. F is continuous at 0. At $x_0 = 2$, condition (a) is fulfilled, but (b) is not. F is not continuous at 2. At $x_0 = 4$, condition (b) is fulfilled, but (a) is not. F is not continuous at 4.

We have said that F is continuous on (a, b) if it is continuous at each $x \in (a, b)$. F is continuous on $[a, b]$ if it is continuous on (a, b), and if the obvious limits hold at the end points. Thus, $\lim_{x \to a^+} F(x) = F(a)$ and $\lim_{x \to b^-} F(x) = F(b)$.

Functions continuous on an interval have been studied intensively, and they play a very important role in mathematics. There are many who feel that the study of calculus requires at this point a proof of some of the basic properties of continuous functions. We shall content ourselves here with a statement of some of the basic properties we need.

Theorems about Continuous Functions   Let F and G be functions that are known to be continuous at x_0.

THEOREM A
H, defined by $H(x) = F(x) + G(x)$, is continuous at x_0.

THEOREM B
H, defined by $H(x) = F(x) \cdot G(x)$, is continuous at x_0.

THEOREM C
H, defined by $H(x) = F(x)/G(x)$, is continuous at x_0, providing that $G(x_0) \neq 0$.

THEOREM D
Let R be a function defined in a neighborhood of $F(x_0)$ which is continuous at $F(x_0)$. Then H, defined by $H = R \circ F$ is continuous at x_0.

THEOREM E

(Intermediate value property)   Suppose F is continuous on $[a, b]$, and that c is a real number between $F(a)$ and $F(b)$. Then there must be at least one value of $x \in [a, b]$, say $\bar{x}$, such that $F(\bar{x}) = c$.

THEOREM F

Suppose F is continuous on $[a, b]$. Then there is at least one $x \in [a, b]$, say x_1, for which $F(x_1) = M$ is the maximum value assumed by F in $[a, b]$. (This means that for any $x \in [a, b]$, $F(x) \leq M$.) Similarly, there is at least one $x \in [a, b]$, say x_2, for which $F(x_2) = m$ is the minimum value assumed by F in $[a, b]$ (that is, if $x \in [a, b]$, then $F(x) \geq m$).

THEOREM G

If $F(x_0) \neq 0$, then $F(x) \neq 0$ for all x in some neighborhood of x_0.

Figure 2.25 enables us to see that Theorems E and F seem reasonable.

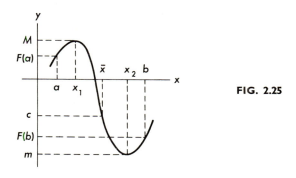

FIG. 2.25

They are easy to believe. It is unfortunately true that they are difficult to prove. They depend on some basic properties of the real number system. For a discussion leading up to their proof, see [2, pages 72–76].

PROBLEMS

A · F is called a constant function on (a, b) if for each $x \in (a, b)$, $F(x)$ is the same number, say c. Prove from the definition that if F is constant on (a, b), then F is continuous on (a, b).

B · Use the definition to prove that the function L defined by $L(x) = x$ is continuous at any point x_0.

C · Conclude, by use of Problems A and B along with any of the theorems you may need, that the polynomial P

$$P(x) = a_n x^n + a_{n-1} x^{n-1} + \cdots + a_1 x + a_0,$$

a_i constant for each i, is continuous at any point x_0.

D · A rational function R is a function for which $R(x)$ may be expressed as the quotient of two polynomials in x; that is

$$R(x) = \frac{P(x)}{Q(x)}$$

where P and Q are polynomials. What can you say about the continuity of such a function R?

E · Use the definition to prove that $G(x) = |2x - 4|$ is continuous at $x_0 = 2$.

F · Suppose we try to replace $[a, b]$ by (a, b) in the statement of Theorem F. Will it remain true?

G · The function M defined in Example 2.1D is continuous on $[\frac{1}{2}, \frac{3}{2}]$ except at one point. Show that Theorem E fails in this case. (Hence, the requirement that F be continuous on $[a, b]$ means just that; continuous at every point of $[a, b]$.)

H · Let $R(x) = \dfrac{32x - 8x^2 - 14}{2x + 1}$.

(a) Show that R is continuous on $[1, 6]$.

(b) By Theorem F, we know there is an x_1 so that if $x \in [1, 6]$, then $R(x) < R(x_1) = M$. You probably cannot find M. However, you can show that for $x \in [1, 6]$, $R(x) < 6$. (Hint: How large can the numerator get if $x \in [1, 6]$? How small can the denominator get?)

(c) Can you find an $\bar{x} \in [1, 3]$ such that $R(\bar{x}) = \sqrt{10}$? Answer yes or no. Give a reason.

We spoke in a deliberately vague way at the beginning of this chapter about well-behaved functions. Though standards of good behavior vary with different situations, we now have the technical background to be specific about at least one measure of good behavior, namely continuity. A function f defined on (a, b) is said to be *piecewise continuous* on (a, b) if it is continuous at all but a finite number of points in the interval. Many of the functions encountered in this chapter were not continuous, but they were piecewise continuous. Most functions met in elementary mathematics are piecewise continuous.

There are other properties soon to be mentioned that we like to require for really good behavior, but we have made a beginning.

2·7 Approximation

Much of mathematics is concerned with approximation. The value of π is approximated by $\frac{22}{7}$ or by 3.14 or by 3.1416 or by some more accurate decimal. But since the decimal expression for π never terminates, we are forced to use an approximation of some kind in any computation that involves π. In the same way we use approximations to $\sqrt{2}$, $\sqrt{3}$, and a host of other numbers. We use them not because we are lazy, but because there is nothing else we can do.

Consider $G(x) = \sqrt[3]{x}$. For this function, we can find $G(8) = 2$ exactly. We cannot find $G(8.4)$ exactly, however. Any answer expressed as a decimal will of necessity be an approximation. Since we wish to know the value of G at the point $8.4 = 8 + 0.4$, our problem is one that we may generally describe as follows.

The Approximation Problem

Suppose F is defined and well behaved in some interval containing x_0, and suppose $y_0 = F(x_0)$ is known. Can we devise a method for easily approximating $F(x_0 + h)$ in the event that h is small and that the exact value of $F(x_0 + h)$ is either very difficult or impossible to find?

The reader has no doubt worked on a number of problems of this sort. (Interpolation from a table is a nice example of this type of problem.) There are a number of reasonable ways to attack such problems. The procedure to be used will depend on the accuracy desired, the functions involved (are they well-enough known so that tables of values have been computed?), and the number of methods known to the problem solver. Our purpose here is to suggest a method that is at the heart of the study of calculus.

We begin by pointing out that the problem is extremely easy if the graph of F is a straight line. Since we know $y_0 = F(x_0)$, this line passes through (x_0, y_0) and from the point–slope form of a straight line (Section 1.3), its equation is

$$y = F(x) = y_0 + m(x - x_0).$$

Then
$$F(x_0 + h) = y_0 + mh.$$

Suppose, however, that F does not have a straight line as a graph. We describe our plan of attack in this situation by reference to Fig. 2.26. Let T

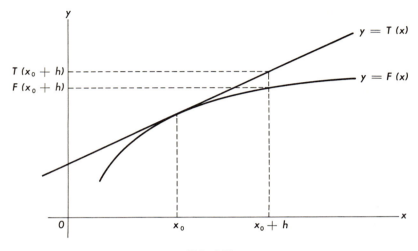

FIG. 2.26

be a function having as its graph a straight line that is tangent to the graph of F at $(x_0, F(x_0))$. (We will define a tangent line more formally later. Let your intuition be your guide here.) We know that $T(x_0 + h)$ is easy to compute. We also know, since the graph of T is a straight line through $(x_0, F(x_0))$, that

$$T(x) = F(x_0) + m(x - x_0). \tag{1}$$

From Fig. 2.26 it seems that if h is not too big, then $T(x_0 + h)$ will approximate $F(x_0 + h)$. We introduce a symbol $\approx$ to express this approximation:

$$T(x_0 + h) \approx F(x_0 + h).$$

What does all this say for the function G considered above? The graph is drawn in Fig. 2.27. Taking $x_0 = 8$, we see that (1) for our example gives

$$T(x) = G(8) - m(x - 8),$$
$$T(x) = 2 + m(x - 8). \tag{2}$$

Our problem is thus reduced to finding the slope m to use in (2). The remarkable thing is, as we shall see in the next chapter, that this problem is not very difficult.

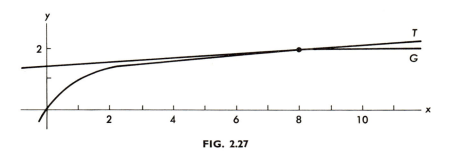

FIG. 2.27

2·7   EXERCISES

In Exercises 1–10, sketch a graph of F in a neighborhood of the indicated point x_0. Then sketch the tangent line at $(x_0, F(x_0))$ and write in the form (1) the formula of the function T having the tangent line as its graph.

1· $F(x) = \sqrt{x}$; $x_0 = 4$.

2· $F(x) = x^3$; $x_0 = 2$.

3· $F(x) = x^3 + 4x + 3$; $x_0 = 2$.

4· $F(x) = x^3 - x^2 + 1$; $x_0 = 2$.

5· $F(x) = \dfrac{5}{x+2}$; $x_0 = 3$.

6· $F(x) = \dfrac{3}{x+1}$; $x_0 = 5$.

7· $F(x) = \dfrac{\sqrt{x}}{x+1}$; $x_0 = 4$.

8· $F(x) = \dfrac{x}{3+\sqrt{x}}$; $x_0 = 4$.

9· $F(x) = \sqrt[3]{x^2} + \sqrt[3]{x}$; $x_0 = 8$.

10· $F(x) = x\sqrt{x} + \sqrt{x}$; $x_0 = 4$.

CHAPTER
3

THE DERIVATIVE

At the close of the last chapter we were confronted with a problem. We wanted to find a function T having as its graph a straight line. This line was to be tangent to the graph of a given function F at a given point $(x_0, F(x_0))$.

In the first section of this chapter we analyze the problem a bit further. We sharpen our understanding of just what is meant by saying that a line is tangent to a curve. Then after introducing the concept of a derivative, we solve our problem.

In Section 3.2 we review the terminology introduced in Section 3.1. Some helpful notations (abbreviations for ideas) are introduced.

In Section 3.3 we introduce a seemingly unrelated problem, that of determining the reading of an automobile speedometer from observations made by someone outside the car. We are quickly led to the same kind of computation we make in Section 3.1.

This computation, which turns up twice, receives our attention in Section 3.4. We learn some rules that relieve us of the algebraic gymnastics required in Sections 3.1 and 3.3. It is here that those who toil, however pessimistically, through the first three sections will be rewarded.

3·1   The Definition

We suppose the function F has a graph described by $y = F(x)$ that passes through $(x_0, y_0) = (x_0, F(x_0))$. We seek a function described by

$$T(x) = F(x_0) + m(x - x_0)$$

that will have as its graph a line tangent to the graph of F at (x_0, y_0). We are motivated by a desire to approximate $F(x_0 + h)$ by $T(x_0 + h)$, since the latter is easier to compute.

Suppose we let the difference between these values be r, or better, $r(x_0, h)$, since the difference depends on both x_0 and h.

$$F(x_0 + h) - T(x_0 + h) = r(x_0, h),$$

$$F(x_0 + h) - [F(x_0) + mh] = r(x_0, h),$$

$$F(x_0 + h) - F(x_0) = mh + r(x_0, h).$$

Multiplication by $1/h$ gives

$$\frac{F(x_0 + h) - F(x_0)}{h} = m + \frac{r(x_0, h)}{h}. \tag{1}$$

The left side of (1) represents the slope of the line k in Fig. 3.1. As h gets closer and closer to zero, line k approaches the graph of T as a limiting

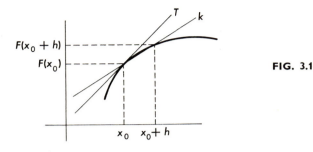

FIG. 3.1

position. This fact indicates how to define formally what we mean by saying that the graph of T is tangent to the graph of F at x_0.

DEFINITION A

The graph of the line $T(x) = F(x_0) + m(x - x_0)$ is said to be tangent to the graph of $y = F(x)$ at x_0 if and only if

$$m = \lim_{h \to 0} \frac{F(x_0 + h) - F(x_0)}{h}.$$

Thus, if T has a graph tangent to that of F at x_0, the left side of (1) approaches m. It follows that as h gets small, $r(x_0, h)/h$ must also be getting close to zero.

Let us pause to emphasize that we have not proved anything with the computation just presented. We have no way of knowing that a tangent line can always be found. Our next step will be to introduce a definition suggested by these computations. Since definitions are never proved, it is of no concern that in the work we just did, we proceeded as if nothing could ever go wrong.

One thing we learned in our preliminary work is that we are concerned with $[r(x_0, h)]/h$ rather than $r(x_0, h)$ itself. This suggests that we replace $r(x_0, h)$ by $hR(x_0, h)$. For reasons to appear later, we also use $|h|$ instead of h. Certainly $|h|$ gets small when h does, whether h is positive or negative.

DEFINITION B

Let F be a real-valued function defined in some interval containing x_0, and suppose there exists a number m for which, when h is small,

$$F(x_0 + h) = F(x_0) + mh + |h| R(x_0, h)$$

where $R(x_0, h)$ approaches zero whenever h approaches zero. Then we say F is differentiable at x_0 and that its derivative is the number m.

We have already admitted that such an m may not exist. In such a case, F is not differentiable at x_0. Suppose, on the other hand, that two such numbers say m_1 and m_2, exist. This would mean our definition left something to be desired, since the same function would have two different derivatives at a point. This cannot happen, however, as we now show. Suppose

$$F(x_0 + h) = F(x_0) + m_1 h + |h| R_1(x_0, h), \tag{2}$$

and

$$F(x_0 + h) = F(x_0) + m_2 h + |h| R_2(x_0, h). \tag{3}$$

It is, of course, possible that $R_1(x_0, h) \neq R_2(x_0, h)$. Subtracting (3) from (2) gives

$$0 = (m_1 - m_2)h + [R_1(x_0, h) - R_2(x_0, h)]|h|.$$

Since $|h| = \pm h$, depending upon whether h is positive or negative,

$$0 = h\{m_1 - m_2 \pm [R_1(x_0, h) - R_2(x_0, h)]\},$$

where the $\pm$ depends on the sign of h. Now h is not zero. (If it were, there would be no approximation problem.) Therefore, the last equation is equivalent to

$$m_2 - m_1 = \pm [R_1(x_0, h) - R_2(x_0, h)].$$

As h gets close to zero, the right side must approach zero, while the left side remains constant. Thus, the constant on the left side is zero; $m_1 = m_2$.

When F is differentiable at x_0, the mechanics of actually finding the number m involve evaluation of the limit indicated in Definition A. To designate this limit by m is not very helpful because m indicates neither the function that was differentiated nor the point. We therefore ι se $F'(x_0)$ in place of m; that is,

$$F'(x_0) = \lim_{h \to 0} \frac{F(x_0 + h) - F(x_0)}{h}. \tag{4}$$

PROBLEMS

A · Suppose R is a constant function (see Problem 2.6A for the definition of a constant function). Use the definition to show that $R'(x) = 0$ for every x.

B · We say the function F has a local maximum at x_0 if F is defined on an interval (a, b) centered at x_0 and if $F(x) \leq F(x_0)$ for each $x \neq x_0$ in (a, b) (Fig 3.2). Suppose F is differentiable at x_0 and has a local maximum there.

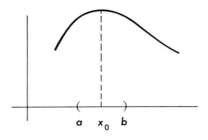

FIG. 3.2

Show that $F'(x_0) = 0$. (Hint: Consider the limit of line (4) as h goes to zero from the right; then from the left. What must be the sign of m in each case?)

C · Define a local minimum for a function F. State and prove a result analogous to Problem B.

At the end of the last chapter we were considering the function

$$G(x) = \sqrt[3]{x}.$$

We were trying to write the equation of a line tangent to the graph of G at $x = 8$. The equation of the tangent line was given in (2) of Section 2.7 as

$$T(x) = 2 + m(x - 8).$$

We needed to determine m. We now know that if T is to have the required approximation properties, then

$$m = G'(8) = \lim_{h \to 0} \frac{G(8 + h) - G(8)}{h}.$$

EXAMPLE A
Determine the required function T and use $T(8.4)$ to approximate $\sqrt[3]{8.4}$.

$$\lim_{h \to 0} \frac{G(8 + h) - G(8)}{h} = \lim_{h \to 0} \frac{\sqrt[3]{8 + h} - \sqrt[3]{8}}{h}.$$

We anticipated this computation in Example 2.4D where we found it to be $\frac{1}{12}$. Thus, $T(x) = 2 + \frac{1}{12}(x - 8)$ (see Fig. 3.3).

Using $T(8.4)$ to approximate $\sqrt[3]{8.4}$ gives

$$T(8.4) = 2 + \frac{1}{12}(0.4) = 2.0333 \ldots$$

(The reader might be interested to know that, correct to four decimals, $\sqrt[3]{8.4} = 2.0328$.) ∎

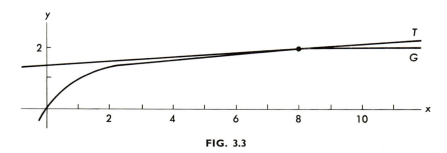

FIG. 3.3

This is a good place to comment on terminology used in conjunction with the accuracy of decimal approximations. In saying that an answer is correct to four decimals, we mean that if the computation were carried out to more decimal places and then rounded off in the usual way, our answer would be obtained. Later on we shall have occasion to ask for answers that are accurate to within one unit in the fourth (or second, or third, or whatever) decimal place. This means that if carried out to more decimal places and then rounded off, the result would not vary from our answer by more than one in the last digit. Thus, in the preceding example, an answer of 2.0327 or 2.0329 would be accurate to within one unit in the fourth decimal place. To be certain of accuracy to within one unit in the fourth decimal place, we must carry our computation to five decimals, have some way to be sure that this five-decimal answer is not in error by more than 0.00005, and then round off to four decimals.

It must be emphasized that not every function is differentiable. A function may even be continuous at a point but not differentiable there.

EXAMPLE B

The function $F(x) = |x|$ is continuous at 0 but not differentiable there.

Given $\varepsilon > 0$, choose $\delta = \varepsilon$. Then if $|x - 0| < \delta$, we certainly have

$$|F(x) - F(0)| < \varepsilon$$

because

$$|F(x) - F(0)| = ||x| - 0| = |x| < \delta = \varepsilon.$$

This establishes continuity. Now notice that

$$\lim_{h \to 0^-} \frac{F(0 + h) - F(0)}{h} = \lim_{h \to 0^-} \frac{|h| - 0}{h} = -1.$$

(This is because $|h| = -h$ when $h < 0$.)

$$\lim_{h \to 0^+} \frac{F(0 + h) - F(0)}{h} = \lim_{h \to 0^+} \frac{|h| - 0}{h} = 1.$$

This means the limit does not exist. F is not differentiable at $x = 0$. ▮

We now have an example of a function that is continuous at a point, but which is not differentiable at that point. Having seen this, anyone endowed with a measure of mathematical curiosity would naturally wonder if this can be turned around. That is, can we find an example of a function that is differentiable at a point, but not continuous there. The next theorem says the answer is no.

THEOREM A

If $F'(x_0)$ exists, then F is continuous at x_0.

PROOF:

We know $F(x_0 + h) - F(x_0) = h[F'(x_0) \pm R(x_0, h)]$. The expression on the left is a measure of how close $F(x_0 + h)$ comes to the target of $F(x_0)$. The right side of the equation (and hence the left side) can be made as close to zero as we please by taking h small. Thus, by restricting the distance of $x_0 + h$ from x_0, we can get $F(x_0 + h)$ as close to $F(x_0)$ as anyone might wish. ▮

This says that a function differentiable on an interval will also be continuous there. Its graph will be a smooth curve as indicated in Fig. 3.4. And, of course, since the function is differentiable, there will be at each point on the graph a line that is tangent to the graph.

Now suppose that, as in Fig. 3.4. the function is zero at each end of the interval. It should seem geometrically obvious that one of the tangent lines to

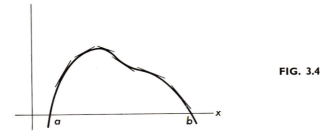

FIG. 3.4

the graph will be horizontal; that is, the slope of the tangent line will be zero. And it so happens that this obvious fact is of great importance. For this reason we will state the result more formally as a theorem and give a proof to satisfy those who understand that " obvious " is often used as a substitute for what is otherwise a difficult argument.

THEOREM B   (Rolle's Theorem)
Suppose F is defined and continuous on $[a, b]$, and that it has a derivative at each point of (a, b). Suppose further that $F(a) = F(b) = 0$. Then there is an $x_0 \in (a, b)$ so that $F'(x_0) = 0$.

PROOF:

If F is constant on $[a, b]$, then the result follows in a trivial way ($F'(x) = 0$ for all x) from Problem A.

If there is an $x_1 \in (a, b)$ so that $F(x_1) \neq F(a)$, then either $F(x_1) > 0$ or $F(x_1) < 0$. If $F(x_1) > 0$, then it is clear that F does not assume its maximum value at $x = a$ or $x = b$. But according to Theorem 2.6F it does assume a maximum value somewhere, say at x_0. And at this point, according to Problem B, $F'(x_0) = 0$.

If $F(x_1) < 0$, the result follows from a consideration of minimum values and use of Problem C. ∎

3 · 1   EXERCISES

In each of Exercises 1–10, find $\lim_{h \to 0}[F(x_0 + h) - F(x_0)]/h$ at the indicated x_0.

1 · $F(x) = x^2 + 3$; $x_0 = 2$.   **2 ·** $F(x) = 2x^2 - 3x$; $x_0 = 1$.

3 · $F(x) = 3x^3 + x^2$; $x_0 = 1$.   **4 ·** $F(x) = x^3 + 2x$; $x_0 = 2$.

5 · $F(x) = \dfrac{1}{x + 3}$; $x_0 = 2$.   **6 ·** $F(x) = \dfrac{2}{x - 2}$; $x_0 = -2$.

7 · $F(x) = \sqrt{x}$; $x_0 = 9$.   **8 ·** $F(x) = \sqrt{x + 1}$; $x_0 = 3$.

9 · $F(x) = \dfrac{x}{x + 4}$; $x_0 = 1$.   **10 ·** $F(x) = \dfrac{x - 1}{x + 2}$; $x_0 = 1$.

In Exercises 11–16, find an expression for

$$\lim_{h \to 0} \frac{F(x + h) - F(x)}{h}$$

at any point x.

11 • $F(x) = 2x^2 + 3x - 7.$ **12 •** $F(x) = 3x^2 - x + 2.$

13 • $F(x) = \dfrac{1}{x^2}.$ **14 •** $F(x) = \dfrac{1}{x}.$

15 • $F(x) = \sqrt{x}.$ **16 •** $F(x) = x - \sqrt{x}.$

17 • Write the equation of the line tangent to

$$F(x) = 3x^3 + x^2$$

at $x = 1$ (note Exercise 3 above). Sketch a graph of F and the tangent line at $x = 1.$

18 • Write the equation of the line tangent to

$$F(x) = x^3 + 2x$$

at $x = 2$ (note Exercise 4 above). Sketch a graph of F and the tangent line at $x = 2.$

19 • Write the equation of the line tangent to

$$F(x) = \sqrt{x}$$

at $x = 9$ (note Exercise 7 above). Sketch a graph of F and the tangent line at $x = 9.$

20 • Write the equation of the line tangent to

$$F(x) = \sqrt{x + 1}$$

at $x = 3$ (note Exercise 8 above). Sketch a graph of F and the tangent line at $x = 3.$

3 · 2 Notation

The point–slope form of a straight line gives us

$$y = y_0 + m(x - x_0).$$

Thus, when seeking the equation of a line tangent to the graph of F at $(x_0, F(x_0))$, we naturally wrote it in the form

$$T(x) = F(x_0) + m(x - x_0). \qquad (1)$$

We saw in the last section that the slope m is given by

$$m = \lim_{h \to 0} \frac{F(x_0 + h) - F(x_0)}{h}.$$

This choice of m enables us to write

$$F(x_0 + h) = F(x_0) + mh + |h| R(x_0, h). \qquad (2)$$

The number m is called the derivative of F at x_0, and we adopted the more suggestive notation of

$$m = F'(x_0).$$

There are other standard symbols related to the derivative with which the reader should be acquainted. Refer to Fig. 3.5. The difference between $x_0 + h$

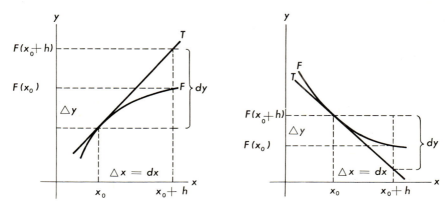

FIG. 3.5

and x_0 is sometimes called Δx (Δ is the capital of the Greek letter delta). The corresponding change in the values of F at $x_0 + h$ and x_0 is then called Δy:

$$\Delta x = (x_0 + h) - x_0 = h,$$

$$\Delta y = F(x_0 + h) - F(x_0).$$

Similarly, with attention focused on T instead of F, we use

$$dx = (x_0 + h) - x_0 = h,$$

$$dy = T(x_0 + h) - T(x_0).$$

It is clear that $\Delta x = dx$, but that $\Delta y \neq dy$. From Fig. 3.5, it appears that Δy will be close to dy when h is small. This is easily verified by noting from (1) that $T(x_0) = F(x_0)$ and $T(x_0 + h) = F(x_0) + mh$, so

$$dy = mh = m\,dx.$$

Then from (2),

$$\Delta y = mh + |h|\,R(x_0, h),$$

and   (3)

$$\Delta y = dy + |h|\,R(x_0, h).$$

Since $|h| R(x_0, h)$ goes to 0 as h gets small, we are justified in asserting that for small h

$$\Delta y \approx dy.$$

This approximation to Δy is of course useful because dy is easily calculated from (4):

$$dy = m\, dx = F'(x_0)\, dx. \qquad (4)$$

Finally, note from (4) (or directly from the definition) that

$$\frac{dy}{dx} = m = F'(x_0),$$

while from (3), remembering that $\Delta x = dx = h$,

$$\frac{\Delta y}{\Delta x} = \frac{dy}{dx} \pm R(x_0, h),$$

from which we have

$$\lim_{\Delta x \to 0} \frac{\Delta y}{\Delta x} = \frac{dy}{dx}.$$

In summary,

$$(x_0 + h) - x_0 = h = \Delta x = dx;$$

$$\Delta y = F(x_0 + \Delta x) - F(x_0);$$

$$F'(x_0) = \left(\frac{dy}{dx}\right) = \lim_{h \to 0} \frac{F(x_0 + h) - F(x_0)}{h};$$

$$dy = \left(\frac{dy}{dx}\right) dx;$$

dy is of interest because it approximates Δy.
The numbers dx and dy are called *differentials*.

If F is differentiable at each point of an interval, then we may think of F' as a function defined on the interval. For each x, there is a value $F'(x)$. In Exercises 11–16 following Section 3.1, the student was asked to find a formula giving the value of $F'(x)$ for an arbitrary x.

PROBLEM

A · Let $F(x) = 6x - x^2 - 5$. Find $F'(2)$. Then find Δy and dy if $\Delta x = \frac{1}{2}$.

Exercises 1–10 are actually continuations of Exercises 1–10 of Section 3.1. For each function you are given the points x_0 and $x_0 + \Delta x$. You are to find Δy and dy in each case.

1. $F(x) = x^2 + 3$,
$x_0 = 2; x_0 + \Delta x = 2.2$.

2. $F(x) = 2x^2 - 3x$,
$x_0 = 1; x_0 + \Delta x = 1.2$

3. $F(x) = 3x^3 + x^2$,
$x_0 = 1; x_0 + \Delta x = 1.25$.

4. $F(x) = x^3 + 2x$,
$x_0 = 2; x_0 + \Delta x = 2.25$.

5. $F(x) = \dfrac{1}{x+3}$,
$x_0 = 2; x_0 + \Delta x = 1.9$.

6. $F(x) = \dfrac{2}{x-2}$,
$x_0 = -2; x_0 + \Delta x = -1.8$.

7. $F(x) = \sqrt{x}$,
$x_0 = 9; x_0 + \Delta x = 8.7$.

8. $F(x) = \sqrt{x+1}$,
$x_0 = 3; x_0 + \Delta x = 2.8$.

9. $F(x) = \dfrac{x}{x+4}$,
$x_0 = 1; x_0 + \Delta x = 1.2$.

10. $F(x) = \dfrac{x-1}{x+2}$,
$x_0 = 1; x_0 + \Delta x = 1.3$.

Exercises 11–16 are continuations of Exercises 11–16 of Section 3.1. Again, x_0 and $x_0 + \Delta x$ are given; you are to find Δy and dy in each case.

11. $F(x) = 2x^2 + 3x - 7$,
$x_0 = 3; x_0 + \Delta x = 2.9$.

12. $F(x) = 3x^2 - x + 2$,
$x_0 = 2; x_0 + \Delta x = 1.8$.

13. $F(x) = \dfrac{1}{x^2}$,
$x_0 = 3; x_0 + \Delta x = 3.3$.

14. $F(x) = \dfrac{1}{x}$,
$x_0 = 3; x_0 + \Delta x = 3.2$.

15. $F(x) = \sqrt{x}$,
$x_0 = 16; x_0 + \Delta x = 16.3$.

16. $F(x) = x - \sqrt{x}$,
$x_0 = 9; x_0 + \Delta x = 9.3$.

Compare Exercises 17–20 with the corresponding exercises of Section 3.1. For each of the indicated functions you are given the points x_0 and $x_0 + \Delta x$. Sketch a graph of the function; show the graph of the line tangent to the curve at x_0. Finally, indicate Δy and dy on the graph.

17. $F(x) = 3x^3 + x^2$ (Exercise 3),
$x_0 = 1; x_0 + \Delta x = 1.25$.

18. $F(x) = x^3 + 2x$ (Exercise 4),
$x_0 = 2; x_0 + \Delta x = 2.25$.

19. $F(x) = \sqrt{x}$ (Exercise 7),
$x_0 = 9; x_0 + \Delta x = 8.7$.

20. $F(x) = \sqrt{x+1}$ (Exercise 8),
$x_0 = 3; x_0 + \Delta x = 2.8$.

3 · 3   The Speedometer Reading

Let y represent the number of miles a car travels in x hours. Suppose

$$y = F(x) = 2x^3 - 12x^2 + 60x \qquad \text{for} \qquad x \in [0, 5].$$

We pose the following question: For a car traveling according to this rule, what does the speedometer read when $x = 3$?

As a first step, we might find the average speed of the car during the fourth hour by computing

$$\text{rate} = \frac{\text{distance}}{\text{time}} = \frac{F(4) - F(3)}{4 - 3} = \frac{176 - 126}{1} = 50.$$

But this is not necessarily the speedometer reading at the exact time $x = 3$. Computing the average speed over the quarter hour following $x = 3$ seems likely to come closer to giving the correct speedometer reading.

$$\frac{F(\frac{13}{4}) - F(3)}{(\frac{13}{4}) - 3} = \frac{136(\frac{29}{32}) - 126}{\frac{1}{4}} = 43\frac{5}{8}.$$

It would be better yet to evaluate this quotient for an even smaller amount of elapsed time. Rather than compute this quotient over and over, however, let us compute the numerator (for this is where the work is) for an arbitrary time interval h:

$$F(3 + h) - F(3) = 2(3 + h)^3 - 12(3 + h)^2 + 60(3 + h) - 126. \qquad (1)$$

Some algebraic computation (which the reader should verify) gives

$$F(3 + h) - F(3) = 42h + 6h^2 + 2h^3.$$

The quotient in which we are interested then is

$$\frac{F(3 + h) - F(3)}{h} = 42 + 6h + 2h^2. \qquad (2)$$

We now verify our earlier results. For $h = 1$, $42 + 6(1) + 2(1)^2 = 50$. For $h = \frac{1}{4}$, $42 + 6(\frac{1}{4}) + 2(\frac{1}{4})^2 = 43\frac{5}{8}$. From (2) it is easily seen now that as h gets small, the quotient approaches 42. This we take to be the instantaneous velocity at $x = 3$.

A further economy of effort could have been achieved by computing, for arbitrary x, the quotient

$$\frac{F(x + h) - F(x)}{(x + h) - x} = \frac{6x^2h + 6xh^2 - 24xh + 2h^3 - 12h^2 + 60h}{h},$$

$$\frac{F(x + h) - F(x)}{(x + h) - x} = 6x^2 - 24x + 60 + 6xh + 2h^2 - 12h.$$

As h gets close to zero, the instantaneous velocity appears as

$$6x^2 - 24x + 60. \tag{3}$$

When $x = 3$, this gives $6(9) - 72 + 60 = 42$, in agreement with the result obtained in the previous paragraph.

We thus see that if $F(x)$ gives the distance a body has moved in time x, then the instantaneous velocity (the reading on the speedometer) is given by

$$\lim_{h \to 0} \frac{F(x + h) - F(x)}{h}.$$

We have seen this expression before. It is the number $F'(x)$. The problem of finding this limit has thus come before us twice. There are many other contexts in which we are led to the same problem. For this reason we next discuss some rules for computing the limit quickly.

3·4   Computational Rules

Through the use of several examples, we have tried to convey the idea that much of our work will involve the expression

$$\lim_{h \to 0} \frac{F(x + h) - F(x)}{h}. \tag{1}$$

This is a limit. Limits can be tricky (Example 2.4A). Even when we have successfully computed such limits (Example 2.4D), the work has required algebraic skill and, the reader may feel, a measure of algebraic inspiration.

We are sorry to say that we cannot entirely skirt the need for some algebraic know-how on the student's part, but there is some relief available. It comes by making use of the theorems about limits which we have at our disposal (Section 2.4).

We begin by finding the derivative of some frequently occurring functions. In expressing these results, we shall indulge in some further abuse of language. We shall write x^n where we really mean "the function F described by $F(x) = x^n$." We also introduce new notation here. We have designated by $F'(x)$ the number obtained as the limit (1). We now use $DF(x)$ for this same number, and of course DF is the function having value $DF(x)$ at x. We also write $DF(x_0)$ to represent the derivative of F evaluated at x_0; that is,

$$DF(x_0) = F'(x_0).$$

D is called the *differential operator*. Thus, instead of saying: Let

$$y = F(x) = x^n; \qquad \text{then} \qquad \frac{dy}{dx} = nx^{n-1};$$

or, let

$$F(x) = x^n; \qquad \text{then} \qquad F'(x) = nx^{n-1};$$

we merely have to write

$$Dx^n = nx^{n-1}.$$

The proof of this assertion is the content of our next theorem.

THEOREM A

For any positive integer n, $Dx^n = nx^{n-1}$.

PROOF:

$Dx^n = \lim_{h \to 0} [(x + h)^n - x^n]/h$. Expanding $(x + h)^n$ by the binomial theorem, we have

$$\frac{1}{h}[(x + h)^n - x^n] = \frac{1}{h}\left[x^n + nx^{n-1}h + \frac{n(n - 1)}{2} x^{n-2}h^2 + \cdots + h^n - x^n \right],$$

$$= nx^{n-1} + \frac{n(n - 1)}{2} x^{n-2}h + \cdots + h^{n-1}.$$

According to Theorem 2.4B, the limit as h approaches 0 of this sum is equal to the sum of the individual limits. The first term of the sum is not affected by h going to zero. All the other terms, however, clearly go to 0 as h does. The theorem is therefore proved. ∎

By way of illustration, this theorem means that if $F(x) = x^4$, then $DF(x) = F'(x) = 4x^3$.

The development of some general rules for differentiation is the next order of business.

Rule 1 If F and G both have derivatives at the point x, then

$$D[F(x) + G(x)] = DF(x) + DG(x).$$

PROOF:

$$D[F(x) + G(x)] = \lim_{h \to 0} \frac{[F(x + h) + G(x + h)] - [F(x) + G(x)]}{h},$$

$$= \lim_{h \to 0} \frac{F(x + h) - F(x)}{h} + \lim_{h \to 0} \frac{G(x + h) - G(x)}{h}.$$

This uses Theorem 2.4B. ∎

Rule 2 If F and G both have derivatives at the point x, then

$$D[F(x) \cdot G(x)] = F(x)[DG(x)] + [DF(x)]G(x),$$

PROOF:

$$D[F(x) \cdot G(x)]$$

$$= \lim_{h \to 0} \frac{F(x + h)G(x + h) - F(x)G(x)}{h},$$

$$= \lim_{h \to 0} \left[\frac{F(x + h)G(x + h) - F(x + h)G(x)}{h} + \frac{F(x + h)G(x) - F(x)G(x)}{h} \right],$$

$$= \lim_{h \to 0} \frac{F(x + h)[G(x + h) - G(x)]}{h} + \lim_{h \to 0} \frac{[F(x + h) - F(x)]G(x)}{h}.$$

We have just used Theorem 2.4B. We now use Theorem 2.4C to write

$$D[F(x) \cdot G(x)]$$

$$= \lim_{h \to 0} F(x + h) \cdot \lim_{h \to 0} \frac{G(x + h) - G(x)}{h} + \lim_{h \to 0} \frac{F(x + h) - F(x)}{h} \lim_{h \to 0} G(x).$$

According to Theorem 3.1A, F is continuous at x. Therefore, Theorem 2.4A says

$$\lim_{h \to 0} F(x + h) = F(x + 0) = F(x). \quad \blacksquare$$

As an elementary (but extremely important) application of these rules, we derive a corollary to Theorem A; this corollary tells us how to differentiate a polynomial. We need to recall in this corollary what we learned in Problem 3.1A namely, that if c is a constant, then $Dc = 0$.

COROLLARY
$$D[a_n x^n + \cdots + a_1 x + a_0] = na_n x^{n-1} + \cdots + a_1.$$

PROOF:

For any k, by Rule 2, $Da_k x^k = a_k Dx^k + (Da_k)x^k$. Therefore, $Da_k x^k = a_k(kx^{k-1}) + 0x^k = ka_k x^{k-1}$. The result now follows from Rule 1. ∎

In Section 3.3 we considered a function

$$F(x) = 2x^3 - 12x^2 + 60x.$$

After considerable computation, we found in (3) of Section 3.3 that

$$F'(x) = 6x^2 - 24x + 60.$$

Notice that with the aid of the Corollary to Theorem A, we can write this result down directly; no pencil work is necessary. In a similar way, Exercises 1, 2, 3, 4, 11, and 12 at the end of Section 3.1 can all be worked much more easily with the help of the Corollary.

Rule 3 If F and G are both differentiable at a point x, then

$$D\frac{F(x)}{G(x)} = \frac{G(x)DF(x) - F(x)DG(x)}{[G(x)]^2} \qquad \text{whenever } G(x) \neq 0.$$

PROOF:

$$D\frac{F(x)}{G(x)} = \lim_{h \to 0} \frac{[F(x + h)]/[G(x + h)] - (F(x)/G(x))}{h}$$

Multiplying the numerator and denominator by $G(x)G(x + h)$,

$$D\frac{F(x)}{G(x)} = \lim_{h \to 0} \frac{F(x + h)G(x) - F(x)G(x + h)}{hG(x)G(x + h)},$$

$$= \lim_{h \to 0} \frac{1}{G(x)G(x + h)}$$

$$\cdot \lim_{h \to 0} \frac{[F(x + h) - F(x)]G(x) - F(x)[G(x + h) - G(x)]}{h}.$$

We have used Theorem 2.4C and some tricky algebra so far. On the first factor, use Theorem 2.4C, continuity (Theorem 3.1A) of G at x, and Theorem 2.4A to get $1/[G(x)]^2$. On the second factor, use Theorem 2.4B and then Theorem 2.4C again. ∎

The proof of our last rule is more involved. To prepare the way we recall that if G is differentiable at x_0 and F is differentiable at y_0, then

$$G(x_0 + h) = G(x_0) + G'(x_0)h + |h| R_1(x_0, h) \quad \text{where} \quad \lim_{h \to 0} R_1(x_0, h) = 0. \quad (2)$$

$$F(y_0 + k) = F(y_0) + F'(y_0)k + |k| R_2(y_0, k) \quad \text{where} \quad \lim_{k \to 0} R_2(y_0, k) = 0. \quad (3)$$

Rule 4   Let $H(x) = F \circ G(x) = F[G(x)]$. Suppose G is differentiable at x_0 and F is differentiable at $y_0 = G(x_0)$. Then

$$H'(x_0) = F'(y_0)G'(x_0).$$

PROOF:

$$\lim_{h \to 0} \frac{H(x_0 + h) - H(x_0)}{h} = \lim_{h \to 0} \frac{F[G(x_0 + h)] - F[G(x_0)]}{h}.$$

Now using (2), this gives

$$H'(x_0) = \lim_{h \to 0} \frac{F[y_0 + G'(x_0)h + |h| R_1(x_0, h)] - F(y_0)}{h}.$$

Set $k = G'(x_0)h + |h| R_1(x_0, h)$. It is clear that as h changes, k changes. In particular, as h gets small, we see that

$$\lim_{h \to 0} h[G'(x_0) \pm R_1(x_0, h] = 0,$$

according to Theorem 2.4C. Hence, when h approaches 0, k approaches 0. With this choice of k, we have, from (3),

$$H'(x_0) = \lim_{h \to 0} \frac{F(y_0) + F'(y_0)(k) + |k| R_2(y_0, k) - F(y_0)}{h}.$$

Replacing k by its value,

$$H'(x_0) = \lim_{h \to 0} \left[\frac{F'(y_0)[G'(x_0)h + |h| R_1(x_0, h)]}{h} \right.$$

$$\left. + \frac{|G'(x_0)h + |h| R_1(x_0, h)| R_2(y_0, k)}{h} \right].$$

We now use Theorem 2.4B to write

$$H'(x_0) = \lim_{h \to 0} F'(y_0) \frac{1}{h} [G'(x_0)h + |h| R_1(x_0, h)]$$

$$+ \lim_{h \to 0} \frac{1}{h} |G'(x_0)h + |h| R_1(x_0, h)| R_2(y_0, k).$$

Since k goes to 0 as h does, $\lim_{h\to 0} R_2(y_0, k) = 0$ and Theorem 2.4C lets us conclude that the last term is 0. Hence, by Theorem 2.4B

$$H'(x_0) = \lim_{h\to 0} F'(y_0)G'(x_0) \pm \lim_{h\to 0} F'(y_0)R_1(x_0, h)$$

$$= F'(y_0)[G'(x_0)]. \ \blacksquare$$

Rule 4 is more easily remembered if we use different notation in stating it. Set

$$y = G(x); \qquad w = F(y) = F(G(x)) = H(x).$$

Then Rule 4 says

$$\left(\frac{dw}{dx}\right) = \left(\frac{dw}{dy}\right)\left(\frac{dy}{dx}\right). \tag{4}$$

This rule is known as the *chain rule*.

The student should commit these rules to memory; in fact, he must commit these rules to memory. The first rule is natural enough so that it is learned almost automatically. The fourth rule is probably best learned by relying on the suggestive formula (4). The second and third rules should be memorized in words, almost in sing-song fashion.

Rule 2 The derivative of a product is the first times the derivative of the second, plus the second times the derivative of the first.

Rule 3 The derivative of a quotient is the denominator times the derivative of the numerator minus the numerator times the derivative of the denominator all over the denominator squared.

EXAMPLE A
Find $D[(3x^2 + 4x - 1)(x^4 + 3x^2 - 7)]$.

$D[(3x^2 + 4x - 1)(x^4 + 3x^2 - 7)]$

$\qquad = (3x^2 + 4x - 1)D[(x^4 + 3x^2 - 7)] + D[(3x^2 + 4x - 1)](x^4 + 3x^2 - 7),$

$\qquad = (3x^2 + 4x - 1)(4x^3 + 6x) + (6x + 4)(x^4 + 3x^2 - 7).$

Step one here uses Rule 2. Then the Corollary to Theorem A was used. (Verify that if we first multiply the original polynomials, then use the Corollary immediately, the result is the same.) $\blacksquare$

EXAMPLE B

$$D\frac{x-1}{x^2+1} = \frac{(x^2+1)D(x-1) - (x-1)D(x^2+1)}{(x^2+1)^2}$$

$$= \frac{(x^2+1) - (x-1)(2x)}{(x^2+1)^2}$$

$$= \frac{-x^2+2x+1}{(x^2+1)^2}. \quad \blacksquare$$

EXAMPLE C

Find $D(x^2+3x+1)^5$.

We could raise the expression to the fifth power and then use the Corollary. We prefer less work. Set

$$w = (x^2+3x+1)^5 \quad \text{and} \quad y = x^2+3x+1.$$

Then $w = y^5$, and the chain rule says

$$\frac{dw}{dx} = \frac{dw}{dy} \cdot \frac{dy}{dx} = (5y^4)(2x+3),$$

$$\frac{dw}{dx} = 5(x^2+3x+1)^4(2x+3).$$

Look at the original problem; look at the answer. After a few problems of this sort, you will have no difficulty in omitting the writing down of the substitutions. ∎

We have seen that $Dx^n = nx^{n-1}$ for any positive integer n. Can we use the same formula if n is not an integer? For example, can we say $Dx^{1/2} = \frac{1}{2}x^{-1/2}$? From the following theorem we get a corollary that answers the question.

THEOREM B

Let F have G as its inverse. (Thus, if $y = F(x)$, then $x = G(y)$.) Suppose F has a continuous nonzero derivative at x. Then the derivative of G at the corresponding y is

$$G'(y) = \frac{1}{F'(x)} \quad \text{where} \quad y = F(x).$$

PROOF:

Set
$$H(x) = G(Fx)) = x.$$
Then
$$H'(x) = 1.$$

By the chain rule,
$$H'(x) = G'(F(x))F'(x).$$
Hence,
$$G'(y)F'(x) = 1.$$

Division by $F'(x)$ gives the result. ∎

When this result is expressed by using the differential operator D, it is customary to use a subscript on the D to indicate the variable with respect to which the derivative is taken. Hence,

$$D_y G(y) = \frac{1}{D_x F(x)} \qquad \text{where} \qquad y = F(x).$$

Strictly speaking, for the proof just given, we should have included in our hypothesis the assumption that G is differentiable at $F(x)$. This can be proved, however, always to be the case under the conditions we have stated [2, page 272].

COROLLARY
$$Dx^r = rx^{r-1} \text{ for any rational number } r.$$

PROOF:

Let $F(x) = x^{1/q}$ where q is a positive integer. Then G, defined by $G(x) = x^q$, is the inverse of F since
$$G(F(x)) = G(x^{1/q}) = (x^{1/q})^q = x.$$

We know, since q is a positive integer, that
$$G'(x) = qx^{q-1}.$$

By the chain rule,
$$G'(F(x)) \cdot F'(x) = 1,$$

so

$$F'(x) = \frac{1}{G'(F(x))}$$

This means that

$$F'(x) = Dx^{1/q} = \frac{1}{q(F(x))^{q-1}} = \frac{1}{qx^{(q-1)/q}} = \frac{1}{q}x^{(1-q)/q} = \frac{1}{q}x^{(1/q)-1}.$$

Now if r is a positive rational, then $r = p/q$, p, q being positive integers.

$$Dx^{p/q} = D(x^{1/q})^p = p(x^{1/q})^{p-1}\frac{1}{q}x^{(1/q)-1} = \frac{p}{q}x^{(p-1)/q}x^{(1-q)/q} = \frac{p}{q}x^{(p-q)/q}.$$

Here we have used the chain rule and some algebra.

The corollary is now proved for positive rationals. If r is a negative rational, then $x^r = x^{-k} = 1/x^k$ where k is a positive rational. The result follows by using Rule 3. ∎

EXAMPLE D

Find $DG(8)$ when $G(x) = \sqrt[3]{x}$. (Notice that this is the same problem we solved in Example 2.4.D.

$$DG(x) = Dx^{1/3} = \frac{1}{3}x^{-2/3};$$

$$DG(8) = \frac{1}{3}8^{-2/3} = \frac{1}{3}\cdot\frac{1}{4} = \frac{1}{12}. \quad ∎$$

3 · 4 EXERCISES

In Exercises 1–4, use the Corollary to Theorem A to find $F'(x)$.

1. $F(x) = 3x^4 - 5x + 3$.
3. $F(x) = 5x^3 - 10x^2 + 9$.

2. $F(x) = 3x^4 - 5x^2 + 9$.
4. $F(x) = 2x^3 - 4x + 3$.

In Exercises 5–8, use Rule 2 along with the Corollary to Theorem A to find $F'(x)$.

5. $F(x) = (x^2 + 3x)(x^3 - 2x^2 + 5)$.
7. $F(x) = (x^3 + 5x - 7)(x^2 - 7x)$.

6. $F(x) = (x^3 - 3)(x^2 + 5x + 6)$.
8. $F(x) = (x^5 - 2x + 1)(x^2 + 3x)$.

In Exercises 9–14, use Rule 3 along with the Corollary to Theorem A to find $F'(x)$.

9. $F(x) = \dfrac{x}{x^2 + 1}$.

10. $F(x) = \dfrac{x^2}{x - 3}$.

11. $F(x) = \dfrac{5}{x - 3}$.

12. $F(x) = \dfrac{4}{x^2 + 7}$.

13. $F(x) = \dfrac{x^2 + 1}{x^3 + 3x}$.

14. $F(x) = \dfrac{3x^2 - x}{x^2 + 4}$.

In Exercises 15–16, use the Corollary to Theorem B along with Rule 1 to find $F'(x)$.

15. $F(x) = 3\sqrt{x} + \sqrt[3]{x}$.

16. $F(x) = 4\sqrt[3]{x} - 3\sqrt{x}$.

In Exercises 17–20, use Rule 4 to find $F'(x)$.

17. $F(x) = \sqrt{x^2 + 3x}$.

18. $F(x) = \sqrt[3]{3x^2 - 9x}$.

19. $F(x) = (4x^3 + 3x - 1)^7$.

20. $F(x) = (2x^2 - 4x + 3)^6$.

In Exercises 21–30, find $F'(x)$ and an expression for dy in terms of x and dx.

21. $y = F(x) = (x^3 + 3x^2 - 7x)$
$\qquad \cdot (x^2 + 4x + 3)$.

22. $y = F(x) = (3x^4 + 7x^2 - 9x + 2)$
$\qquad \cdot (x^2 - 7)$.

23. $y = F(x) = \dfrac{x - 1}{x + 2}$.

24. $y = F(x) = \dfrac{x + 1}{x - 1}$.

25. $y = F(x) = x^{1/3}(x^2 + 1)$.

26. $y = F(x) = (x - 1)^{1/3}(x^2 + 4)^{1/2}$.

27. $y = F(x) = \dfrac{x^{1/2}}{3x + 1}$.

28. $y = F(x) = \dfrac{2x^2}{1 + \sqrt{x}}$.

29. $y = F(x) = (3x^2 + 4)^5$.

30. $y = F(x) = (3\sqrt{x} + 1)^6$.

In Exercises 31–40, find dy/dx.

31. $y = \dfrac{\sqrt{x + 1}}{\sqrt{x - 1}}$.

32. $y = \dfrac{\sqrt{x + 1}}{\sqrt{x - 1}}$.

33. $y = \dfrac{x^2 + 1}{\sqrt{x}}$.

34. $y = \dfrac{x + 2}{(x - 1)^3}$.

35. $y = \dfrac{x^{-1} + x}{(x + 1)^{-1}}$.

36. $y = \dfrac{x^{1/2} + x^{-1/2}}{(x + 1)^{1/2}}$.

37. $y = \dfrac{(x^2 + 1)^{1/2}}{x}$.

38. $y = \dfrac{x}{(x^2 + 1)^{1/2}}$.

39. $y = \dfrac{1}{x^3 + 4x^2}$.

40. $y = \dfrac{1}{(x^2 + 3x - 7)^3}$.

In Exercises 41–46, find dy/dt in two ways. First express y as a function of t and differentiate; second, use the chain rule.

41 . $y = x^2; x = t^{1/2} + 1.$

42 . $y = \dfrac{x}{x+1}; x = \sqrt{t-1}.$

43 . $y = (x^3 + 3x)^2; x = t^2 + 1.$

44 . $y = (2x^3 - 1)^{1/3}; x = \dfrac{1}{t+1}.$

45 . $y = \sqrt{x^2 + x}; x = \dfrac{t}{t+1}.$

46 . $y = \dfrac{x^2 + 1}{\sqrt{x}}; x = t^2 - 3.$

In Exercises 47–54, find $H'(t_0)$ if $H = G \circ F$, the composition of the indicated functions.

47 . $x = F(t) = t^2 + 1, t_0 = 1;$

$y = G(x) = \dfrac{1}{x-1}.$

48 . $x = F(t) = (t^2 - 1)^{1/2}, t_0 = 1;$

$y = G(x) = (x + 1)^2.$

49 . $x = F(t) = \dfrac{t}{t+1}, t_0 = 1;$

$y = G(x) = x^2 + 3.$

50 . $x = F(t) = (t^2 + t)^{1/2}, t_0 = 0;$

$y = G(x) = x^{1/3}.$

51 . $x = F(t) = t^2 + t, t_0 = -1;$

$y = G(x) = [x^2 + x]^{1/2}.$

52 . $x = F(t) = \dfrac{1}{t+1}, t_0 = 0;$

$y = G(x) = \dfrac{x-1}{x^{1/2}}.$

53 . $x = F(t) = \sqrt{t} - \sqrt[3]{t}, t_0 = 64;$

$y = G(x) = \dfrac{x}{x^2 + 1}.$

54 . $x = F(t) = \dfrac{t-1}{\sqrt{t+1}}, t_0 = 3;$

$y = G(x) = x^2 - 1.$

In Exercises 55–62, find, in two ways, $DF^{-1}(y_0)$ where $y_0 = F(x_0)$. First find the appropriate inverse, and compute $DF^{-1}(y_0)$ directly; then find $DF^{-1}(y_0)$ by use of Theorem B.

55 . $F(x) = \dfrac{1}{x-1}, x_0 = 2$

56 . $F(x) = -\sqrt{x}, x_0 = 4.$

57 . $F(x) = 1 - x^2, x_0 = 2.$

58 . $F(x) = \dfrac{3}{x+2}, x_0 = 1.$

59 . $F(x) = \dfrac{x}{x+1}, x_0 = -2.$

60 . $F(x) = \dfrac{2x+1}{x}, x_0 = 1.$

61 . $F(x) = \dfrac{x^2}{x+1}, x_0 = -2.$

62 . $F(x) = \dfrac{x}{x^2+1}, x_0 = 3.$

For the function F given in each of Exercises 63–70, find Δy and dy when x_0 and Δx are as indicated.

63 · $F(x) = x^2 + 1$, $x_0 = 0$,
$\Delta x = 0.1$.

64 · $F(x) = 3x^2 - 1$, $x_0 = 2$,
$\Delta x = 0.1$.

65 · $F(x) = \sqrt{x+1}$, $x_0 = 3$,
$\Delta x = 0.2$.

66 · $F(x) = \sqrt[3]{x^2 + 1}$, $x_0 = 0$,
$\Delta x = 0.2$.

67 · $F(x) = \dfrac{3x}{2x-1}$, $x_0 = 1$,
$\Delta x = -0.1$.

68 · $F(x) = \dfrac{x}{x+1}$, $x_0 = 0$,
$\Delta x = -0.1$.

69 · $F(x) = \dfrac{\sqrt{x}}{x+1}$, $x_0 = 0$,
$\Delta x = \frac{1}{4}$.

70 · $F(x) = \dfrac{x^{-1} + x}{\sqrt{x}}$, $x_0 = 1$,
$\Delta x = \frac{1}{4}$.

As Exercises 71–78, use each of the functions given in Exercises 63–70, and write the equation of the line tangent to the graph of F at the indicated point x_0. Sketch the graph of F and the graph of the tangent line.

3 · 5   The Derivative as a Linear Transformation

There is a second way to define the derivative that ultimately turns out to be of great help. In order to understand this definition, we first need to understand the concept of a *linear* function. A function L is called *linear* if it satisfies

$$L(ax) = aL(x) \text{ for any real number } a; \tag{1}$$
$$L(x_1 + x_2) = L(x_1) + L(x_2). \tag{2}$$

It is immediately clear that $L(0) = 0$, since $L(0) = L(0x) = 0L(x) = 0$. For functions of a single real variable such as we are studying,

$$L(x) = L(x1) = xL(1),$$

so the function is completely determined by knowing $L(1)$. Given a *linear* function, let $L(1) = m$. Then

$$L(x) = mx. \tag{3}$$

Appropriately, the graph of the *linear* function L is a line (Fig. 3.6); $L(1) = m$ is the slope of the line.

There is for each *linear* function L a real number m for which (3) holds. Conversely, for each real number m, (3) defines a *linear* function. There is therefore a one-to-one correspondence between *linear* functions and real numbers.

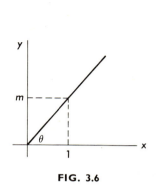

FIG. 3.6

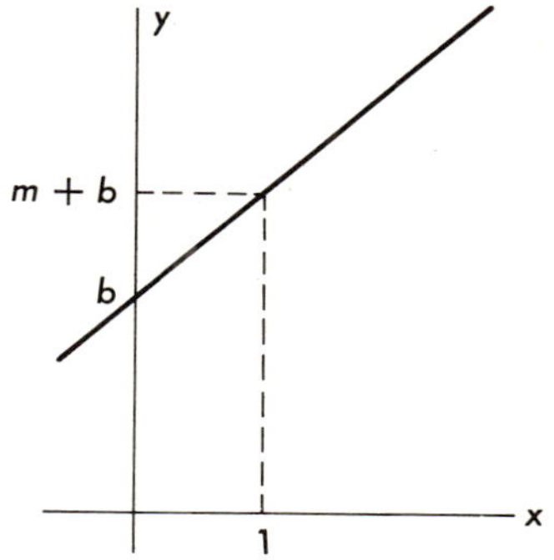

FIG. 3.7

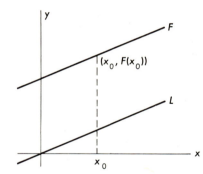

FIG. 3.8

Now consider the function H defined by $H(x) = mx + b$. We know that the graph is again a straight line (Fig. 3.7), and most books written at an elementary level call such functions linear even though they do not satisfy (1) or (2). We prefer not to adopt ideas that need to be relearned later, so we shall italicize the word *linear* wherever we use it as a reminder that we refer to conditions (1) and (2). Functions $H(x) = mx + b$, referred to as linear in other textbooks, are thus *linear* if and only if $b = 0$.

A straight line passing through an arbitrary point (x_0, y_0) with slope m is the graph of the function described by

$$y = y_0 + m(x - x_0).$$

If L is the *linear* function for which $L(x) = mx$, then $L(x - x_0) = m(x - x_0)$ and the line through (x_0, y_0) with slope m may be written

$$y = y_0 + L(x - x_0).$$

Any function F that has a nonvertical straight line as its graph must be described, therefore, by a formula of the form

$$F(x) = y_0 + L(x - x_0). \tag{4}$$

Clearly, $y_0 = F(x_0)$ since $L(x_0 - x_0) = L(0) = 0$. (See Fig. 3.8).

3 · 5A EXERCISES

In each of Exercises 1–6 you are given a *linear* function and a point. Write the equation of a line passing through the point and running parallel to the graph of the *linear* function.

1 · $P_0(3, 7)$; $L(x) = 2x$.   **2 ·** $P_0(-4, 1)$; $L(x) = 3x$.

3 · $P_0(1, -2)$; $L(x) = 4x$.   **4 ·** $P_0(3, -2)$; $L(x) = -x$.

5 · $P_0(-3, -1)$; $L(x) = 0$.   **6 ·** $P_0(4, -3)$; $L(x) = 0$.

In each of Exercises 7–12 you are given the equation of a straight line. Write the equation of a *linear* function having its graph parallel to the given line.

7 · $2y = 3x + 4$.   **8 ·** $5y = 7x - 9$.

9 · $2x + 3y - 7 = 0$.   **10 ·** $4x - 3y - 2 = 0$.

11 · $3x - 4y + 5 = 0$.   **12 ·** $3x + 5y - 10 = 0$.

We now refer back to Section 2.7 where we gave motivation for the definition of the derivative. We were trying to determine a number m so that the straight line graph of

$$T(x) = F(x_0) + m(x - x_0)$$

would be tangent to the graph of F at x_0. But we have now observed that the numbers m are in one-to-one correspondence with the linear functions L, so we might rephrase our problem as follows. Determine a *linear* function L so that the graph of

$$T(x) = F(x_0) + L(x - x_0)$$

is tangent to the graph of F at x_0. Or, phrasing the same idea analytically, we are led to the following modification of Definition 3.1B.

DEFINITION A

Let F be a real-valued function defined in some interval containing x_0, and suppose there exists a *linear* function L for which, when h is small,

$$F(x_0 + h) = F(x_0) + L(h) + |h| R(x_0, h),$$

where $R(x_0, h)$ approaches zero whenever h approaches zero. Then we say F is differentiable at x_0 and that its derivative is the *linear* function L.

Note that according to this definition, $F'(x_0) = L$ is a function, not a number. The graph of the *linear* function $F'(x_0)$ is indicated in Fig. 3.9. The number m

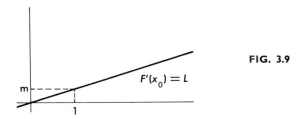

FIG. 3.9

previously defined to be the derivative is obtained as the value of $F'(x_0) = L$ at $x = 1$. A useful notational convenience is to indicate the number $F'(x_0)(1)$ by $[F'(x_0)]$.

In the case of real-valued functions of a real variable where we have a one-to-one correspondence between the numbers m and the functions L, this new definition is simply a matter of terminology. But in the study of functions of several variables and in more abstract settings, the concept of the derivative as a *linear* function (or linear transformation) still makes sense whereas the derivative as a number does not.

PROBLEMS

A · Suppose F is a *linear* function. Use Definition A to show that $F'(x) = F$ for every x.

B · Suppose F is a constant function (as defined in Problem 2.6A). Use Definition A to show that $F'(x) = \mathcal{O}$ for every x. Note that $\mathcal{O}$ represents the *linear* function that maps every number into the number 0.

C · Let $F(x) = x^3$. Find $[F'(2)]$ and $F'(2)(3)$.

MORE ABOUT THE DERIVATIVE

Suppose F is a function that is defined on $[a, b]$ and is differentiable on (a, b). We know that this means that if $x_0 \in (a, b)$, then we may define a function T,

$$T(x) F = (x_0) + F'(x_0)(x - x_0) \tag{1}$$

having as its graph a line that is tangent to the graph of F at x_0. Look at Fig. 4.1.

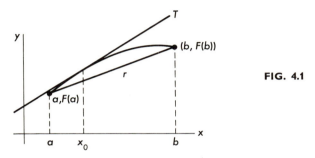

FIG. 4.1

The slope of the tangent line naturally depends on the choice of x_0. It is given by the number

$$m(x_0) = F'(x_0).$$

In the first two sections of this chapter we regard m as a function defined on (a, b). Its properties are related to the function F.

The function T is described by a first-degree polynomial. For values of x near x_0, $T(x)$ approximates $F(x)$. In the third (and final) section of this chapter, we use the results of the first two sections to define polynomials of higher degree, which give better approximations to the values of F than T does.

4·1   The Mean Value Theorem

Here is a question. In Fig. 4.1 we have drawn a line segment r connecting $(a, F(a))$ and $(b, F(b))$. In which direction would you move x_0 so that the tangent line at $(x_0. F(x_0))$ would be parallel to the line segment r?

Hopefully, you would move it to the right of its present position. (If you carefully considered the question and opted for moving x_0 to the left, you should take a short rest and then begin this chapter again.) Actually your answer is not important. The important thing is for you to agree that there is an answer; that is, that we can always find an x_0 so that the tangent line at $(x_0, F(x_0))$ will be parallel to r. Stated another way, we can always find an x_0 so that the slope $m(x_0) = F'(x_0)$ of the tangent line will equal the slope of r, which is

$$\frac{F(b) - F(a)}{b - a}.$$

If you agree that this is reasonable, then you have agreed to the truth of a very important theorem.

THEOREM A (Mean Value Theorem)

Let F be defined and continuous on $[a, b]$, and suppose it has a derivative at every point of (a, b). Then it is always possible to find at least one $x_0 \in (a, b)$ such that

$$\frac{F(b) - F(a)}{b - a} = F'(x_0).$$

The proof is deferred to Problem D. We shall occupy ourselves here with two examples intended to warn the reader that when he uses Theorem A, he must be certain that the hypothesis is fully satisfied.

EXAMPLE A

Let A be defined on $[-1, 1]$ by $A(x) = |x|$. Then

$$A'(x) = \begin{cases} 1, & x \in (0, 1), \\ -1, & x \in (-1, 0). \end{cases}$$

$$\frac{A(1) - A(-1)}{1 - (-1)} = \frac{1 - 1}{2} = 0 \neq A'(x) \qquad \text{for any } x.$$

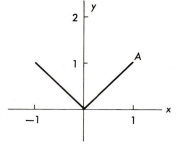

FIG. 4.2

The problem here is that A is not differentiable on all of $(-1, 1)$. One disagreeable point messes up the theorem. ▋

EXAMPLE B
Define B by

$$B(x) = \begin{cases} x, & x \in [0, 1), \\ 2, & x = 1. \end{cases}$$

Then $B'(x) = 1$ exists for all $x \in (0, 1)$.

$$\frac{B(1) - B(0)}{1 - 0} = \frac{2 - 0}{1} = 2 \neq B'(x) \qquad \text{for any } x.$$

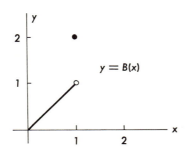

y = B(x)

FIG. 4.3

Now what is wrong ? ∎

PROBLEMS

A · The function $P(x) = \sqrt{x}$ is continuous and differentiable on $[1, 4]$. Find x_0 so that

$$\frac{P(4) - P(1)}{4 - 1} = P'(x_0).$$

B · Suppose you know that F is continuous on $[a, b]$ and that it has a derivative on all of (a, b) for which $F'(x) = 0$ for every x. Show that F is constant. (Hint: Take any $x \in [a, b]$. Show $F(x) = F(a)$.)

C · Suppose that in the preceding problem we had said that for every $x \in (a, b)$, $F'(x) = c$ where c is a nonzero constant. Show that the graph of F is a straight line passing through $(a, F(a))$. (Hint: Show that $F(x) = F(a) + c(x - a)$.)

D · Prove the mean value theorem. (Hint: Use Rolle's theorem (Theorem 3.1B) on the function G defined by

$$G(x) = F(x) - F(a) - \frac{F(b) - F(a)}{b - a}(x - a).)$$

The function G is suggested (or remembered) by taking the difference between the function F and the line passing through $(a, F(a))$ with a slope of $[F(b) - F(a)]/(b - a)$.

In each of Exercises 1–6 you are given a function F and an interval $[a, b]$ on which F satisfies the hypotheses of the mean value theorem. Find, in each case, the point $x_0 \in (a, b)$ for which

$$\frac{F(b) - F(a)}{b - a} = F'(x_0).$$

1. $F(x) = x^{1/2}$, $a = 0$, $b = 4$.

2. $F(x) = x^{1/3}$, $a = 1$, $b = 8$.

3. $F(x) = \dfrac{x}{x + 1}$, $a = -\frac{1}{2}$, $b = \frac{1}{2}$.

4. $F(x) = \dfrac{x}{x - 1}$, $a = 2$, $b = 3$.

5. $F(x) = x^3$, $a = -1$, $b = 1$.

6. $F(x) = x^3 - 4x^2 + 4x$, $a = 0$, $b = 3$.

4 · 2 Higher Derivatives

In the introduction to this chapter we called attention to the real-valued function of a real variable,

$$m(x) = F'(x).$$

This function is, of course, defined on (a, b). As such, it may itself be differentiable at some (or all) of the points in (a, b). At such points we say that F is twice differentiable.

DEFINITION A (The Second Derivative)
Let F be differentiable in a neighborhood of x_0. Set

$$m(x) = F'(x).$$

Then if $m'(x_0)$ exists, it is called the *second derivative of F at x_0*. We write

$$F''(x_0) = m'(x_0).$$

EXAMPLE A

Find all the derivatives of

$$F(x) = \tfrac{1}{3}x^3 - 2x^2 + 3x + 3.$$

From our knowledge of computing the derivative, we know that

$$F'(x) = x^2 - 4x + 3,$$
$$F''(x) = 2x - 4,$$
$$F'''(x) = 2,$$
$$F''''(x) = 0.$$

It is clear that all further derivatives must be zero. ▋

The alternative notations for the derivative have their counterparts for derivatives of higher order. Let

$$y = F(x),$$

$$\frac{dy}{dx} = F'(x),$$

Then

$$\frac{d}{dx}\frac{dy}{dx} = \frac{d}{dx}F'(x) = F''(x).$$

This suggests adoption of the notation

$$\frac{d^2y}{dx^2} = F''(x).$$

Similarly, we write

$$\frac{d^n y}{dx^n} = F^{(n)}(x)$$

for the nth derivative. We also have the notation

$$DF(x) = F'(x),$$

so

$$D\{DF(x)\} = DF'(x) = F''(x).$$

This suggests

$$D^2F(x) = F''(x)$$

and, more generally,

$$D^n F(x) = F^{(n)}(x).$$

PROBLEMS

A · On the same sheet of paper, draw three sets of coordinate axes, situated as in Fig. 4.4. On the top set of axes, graph $y = F(x) = \frac{1}{3}x^3 - 2x^2 + 3x + 3$,

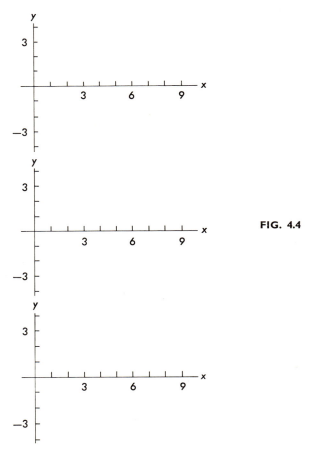

FIG. 4.4

F being the function of Example A. Using the same function, graph $y = F'(x)$ on the second set of axes and $y = F''(x)$ on the third set. Draw the tangent lines to $F(x)$ and to $F'(x)$ at $x = 2$.

B · (This is a continuation of Problem A.) Draw the tangent line to $F(x)$ at $x = 2.5$. Has $m(x) = F'(x)$ increased or decreased as x moved from 2 to 2.5 ? How can you tell the answer from a glance at your graph of $F'(x)$? If the graphs of $F(x)$ and $F'(x)$ were missing, could you tell the answer from looking at the graph of $F''(x)$?

C· $P(x) = \sqrt{x}$. We saw, in Example A, that $D^4 F(x) = 0$. Can you find an n so that $D^n P(x) = 0$?

In Exercises 1–12, find the second derivative.

1· $F(x) = 4x^{3/2}$.

2· $F(x) = 9x^{4/3}$.

3· $F(x) = \dfrac{x}{2(x+1)}$.

4· $F(x) = \dfrac{3x}{1-x}$.

5· $F(x) = \sqrt{x^2 + 1}$.

6· $F(x) = (x^3 + 3x - 7)^4$.

7· $F(x) = \dfrac{x}{x^2 + 1}$.

8· $F(x) = \dfrac{x}{x^3 + 4}$.

9· $F(x) = \left(\dfrac{x}{x+1}\right)^5$.

10· $F(x) = \dfrac{x}{\sqrt{x-1}}$.

11· $F(x) = \dfrac{x-a}{x-b}$.

12· $F(x) = \dfrac{x^2 + a}{x^2 + b}$.

In Exercises 13–18, find the third derivative.

13· $y = x^{2/3}$.

14· $y = \dfrac{x}{x+4}$.

15· $y = \dfrac{x^2}{x+1}$.

16· $y = \dfrac{x^2 + 1}{x + 1}$.

17· $y = (x^2 + 1)^{1/2}$.

18· $y = (x^2 + 4)^{3/2}$.

Note that:

$$D\frac{1}{x} = Dx^{-1} = -x^{-2},$$

$$D^2 \frac{1}{x} = D(-x^{-2}) = 2x^{-3},$$

$$\vdots$$

$$D^n \frac{1}{x} = (-1)^n n! x^{-(n+1)} \quad \text{where} \quad n! = n(n-1)\cdots(2)(1).$$

We have an expression for the nth derivative. Find such an expression for each of the following.

19· $F(x) = \dfrac{x+1}{x-1}$.

20· $F(x) = \sqrt{x}$.

21· $F(x) = \sqrt{x+1}$.

22· $F(x) = \dfrac{1}{3-x}$.

4 · 3 Taylor's Formula

In section 2.7 we raised an important question to which we shall return several times.

The Approximation Problem

Suppose F is defined and well behaved in some interval containing x_0, and suppose $F(x_0)$ is known. Can we devise a method for easily approximating $F(x_0 + h)$ in the event that h is small and that the exact value of $F(x_0 + h)$ is either very difficult or impossible to find ?

In Chapter 3 we found such a method. It hinged on understanding the phrase " F is ... well behaved in some interval containing x_0" to mean that F was differentiable at x_0. Thus, we assumed that we could write

$$F(x_0 + h) = F(x_0) + F'(x_0)(h) + |h| R(x_0, h).$$

We set $x = x_0 + h$;

$$F(x) = F(x_0) + F'(x_0)(x - x_0) + |x - x_0| R(x_0, x - x_0)$$

and we set

$$T(x) = F(x_0) + F'(x_0)(x - x_0).$$

Then we could write

$$F(x) = T(x) + |x - x_0| R(x_0, x - x_0). \tag{1}$$

Because $R(x_0, x - x_0)$ gets small as x gets close to x_0, it was clear that

$$F(x) \approx T(x)$$

when x was taken close to x_0. With our knowledge of higher derivatives and with the mean value theorem available as a tool, we can now improve our method by proving another theorem about approximation. This theorem will include (1) as a special case.

We begin by considering the approximation problem for the case where F is a polynomial. Now the reader may object that when F is a polynomial, it is no work at all to find $F(x_0 + h)$ exactly by direct substitution. We agree. But suppose you have a project going that requires you to find the value of

$F(x_0 + h)$ for many choices of h. We faced such a project, in fact, in Section 3.3, with

$$F(x) = 2x^3 - 12x^2 + 60x \tag{2}$$

and

$$x_0 + h = 3 + h.$$

Suppose we had known back there that (2) could be written in the form

$$F(x) = 126 + 42(x - 3) + 6(x - 3)^2 + 2(x - 3)^3. \tag{3}$$

Then

$$F(3 + h) - F(3) = 42h + 6h^2 + 2h^3$$

could have been easily obtained from (3) by subtracting $F(3) = 126$ from both sides and substituting $3 + h$ for x. See (1) of Section 3.3.

The problem seems to be to find, without devoting an inordinate amount of time to the job, the coefficients 126, 42, 6, and 2. It happens that there is a very easy way. Suppose we want to write, for the reasons suggested,

$$F(x) = a_0 + a_1(x - 3) + a_2(x - 3)^2 + a_3(x - 3)^3 \tag{4}$$

where the trick is to find a_0, a_1, a_2, a_3.

The reader may wonder at this point why we stopped when we did in writing down (4). Why did we not include $a_4(x - 3)^4$, for instance ? We offer the following explanation. If we replace the left side of (4) by the expression given in (2), our goal in choosing a_0, a_1, a_2, a_3 is to make the resulting expression an identity; it must hold for every choice of x. Thus, when the right side is multiplied out and like terms are collected, the coefficients of like powers of x must be the same on each side of the equal sign. (Try this for the choices of the a_i that we said work.) Now if the right side had a term $a_4(x - 3)^4$, then after multiplication, x^4 would have a coefficient of a_4. A look at the left side quickly shows then that $a_4 = 0$. That is why we do not include in (4) any factors of power higher than three.

Returning to the problem of determining the a_i, we see immediately, by substituting in (4), that

$$a_0 = F(3) = 126.$$

the latter equality being obtained from (2). The key idea to further progress is to think of differentiating (4).

$$DF(x) = a_1 + 2a_2(x - 3) + 3a_3(x - 3)^2.$$

From (2) we learn that $DF(3) = 42$, so substitution in our present expression gives

$$a_1 = DF(3) = 42.$$

With the pattern of our procedure now established, we continue.

$$D^2 F(x) = 2a_2 + 3(2)a_3(x - 3);$$

$$2a_2 = D^2 F(3) = 12 \quad \text{from which} \quad a_2 = \frac{12}{2} = 6$$

$$D^3 F(x) = 3(2)a_3;$$

$$3(2)a_3 = D^3 F(3) = 12 \quad \text{and so} \quad a_3 = \frac{12}{3(2)} = 2.$$

In exactly the same way, any polynomial F of degree n may be written in the form

$$F(x) = F(x_0) + DF(x_0)(x - x_0) + \frac{D^2 F(x_0)}{2!} (x - x_0)^2$$

$$+ \cdots + \frac{D^n F(x_0)}{n!} (x - x_0)^n \tag{5}$$

where we have used the notation $n! = n(n - 1) \cdots (3)(2)(1)$. Understand $0! = 1$.

PROBLEM

A · Let F be the polynomial $F(x) = x^3 - 3x^2 + 7x + 4$. Choose $x_0 = 2$ and express F in the form (5).

This is all well and good for polynomials of course, but what has it to do with approximating an arbitrary function F in a neighborhood of x_0 ? We hope it has been suggestive. Suppose the function F we wish to approximate is n times differentiable at x_0. There is nothing to prevent us from obtaining a polynomial of degree n in exactly the same way. Of course, this polynomial will not be equal to F if F is not a polynomial. In many cases of interest, however, it will closely approximate the given function. Let us first define the polynomial we are associating with F. Then we will look at some examples to convince the reader that what we are proposing does in fact work. Finally we will prove that it works.

DEFINITION A   (The Taylor Polynomial of F at x_0)

Let F be defined and n times differentiable at x_0. The Taylor polynomial of degree n associated with F at x_0, which we shall designate with $T_{F(x_0),n}$, is defined by

$$T_{F(x_0),n}(x) = F(x_0) + DF(x_0)(x - x_0) + \frac{D^2F(x_0)}{2!}(x - x_0)^2$$

$$+ \cdots + \frac{D^nF(x_0)}{n!}(x - x_0)^n.$$

Whenever the function F and the point x_0 are clear from the context, we will, for obvious reasons, prefer to abbreviate the subscript, writing T_n.

EXAMPLE A

$G(x) = \sqrt[3]{x}$. This is an old friend. We have worked before at approximating $G(8.4)$. Let us try it again, using the Taylor polynomial of degree 1 corresponding to G at $x_0 = 8$.

$$T_1(x) = G(8) + DG(8)(x - 8).$$

$$DG(x) = \tfrac{1}{3}x^{-2/3};$$

$$DG(8) = \tfrac{1}{3}8^{-2/3} = \tfrac{1}{12};$$

$$T_1(x) = 2 + \tfrac{1}{12}(x - 8).$$

This is exactly the function T obtained in Example 3.1A. ∎

The result of the preceding example is not surprising. It is evident from an inspection of the definition of the Taylor polynomial at x_0 that if we take $n = 1$, then this polynomial is simply the function $T(x)$ described in (1) of the Introduction to this chapter. The Taylor polynomial $T_{G(x_0),1}$ thus describes the function having as its graph a line tangent to the graph of the function G at x_0.

EXAMPLE B

Again let $G(x) = \sqrt[3]{x}$. This time, find the Taylor polynomial of degree 2; that is, find $T_{G(8),\,2}$.

We know from the preceding example that

$$DG(x) = \tfrac{1}{3}x^{-2/3};$$

$$D^2 G(x) = \tfrac{1}{3}(-\tfrac{2}{3})x^{-5/3};$$

$$D^2 G(8) = -\frac{2}{9}\frac{1}{2^5} = -\frac{1}{144}.$$

Substituting in the formula defining the Taylor polynomial, we get

$$T_2(x) = 2 + \frac{1}{12}(x - 8) + \left(-\frac{1}{288}\right)(x - 8)^2.$$

From this we learn that

$$T_2(8.4) = 2 + 0.0333 - 0.0005 = 2.0328.$$

Compare this approximation of $\sqrt[3]{8.4}$ with the answer correct to four decimal places as given at the end of Example 3.1A. The graphs of the functions T_1, T_2, and G are shown in Fig. 4.5. ∎

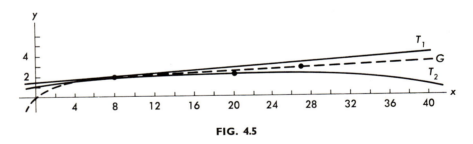

FIG. 4.5

It seems that $T_{G(x_0),n}$ does, at least in this instance, approximate G in a neighborhood of x_0. It also seems that if we get too far from x_0, we will not have very good luck in our approximation attempts. Clearly, we need to know the answer to the following question.

If F is n times differentiable at a point x_0, so that we can form T_n, how closely does $T_n(x_0 + h)$ approximate $F(x_0 + h)$ if h is restricted to some certain neighborhood of 0; that is, $x_0 + h$ is restricted to some certain neighborhood of x_0?

The following theorem gives an answer to this question. This is a well-known result which has been proved in a variety of ways.

THEOREM A

Let F have n continuous derivatives on $[a, b]$ and suppose that $D^{n+1}F$ exists on (a, b). For any $x_0 \in [a, b]$, the Taylor polynomial $T_{F(x_0),n}$ may be formed. Then for any $x \in [a, b]$,

$$F(x) = T_n(x) + \frac{(x - x_0)^{n+1}}{(n+1)!} D^{n+1}F(\bar{x}) \qquad (6)$$

where $\bar{x}$ is some point between x_0 and x. The last term, being the difference between $F(x)$ and $T_n(x)$, is called the *remainder term*. It is denoted by $R_n(x_0, x)$; that is,

$$R_n(x_0, x) = \frac{(x - x_0)^{n+1}}{(n+1)!} D^{n+1}F(\bar{x})$$

where $\bar{x}$ is between x_0 and x.

PROOF:

We will establish (6) by showing that it holds for any arbitrarily chosen point $x_1 \in [a, b]$. Given such a point, we first choose a number c so that

$$F(x_1) + c(x_1 - x_0)^{n+1} - T_n(x_1) = 0. \qquad (7)$$

With c thus determined, we define a function G on $[a, b]$ by

$$G(x) = F(x) + c(x - x_0)^{n+1} - T_n(x). \qquad (8)$$

By (7), we see immediately that $G(x_1) = 0$. From (8), G shares the differentiability properties of F, and

$$DG(x) = DF(x) + (n+1)c(x - x_0)^n - DT_n(x)$$
$$\vdots$$
$$D^nG(x) = D^nF(x) + (n+1)! \, c(x - x_0) - D^nT_n(x).$$

Recalling that $D^rF(x_0) = D^rT_n(x_0)$ for $r = 0, 1, \ldots, n$ we see from these equations that $D^rG(x_0) = 0$, $r = 0, 1, \ldots, n$. Now G vanishes for x_0 and x_1. Rolle's theorem (Theorem 3.1B) guarantees the existence of $\bar{x}_1$ between x_0 and x_1 such that $DG(\bar{x}_1) = 0$. Now DG is zero at $\bar{x}_1$ and x_0. Another appeal to Rolle's theorem gives us an $\bar{x}_2$ between $\bar{x}_1$ and x_0 (so certainly between x_0 and x_1) for which $D(DG)(\bar{x}_2) = D^2G(\bar{x}_2) = 0$. Continuing in this fashion, we ultimately find that there must be an $\bar{x}$ between x_0 and $\bar{x}_n$, hence between x_0 and x_1 for which $D^{n+1}G(\bar{x}) = 0$.

$$D^{n+1}G(x) = D^{n+1}F(x) + (n+1)!c.$$

Substituting $\bar{x}$, which we know makes the left side 0, gives

$$c = -\frac{D^{n+1}F(\bar{x})}{(n+1)!}.$$

Replacing c in (7) by this value gives

$$F(x_1) - \frac{D^{n+1}F(\bar{x})}{(n+1)!}(x_1 - x_0)^{n+1} - T_n(x_1) = 0. \quad \blacksquare$$

The typical use of this theorem is as follows. We are trying to estimate $F(x)$ by using $T_n(x)$. We find $T_n(x)$ and reason that the magnitude of our error is

$$|F(x) - T_n(x)| = |R_n(x_0, x)|.$$

Now the term on the right depends on $\bar{x}$, which is generally not known. It is restricted, however, in that it lies between x and x_0. Using this information and the expression for $R_n(x_0, x)$ given by the theorem, we are often able to estimate an upper bound for $|R_n(x_0, x)|$. Thus, while we will probably not know the exact error, we may be able to say the error is no greater than some number. If, for example, the error is known to be less than 0.0005, then we know that $T(x)$ approximates $F(x)$ accurately to within one unit in the third decimal place.

A practical reminder of a matter first called to your attention in Problem 2.6H. If for $x \in [a, b]$, we know that $|N(x)| \leq r$ and $|D(x)| \geq s > 0$, then it follows that for $x \in [a, b]$,

$$\left|\frac{N(x)}{D(x)}\right| < \frac{r}{s}.$$

The use of Taylor's theorem is illustrated in the following two examples.

EXAMPLE C

Suppose we decide to use a Taylor polynomial to approximate $F(x) = (1 + x)/(1 - x)$ for $x \in (4.5, 5.5)$. How large must we take n if we want our answer accurate within one unit in the second decimal place ?

We are obviously going to use the polynomials $T_{F(5),n}$. Taylor's theorem gives

$$R_n(5, x) = \frac{(x - 5)^{n+1}}{(n+1)!} D^{n+1}F(\bar{x})$$

where $\bar{x}$ is between x and 5. Since $x \in (4.5, 5.5)$, $\bar{x}$ is also in this interval.

$$DF(x) = \frac{(1 - x) - (1 + x)(-1)}{(1 - x)^2} = 2(1 - x)^{-2};$$

$$D^2F(x) = 2(2)(1 - x)^{-3}$$

$$\vdots$$

$$D^nF(x) = 2(n!)(1 - x)^{-(n+1)}.$$

Therefore,

$$|R_n(5, x)| = \frac{|x - 5|^{n+1}}{(n + 1)!} \frac{2(n + 1)!}{|1 - \bar{x}|^{n+2}}$$

$$= \frac{2|x - 5|^{n+1}}{|1 - \bar{x}|^{n+2}} = \frac{N(x)}{D(\bar{x})}.$$

Since $x \in (4.5, 5.5)$, it is clear that $|N(x)| < 2(\frac{1}{2})^{n+1}$. And since $\bar{x} \in (4.5, 5.5)$, $|D(\bar{x})| \geq |1 - 4.5|^{n+2}$. Hence, making use of the practical reminder given just before this example, we find that

$$|R_n(5, x)| \leq \frac{2(\frac{1}{2})^{n+1}}{|1 - 4.5|^{n+2}} = \frac{4}{7^{n+2}}$$

By trial and error, we choose n large enough so that this expression is less than 0.005. For $n = 2$,

$$|R_2(5, x)| \leq \frac{4}{7^4} < \frac{4}{(48)(48)} < \frac{1}{500} = 0.002.$$

Thus,

$$T_2(x) = -\tfrac{3}{2} + \tfrac{1}{8}(x - 5) - \tfrac{1}{32}(x - 5)^2$$

may be used to approximate $F(x)$. For $x \in (4.5, 5.5)$, the approximation so obtained will be accurate to within one unit in the second decimal place.

By way of illustration,

$$T_2(5.4) = -1.5000 + 0.0500 - 0.0050 = -1.455.$$

Depending on whose rules are being used, this approximation rounded off to two decimal places will be given as -1.45 or -1.46. Direct computation, which we carry out here for purposes of comparison, given

$$F(5.4) = \frac{6.4}{-4.4} = -1.454.$$

Rounded off to two decimals (by anybody's rules), the correct answer is -1.45. Our approximation (either one) is thus within one unit in the second decimal place of the correct answer. ∎

EXAMPLE D
We use the theorem to estimate the error made in Example B above, where we used $T_2(8.4)$ to approximate $G(8.4) = \sqrt[3]{8.4}$.

In Example B we found

$$D^2G(x) = \tfrac{1}{3}(-\tfrac{2}{3})x^{-5/3}.$$

Therefore

$$D^3G(x) = \tfrac{1}{3}(-\tfrac{2}{3})(-\tfrac{5}{3})x^{-8/3}.$$

Using the formula provided by Taylor's theorem, we find that

$$R_2(8, 8.4) = \frac{(8.4-8)^3}{3!} D^3G(\bar{x})$$

where $\bar{x} \in (8, 8.4)$. Since

$$D^3G(\bar{x}) = \frac{10}{27} \cdot \frac{1}{\bar{x}^{8/3}},$$

$$R_2(8, 8.4) = \frac{(0.4)^3}{6} \cdot \frac{10}{27} \cdot \frac{1}{\bar{x}^{8/3}}.$$

The restriction that $\bar{x} \in (8, 8.4)$ means

$$R_2(8, 8.4) \le \frac{(0.4)^3}{6} \cdot \frac{10}{27} \cdot \frac{1}{2^8} < 0.00002.$$

This means our answer of 2.0328 obtained in Example B is accurate to within one unit in the fourth decimal place. (We have already seen, of course, that in this case our answer is actually exactly correct to four decimal places.) ∎

PROBLEMS

B · Let F be defined by $F(x) = 1/(1 - x)$. Write the indicated Taylor polynomials.
    (a) $T_{F(0),0}$.   (c) $T_{F(0),2}$.
    (b) $T_{F(0),1}$.   (d) $T_{F(0),3}$.
Now on the same axes, graph all four of these polynomials and the function F for $x \in [-2, 2]$.

C · (The purpose of this rather long problem is to help you understand the proof of the Taylor theorem.) Let $F(x) = 1/(1 - x)$ and suppose we take $n = 2$, $x_0 = 0$. Choose $x_1 = \frac{1}{4}$. Taylor's theorem says, in this case,

$$\frac{1}{1 - \frac{1}{4}} = 1 + \frac{1}{4} + (\frac{1}{4})^2 + \frac{(\frac{1}{4})^3}{3!} \cdot \frac{3!}{(1 - \bar{x})^4}, \qquad \bar{x} \in (0, \tfrac{1}{4}). \qquad (6')$$

Prove this by tracing the steps of the proof. Hint: Begin by determining c from (7); that is, solve

$$\tfrac{4}{3} + c(\tfrac{1}{4})^3 - [1 + \tfrac{1}{4} + \tfrac{1}{16}] = 0.$$

Using the resulting value of c, substitute in (8) to get

$$G(x) = (1 - x)^{-1} + cx^3 - [1 + x + x^2].$$

(a) Verify $G(0) = 0$.   (c) Verify $DG(0) = 0$.
(b) Verify $G(\tfrac{1}{4}) = 0$.   (d) Verify $D^2G(0) = 0$.

Since $G(0) = G(\tfrac{1}{4})$, we use Rolle's theorem to conclude that there is an $\bar{x}_1$ for which $DG(\bar{x}_1) = 0$; $\bar{x}_1 \in (0, \tfrac{1}{4})$.
Since $DG(0) = DG(\bar{x}_1)$, use Rolle's theorem to conclude that there is an $\bar{x}_2$ for which $D^2G(\bar{x}_2) = 0$; $\bar{x}_2 \in (0, \bar{x}_1)$.
Finally conclude there is an $\bar{x}$ such that $D^3G(\bar{x}) = 0$, $\bar{x} \in (0, \tfrac{1}{4})$. Actually find $\bar{x}$ for this example.

D · Show from (6′) in Problem C that if we use $T_2(\tfrac{1}{4})$ to approximate $F(\tfrac{1}{4})$, the error is no more than $(4/3^4) \approx 0.049$. Then find the actual error. Finally, using

$$R_n(0, \tfrac{1}{4}) = \frac{(\tfrac{1}{4})^{n+1}}{(n + 1)!} D^{n+1}F(\bar{x}), \qquad \bar{x} \in (0, \tfrac{1}{4}),$$

find out which Taylor polynomial (that is, which n) may be used if we want to be sure our answer is correct to within one unit in the second decimal place. (This means the error term must be less than 0.005.)

===   **4·3 EXERCISES**

Compute, for each of the functions in Exercises 1–10, the Taylor polynomial $T_{F(x_0), 2}$.

1 · $F(x) = x^4 - 3x^3 + 2x^2 - x + 1$ at $x_0 = 2$.
2 · $F(x) = x^4 - 3x^3 + 2x^2 - x + 1$ at $x_0 = -1$.
3 · $F(x) = x^4 - 3x^3 + 2x^2 - x + 1$ at $x_0 = 3$.
4 · $F(x) = x^4 - 3x^3 + 2x^2 - x + 1$ at $x_0 = -2$.

5. $F(x) = \dfrac{1}{1 + x^2}$ at $x_0 = 0$.

6. $F(x) = \dfrac{1}{1 - x^2}$ at $x_0 = 0$.

7. $F(x) = \dfrac{x + 2}{-(1 + x)}$ at $x_0 = -2$.

8. $F(x) = \dfrac{x - 2}{3 - x}$ at $x_0 = 2$.

9. $F(x) = \dfrac{1 - x}{1 + x}$ at $x_0 = 0$.

10. $F(x) = \dfrac{x}{2x + 3}$ at $x_0 = 3$.

11. Find $R_2(2, \frac{9}{4})$ for Exercise 1, and $R_2(3, \frac{9}{4})$ for Exercise 3. Estimate the maximum value of $|R_2(2, \frac{9}{4})|$, remembering that $\bar{x} \in [2, \frac{9}{4}]$. Also, estimate the maximum of $|R_2(3, \frac{9}{4})|$, $\bar{x} \in [\frac{9}{4}, 3]$.

12. Find $R_2(-1, -\frac{5}{4})$ for Exercise 2, and $R_2(-2, -\frac{5}{4})$ for Exercise 4. Estimate the maximum value of $|R_2(-1, -\frac{5}{4}),|$ remembering that $\bar{x} \in [-\frac{5}{4}, -1]$. Also estimate the maximum of $|R_2(-2, -\frac{5}{4})|$, $\bar{x} \in [-2, -\frac{5}{4}]$.

13. In Exercise 1, compute the actual difference $F(\frac{9}{4}) - T_2(\frac{9}{4})$. Compute the same difference for Exercise 3. Compare these numbers with the estimates of Exercise 11.

14. For Exercise 2, compute the actual difference $F(\frac{5}{4}) - T_2(\frac{5}{4})$. Compute the same difference for Exercise 4. Compare these numbers with the estimates of Exercise 12.

15. Use the T_2 computed in Exercise 9 to estimate $F(\frac{1}{4})$. Use $|R_2(0, \frac{1}{4})|$ to get an upper bound on the possible error of this estimate.

16. Use the T_2 computed in Exercise 10 to estimate $F(\frac{13}{4})$. Use $|R_2(3, \frac{13}{4})|$ to get an upper bound on the possible error of this estimate.

CHAPTER

5

APPLICATIONS

HE REALLY SEEMS TO BELIEVE THIS KIND OF AN
EXPLANATION, AND I FIND NO EXCUSE FOR HIM
EXCEPT THAT A MATHEMATICIAN, HOWEVER
GREAT, WITHOUT THE HELP OF A GOOD DRAW-
ING IS NOT ONLY HALF A MATHEMATICIAN, BUT
ALSO A MAN WITHOUT EYES.

—*Ludovico Cigoli*
from a letter to Galileo as quoted by
Santillana, The Crime of Galileo.

In Section 2.2 we saw that the graph is an aid to understanding the behavior of a real-valued function of a real variable. We tried to acquire some skill in drawing such graphs in Section 2.3. At that point there was no great interest in producing a terribly accurate graph. In many cases the reader was encouraged simply to plot a few points and sketch a curve through them.

Although the graphs we drew earlier are adequate for many purposes, there are applications of mathematics in which a more refined picture is needed. We now have at our disposal the tools necessary for such refinement. Section 5.1 is devoted to the application of these tools to the problem of drawing as accurate a graph as could be desired. The reader should find this section esthetically satisfying as well as practical. It is properly viewed as a continuation of Section 2.3.

One consequence of being able to draw an accurate graph is that you are able to find precisely the relative maximum and relative minimum points. This ability can in turn be used to solve a host of interesting problems drawn from different fields. Such problems are the subject of Section 5.2.

We have already seen (Section 3.3) that the derivative may be interpreted as the instantaneous velocity of a moving object. Further exploration of this idea is the purpose of the last section in this chapter.

5 · 1   Graphs; Refining the Picture

Almost any daily newspaper will contain a number of graphs. They are used to indicate population trends, price changes on the stock market, changes in yesterday's temperature, etc. In almost every case, they are used because they show at a glance the intervals in which there was an increase and the intervals in which there was a decrease.

We are similarly interested, when drawing a graph of $y = F(x)$, to know the intervals on which F is increasing and the intervals on which F is decreasing. This is particularly easy to determine if the function is differentiable in the interval on which it is being studied. One would expect in this case that since $F'(x)$ gives the slope of the line tangent to the graph of F at x, a positive value of $F'(x)$ on an interval would indicate an increasing function; and negative values of $F'(x)$ on the interval would indicate a decreasing function. We now show that this expectation is correct.

THEOREM A
> Suppose F is continuous on $[a, b]$, differentiable on (a, b). Then $F'(x) \geq 0$ on (a, b) implies that F is increasing on $[a, b]$; and $F'(x) \leq 0$ on (a, b) implies that F is decreasing on $[a, b]$.

PROOF:

To show that F is increasing on $[a, b]$, we need to show that if $a \leq x_1 < x_2 \leq b$, then $F(x_1) \leq F(x_2)$. We appeal to the mean value theorem (Theorem 4.1A) to obtain

$$\frac{F(x_2) - F(x_1)}{x_2 - x_1} = F'(x_0),$$

for some $x_0 \in (a, b)$. And since $F'(x_0) \geq 0$,

$$F(x_2) - F(x_1) = F'(x_0)(x_2 - x_1) \geq 0,$$

which proves the theorem for increasing functions. The proof for decreasing functions is similar. ∎

Both the statement of the above theorem and its proof would remain valid if we replaced all the inequalities by strict inequalities. In such cases, we refer to F as *strictly increasing* or *strictly decreasing*.

With the intervals on which $F'(x) \neq 0$ thus characterized, we now wish to turn attention to those points for which $F'(x) = 0$. In order to talk intelligently about such points, it will be a good idea to review some terms we have already met and to introduce a few new ones.

A function F is said to have a *local maximum* at x_0 if F is defined on an open neighborhood N of x_0, and if $F(x) \leq F(x_0)$ for each $x \neq x_0$ in N. A *local minimum* is similarly defined. As usual, our definitions will make more sense if we think about them while looking at a picture. The function F graphed in Fig. 5.1 has a local maximum at the points x_0 and x_2 and a local minimum at x_1. Local maximums and local minimums of F are called *extreme points*.

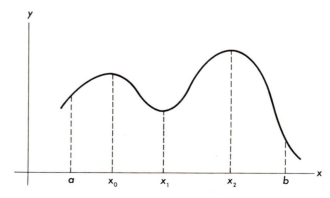

FIG. 5.1

We also see from Fig. 5.1 the wisdom of referring to *local* maximums and *local* minimums (sometimes called *relative* maximums or minimums). There is a local maximum at x_0 because $F(x_0)$ is bigger than $F(x)$ for any x in a neighborhood of x_0. But as often happens to the biggest kid in the neighborhood, so it happens to x_0 that if we look outside of the immediate neighborhood, there is a point (in fact, there are a lot of points) at which $F(x)$ is larger than $F(x_0)$. To push this analogy a bit further, it is true that if we restrict ourselves to one district having a well-defined boundary, there will be one kid who is absolutely the biggest. And so it happens in Fig. 5.1. If we restrict our consideration of F to the domain $[a, b]$, then there is a maximum value for F, attained at x_2.

Still looking at Fig. 5.1, the maximum of F on $[a, b]$ occurs at x_2, which is also a point where F has a local maximum. Note, however, that on $[a, b]$, the minimum value of F occurs at the point $x = b$. Yet $x = b$ is not a local minimum since there is no open neighborhood of $x = b$ contained in the interval $[a, b]$ under consideration. Finally we note by observing the graph of $y = 1/x$ on the interval $(0, 1)$ that a function may have no maximum or minimum value in an interval. (Why is 1 not a minimum value for $1/x$ in $(0, 1)$?)

The example just given of a function with neither maximum nor minimum was defined on an open interval. There is a very good reason for this. We have already seen (Theorem 2.6F) that if a function F is continuous on a closed interval $[a, b]$, then there is a point x_1 for which $F(x_1) = M$ is the maximum value of F in $[a, b]$ and there is a point x_2 for which $F(x_2) = m$ is the minimum value of F in $[a, b]$.

If one thinks of $F'(x)$ as the slope of the line tangent to the graph of F at x, then it seems geometrically obvious that $F'(x)$ will be zero at points where F has extreme points. Not everything that seems geometrically obvious turns out to be true, of course, but this is one of those happy situations in which our intuition is a reliable guide. We have, in fact, already proved this (Problems 3.1B and 3.1C).

To summarize, if F is differentiable on (a, b) and continuous on $[a, b]$, then

(1) F assumes both a maximum and a minimum value on $[a, b]$. These points may or may not occur at the end of the interval.

(2)   There may be points in (a, b) at which F has a local maximum or minimum. If x_0 is such a point, then $F'(x_0) = 0$.

DEFINITION A

Let F be differentiable on (a, b). If $F'(\bar{x}) = 0$ for $\bar{x} \in (a, b)$, then $\bar{x}$ is called a *critical point* of F.

Mathematics is supposed to teach one to think clearly, so let's see how we're doing. Suppose F has a critical point at x_0. Does it follow that F has either a local maximum or a local minimum at x_0 ?

The answer is no. Those who answered yes suffer from a common impediment to clear thinking known as assuming the converse of a true statement to be true. For this group we offer the following example as an antidote.

EXAMPLE A
Find the critical points of $C(x) = x^3$.

$C'(x) = 3x^2$, so the only critical point occurs at $x = 0$. The graph (Fig. 5.2) indicates that $x = 0$ is neither a local maximum nor a local minimum. ∎

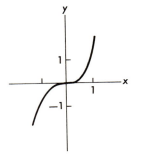

FIG. 5.2

Thus, a critical point may be a local maximum, a local minimum, or neither. In the paragraphs below we develop some tests to classify critical points in the case of functions that are twice differentiable. But the reader will do well to remember that critical points can very often be classified by merely looking at a rough sketch of the graph of a function. As usual, then, we urge the use of a picture along with analytical calculations.

EXAMPLE B
A graph of $G(x) = (x - 1)^2/(x + 3)$ was sketched in Example 2.3B. Refine this picture by locating and classifying the critical points:

$$G'(x) = \frac{(x + 3)2(x - 1) - (x - 1)^2}{(x + 3)^2}$$

$$= \frac{(x - 1)(x + 7)}{(x + 3)^2}.$$

There are critical points at $x = 1$ and $x = -7$. The one at $x = 1$ was, of course, located and identified in Fig. 2.10 as a local minimum. From this same sketch it is clear that $x = -7$ is a local maximum. Using our added information about $x = -7$, we refine Fig. 2.10 as indicated in Fig. 5.3. ▉

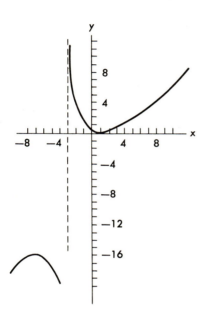

FIG. 5.3

We emphasize our assumption that F is differentiable throughout the interval on which we are working. Thus, although the function $S(x) = 4(x - 2)^{2/3} /x^2$ has a local minimum at $x = 2$ (see Problem 2.4G), we will not find it by studying the derivatives of S, since the derivative of S does not exist at $x = 2$.

EXAMPLE C

In Example 2.3A we had the problem of plotting the curve defined by

$$4x^2 - 24x - y^2 - 4y + 16 = 0. \qquad (1)$$

Suppose we wish to find the slope of this curve corresponding to some point x.

We know that (1) determines not one but two functions of x, namely

$$y = H_1(x) = -2 + 2\sqrt{(x - 3)^2 - 4},$$

$$y = H_2(x) = -2 - 2\sqrt{(x - 3)^2 - 4}.$$

Corresponding to $x = 6$, the slope depends on whether we compute $DH_1(6)$ or $DH_2(6)$.

$$DH_1(x) = [(x-3)^2 - 4]^{-1/2}(2)(x-3) \qquad \text{so} \qquad DH_1(6) = 6/\sqrt{5};$$

$$DH_2(x) = -[(x-3)^2 - 4]^{-1/2}(2)(x-3) \qquad \text{so} \qquad DH_2(6) = -6/\sqrt{5}.$$

These values are consistent with Fig. 2.8 of Section 2.3 where we see that the graph of H_1, lying above the line $y = -2$, has a positive slope at $x = 6$, while H_2, lying below the line $y = -2$, has a negative slope at $x = 6$.

It is instructive to derive this result by a second method, called *implicit differentiation*. In this process, we think of y as a function of x, writing $y = H(x)$ without specifying which function we have in mind. By the chain rule, the derivative of y^2 as a function of x is

$$\frac{d}{dx}y^2 = 2y\frac{dy}{dx}.$$

Differentiation of (1) then gives

$$8x - 24 - 2y\frac{dy}{dx} - 4\frac{dy}{dx} = 0,$$

$$\frac{dy}{dx} = \frac{8x - 24}{2y + 4} = \frac{4x - 12}{y + 2}.$$

(2)

Now to find $\left.\dfrac{dy}{dx}\right|_{x=6}$ (read this "the value of dy/dx when $x = 6$") we must know y as well as x. Setting $x = 6$ in (1),

$$144 - 144 - y^2 - 4y + 16 = 0,$$

$$y = \frac{-4 \pm \sqrt{16 + 64}}{2} = -2 \pm 2\sqrt{5}.$$

Thus, (2) will give two answers, depending on the value we use for y. If we use $-2 + 2\sqrt{5}$, which is above the line $y = -2$, we get

$$\frac{dy}{dx} = \frac{4(6) - 12}{(-2 + 2\sqrt{5}) + 2} = \frac{12}{2\sqrt{5}} = \frac{6}{\sqrt{5}}.$$

If we use $-2 - 2\sqrt{5}$, which is below $y = -2$, we get

$$\frac{dy}{dx} = \frac{4(6) - 12}{(-2 - 2\sqrt{5}) + 2} = \frac{-6}{\sqrt{5}}. \quad \blacksquare$$

PROBLEMS

A· Refine the graph you drew of $R(x) = (x^2 + 4x - 3)/(x^2 - 4x + 3)$ defined in Problem 2.4D by finding and identifying any critical points.

B· Refine your graph of $S(x) = [4(x - 2)^{2/3}]/x^2$ defined in Problem 2.4G by considering $\lim_{x \to 2^-} S'(x)$ and $\lim_{x \to 2^+} S'(x)$.

C· Differentiate $3y^2 - 12y - 2x^2 - 12x - 6 = 0$ implicitly and use the result to find the slopes of the corresponding curve at the points $x = -2, -1, 0, 1, 2$. Explain your result.

D· In Problem 2.6H you studied $R(x) = (32x - 8x^2 - 14)/(2x + 1)$. You showed that for $x \in [1, 6]$, $R(x) < 6$. Reread this problem. Can you now find the number x_1 so that if $x \in [1, 6]$, then $R(x) \le R(x_1)$? Also, can you find x_2 so that if $x \in [1, 6]$, then $R(x) \ge R(x_2)$?

=== **5.1A   EXERCISES**

In Exercises 1–18, sketch a graph of the indicated function. Then refine your graph by locating all extreme points exactly.

1· $y = x^3 - 5x^2 + 3x + 2.$ **2·** $y = x^3 - x^2 - 8x + 4.$

3· $y = x^4 - 6x^2 + 8x + 3.$ **4·** $y = x^4 + 4x^3 - 16x - 5.$

5· $y = x^4 + 2x^3 - 2x^2 + 1.$ **6·** $y = x^4 + 3x^3 - \frac{8}{3}x^2 + 1.$

7· $y = 4x^5 + 7x^4 - 8x^3 - 2.$ **8·** $y = 3x^5 + x^4 - x^3 - 4.$

9· $y = \dfrac{x - 1}{x^2}.$ **10·** $y = \dfrac{x + 3}{(x - 2)^2}.$

11· $y = \dfrac{x^2}{x^2 - 6x + 8}.$ **12·** $y = \dfrac{x^2 - 9x + 20}{x - 3}.$

13· $y = x^{3/2}(x - 4).$ **14·** $y = \dfrac{x^{3/2}}{x - 2}.$

15· $y = x^{2/3}(x - 4).$ **16·** $y = x^{1/3}(x - 4).$

17· $y = x^{4/3}(x^2 - 4).$ **18·** $y = x^{3/2}(x^2 + 2).$

In Exercises 19–28 find dy/dx by implicit differentiation.

19· $y^3 + 3y - 2x^2 + 4x = 0.$ **20·** $3y^2 - y^3 - x^6 + x^3 = 0.$

21· $x^2y + 3x^4 = 0.$ **22·** $x^3y - 2x^2 = 0.$

23· $xy^2 + 4y - x^3 = 0.$ **24·** $x^2y^2 + 3y - x^2 = 0.$

25· $\dfrac{x}{y} + xy = 0.$ **26·** $\dfrac{y}{x} - \dfrac{x}{y} = 0.$

27· $\dfrac{xy + 3}{y^2} = x^3.$ **28·** $\dfrac{x^2y + xy^2}{x + y} = 4.$

Using the Second Derivatives

We now turn to the use of the second derivative in classifying critical points. We might anticipate the facts by thinking a moment about Fig. 5.4 in which

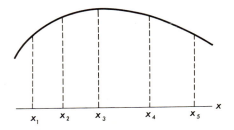

FIG. 5.4

we have pictured a local maximum point. Suppose you drew lines tangent to the graph at x_1, x_2, x_3, x_4, and x_5. The slope at each point is less than the slope at the preceeding point. But these slopes are precisely the values

$$F'(x_1) > F'(x_2) > F'(x_3) > F'(x_4) > F'(x_5).$$

Thus F' is a decreasing function. But if F' is strictly decreasing on an interval, then its derivative, F'', must be negative on the entire interval. So we anticipate that if the second derivative is negative in a neighborhood of a critical point, then F will have a local maximum at that point. Similarly, one sees that if $F''(x) > 0$ in a neighborhood of a critical point x_0, then F ought to have a local minimum at x_0.

Let us prove that once again our geometric intuition is indeed correct. We consider three cases for a function F that is twice differentiable on (a, b) and has a critical point at $x_0 \in (a, b)$;

    i. $F''(x_0) < 0$,
   ii. $F''(x_0) > 0$,
  iii. $F''(x_0) = 0$.

According to Theorem 2.6G, in cases i and ii, there will be a neighborhood N of x_0 on which $F''(x)$ will remain negative (in case i) or positive (in case ii). Now Taylor's theorem (Theorem 4.3A) tells us that for any $x \in N$.

$$F(x) = F(x_0) + F'(x_0)(x - x_0) + \tfrac{1}{2}F''(\bar{x})(x - x_0)^2,$$

where $\bar{x}$ is between x and x_0. Since $F'(x_0) = 0$, we see that

$$F(x) - F(x_0) = \tfrac{1}{2}F''(\bar{x})(x - x_0)^2.$$

Now $(x - x_0)^2$ is always positive, so if $F''(\bar{x}) < 0$ as in case i, $F(x) < F(x_0)$, and F has a relative maximum at x_0. And if $F''(\bar{x}) > 0$ as in case ii, $F(x) > F(x_0)$, and F has a local minimum at x_0.

Finally, in the case where $F''(x_0) = 0$, it will still be true that if $F''(x) < 0$ for $x \neq x_0$, $x \in N$, then x_0 will be a local maximum point, and if $F''(x) > 0$ for $x \neq x_0$, $x \in N$, then x_0 will be a local minimum point. The only remaining possibility is that $F''(x)$ changes sign at x_0, being positive on one side of x_0 and negative on the other. In this case we say F has a point of *inflection* at x_0.

DEFINITION B

The function F, twice differentiable on (a, b), is said to have a point of inflection at $x_0 \in (a, b)$ if $F''(x)$ changes sign at x_0.

The function $C(x) = x^3$ of Example A has $C''(x) = 6x$, which is negative for $x < 0$, positive for $x > 0$. This is why $x = 0$ is neither a local maximum nor minimum. Note carefully that the condition $F''(x_0) = 0$ is not enough to guarantee that x_0 is a point of inflection. Study the behavior of $F(x) = x^4$ at $x_0 = 0$. Finally, note that a point may be an inflection point but not a critical point. In summary, a function F for which $F''(x)$ exists and is continuous on (a, b) has $x_0 \in (a, b)$ as a

i.   local minimum if $F'(x_0) = 0$ and $F''(x) \geq 0$ in a neighborhood of x_0,

ii. local maximum if $F'(x_0) = 0$ and $F''(x) \leq 0$ in a neighborhood of x_0,

iii. point of inflection if $F''(x_0) = 0$ and $F''(x)$ changes sign at x_0.

In Problem 4.2A the reader was asked to sketch graphs of F, F', and F'' for $F(x) = \frac{1}{3}x^3 - 2x^2 + 3x + 3$. The desired graphs are shown in Fig. 5.5. They should be studied in the light of this section.

The graph of F has critical points at $x = 1$ and $x = 3$ where $F'(x) = 0$. $F''(1) < 0$, so F has a local maximum at $x = 1$, and $F''(3) > 0$, so F has a local minimum at $x = 3$.

The graph of F' is positive on $(-\infty, 1)$ and $(3, \infty)$, corresponding to the intervals on which F increases. And F' is negative on $(1, 3)$ where F decreases. F' decreases on $(-\infty, 2)$ corresponding to the decreasing slope of the tangent to $y = F(x)$ on this interval. F' has a critical point (where $F''(x) = 0$) at $x = 2$; this is as small, that is as far negative, as the slope of the tangent line to F will go. F' increases on $(2, \infty)$.

Slide a pencil along the graph of F, keeping it tangent to the curve at all points. The pencil turns clockwise (decreasing slope) on $(-\infty, 2)$ where F'' is negative. It turns counterclockwise (increasing slope) on $(2, \infty)$ where F'' is

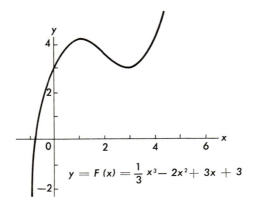

$$y = F(x) = \frac{1}{3}x^3 - 2x^2 + 3x + 3$$

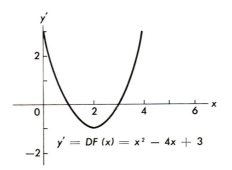

$$y' = DF(x) = x^2 - 4x + 3$$

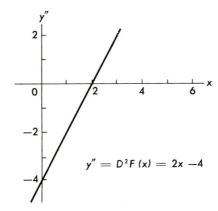

$$y'' = D^2F(x) = 2x - 4$$

FIG. 5.5

positive. The point of inflection at $x = 2$ is the point at which the tangent quits turning one way and starts to turn another.

PROBLEM

E · Suppose the graph of $F(x) = \frac{1}{3}x^3 - 2x^2 + 3x + 3$ studied in this section represented on some economic scale the profits of a company at time x. Interpret the second derivative.

═══ 5·1B EXERCISES

Sketch careful graphs of the following functions. Locate extreme points and identify them as local maximums, minimums, or inflection points. Locate all points of inflection.

1 · $y = 3x^5 - 6x^4 + x^3 + 3x^2 - 2.$

2 · $y = 2x^5 + 5x^4 - 10x^3 - 40x^2 - 40x.$

3 · $y = \dfrac{x^2 - 3x - 4}{x^2 - 3x}.$

4 · $y = \dfrac{x^2 + x - 6}{x^2 + x - 2}.$

5 · $y = x + \dfrac{1}{x}.$

6 · $y = x^2 - \dfrac{1}{x^2}.$

7 · $y = (x - a)^2(x - b)^3.$

8 · $y = (x - a)^3(x - b)^4.$

In Exercises 9–12, sketch smooth graphs consistent with the information given.

9 · $F'(x) > 0$ on $(3, 5)$; $F'(x) \le 0$ elsewhere. $F'(3) = F'(5) = F''(4) = F''(6) = 0.$ $F''(x) > 0$ for $x > 6.$

10 · $F'(x) < 0$ on $(3, 5)$; $F'(x) \ge 0$ elsewhere. $F'(3) = F'(5) = F''(4) = F''(6) = 0.$ $F''(x) < 0$ for $x > 6.$

11 · $F''(x) = (x - 1)(x - 3)(x - 4)^2 g(x)$, where $g(x) \ge 0$ for all x. $F'(4) = F'(1) = 0.$

12 · $F''(6) = x(x - 3)^2(x + 1)^2(x - 1)g(x)$, where $g(x) \ge 0$ for all x. $F'(-1) = F'(3) = 0.$

Consider the functions of Exercises 13–16 to be defined on the closed interval $[0, 3]$. According to Theorem 2.6F, each function assumes a maximum M and a minimum m on the interval. For each function, find any local maximums and minimums and also the maximum M and minimum m.

13 · $F(x) = 4x^3 - 18x^2 + 15x + 5.$

14 · $F(x) = 2x^3 - 9x^2 + 12x - 5.$

15 · $F(x) = 2x^3 - 3x^2 - 12x + 5.$

16 · $F(x) = x^3 - 6x^2 + 9x - 2.$

5 · 2 Maxima–Minima Problems

An almost endless variety of problems exist in which it would be helpful to know how big or how small a real-valued function of one variable can become. The technique for solving such problems, a standard one, is illustrated in Example A. We then summarize our method, and finally use the steps of the summary to lead us through the solution of a second problem in Example B.

EXAMPLE A

A certain company wants to manufacture a can that will have a volume of 64 cubic inches. They also want to use as little material as possible. What ought to be the radius of the can ?

Let the radius of the can be x and the height h (Fig. 5.6). We know

$$64 = \pi x^2 h; \tag{1}$$

$$\text{area } A = 2\pi x^2 + 2\pi x h. \tag{2}$$

A depends on x and h. Using (1) enables us to express the area as a function of the single variable x.

$$A(x) = 2\pi x^2 + 2\pi x \frac{64}{\pi x^2} = 2\pi x^2 + 128 x^{-1}. \tag{3}$$

The graph of $y = A(x) = (2\pi x^3 + 128)/x$ may be sketched quickly, as indicated in Fig. 5.7.

We are not interested in negative values of x, and we see that, in accord with what we would expect from Fig. 5.6 the area can be made very large by using either very small or very large values of x. We seek the can that has the least area (uses the least material). It is geometrically obvious from Fig. 5.7 that we should use what we know about local minimums, namely, that at a point where a local minimum occurs, the derivative is zero. From (3), we get

$$A'(x) = 4\pi x - 128 x^{-2} = x^{-2}[4\pi x^3 - 128].$$

This will be zero only if $x = \sqrt[3]{128/4\pi} = 4/\sqrt[3]{2\pi} \approx 2.2$. Of course, not every critical point (which is technically what we have found) is a local minimum, but Fig. 5.7 convinces us that we have found a local minimum in this case. We are thus relieved of finding the second derivative.

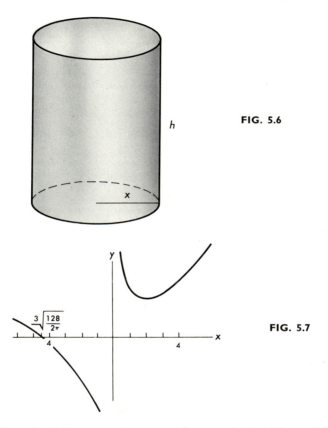

FIG. 5.6

FIG. 5.7

In (2) we had A expressed in terms of two variables. We used the additional information from (1) to eliminate one variable. Sometimes the additional information is not so easily solved for one variable in terms of the other. Then the chain rule may prove a handy device. We illustrate this by working this example a second time. Keep in mind that h is a function of x, and that we seek the derivative of A with respect to x.

From (1)

$$0 = \pi \left[x^2 \frac{dh}{dx} + 2xh \right].$$

From (2)

$$\frac{dA}{dx} = 4\pi x + 2\pi \left[x \frac{dh}{dx} + h \right].$$

Solving the first equation for dh/dx, we get $dh/dx = -2h/x$. Substituting this for dh/dx in the second equation and at the same time setting $dA/dx = 0$,

$$0 = 4\pi x + 2\pi \left[x\left(\frac{-2h}{x}\right) + h \right].$$

Solving this gives $h = 2x$, and using this in (1) gives $64 = 2\pi x^3$; $x = \sqrt[3]{64/2\pi}$. ▮

We now summarize our method.

(a)   We made certain simplifying assumptions. We assumed the material to be the same thickness on the ends as on the sides. We neglected waste. We also paid no attention to the company's other problems, such as whether the item they want to pack will fit in a can with the radius we recommend.

(b)   We drew a picture of the can and indicated our variables clearly. Every mathematician does this. Only students seem to be in too much of a hurry for this step.

(c)   We thought of the function to be minimized as a function of one variable and derived an equation expressing this. Even when working our problem by the second method, we were thinking of h, and therefore of A, as a function of the variable x.

(d)   We sketched a quick graph of the function to help us interpret our results, and we noted that there were physical limitations imposed on the variable x (it could not be negative). There are cases where the graph of the relevant function is more work to draw than is justified by its usefulness. In such cases, the second derivative can be analyzed. In all cases, the physical limitations on the variable must be noted.

(e)   We found the critical points by equating the derivative to 0, and then identified the critical point that was the object of our inquiry. We keep in mind that if we seek an extreme value of a function $F(x)$ on $[a, b]$, it may occur at an end point, in which case the derivative need not be 0.

PROBLEM

A · It would be more realistic in Example A to suppose that the bases of radius x were cut from squares of material with sides $2x$. Assuming that the remainder of these squares is discarded as waste, how should the can be constructed to minimize the material used?

EXAMPLE B

A man is in a boat three-quarters of a mile from shore and wishes to reach a cottage one-half of a mile down the shore. He can row his boat 3 mph, and he can cover 5 mph on land. Where should he land to minimize his time to the cottage ?

(a)   We assume the shore line is straight, that the man can row his boat in a straight line, and that he is willing to abandon his boat anywhere along the shore.

(b)   The boat is at point B, the cottage at C (see Fig. 5.8). We let x be the distance along the shore that is not covered on foot.

(c)   Time in water $= \dfrac{\sqrt{x^2 + (\frac{9}{16})}}{3}$;

time on land $= \dfrac{\frac{1}{2} - x}{5}$;

total time $T = \frac{1}{3}(x^2 + \frac{9}{16})^{1/2} + \frac{1}{10} - \frac{1}{5}x.$

(d)   We get the desired rough sketch by graphing

$$T_1 = \frac{1}{3}\sqrt{x^2 + \frac{9}{16}}; \qquad T_2 = \frac{1}{10} - \frac{1}{5}x$$

and then adding ordinates (Fig. 5.9). We observe that we are only interested in $x \in [0, \frac{1}{2}]$.

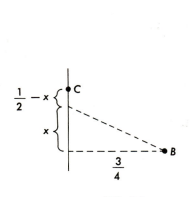

FIG. 5.8

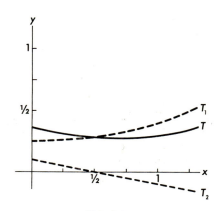

FIG. 5.9

(e) $dT/dx = \frac{1}{3}x(x^2 + \frac{9}{16})^{-1/2} - \frac{1}{5};$

$$0 = (x^2 + \tfrac{9}{16})^{-1/2}[\tfrac{1}{3}x - \tfrac{1}{5}\sqrt{x^2 + \tfrac{9}{16}}].$$

The first factor is never 0. Setting the second equal to 0 and performing some algebra gives as the only solutions $x = \pm\frac{9}{16}$. Neither of these values falls in the domain of x in which we are interested. It is thus clear from the graph that for $x \in [0, \frac{1}{2}]$, T is at a minimum when $x = \frac{1}{2}$. He should row all the way. ∎

PROBLEMS

B · Work the problem of Example B if the cottage is 1 mile down the shore.

C · Find the maximum and minimum values of the function $G(x) = x^3 - 5x^2 + 3x + 10$ for $x \in [0, \frac{9}{2}]$.

$$\equiv\! \quad 5 \cdot 2 \quad \text{EXERCISES}$$

For Exercises 1–6 refer to the accompanying illustrations and give: (a) An equation relating the variable to be maximized or minimized to the variables x and y indicated in the appropriate figure; (b) a second equation relating x and y; (c) the required solution.

1 · A box is to have a square base and is to contain 48 ft³. The bottom and the sides are to be lined with material making them twice as expensive as the top (Fig. 5.10). Find the dimensions of the least expensive box.

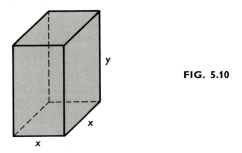

FIG. 5.10

2 · A rectangular building is to be placed on a triangular lot as indicated in Fig. 5.11. What is the maximum size of the building in square feet?

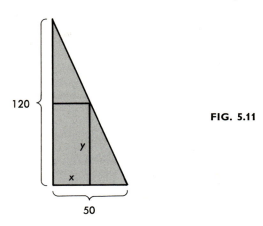

FIG. 5.11

120

50

3. A right circular cone has as its base a circle of radius 6. Its height is 12. Find the volume of the largest right circular cylinder that can be inscribed (Fig. 5.12).

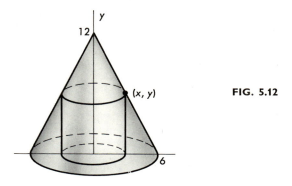

FIG. 5.12

4. Find the radius of the right circular cone of largest volume that can be inscribed in a sphere of radius 4 in (Fig. 5.13).

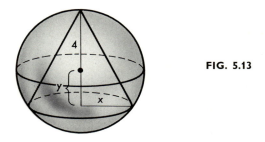

FIG. 5.13

5. A boy rests his 8-ft fishing pole on a $2\sqrt{2}$-ft-high rail at the end of a pier. He holds the end of the pole with his foot against the pier. How should he place the pole in order that the line be as far as possible from the end of the pier (Fig. 5.14)?

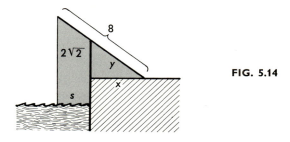

FIG. 5.14

6 · A drinking fountain in a city park is 256 ft from the sidewalk along one side of the park and 108 ft from the sidewalk along a side of the park perpendicular to the first (Fig. 5.15). It is decided to put a walk diagonally across the park

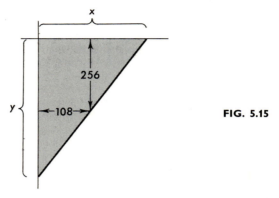

FIG. 5.15

going past the fountain. How should the walk be put in to minimize its length ?

It is not advantageous in all problems, of course, to express the variable to be maximized or minimized in terms of two variables. In Exercises 7 and 8, it is convenient to use just one variable x, as indicated in the figures. Solve these problems.

7 · A gutter is to be made from a piece of tin by bending it so that the cross section is shaped as indicated in Fig. 5.16. What should be the width across the top to maximize the capacity of the gutter ?

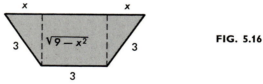

FIG. 5.16

8 · A box is to be formed from a piece of metal that is 4 in. by 8 in. Squares are to be cut out of each corner, and the sides then bent up (see Fig. 5.17). How long should the edge of the squares be if the box is to have maximum volume ?

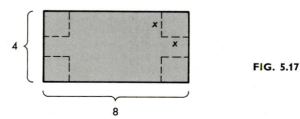

FIG. 5.17

Solve the remaining exercises by whatever method appeals (or appears) to you.

9. A farmer has 1500 ft of fencing. He wants to enclose as large a pasture as possible. One side of the pasture is to be bounded by the bank of a (straight, naturally) river, and no fence will be required on this side. How large a pasture can he enclose ?

10. A piece of wire 60 in. long is bent into a rectangle. Find the dimensions of the rectangle having a maximum area.

11. The combined resistance R of two resistors of resistance R_1 and R_2, respectively, is given by

$$\frac{1}{R} = \frac{1}{R_1} + \frac{1}{R_2}.$$

If $R_1 > 0$, $R_2 > 0$, and the sum $R_1 + R_2 = a$ is fixed, show that the maximum resistance R is obtained by choosing resistors for which $R_1 = R_2$.

12. The strength of a rectangular beam is proportional to the width of the beam and to the square of the depth. Show that the strongest beam that may be cut from a circular log of diameter b has a width of $b/\sqrt{3}$.

13. A right circular cylinder is to be inscribed in a sphere of radius a. Find the height of the cylinder having maximum volume.

14. A rectangle is to be inscribed in the ellipse

$$\frac{x^2}{a^2} + \frac{y^2}{b^2} = 1.$$

Find the area of the rectangle for which the area is maximum.

15. An open cylindrical tank of fixed volume V is to be constructed. If the material for the sides costs \$3 per square foot and the material for the bottom costs \$9 per square foot, find the most economical dimensions for the tank.

16. A cylindrical silo is to be covered by a hemispherical roof. The roof costs three times as much per square foot as the sides. For a fixed volume, what should be the ratio of the radius to the height of the cylinder to build the most economical silo? (Ignore the bottom.)

17. A craftsman finds that he can make about 30 items per day, but if he works more slowly, the quality is better and he is able to charge more for his products. He finds, in fact, that if he makes x items each day, he can sell each one for

$$20\sqrt{231 - 7x} \text{ cents.}$$

How many should he make each day to make the most money?

18. A builder owns 1 mile of lake front property. Local ordinances require that any building must be put on a lot having at least 50 ft of lake frontage. People naturally prefer bigger lots, but the builder finds that the price they are willing to pay for bigger lots reaches a point of diminishing returns. More specifically, he concludes that they will pay, for a lot of x lake-front footage,

$$L(x) = 16,000 \frac{\sqrt{x-50}}{x}.$$

Into how many equal-sized lots should the builder subdivide his property to realize the maximum profit?

19 · A wire 100 ft long is to be cut into two pieces. One piece will be shaped into a square; the other a circle. How should the wire be cut if the sum of the enclosed areas is to be a maximum; a minimum?

20 · A wire 100 ft long is to be cut into two pieces. One piece will be shaped into an equilateral triangle; the other a circle. How should the wire be cut if the sum of the enclosed areas is to be a maximum; a minimum?

21 · A radio newscaster is to interview people on Main Street between Sixth and Seventh Streets. Parked at Sixth and Main is a sound truck playing an advertising slogan, and at Seventh and Main a work crew is using a pneumatic hammer. If the hammer is eight times as loud as the sound truck, where along Main Street should the newscaster stand to minimize noise interference? (Recall that the intensity of sound varies inversely with the square of the distance from the source.)

22 · A power station located on one bank of a river wishes to run a cable to a factory located on the opposite bank and 1000 feet upstream. If the river is 300 feet wide and it costs twice as much to lay the cable under water as it does to run it along the shore, how should the cable be put in?

23 · Work Exercise 22 assuming the factory is 150 feet upstream from the power station.

24 ·* A circle of radius 1 is given. Find the altitude of the circumscribed isosceles triangle having the least area.

25 ·* A sphere of radius 1 is given. Find the altitude of the circumscribed right circular cone having the least volume.

26 ·* Find the point on the curve $y^2 - 4y - 4x = 16$ that is closest to the origin.

27 ·* Find the point on the curve $x^2 - 4x - 4y = 16$ that is closest to the origin.

5 · 3   Velocity and Acceleration; Related Rates

In Section 3.3 we considered a function F, the values of which gave the position of a car at time x. We asked for an expression giving the instantaneous velocity at time x. It turned out, in terms of notation developed since then, that the velocity at time x is $DF(x)$. Review Section 3.3.

Suppose we take this interpretation of the function

$$F(x) = \tfrac{1}{3}x^3 - 2x^2 + 3x + 3 \qquad (1)$$

introduced in Example 4.2A and studied more extensively in Section 5.1 (Fig. 5.5). We thus think of (1) as giving the distance of some moving object from a stationary point P. At time $x = 0$, it is 3 units past P. (One unit of time before this, it was 7/3 units behind P.) It moves away from P until time $x = 1$, at which time it begins moving back to P (so for an instant it was stopped). It moved backward until $x = 3$, when it again stopped and began to move forward.

Now DF represents the velocity. Hence its graph is 0 at the times when we noticed that the object had stopped. The backward motion from $x = 1$ to $x = 3$ corresponds to a negative velocity. The velocity, we note, decreased from $x = 0$ until $x = 2$. (We view speeding up in reverse, as our object does from $x = 1$ to $x = 2$, as a decrease in velocity.) Between $x = 2$ and $x = 3$, the object continues to move in reverse, but not so fast; that is, the velocity has begun to increase. Pretty soon it gets back to 0 (the object stops its backing up) and becomes positive (goes forward).

We have another graph; that of D^2F. From it we can tell much of what was said in the last paragraph. The velocity decreased from $x = 0$ to $x = 2$, when it began to increase. A change in velocity is called *acceleration*.

DEFINITION A   (Velocity and Acceleration)
   Let $F(x)$ give the distance of a moving object from a fixed point at time x. Then the number $DF(x)$ is called the *velocity at time* x; $D^2F(x)$ is called the *acceleration*.

When the function $F(x)$ describes something like the height of a liquid in a container at time x, the derivative is sometimes referred to as the *rate of change* rather than the velocity. If the position of one object depends on the position of a second object, then any motion of one naturally affects the other. The rate of change of one may determine the rate of change of the second. Problems in which this principle is used are called *related rates* problems.

EXAMPLE A
   A 6-ft man walks down a street at 4 ft/sec. A street light mounted 14 ft above the ground casts a shadow of the man. How fast is the shadow increasing in length t sec after the man passes the light?

The position of the man t sec after passing the light is

$$x = 4t.$$

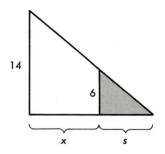

FIG. 5.18

The length of the shadow s satisfies (Fig. 5.18)

$$\frac{s}{6} = \frac{s+x}{14}.$$ (2)

Substitution of $x = 4t$ and solution for s gives

$$s = 3t.$$

Then the rate of increase is

$$\frac{ds}{dt} = 3 \frac{\text{ft}}{\text{sec}}.$$

This means, of course, that the shadow increases its length at a constant rate.

There is a second procedure often used to solve problems of this kind. Returning to (2) for a moment, we observe that both x and s are functions of the time t. If we thus differentiate (2) with respect to t, we get

$$\frac{1}{6}\frac{ds}{dt} = \frac{1}{14}\left(\frac{ds}{dt} + \frac{dx}{dt}\right).$$

Now dx/dt is the velocity of the man; that is, $dx/dt = 4$ ft/sec. Therefore

$$\frac{1}{6}\frac{ds}{dt} = \frac{1}{14}\frac{ds}{dt} + \frac{4}{14}.$$

Solving, we get

$$\frac{ds}{dt} = 3\frac{\text{ft}}{\text{sec}}. \quad \blacksquare$$

PROBLEM

A · An 18-ft ladder leans against a house. If the lower end slips along the level ground at the rate of 3 ft/sec, find the rate of descent of the upper end along the wall when the lower end is 6 ft from the wall (Fig. 5.19).

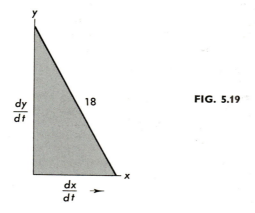

FIG. 5.19

1 · A rope tied to a small boat is wound onto a winch at the end of a pier. If the winch is 12 ft above the surface of the water and the rope is wound up at the rate of 8 ft/sec, how fast is the boat approaching the pier when 20 ft of rope is still out?

2 · A loading platform is 4 ft above ground level, and a pulley is mounted at a height of 12 ft over the edge of the platform. A 54-ft cable is attached to a box on the loading dock, run over the pulley, and attached to a small tractor at a point 1 ft above the ground. If the tractor moves forward at 12 ft/sec, how fast is the box moving across the dock when it is 9 ft from the edge?

3 · Highways A and B are perpendicular to each other. At noon, a truck traveling east on highway A at 50 mph is 13 miles west of the point where A crosses over B. A northbound automobile traveling 70 mph on highway B is at the same time 30 miles south of the junction.
(a) When will the vehicles be closest to each other?
(b) How fast are they moving away from each other at 12:30 P.M.?

4 · Ship A moves east at 12 knots (12 nautical mph). A disabled ship B is drifting south at 1 knot. At noon ship A is 4 (nautical) miles south and 18 miles west of ship B. When is a flare sent up from B most likely to be seen by A? How fast are the ships separating at 2:00 P.M.?

5 · A conical funnel is 8 in. across the top and 12 in. deep. If a liquid runs out of the funnel at π cubic inches per second, how fast is the level of the liquid falling 8 sec after the funnel was filled?

6 · Sand is poured onto the ground in such a way that the pile forms a right circular cone in which the diameter of the base is always three times the altitude. If the sand is poured on at the rate of 12 ft³/min, find the rate at which the apex of the pile is moving up when the diameter of the pile is 15 ft.

7 · In a flush box, the float is attached to a 14-in. straight rod. This rod is pivoted 2 in. from the other end on a support 10 in. above the bottom of the box. If the level of the water rises 1 in. each 3 sec, how fast is the upper end of the rod moving down when the float is 6 in. from the bottom of the tank?

8 · A seesaw is made from a 14-ft plank balanced in the middle. Two boys of different weights ride it, one sitting at the very edge, the other sitting 3 ft from the edge. If the boy closer to the center goes down 6 ft/sec, how fast does the other boy go up?

9 · Air is pumped into a spherical balloon so that its volume increases by 4 cubic inches per second. How fast is the diameter expanding at the end of the eighth second?

10 · The capacity of a cylindrical tank with a radius of 16 ft may be altered by raising or lowering the top. If the top can be raised 3 ft/min, how fast can the volume be changed?

Antiderivatives and Free-Falling Bodies

Closely related to a discussion of velocity and acceleration is a second idea, which we will introduce by returning to a consideration of the instantaneous speed of an automobile.

Sometimes a passenger in a car watches the speedometer instead of the scenery. Such a passenger in the car described in Section 3.3, if he kept a record, could report that at time x, the speedometer reading was given by

$$V(x) = 6x^2 - 24x + 60.$$

He would not know, of course, where he was at time x. Could you tell him? Your reasoning might go like this. The velocity is obtained by differentiating the function giving the distance. Can I find a function F for which $DF(x) = V(x)$? A little experimenting will lead you to

$$F(x) = 2x^3 - 12x^2 + 60x + C$$

where C is a constant. You have no hope of telling what C is, of course, since any constant will drop out when we differentiate. If the rider tells you that he started three units from the fixed point, then of course we can find $F(0) = C = 3$.

We have recovered F by "undoing" the differentiation that was done to get V. The best way to verify that you have gone backward correctly is to try going forward again from F; that is, check to see that

$$DF(x) = V(x). \tag{3}$$

The process of undoing differentiation is an important one. It is called *antidifferentiation*. The symbol used to indicate that we seek the antiderivative of V is inspired by (3). We want to express the idea that something is done to V and that F is the answer. Though D^{-1} has no meaning as yet, our algebraic instinct tells us that it ought to be given a meaning, so we could write $D^{-1}(DF(x)) = F(x)$; then (3) would give

$$F(x) = D^{-1}V(x).$$

With this guiding idea, we formalize our definition.

DEFINITION B (Antiderivative)

Let V be defined on $[a, b]$. An antiderivative of V is any function F for which $DF(x) = V(x)$ for each $x \in [a, b]$. An antiderivative of V is denoted by $D^{-1}V$.

We defined *an* antiderivative rather than *the* antiderivative since antiderivatives are not unique. You will be asked to prove, in the next group of problems, that if F_1 and F_2 are both antiderivatives of V, continuous on $[a, b]$, then there is a constant C such that

$$F_1(x) = F_2(x) + C \qquad \text{for all} \qquad x \in [a, b]. \tag{4}$$

If the antiderivative of a function A again has an antiderivative, we abbreviate $D^{-1}[D^{-1}A]$ to $D^{-2}A$. Thus if $D^{-2}A = F$, then $A = D^2F$.

PROBLEMS

B · Prove (4). (Hint: Set $G(x) = F_1(x) - F_2(x)$. Use problem 4.1B.)

C · It is to be emphasized that not all functions have antiderivatives that can be expressed in the form of a familiar algebraic function. Moreover, some functions do have antiderivatives, but they are not likely to be found by the beginning student. Find as many of the indicated antiderivatives as you can.

(a) $D^{-1}(x^{-1/3} + x^{1/3})$. (b) $D^{-1}(x^{-2})$.
(c) $D^{-1}(x^{-1})$. (d) $D^{-2}(x^{1/4})$.
(e) $D^{-1}(x\sqrt{x^2 + 1})$. (f) $D^{-1}(x^2 + 1)^{-3/2}$.

D · Can you give a general formula for $D^{-1}x^p$ where p is rational? (Note that parts (a)–(d) of Problem C involve expressions of this sort.)

Physicists tell us that the acceleration due to gravity is a constant that they designate by g (about 32 ft/sec/sec). Suppose we let S be a function of t (for time) that gives the position of a free-falling body at time t. Then the constant acceleration due to gravity is expressed by writing

$$D^2 S(t) = g \qquad \text{(all } t\text{)},$$
$$DS(t) = D^{-1}g = gt + C_1.$$

(5)

Now $DS(t)$ is the velocity of the object at time t, so if we set $t = 0$, (5) becomes

$$V(0) = C_1.$$

C_1 must therefore be the initial velocity, so we will rename it v_0. Now operating on (5) with D^{-1} gives

$$S(t) = D^{-1}(gt + v_0) = \tfrac{1}{2}gt^2 + v_0 t + C_2.$$

Since $S(0) = C_2$ is the initial position, we call it s_0.

$$S(t) = \tfrac{1}{2}gt^2 + v_0 t + s_0.$$

This formula can be memorized, but we urge the reader instead to try to solve a few problems by working from first principles.

EXAMPLE B

A ball is thrown up from the top of a 224-ft building with a speed of 80 ft/sec.

(a) How high does the ball get above the ground?
(b) How much time elapses before it hits the ground?
(c) Find the ball's velocity when (just before; not just after) it hits the ground.

Following our own advice, we start with

$$D^2 S(t) = -32.$$

We use -32 because gravity pulls down, and we like to think of up as positive, a personal preference. The important thing, of course, is to be consistent.

$$DS(t) = -32t + C_1 = V(t).$$

Now we know that $V(0) = 80$ (positive; it was thrown up).

$$DS(t) = -32t + 80.$$

(6)

We can also see immediately, since the ball has velocity 0 at its highest point, that this occurs when

$$-32t + 80 = 0; \qquad t = \tfrac{5}{2}$$
$$S(t) = -16t^2 + 80t + C_2 = -16t^2 + 80t + 224.$$

(7)

We determined C_2 by knowing the height at time 0 was $S(0) = 224$. The maximum height, attained at $t = 5/2$, is

$$S(\tfrac{5}{2}) = 324.$$

To find the time when the object hits the ground, we set $S(t) = 0$.

$$-16t^2 + 80t + 224 = 0$$

$$t = 7, -2.$$

We are interested in the positive root. The object hits the ground 7 sec after being thrown. What meaning would you give the root $t = -2$?

If we had not stopped at (6) to find the time when $DS(t)$ was 0, we would have gone on to obtain (7) and would have then asked ourselves when this function obtains its maximum. This would suggest finding $DS(t)$ and setting it equal to 0, as we did. To find the velocity at the time the ball hits the earth, we use

$$V(t) = -32t + 80$$

so

$$V(7) = -144 \text{ ft/sec.}$$

Are you surprised that the velocity is negative? ▮

PROBLEMS

E · Refer to Example B. With what velocity would a ball have to be thrown from the ground in order to attain the same maximum height?

F · A projectile is fired from a gun tilted to make a 30° angle with the ground. The muzzle velocity is 400 ft/sec. How far will the projectile travel? (Hint: Resolve the initial velocity into vertical and horizontal components.)

=================================== EXERCISES 5 · 3B

In the following problems, use 32 ft/sec/sec as the acceleration due to gravity.

1 · The initial velocity of a body is 36 ft/sec upward. How far does the body rise?

2 · Three seconds after being thrown upward, a body is rising at 200 ft/sec. What was the initial velocity?

3 · An object propelled upward from the ground attains a height of 64 ft in 4 sec. When will it reach its maximum height?

4 · An object is dropped from the top of a building, and it hits the ground at 112 ft/sec. How high is the building?

5. A ball is thrown upward from the ground. An observer in a building sees the ball going up, and again as it comes down. Prove that its speed is the same each time it crosses his line of vision.

6. An object propelled up from the ground attains its maximum height in t_0 sec. Prove that it hits the ground again in $2t_0$ sec, and that it returns at the same speed with which it left.

7. There is a 60 mph speed limit posted on a road that runs along the ridge of a bluff. The driver of a car loses control, breaking through a guard rail. The car lands in the ravine 36 ft below the road and 150 feet from where it went through the rail. Was the driver speeding?

8. A balloon rises 24 ft/sec. A stone is released from the balloon and hits the ground in 12 sec. What is the maximum height attained by the stone?

9. An automobile traveling 60 mph decelerates at 16 ft/sec/sec. How far will the car travel before coming to rest? Answer the question if the car is traveling 30 mph.

10. For a fixed muzzle velocity, suppose a gun is tilted at a 30-deg angle with the ground and fired, then at a 45-deg angle with the ground and fired again. Which shell will go farther?

In Exercises 11–20, find as many of the indicated antiderivatives as you can.

11. $D^{-1}(2x^2 - 3x + 4)$.

12. $D^{-1}(5x^3 - 3x + 7)$.

13. $D^{-1}(\frac{2}{3}x^{1/2} + \frac{1}{4}x^{-1/3})$.

14. $D^{-1}(\frac{1}{3}x^{2/3} - \frac{3}{5}x^{-1/3})$.

15. $D^{-1}x^2(4 + x^3)^{1/2}$.

16. $D^{-1}x(x^2 + 1)^{1/2}$.

17. $D^{-1}\dfrac{1}{1 + x^2}$,

18. $D^{-1}\dfrac{x^2}{(1 + x^3)^2}$.

19. $D^{-1}\dfrac{3x}{\sqrt{1 + x^2}}$.

20. $D^{-1}\dfrac{1}{\sqrt{1 - x^2}}$.

THE DEFINITE
INTEGRAL

As we begin this chapter, it will appear that we have decided to study a subject having nothing to do with differentiation. Only those who have had a sneak preview of calculus will realize what is coming, and as with any story, those who know how it comes out should not tell the rest. In fact, since this is an old story for which there is no copyright, the development of the plot varies a bit, depending on the storyteller. Thus, even those who know the conclusion may find that their understanding is increased if they read this chapter as if it were new.

6·1   The Area under a Curve

In Fig. 6.1 we have indicated a region R enclosed by the x and y axes, the line $x = 2$, and the graph of the function G,

$$G(x) = x^2 + 2.$$

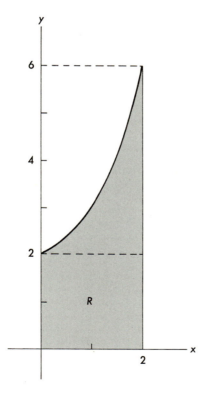

FIG. 6.1

From this figure it is clear that the indicated region R must have an area greater than four, the area of the smaller (square) rectangle indicated by a dotted line. An alert reader may wonder, at this point, just what we mean by the area under a curve. Although this is a good question, we will avoid it for the moment by suggesting that anyone who thought of it should consider becoming a mathematics major.

We can all agree on the area of a rectangle, so our assertion still stands. Any definition we adopt for assigning areas to regions enclosed by curved lines had better assign a number greater than four to the region R. From Fig 6.2 it

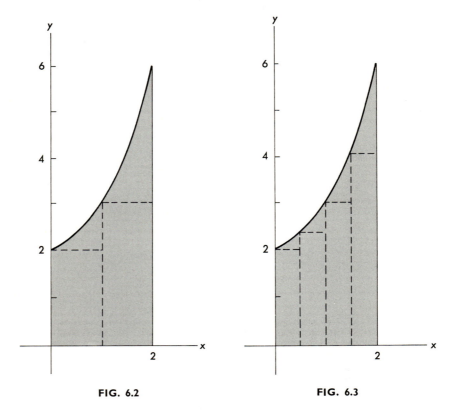

FIG. 6.2   FIG. 6.3

is similarly argued that the area of R had better exceed $2 + 3 = 5$, and Fig. 6.3 indicates that the area of R exceeds

$$\tfrac{1}{2}(2) + \tfrac{1}{2}(\tfrac{9}{4}) + \tfrac{1}{2}(3) + \tfrac{1}{2}(\tfrac{17}{4}) = \tfrac{23}{4}.$$

Let us write down what we are doing in a way that will enable us to see how things look after n steps. The rectangles have, at each step, a base of length 2 divided by 2^n (since we have been doubling the number of rectangles each time). Thus, corresponding to Figs. 6.1, 6.2, and 6.3 we have the numbers

$$\underline{S}_0 = \frac{2}{2^0} G(0) = 4,$$

$$\underline{S}_1 = \frac{2}{2^1} \left[G\left(0\,\frac{2}{2^1}\right) + G\left(1\,\frac{2}{2^1}\right) \right] = 5,$$

$$\underline{S}_2 = \frac{2}{2^2} \left[G\left(0\,\frac{2}{2^2}\right) + G\left(1\,\frac{2}{2^2}\right) + G\left(2\,\frac{2}{2^2}\right) + G\left(3\,\frac{2}{2^2}\right) \right] = \frac{23}{4}.$$

In general,

$$\underline{S}_n = \frac{2}{2^n} \left[G\left(0\,\frac{2}{2^n}\right) + G\left(1\,\frac{2}{2^n}\right) + \cdots + G\left([2^n - 1]\,\frac{2}{2^n}\right) \right].$$

It is reasonable to wonder at this point why we chose to put a bar under each S. We use this to remind us later that we have a *lower sum*; that is a sum that is certainly smaller (lower) than the desired area. We aim now at getting a more compact expression for $\underline{S}_n$. For any integer k,

$$G\left(k\,\frac{2}{2^n}\right) = \left(\frac{k}{2^{n-1}}\right)^2 + 2 = \frac{k^2}{2^{2n-2}} + 2,$$

$$\underline{S}_n = \frac{2}{2^n} \left\{ [0 + 2] + \left[\frac{1}{2^{2n-2}} + 2\right] + \left[\frac{4}{2^{2n-2}} + 2\right] + \cdots + \left[\frac{(2^n - 1)^2}{2^{2n-2}} + 2\right] \right\};$$

$$\tag{1}$$

$$\underline{S}_n = \frac{1}{2^{n-1}} \left\{ \frac{1}{2^{2n-2}} [0 + 1 + 4 + \cdots + (2^n - 1)^2] + 2^n(2) \right\}. \tag{2}$$

The last term of $2^n(2)$ comes from the fact that there are 2^n bracketed terms in (1), each of which contributes a 2. To be certain of understanding (2), the reader should set $n = 2$ in this expression and verify that he gets $\frac{23}{4}$.

Sums of the form (1) and (2) occur often enough in mathematics to justify a method of abbreviating them. We introduce for this purpose the Greek letter $\sum$ (sigma).

DEFINITION A (Summation Notation)

Given $n + 1$ numbers, $a_0, a_1, \ldots, a_m, \ldots, a_n$, we define

$$\sum_{k=m}^{n} a_k = a_m + a_{m+1} + \cdots + a_n,$$

That is all there is to it; and a little practice is all that is needed to use this notation with facility.

PROBLEMS

A · Let $a_k = k^2$. Show: $\sum_{k=1}^{6} a_k = 91$, $\sum_{k=4}^{7} a_k = 126$. (Rather than write the formula for a_k separately, we often write $\sum_{k=1}^{6} k^2 = 91$. Note that $\sum_{k=0}^{6} k^2 = \sum_{k=1}^{6} k^2$.)

B · (a) Show that if $b_k = k^2 + 3$, then $\sum_{k=0}^{6} b_k = 112$.
(b) Note that $112 = 91 + (6+1)(3)$.   Suppose $b_k = ca_k + d$ where c and d are constants. Show that $\sum_{k=0}^{n} b_k = c \sum_{k=0}^{n} a_k + (n+1)d$.

C · Use induction (you may have to review a page from your algebra book) to show

$$\sum_{k=0}^{n} k^2 = \frac{n(n+1)(2n+1)}{6}.$$

(Note that, in accord with what we found in Problem A,

$$\sum_{k=0}^{6} k^2 = \frac{6(7)(13)}{6} = 91.)$$

Equipped with our new tool, we rewrite Eq. (1).

$$\underline{S}_n = \frac{2}{2^n} \left\{ \sum_{k=0}^{2^n-1} \left[\frac{k^2}{2^{2n-2}} + 2 \right] \right\}. \tag{1'}$$

We refer to Problem B setting $c = 1/2^{2n-2}$ and $d = 2$. Then directly from (1') we get

$$\underline{S}_n = \frac{1}{2^{n-1}} \left\{ \frac{1}{2^{2n-2}} \sum_{k=0}^{2^n-1} k^2 + 2^n(2) \right\}, \tag{2'}$$

which is simply (2) written in terms of sigma notation. We may now refer to Problem C to get

$$\sum_{k=0}^{2^n-1} k^2 = \frac{(2^n - 1)2^n(2^{n+1} - 1)}{6}.$$

We substitute this expression into (2') to get

$$\underline{S}_n = \frac{1}{2^{n-1}} \left\{ \frac{1}{2^{2n-2}} \frac{(2^n - 1)2^n(2^{n+1} - 1)}{6} + 2^{n+1} \right\}.$$

We then exercise our algebraic muscles a bit to simplify this expression for $\underline{S}_n$. We get

$$\underline{S}_n = \frac{20}{3} - \frac{1}{2^{n-2}} + \frac{1}{3 \cdot 2^{2n-2}}. \tag{3}$$

(Doubters may restore some of their confidence by using (3) to compute $\underline{S}_1$ and $\underline{S}_2$.)

It is our idea that for any n, the area of the region R is larger than the number S_n computed from (3). Now let us consider the problem from another point of view. A glance at the larger rectangle in Fig. 6.1 convinces us that the area of R must be less than $\bar{S}_0 = 12$. Here the bar above the S is to remind us that we have an *upper sum*. It is bigger than the desired area. Consideration of Figs. 6.4 and 6.5 lead us in a similar way to the upper sums $\bar{S}_1$ and $\bar{S}_2$.

Explain each statement that follows:

$$\bar{S}_0 = \frac{2}{2^0} G\left(1\frac{2}{2^0}\right) = 12,$$

$$\bar{S}_1 = \frac{2}{2^1}\left[G\left(1\frac{2}{2^1}\right) + G\left(2\frac{2}{2^1}\right)\right] = 3 + 6 = 9,$$

$$\bar{S}_2 = \frac{2}{2^2}\left[G\left(1\frac{2}{2^2}\right) + G\left(2\frac{2}{2^2}\right) + G\left(3\frac{2}{2^2}\right) + G\left(4\frac{2}{2^2}\right)\right] = \frac{31}{4},$$

$$\vdots$$

$$\bar{S}_n = \frac{2}{2^n} \sum_{k=1}^{2^n} G\left(k\frac{2}{2^n}\right) = \frac{2}{2^n}\left\{\sum_{k=1}^{2^n}\left[\frac{k^2}{2^{2n-2}} + 2\right]\right\}, \tag{4}$$

$$\bar{S}_n = \frac{1}{2^{n-1}}\left\{\frac{1}{2^{2n-2}}\sum_{k=1}^{2^n} k^2 + 2^n(2)\right\},$$

$$\bar{S}_n = \frac{1}{2^{n-1} \cdot 2^{2n-2}} \cdot \frac{2^n(2^n+1)(2^{n+1}+1)}{6} + 4,$$

$$\bar{S}_n = \frac{20}{3} + \frac{1}{2^{n-2}} + \frac{1}{3 \cdot 2^{2n-2}}. \tag{5}$$

Readers may again buoy up their confidence by computing $\bar{S}_1$ and $\bar{S}_2$ from this formula.

Now we still have not defined what we will mean by the area of R, but we recall that this area is greater than S_n and less than $\bar{S}_n$. Hence, from (3) and (5), we see that however we define area, we must have

$$\frac{20}{3} - \frac{1}{2^{n-2}} + \frac{1}{3 \cdot 2^{2n-2}} < \text{Area of } R < \frac{20}{3} + \frac{1}{2^{n-2}} + \frac{1}{3 \cdot 2^{2n-2}}$$

for any integer n. As n gets large, both the upper bound and lower bound of the area approach 20/3. We may not have defined the area of R, but we know what the answer ought to be.

A reasonable question has been put to the author about the computations above. Why was the interval divided into two segments, then four, then eight, and so on? The questioner would have divided the interval into two equal segments, then three, then four, etc. This would simplify the computation, or at least the notation, in that it would eliminate all the powers of two.

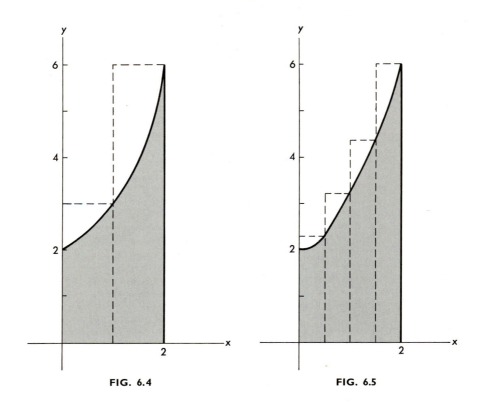

FIG. 6.4   **FIG. 6.5**

There are two reasons for retaining our procedure. One is that the method we have followed best illustrates the way an actual computation would be made in a case where we could not find nice formulas like (3) and (5) for $\underline{S}_n$ and $\bar{S}_n$. Note that at any step, one can use half the computations from the previous step. For example, in computing $\underline{S}_2$ we needed only to compute $G(\frac{1}{2})$ and $G(\frac{3}{2})$, because $G(0)$ and $G(1)$ had been computed previously. If we had partitioned [0, 2] into three equal segments, all points of division except 0 would have been new, requiring new calculations. And this problem intensifies with increasing n. We have therefore followed a method that would be far more efficient in setting up a computer program, for example.

Secondly, it is true that for a beginner trying to get the idea, especially in cases where it is possible to obtain a formula for $\underline{S}_n$, there is less confusion if one uses just n segments instead of 2^n. So we have done the more confusing computation, leaving the easier one to be worked out by the reader. This is an intentional deviation from the more common practice of working out the easy case as an illustration, leaving the more involved case for the reader.

PROBLEM

D · In the illustration above, corresponding to the sum $\underline{S}_n$, we partitioned $[0, 2]$ into 2^n segments. Work through the same example, this time obtaining $\underline{S}_n$ by partitioning $[0, 2]$ into n equal segments. As a starter, this procedure gives

$$\underline{S}_1 = \tfrac{2}{1}G(0) = 4,$$

$$\underline{S}_2 = \tfrac{2}{2}[G(0 \cdot \tfrac{2}{2}) + G(1 \cdot \tfrac{2}{2})] = 5,$$

$$\underline{S}_3 = \tfrac{2}{3}[G(0 \cdot \tfrac{2}{3}) + G(1 \cdot \tfrac{2}{3}) + G(2 \cdot \tfrac{2}{3})] = \tfrac{148}{27}.$$

(This last sum, of course, corresponds to Fig. 6.6) Show that

$$\underline{S}_n = \frac{2}{n} \sum_{k=0}^{n-1} \left[\left(\frac{2k}{n}\right)^2 + 2 \right],$$

and then simplify as in the text. Finally, find $\lim_{n \to \infty} \underline{S}_n$.

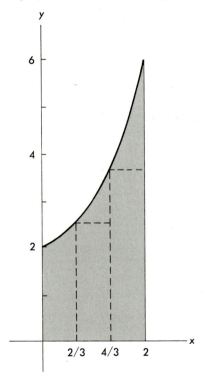

FIG. 6.6

2/3   4/3   2

1· Find numerical values for the following expressions.

(a) $\displaystyle\sum_{k=4}^{11} (3k + 2)$.   (c) $\displaystyle\sum_{k=1}^{10} (-1)^k(4k + 1)$.

(b) $\displaystyle\sum_{k=1}^{7} (k^2 + 3k)$.   (d) $\displaystyle\sum_{k=0}^{25} k^2$.

2· Solve $\displaystyle\sum_{k=1}^{3} (k^2 + 1)x^k = 0$.

3· Prove $\displaystyle\sum_{k=1}^{n} k = \frac{(n + 1)n}{2}$.

4· Use the method of this section, dividing [0, 1] into halves, then quarters, and so on, and then inscribing rectangles to show that the area bounded by the coordinate axes, the line $x = 1$, and the line $y = x + 1$ ought to be $\frac{3}{2}$. (Use the result of Exercise 3.)

5· Use the method of this section, dividing [0, 1] into halves, then quarters, and so on, and then inscribing rectangles to show that the area bounded by the coordinate axes, the line $x = 1$ and the curve $y = 2 - x^2$ ought to be $\frac{5}{3}$. (Use the result of Problem C.)

6·2 The Integral Defined

The division of an interval $[a, b]$ into nonoverlapping subintervals that cover $[a, b]$ is called a partition of $[a, b]$. At the nth step of the last section, we constructed a partition of [0, 2] using the intervals

$$\left[0, \frac{2}{2^n}\right], \quad \left[\frac{2}{2^n}, 2\frac{2}{2^n}\right], \quad \ldots, \quad \left[(2^n - 1)\frac{2}{2^n}, 2^n\frac{2}{2^n}\right].$$

Note that it is the collection of all these intervals that make up the partition. It makes sense to talk about the kth subinterval, in this case $[(k - 1)(2/2^n), k(2/2^n)]$, of the nth partition.

In the example just cited, the intervals of the partition all have the same length of $1/2^{n-1}$. In general, the intervals of a partition need not be of the same length. The maximum length of the subintervals of a partition P is called the *mesh* of P, written $\delta(P)$. (See Fig. 6.7.) The mesh of the nth partition used in Problem 6.1D was $2/n$.

After we had chosen our partition in Section 6.1, we computed $\underline{S}_n$ and $\bar{S}_n$. In the first case we selected the left end point of each subinterval (we shall

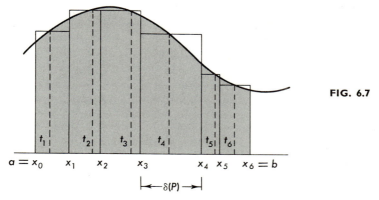

FIG. 6.7

call the left end point of the kth subinterval l_k here) and formed the sum (1′). We now write this sum in the form

$$\underline{S}_n = \sum_{k=1}^{2^n} G(l_k) \frac{1}{2^{n-1}} .$$

Designating the right end point of the kth subinterval by r_k, we similarly write (4) in the form

$$\overline{S}_n = \sum_{k=1}^{2^n} G(r_k) \frac{1}{2^{n-1}} .$$

We found that $\lim_{n \to \infty} \underline{S}_n = \lim_{n \to \infty} \overline{S}_n = \frac{20}{3}$. Whether we used the left end point or the right end point of each subinterval of the partition seemed to make no difference in the result. Since these extreme choices (left end point, then right end point) give the same result, we could get the idea that any point in between would also give the same result. That is, suppose we pick any point t_k in the kth subinterval of the partition and form

$$\sum_{k=1}^{2^n} G(t_k) \frac{1}{2^{n-1}} .$$

When t_k is chosen to be l_k, this is of course $\underline{S}_n$; and when t_k is chosen to be r_k, then this sum is just $\overline{S}_n$. We might reasonably expect that the limit of this sum as $n \to \infty$ would again be $\frac{20}{3}$. For the function G, this is actually the case.

There is still another observation. Problem 6.1D indicates that the particular sequence of partitions used in forming our sums does not affect the result either. We can divide the original interval into three segments, then four, etc., instead of halves, then quarters, and so on. So long as the mesh of the partitions gets smaller as n increases, the particular partitions used do not affect the value of the final limit.

Any of the sums we considered in the last section, including the ones used in Problem D are examples of the following sum, which we construct for an arbitrary function G defined on $[a, b]$.

DEFINITION A (Riemann Sums $S(G, P, \{t_k\})$)

Let P be a partition of $[a, b]$ determined by the consecutive points $a = x_0, x_1, x_2, \ldots, x_n = b$. In each subinterval $[x_{k-1}, x_k]$ choose a point t_k. Then form

$$S(G, P, \{t_k\}) = \sum_{k=1}^{n} G(t_k)(x_k - x_{k-1}).$$

This is called a *Riemann sum*.

PROBLEM

A · In Example 2.1D, we defined a function M as follows.

$$M(x) = \begin{cases} 0 & \text{if } x = 0, \\ 8 & \text{if } x \in (0, 1], \\ 16 & \text{if } x \in (1, 2]. \\ \vdots & \\ 8n & \text{if } x \in (n-1, n]. \end{cases}$$

Consider this function on $[0, 3]$. Let P be a partition of this interval into 12 equal subintervals. Choose t_k as the midpoint of the kth subinterval. Find the corresponding Riemann sum

$$S(M, P, \{t_k\}).$$

If the function G is positive on $[a, b]$, and continuous there (we have not required either), then we can picture $S(G, P, \{t_k\})$ as representing the sums of the areas of rectangles, as indicated in Fig. 6.7.

For the function G in Section 6.1 we had the feeling that no matter how we chose our partitions P and no matter how we selected the points t_k in the kth subinterval, the limit of the sums $S(G, P, \{t_k\})$ would approach $\frac{20}{3}$ as the mesh of the partitions went to 0. This feeling was based on only three trials plus our faith that things often go as we hope they will, so we realize that there remains much to prove. We can, however, make a definition.

DEFINITION B (The Riemann Integral of G Defined on $[a, b]$)

Suppose we are given an arbitrary neighborhood N of some number r, and that we are then able to guarantee that for any partition P with a small enough mesh (we specify how small, depending on the neighborhood of r) the number $S(G, P, \{t_k\})$ will fall in the neighborhood N, no matter how the t_k's are chosen in the subintervals $[x_{k-1}, x_k]$. Then the function G is said to be *Riemann integrable* on $[a, b]$, and the number r is called the *Riemann integral of G on* $[a, b]$.

The Riemann integral r depends only on G and the interval $[a, b]$. We thus denote r by stretching out our S (for sums) a bit and writing

$$r = \int_a^b G(x) \, dx.$$

More will be said later about inclusion of the symbol dx in the expression at the right. For now, let it serve as a reminder that the sums were formed by using the function G and partitions $[x_{k-1}, x_k]$ of the x axis.

It is also appropriate to mention that several modifications of the symbol $\int_a^b G(x) \, dx$ are used to advantage in certain situations. One can be pointed out immediately. When G is described by $G(x) = x^2 + 2$, the Riemann integral of G over $[a, b]$ is often written

$$\int_a^b (x^2 + 2) \, dx.$$

Inspection of the definition of the Riemann integral makes it easy to believe that there are functions which are not integrable. As the sudent might guess however, we have given a definition for which many functions will be integrable.

Since it is true that many functions are not Riemann integrable, we wish to mention here that other definitions of the integral have been devised so that more functions will turn out to be integrable. Thus there is the Lebesgue integral, for instance. Care is always taken in forming new definitions of finite integrals to see that a function which is Riemann integrable is integrable by the new definition, and that the integrals have the same value according to either definition. In this book we shall be concerned exclusively with the Riemann integral; therefore, we shall customarily drop the descriptive name Riemann, expecting this to be understood.

We have reason to think $\int_0^2 (x^2 + 2) \, dx = \frac{20}{3}$. We emphasize that we do not know this. After all, we have only computed the sums $S(G, P, \{t_k\})$ for three kinds of partitions, and for very special choices of t_k. Since we will not live long enough to try all possible partitions, it would, of course, be helpful to have some theorem that assures us that an integral exists. Once we know it exists, we have some hope of finding the value. The variety of theorems dealing with this problem is truly staggering. For our purposes, we select three modest versions that will meet our needs. We will not prove Theorem A.† Problems B and C suggest proofs for Theorems B and C, respectively.

THEOREM A

If a function F is continuous at each point of $[a, b]$, it is integrable on that interval.

The statement of Theorem B requires that we formalize a bit our notion of a function that is steadily increasing (or decreasing). The function F defined on $[a, b]$ is said to be *monotone increasing* on $[a, b]$ if for $x_1, x_2, \in [a, b]$, $x_1 \leq x_2$ implies $F(x_1) \leq F(x_2)$.

THEOREM B

If a function F is defined and monotone increasing (decreasing) on $[a, b]$, then F is integrable on that interval.

THEOREM C

If a function F is piecewise continuous (Section 2.6) on $[a, b]$, and if there is an M such that $|F(x)| \leq M$ for all $x \in [a, b]$ at which F is defined, then F is integrable on that interval.

From Theorem A it is clear that $\int_0^2 (x^2 + 2) \, dx$ exists. Since the limit of the sums $S(G, P, \{t_k\})$ for any partition P and choice of t_k must, therefore, give the same result, any one of the three computations we have made would in fact suffice to show the value of this integral to be $\frac{20}{3}$.

The function M defined in Problem A is not continuous, so Theorem A says nothing about the integrability of M. It is monotone increasing, however, so it is integrable on $[0, 3]$ according to Theorem B. Once the existence of the integral is established, it is easy to see that $\int_0^3 M(x) \, dx = 48$.

The function M also satisfies the hypothesis of Theorem C. It is continuous on $[0, 3]$ except at the points $0, 1, 2, 3$; and the function is bounded by 24 on the interval. Hence, we could also have used Theorem C to conclude that the integral existed.

---

† The proof of this theorem is not too hard for anyone who has read this book this far. We omit it because it is long and interrupts our continuity of thought. A very adequate proof, given from our point of view, is to be found in A. Taylor, "Advanced Calculus," pp. 504–509, Blaisdell, 1955. Proofs of the properties of the integral described in Section 6.3 may also be found in this reference.

Notice the importance of having the function bounded in Theorem C. $F(x) = 1/x^2$ is continuous on $[-1, 1]$ except at $x = 0$. F, however, is not bounded, so we are not able to conclude that $\int_{-1}^{1} F(x) \, dx$ exists. This is fortunate since, as we shall see later, this integral does not exist.

PROBLEMS

B · Prove Theorem B for monotone increasing functions. Hint: For any $t_k \in [x_{k-1}, x_k]$, we have the inequality

$$\sum_{k=1}^{n} F(x_{k-1})(x_k - x_{k-1}) \leq \sum_{k=1}^{n} F(t_k)(x_k - x_{k-1}) \leq \sum_{k=1}^{n} F(x_k)(x_k - x_{k-1}).$$

Now as the mesh of the partition goes to zero, the difference between the last sum and the first sum goes to zero. Hence, both sums must be tending to the same limit, as must be the arbitrary Riemann sum trapped between them.

C · Prove Theorem C. Hint: Choose a partition that includes each of the points of discontinuity $c_1, c_2, \ldots, c_r$ interior to a subinterval of length δ. The contribution of each of the corresponding terms in the Riemann sum is less than or equal to $M\delta$.

Now observe something. Our definitions were motivated by considerations of area. But in neither Definition A nor Definition B does the word "area" appear. Our definition circumvents the need to define what we mean by the area under a curve. In fact, the best way to proceed in the case of our region R in Section 6.1 is simply to define the area of R to be the value of the integral

$$\int_0^2 (x^2 + 2) \, dx = \tfrac{20}{3}.$$

There are other advantages to defining the integral independent of any notion of area. It makes it a much more versatile tool, as we now illustrate in a series of examples.

EXAMPLE A

Let the region R of Section 6.1 be rotated about the x axis, thereby generating the solid figure indicated in Fig. 6.8. What volume shall we assign to this solid figure?

We proceed as follows. Let P be a partition of $[0, 2]$; $0 = x_0, x_1, \ldots,$ $x_n = 2$. Pass two planes perpendicular to the X axis at x_{k-1} and x_k. Choose $t_k \in [x_{k-1}, x_k]$ and form

$$\pi[G(t_k)]^2 (x_k - x_{k-1}).$$

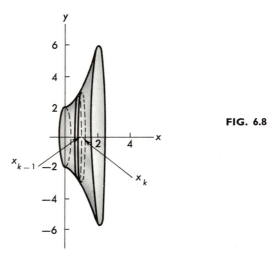

FIG. 6.8

This is approximately the volume of the disk cut out of the solid by the two planes. (Why approximately?) Now form

$$\sum_{k=1}^{n} \pi[G(t_k)]^2(x_k - x_{k-1}).\tag{1}$$

If we define $F(x) = \pi[G(x)]^2$, then (1) becomes the sum $S(F, P, \{t_k\})$. Since G is continuous, F is continuous; so these sums have a limit as the mesh of the partitions goes to zero.

$$\int_0^2 F(x)\,dx = \int_0^2 \pi(x^2 + 2)^2\,dx = \int_0^2 \pi(x^4 + 4x^2 + 4)\,dx.\tag{2}$$

In the same way that we used an integral to define the area of R, we now use this integral to define the volume generated by rotating R about the x axis. This is known as finding volumes by *the method of disks*. ∎

Before going on to another example, the reader should note that if he were asked what the integral $\int_0^2 \pi(x^4 + 4x^2 + 4)\,dx$ stood for, he might respond by saying it is the area under the curve $F(x) = \pi(x^4 + 4x^2 + 4)$. (See Fig. 6.9.) Indeed, if required to find this area, the integral of (2) is precisely what you would set up (we hope). But this is not the consideration that led us to (2). We have an illustration now of why it was a point of honor not to mention area in defining the integral. We are free to interpret our definition as an area if we please, but we are also free to interpret it in many other ways.

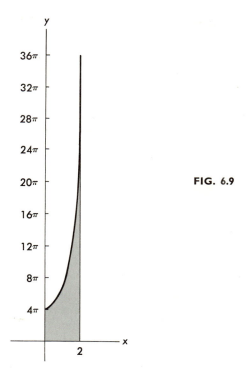

FIG. 6.9

EXAMPLE B

Let the region R of Section 6.1 be rotated about the y axis, generating the "scooped-out cylinder" of Fig. 6.10. Find the volume.

We could use a modification of the disk method, but we choose to illustrate a second method. Suppose the solid is made of clay. Take a tin can of radius 1 without top or bottom and push it through our solid of clay, keeping the respective centers lined up. When the can has been pushed all the way through, a certain volume of clay is missing.

Now let us formalize this. Partition $[0, 2]$ as in Example A. The can is to have sides thick enough so the outside of the can passes through x_k, the inside through x_{k-1}. If we choose t_k in $[x_{k-1}, x_k]$, the volume of clay removed is approximately

$$2\pi t_k\, G(t_k)(x_k - x_{k-1}).$$

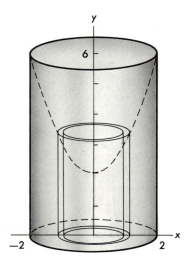

FIG. 6.10

If we set $H(x) = 2\pi x G(x)$, then

$$S(H, P, \{t_k\}) = \sum_{k=1}^{n} H(t_k)(x_k - x_{k-1}) = \sum_{k=1}^{n} 2\pi t_k\, G(t_k)(x_k - x_{k-1}).$$

The limit of this exists since H is continuous, and we recognize that the sum on the right represents the volume we seek. Hence, the volume is

$$V = \int_0^2 2\pi x(x^2 + 2)\, dx.$$

This method of finding volumes is known as the *method of cylindrical shells*. To be complete, we should now show that the method of cylindrical shells produces the same answer as does the method of disks. We omit this verification. ▮

EXAMPLE C

Suppose the familiar region R has now been cut out of stiff sheet metal. Where would you put your finger in order to balance it?

Problems of this sort are solved by considering moments. We digress to solve a similar but easier problem first. Consider three plates of equal weight ρ placed on a tray conveniently provided with coordinate axes marked on its edges. (See Fig. 6.11.) Our problem is to pile the plates up and locate the pile on the tray in such a way that the waiter will notice no change in balancing his load. The first simplification in such a problem is to assume all the weight

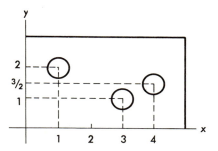

FIG. 6.11

of each plate to be concentrated at its center. Then the tendency of the plates to tip the tray if it were hinged along the y axis would be

$$\rho(1) + \rho(3) + \rho(4) = 8\rho. \qquad (3)$$

The expression (3) is called the *moment of the system about the y axis*, designated M_y. If the pile of three plates is centered at $(\bar{x}, \bar{y})$, then its moment about the y axis would be $3\rho\bar{x}$. Since the effect is to be the same, we have $3\rho\bar{x} = M_y = 8\rho$.

$$\bar{x} = \frac{M_y}{3\rho} = \frac{8\rho}{3\rho} = \frac{8}{3}. \qquad (4)$$

The reader may now verify his understanding by imagining the tray hinged along the x axis, finding first the moment M_x and then

$$\bar{y} = \frac{M_x}{3\rho} = \frac{3}{2}. \qquad (5)$$

(Explain why we could have guessed $\bar{y} = \frac{3}{2}$.)

We now return to our problem. If the uniform material from which R is cut has a density of ρ units per square unit of area, then the mass of the body is known from the work of Section 6.1 to be $\frac{20}{3} \rho$. Using the relations (4) and (5), we have

$$\bar{x} = \frac{M_y}{\frac{20}{3}\rho}, \qquad \bar{y} = \frac{M_x}{\frac{20}{3}\rho},$$

where M_y and M_x represent the moments of R about the y and x axes, respectively. The point $(\bar{x}, \bar{y})$ is the center of gravity, also called the *centroid* in the case of a plane area of uniform density.

Consider an arbitrary rectangle used in finding one of the sums $S(G, P, \{t_k\})$. This rectangle may be assumed to have its mass centered at $(t_k, G(t_k)/2)$. It has a moment about the y axis of about $t_k G(t_k)(x_k - x_{k-1})\rho$ and a moment about the x axis approximated by $(G(t_k)/2)G(t_k)(x_k - x_{k-1})\rho$ (Fig. 6.12).

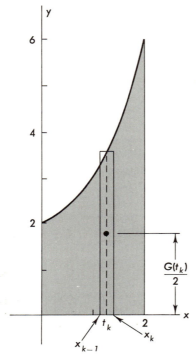

FIG. 6.12

Summed up for all the rectangles of the partition P,

$$M_y(P) = \sum_{k=1}^{n} [t_k G(t_k)\rho](x_k - x_{k-1});$$

$$M_x(P) = \sum_{k=1}^{n} [\tfrac{1}{2}G^2(t_k)\rho](x_k - x_{k-1}).$$

These give rise in the usual way to the integrals

$$M_y = \int_0^2 x(x^2 + 2)\rho \, dx;$$

$$M_x = \int_0^2 \tfrac{1}{2}(x^2 + 2)^2 \rho \, dx. \quad \blacksquare$$

Before leaving Example C, we wish to introduce a mnemonic device that is often useful. By setting $\Delta x_k = x_k - x_{k-1}$ and then identifying $dx = \Delta x$ (see Section 3.2), we see that Fig. 6.12 could be labeled as in Fig. 6.13. Looking at

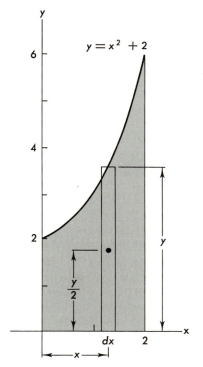

FIG. 6.13

the one element sketched in this figure, we write down:

$$\text{Area} = y \, dx;$$
$$M_y = \rho x y \, dx;$$
$$M_x = \rho \frac{y}{2} y \, dx.$$

The respective quantities for the region R are obtained by summing these ($\int$) between 0 and 2.

$$\text{Area} = \int_0^2 y \, dx = \int_0^2 (x^2 + 2) \, dx;$$

$$M_y = \int_0^2 \rho x y \, dx = \int_0^2 \rho(x^3 + 2x) \, dx;$$

$$M_x = \int_0^2 \tfrac{1}{2}\rho y^2 \, dx = \int_0^2 \frac{\rho}{2}(x^2 + 2)^2 \, dx$$

Used in this way, the symbol dx has utility in setting up the integral corresponding to a given problem. It has other assets as well. It serves as an indicator that tells us, in the first integral set up for the area, $\int_0^2 y\, dx$, that we are to consider this as an integral of a function of x, not of y. This is important because by itself, the variable under the integral sign is a dummy; that is,

$$\int_0^2 y\, dy = \int_0^2 z\, dz = \int_0^2 \xi\, d\xi. \tag{6}$$

Hence, if we want $\int_0^2 (x^2 + 2)\, dx$, it will not suffice to write down $\int_0^2 y$ unless we have some way to warn the reader that we want him to substitute $y = x^2 + 2$. The notation $\int_0^2 y\, dx$ does just this. Another asset of this notation will appear in the next section.

Each of the three Examples A, B, and C has led to an integral. In Example A we showed how the resulting integral could have arisen from a different problem. This could have been done with all the other integrals. The problem for us, however, is not so much "Where did it come from?" as "What do we do with it?" In Section 6.1 we actually evaluated $\int_0^2 (x^2 + 2)\, dx$. It was a lot of work. The prospect of evaluating the other integrals that have now come up seems to promise a week or more of computation. Computers can be (and, in fact, are) used for such work. But it is also true that if we know how, all of the integrals of our examples can be evaluated very quickly without the use of a computer. It is toward this goal that we move in the next two sections.

PROBLEMS

D · By interpreting the following integrals as areas, you can evaluate them. Do so.

(a) $\displaystyle\int_0^5 \sqrt{25 - x^2}\, dx.$

(b) $\displaystyle\int_{-1}^2 (2x + 3)\, dx.$

(c) $\displaystyle\int_0^2 (4 - x^2)\, dx.$ (Hint: $4 = 6 - 2$.)

E · Make up three problems, one about area, one about volume, and one about finding a moment M_x, for which the solution leads to the integral $\int_0^2 \pi x^2\, dx$.

F · Let G be a function constant on $[a, b]$; $G(x) = c$ for each $x \in [a, b]$. Show $\int_a^b G(x)\, dx = c(b - a)$.

In each of the following exercises, set up the integral you would evaluate to find the indicated number. Do not try to evaluate the integral.

1 · The area bounded by the positive coordinate axes, the line $x = 1$, and the curve $y = 2 - x^2$.

2 · The area bounded by the x axis, the line $x = 1$, and the curve $y = x^3$.

3 · The area bounded by the x axis, the lines $x = 2$ and $x = 4$, and the curve $y = x^2 - 4x + 7$.

4 · The area bounded by the x axis, the lines $x = 1$ and $x = 3$, and the curve $y = -x^2 + 6x - 4$.

5 · The volume generated by rotating the region of Exercise 1 about the x axis. (Use the method of disks.)

6 · The volume generated by rotating the region of Exercise 2 about the x axis. (Use the method of disks.)

7 · The volume generated by rotating the region of Exercise 3 about the x axis. (Use the method of disks.)

8 · The volume generated by rotating the region of Exercise 4 about the x axis. (Use the method of disks.)

9 · The volume generated by rotating the region of Exercise 1 about the y axis. (Use the method of cylindrical shells.)

10 · The volume generated by rotating the region of Exercise 2 about the y axis. (Use the method of cylindrical shells.)

11 · The volume generated by rotating the region of Exercise 3 about the y axis. (Use the method of cylindrical shells.)

12 · The volume generated by rotating the region of Exercise 4 about the y axis. (Use the method of cylindrical shells.)

13 · The moment about the x axis of the region described in Exercise 1.

14 · The moment about the x axis of the region described in Exercise 2.

15 · The moment about the x axis of the region described in Exercise 3.

16 · The moment about the x axis of the region described in Exercise 4.

17 · The moment about the y axis of the region described in Exercise 1.

18 · The moment about the y axis of the region described in Exercise 2.

19 · The moment about the y axis of the region described in Exercise 3.

20 · The moment about the y axis of the region described in Exercise 4.

6·3 Properties of the Integral

We collect here some facts we need about the definite integral. It should be remembered that because we wish the integral to be a tool for problems other than finding area, appeals to sketches of areas that can be associated with an integral do not constitute proofs. They are very helpful, however, in picturing what a statement about integrals really says. Thus, while the brief discussions and pictures accompanying the theorems that follow do indicate the method of proof, they are intended primarily to make clear the statements of fact. It is assumed in all the following theorems that the functions are such that the integrals involved do exist.

If the partition is indexed as in Fig. 6.14, then each term of the correspond-

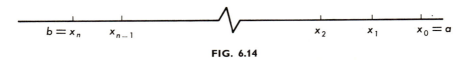

FIG. 6.14

ing sums $S(F, P, \{t_k\})$ has a negative factor $x_k - x_{k-1}$. The integral defined by such partitions is represented by $\int_b^a F(x)\,dx$. Naturally, it is the negative of the integral defined by the partitions we previously considered.

THEOREM A

$$\int_a^b F(x)\,dx = -\int_b^a F(x)\,dx.$$

Since $\sum_{k=1}^n cF(t_k)(x_k - x_{k-1}) = c\sum_{k=1}^n F(t_k)(x_k - x_{k-1})$, we have another very useful property.

THEOREM B

$$\int_a^b cF(x)\,dx = c\int_a^b F(x)\,dx \qquad \text{where } c \text{ is a constant.}$$

Now consider Figs. 6.15 and 6.16. If we note that $\int_b^c F(x)\,dx$ is negative in Fig. 6.16, we can convince ourselves that the next theorem is true regardless of whether c is between a and b or not.

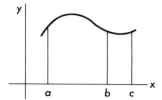

FIG. 6.15

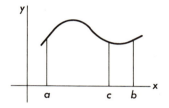

FIG. 6.16

THEOREM C

$$\int_a^b F(x)\, dx + \int_b^c F(x)\, dx = \int_a^c F(x)\, dx.$$

Since $(F + G)(t_k)$ means $F(t_k) + G(t_k)$, the definition of the integral has the additivity of functions as an immediate consequence.

THEOREM D

$$\int_a^b [F(x) + G(x)]\, dx = \int_a^b F(x)\, dx + \int_a^b G(x)\, dx.$$

In Fig. 6.17, $F(x) \le G(x)$ for every $x \in [a, b]$.

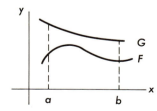

FIG. 6.17

It appears natural then that the area under G is greater than that under F.

THEOREM E

If $F(x) \leq G(x)$ for every $x \in [a, b]$, then

$$\int_a^b F(x) \, dx \leq \int_a^b G(x) \, dx.$$

The triangle inequality for real numbers r and s states that $|r + s| \leq |r| + |s|$. This extends to any finite collection of real numbers, enabling us to write

$$\left| \sum_{k=1}^n F(t_k)(x_k - x_{k-1}) \right| \leq \sum_{k=1}^n |F(t_k)| \, |x_k - x_{k-1}|.$$

This inequality, which holds for any partition, gives rise to an inequality between integrals.

THEOREM F

$$\text{For } a < b, \quad \left| \int_a^b F(x) \, dx \right| \leq \int_a^b |F(x)| \, dx.$$

PROBLEMS

A · We know $\int_0^2 (x^2 + 2) \, dx = \frac{20}{3}$. Use Theorem D and the result of Problem 6.2F to find $\int_0^2 x^2 \, dx$.

B · By interpreting the integral as an area, find the value of $\int_{-1}^2 x \, dx$. Use this result together with Theorem D, Theorem B, and Problem 6.2F to find the value of $\int_{-1}^2 (2x + 3) \, dx$. (Compare your answer with the answer you got the last time you evaluated this integral in Problem 6.2D.)

C · Given $\int_0^3 (x^2 + 2) \, dx = 15$, find $\int_2^3 (x^2 + 2) \, dx$.

D · Show that $\int_0^2 (x^2 - 2\sqrt{2x} + 2) \, dx \geq 0$.

E · If $x > 0$, then the function H defined by $H(t) = 1/t$ is continuous on the interval between 1 and x. Then by Theorem 6.2A, the integral $\int_1^x dt/t$ exists. Naturally the value of the integral depends on the choice of x (Fig. 6.18); that is, we have a function of x. Call it L:

$$L(x) = \int_1^x \frac{dt}{t}.$$

(a) Show that $L(2)$ satisfies $\frac{7}{12} < L(2) < \frac{10}{12}$. (Hint: Partition $[1, 2]$ into two equal subintervals.)

(b) Guess at the value of x for which $L(x) = 1$.

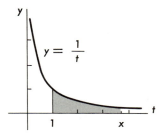

FIG. 6.18

(c)   Sketch as good a graph as you can of L on $(0, \infty)$. (Use only the definition of L as given above.)

(d)   Suppose $DL(x)$ exists for all $x > 0$. By looking at the graph you drew in part (c), sketch the graph of $y = DL(x)$.

6 · 4   The Fundamental Theorem

Let G be continuous on $[a, b]$. We have seen that it is then possible to define a second function F on $[a, b]$ by setting

$$F(x) = \int_a^x G(t)\, dt. \tag{1}$$

The study of this function F turns out to be very rewarding. As usual, the greatest rewards come to those asking the right questions. In this case, the questions to ask are "Does F have a derivative at each point of $[a, b]$? If so, can we find $DF(x)$?"

If the reader still recalls what he learned about finding $DF(x)$, he will begin by computing, for arbitrary $x \in [a, b]$,

$$F(x + h) - F(x) = \int_a^{x+h} G(t)\, dt - \int_a^x G(t)\, dt,$$

$$= \int_x^a G(t)\, dt + \int_a^{x+h} G(t)\, dt = \int_x^{x+h} G(t)\, dt. \tag{2}$$

Theorems A and C of the previous section have been used here.

We now need to reach back for two facts from Section 2.6 about continuous functions. First we recall (Theorem 2.6F) that the continuous function G has a maximum value M and a minimum value m which it attains on $[x, x + h]$. Hence, on $[x, x + h]$, $m \le G(x) \le M$, and so Theorem E of the last section enables us to write

$$\int_x^{x+h} m\, dt \le \int_x^{x+h} G(t)\, dt \le \int_x^{x+h} M\, dt.$$

According to Problem 6.2F

$$\int_x^{x+h} m \, dt = m[(x + h) - x] = mh.$$

Similarly,

$$\int_x^{x+h} M \, dt = Mh.$$

We therefore have

$$mh \leq \int_x^{x+h} G(t) \, dt \leq Mh$$

or

$$m \leq \frac{1}{h} \int_x^{x+h} G(t) \, dt \leq M.$$

The second fact we now need about the continuous function G is that if it takes on the values m and M, then it takes on all values between m and M (Theorem 2.6E); hence, there is an $\bar{x} \in [x, x + h]$ for which

$$G(\bar{x}) = \frac{1}{h} \int_x^{x+h} G(t) \, dt.$$

Substitution in (2) gives

$$F(x + h) - F(x) = hG(\bar{x}),$$

so

$$\frac{F(x + h) - F(x)}{h} = G(\bar{x}), \qquad \bar{x} \in [x, x + h].$$

As $h \to 0$, $\bar{x}$ is forced closer to x. Hence

$$DF(x) = G(x).$$

Since x was arbitrary in $[a, b]$, we have proved the following very important theorem.

THEOREM A

Let G be continuous on $[a, b]$, and suppose F is defined by (1)

$$F(x) = \int_a^x G(t) \, dt.$$

Then $DF = G$; that is, $F = D^{-1}G$.

This is a powerful theorem. For example, it tells us something about the function

$$L(x) = \int_1^x \frac{dt}{t}$$

introduced and studied in Problem 6.3E. It tells us that

$$DL(x) = \frac{1}{x}.$$

Did your graph (part (d) of Problem 6.3E) for $y = DL(x)$ look like $y = 1/x$? For $x > 0$, the slope of the line tangent to the graph of L at x is always positive. Does this agree with the graph you have already drawn of L? Consider also the "gap" that existed in your solution to Problem 5.3D.

There is an easy but extremely important consequence of Theorem A. Given a function G, continuous on $[a, b]$, we have found an antiderivative F defined by (1). Suppose now that H is any other antiderivative, so that $DH(x) = G(x)$. We know (Problem 5.3B) that $F(x) = H(x) + C$, C a constant; that is

$$\int_a^x G(t)\, dt = H(x) + C.$$

Set $x = a$; $0 = H(a) + C$, so $C = -H(a)$. Now if we set $x = b$, we have

$$\int_a^b G(t)\, dt = H(b) - H(a).$$

We have related the integral to the idea of antidifferentiation. The fact that these seemingly diverse ideas are connected in such a simple way is of central importance. We state the result formally.

THEOREM B   (Fundamental Theorem of Integral Calculus)
Let G be continuous on $[a, b]$. Then

$$\int_a^b G(t)\, dt = D^{-1}G(b) - D^{-1}G(a)$$

where $D^{-1}G$ is understood to be any antiderivative of G.

The fundamental theorem thus relates integration, a long and messy computation, to the finding of an antiderivative, sometimes a very simple process. For this reason the actual work of finding an antiderivative has become so closely identified with the work of finding the value of an integral that a slight variation of the integral sign is traditionally used where we have consistently used D^{-1}.

The traditional notation is an integral sign $\int$ with no limits indicated on it. Hence

$$\int F(x)\,dx = D^{-1}F(x).$$

By virtue of definition, $\int F(x)\,dx$ has nothing to do with integration; it has to do with "undoing" differentiation. Integration has to do with summing up rectangles under a curve (roughly speaking). But because the fundamental theorem ties these ideas together, a modified symbol for summing rectangles is used to indicate the finding of antiderivatives.

The symbol $\int F(x)\,dx$ is called the *indefinite integral of F*. It should be called the *antiderivative*. From this point on we shall use the customary symbol $\int F(x)\,dx$ except where we wish to emphasize what is going on; but we shall refer to it as the antiderivative.

EXAMPLE A

Evaluate $\int_0^2 (x^2 + 2)\,dx$ by using the fundamental theorem.
$\int (x^2 + 2)\,dx = (x^3/3) + 2x + C$. The theorem gives us

$$\int_0^2 (x^2 + 2)\,dx = \frac{2^3}{3} + 2(2) + C - \left(\frac{0}{3} + 2(0) + C\right) = \frac{20}{3}.$$

This agrees with the result we got in Section 6.1. It is obviously an easier way to evaluate the integral. ∎

Note in the preceding example that the constant C dropped out in the subtraction. Since this will always happen, we see why the particular antiderivative used is not important. For this reason, it is common to simply choose $C = 0$.

There is a notational device that is sometimes used; we abbreviate the difference $F(b) - F(a)$ by $F(x)\big|_a^b$.

EXAMPLE B

In Example 6.2A we confronted the problem of evaluating $\int_0^2 \pi(x^4 + 4x^2 + 4)\,dx$. Finish this problem by using the fundamental theorem.

$$\int_0^2 \pi(x^4 + 4x^2 + 4)\,dx = \pi\left(\frac{x^5}{5} + \frac{4x^3}{3} + 4x\right)\bigg|_0^2,$$

$$= \pi\left(\frac{32}{5} + \frac{4\cdot 8}{3} + 8\right) = \frac{376}{15}\pi. \; ∎$$

PROBLEMS

A · Use the fundamental theorem to compute $\int_0^2 x^2\,dx$. (You have evaluated this integral using other methods in Problem 6.3A.)

B · Use the fundamental theorem to evaluate $\int_{-1}^2 (2x+3)\,dx$. Compare your answer with the one you obtained in Problem 6.3B.

C · In Problem 6.3C you were told that $\int_0^3 (x^2 + 2)\,dx = 15$. Verify this by using the fundamental theorem.

D · Compute $\int_0^2 (x^2 - 2\sqrt{2x} + 2)\,dx$. Compare your answer with the assertion of Problem 6.3D.

E · Compute $\int_0^2 x\,dx$. Also compute $\int_0^2 y\,dy$. This illustrates line (6) of Section 6.2 which is an important idea. Read it again.

═══════════════════════════════════   6·4A EXERCISES

1–20 · Evaluate the integrals obtained for Exercises 1–20 in Section 6.2.

In Exercises 21–30, evaluate the indicated integral.

21 · $\displaystyle\int_0^1 (3x^2 + x)\,dx.$

22 · $\displaystyle\int_0^2 (\tfrac{1}{2}x^2 - x)\,dx.$

23 · $\displaystyle\int_1^2 \frac{dx}{x^2}.$

24 · $\displaystyle\int_1^2 \left(x - \frac{1}{x^3}\right) dx.$

25 · $\displaystyle\int_0^1 \sqrt{x}\,dx.$

26 · $\displaystyle\int_0^1 x^{3/2}\,dx.$

27 · $\displaystyle\int_2^3 \frac{x-1}{x^3}\,dx.$

28 · $\displaystyle\int_1^2 \frac{2-x}{x^3}\,dx.$

29 · $\displaystyle\int_1^{64} (\sqrt{x} - \sqrt[3]{x})\,dx.$

30 · $\displaystyle\int_0^8 (\sqrt{x} + \sqrt[3]{x})\,dx.$

Substitutions under the Integral

By reason of the fundamental theorem, any formula for the differentiation of a function also gives a formula for the integration of a function. A particularly useful example of this idea is obtained by consideration of the chain rule. Figure 6.19 recalls the composition of functions to which the chain rule applies. A function G takes the x axis into the u axis; that is, $u = G(x)$. $DG(x)$

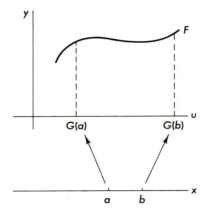

FIG. 6.19

is assumed to exist and be continuous on $[a, b]$, and a function F having a derivative

$$D_u \, F(u) = R(u) \tag{3}$$

is assumed to be continuous on $[G(a), G(b)]$. Then the chain rule says that the composite function $H(x) = F(G(x))$ has the derivative

$$D_x H(x) = [D_u F(u)][D_x G(x)] = R(G(x)) D_x G(x). \tag{4}$$

The fundamental theorem applied to (3) and (4) gives, respectively,

$$\int_{G(a)}^{G(b)} R(u) \, du = F(G(b)) - F(G(a)) \tag{3'}$$

and

$$\int_a^b R(G(x)) D_x G(x) \, dx = H(b) - H(a),$$
$$= F(G(b)) - F(G(a)). \tag{4'}$$

We state our result formally.

THEOREM C

Let G and DG be defined and continuous on $[a, b]$. Let R be continuous on $[G(a), G(b)]$. Then

$$\int_{G(a)}^{G(b)} R(u) \, du = \int_a^b R[G(x)] D_x \, G(x) \, dx. \tag{5}$$

Here is another instance in which the introduction of the dx under the integral sign has utility. Having proved Theorem C without reference to the notion of dx as a differential, it is nevertheless a very useful aid to our memory to note that the right side of (5) is a natural consequence of the substitution $u = G(x)$, $du = D_x G(x)\, dx$.

EXAMPLE C

In Example 6.2B we were confronted with the problem of evaluating $V = \int_0^2 2\pi x(x^2 + 2)\, dx$. Complete this problem.

We complete Example 6.2B in two ways. The first is the more natural; the second illustrates the use of Theorem C in a simple situation.

$$V = \int_0^2 2\pi x(x^2 + 2)\, dx = 2\pi \int_0^2 (x^3 + 2x)\, dx = 2\pi \left(\frac{x^4}{4} + x^2\right)\bigg|_0^2 = 16\pi.$$

We could begin by setting $u = x^2 + 2$; then $du = 2x\, dx$. When $x = 0$, $u = 2$; and when $x = 2$, $u = 6$. Hence, by Theorem C,

$$V = \int_0^2 \pi(x^2 + 2)(2x\, dx) = \pi \int_2^6 u\, du = \pi \frac{u^2}{2}\bigg|_2^6,$$

$$V = \pi[18 - 2] = 16\pi. \quad \blacksquare$$

PROBLEMS

F · Evaluate $\int_0^4 2x\sqrt{x^2 + 9}\, dx$ by means of Theorem C, setting $u = x^2 + 9$.

G · Show that $\int_1^{ab} dt/t = \int_{1/b}^a dt/t$. (Hint: We know $\int_1^{ab} dt/t = \int_1^{ab} du/u$. (See Problem 6.3E.) In the second integral, use Theorem C, setting $u = bt$; a and b are, of course, constants.)

EXAMPLE D

Find the area included between the curves described by

$$4y + x^2 - 8x + 8 = 0,$$

$$x + 2y - 4 = 0.$$

Also set up the integral you would use to find the moment about the x axis, assuming a density of $\rho = 1$ per square unit.

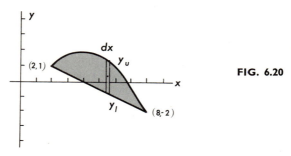

FIG. 6.20

The graph is drawn (Fig. 6.20), with intersections indicated. The area of the element shown is $(y_u - y_l)\,dx$ where

$$y_u = -\tfrac{1}{4}(x^2 - 8x + 8),$$
$$y_l = -\tfrac{1}{2}(x - 4).$$

Hence the required area, obtained by summing such elements from $x = 2$ to $x = 8$, is

$$A = \int_2^8 (y_u - y_l)\,dx = \int_2^8 (-\tfrac{1}{4}x^2 + \tfrac{5}{2}x - 4)\,dx,$$

$$A = (-\tfrac{1}{12}x^3 + \tfrac{5}{4}x^2 - 4x)\Big|_2^8 = 9.$$

The moment of the indicated element about the x axis is

$$(y_u - y_l)\frac{y_u + y_l}{2}\,dx = \tfrac{1}{2}(y_u{}^2 - y_l{}^2)\,dx,$$

$$M_x = \frac{1}{2}\int_2^8 \left(\left[\frac{x^2 - 8x + 8}{4}\right]^2 - \left[\frac{x - 4}{2}\right]^2\right)dx. \ \blacksquare$$

6 · 4B   EXERCISES

Exercises 1–10 are to be evaluated by using Theorem C. We have in each case indicated the proper choice of u.

1. $\displaystyle\int_{-1}^{2} \frac{dx}{(x + 3)^2}$; $u = x + 3$.   **2.** $\displaystyle\int_{2}^{3} \frac{dx}{(x + 1)^2}$; $u = x + 1$.

3. $\displaystyle\int_{1}^{2} -2x\sqrt{4 - x^2}\,dx$;   **4.** $\displaystyle\int_{0}^{1} 2x\sqrt{2 + x^2}\,dx$;

$\quad\quad u = 4 - x^2.$ $\quad\quad u = 2 + x^2.$

5. $\int_{-1}^{1} 3x^2\sqrt{2+x^3}\, dx$;

$\qquad u = 2 + x^3.$

6. $\int_{1}^{4} (2x-1)\sqrt{x^2 - x}\, dx$;

$\qquad u = x^2 - x.$

7. $\int_{-2}^{1} \dfrac{5dx}{(x-2)^2}$; $u = x - 2.$

8. $\int_{1}^{2} -2x\sqrt{4-x^2}\, dx$; $u = 4 - x^2.$

9. $\int_{0}^{1} \dfrac{2x\, dx}{\sqrt{x^2+1}}$; $u = x^2 + 1.$

10. $\int_{0}^{2} \dfrac{6x\, dx}{\sqrt{3x^2+4}}$; $u = 3x^2 + 4.$

In Exercises 11–16, find the area of the region included between the two curves.

11. $x^2 + 4y = 16$,
$\quad y = \frac{3}{2}|x|.$

12. $y = 4$,
$\quad y = x^2.$

13. $y = x^2$,
$\quad y = 2x.$

14. $y = x^3$,
$\quad y = |x|.$

15. $y = x^3 - 6x^2 + 9x - 1$,
$\quad 3y + x^2 - 5x - 5 = 0.$

16. $x^2 + y - 8x + 13 = 0$,
$\quad y = x^2 - 10x + 23.$

17. The region of Exercise 11 is rotated about the y axis. Find the volume generated.

18. The region of Exercise 12 is rotated about the y axis. Find the volume generated.

19. The region of Exercise 11 is rotated about the x axis. Find the volume generated.

20. The region of Exercise 12 is rotated about the x axis. Find the volume generated.

21. Find the centroid of the region of Exercise 11.

22. Find the centroid of the region of Exercise 12.

23. The region of Exercise 13 is rotated about the y axis. Find the volume generated.

24. The region of Exercise 14 is rotated about the y axis. Find the volume generated.

25. The region of Exercise 13 is rotated about the x axis. Find the volume generated.

26. The region pf Exercise 14 is rotated about the x axis. Find the volume generated.

27. Find the centroid of the region of Exercise 13.

28. Find the centroid of the region of Exercise 14.

6 · 5 Improper Integrals

Since $F(x) = 1/\sqrt{x}$ is not defined at the origin, the integral $\int_{0}^{2} dx/\sqrt{x}$ is not defined. The difficulty is caused by the failure of F to be defined at $x = 0$, since $\int_{r}^{2} dx/\sqrt{x}$ clearly exists for every choice of $r > 0$. This leads us to raise the following question. Suppose we just arbitrarily define $F(0) = 0$. Does

$\int_0^2 F(x)dx$ exist now? We cannot appeal to Theorem 6.2B because F is not monotone on the *closed* interval $[0, 2]$ (and no other choice of a value for $F(0)$ would have made it so). We cannot appeal to Theorem 6.2C because F is not bounded on $[0, 2]$. On the other hand, we cannot say that $\int_0^2 F(x)\,dx$ does not exist either.

The way out of our dilemma has already been suggested. We consider

$$\int_r^2 F(x)\,dx = \int_r^2 \frac{dx}{\sqrt{x}} \qquad \text{where} \quad r > 0.$$

Since we now have an interval on which F is continuous, the integral does exist; in fact

$$\int_r^2 \frac{dx}{\sqrt{x}} = 2x^{1/2}\Big|_r^2 = 2\sqrt{2} - 2\sqrt{r}.$$

Now as r gets closer and closer to zero, the value of the integral clearly approaches $2\sqrt{2}$.

DEFINITION A

Let F be continuous on $(a, b]$, and suppose $\lim_{x \to a^+} F(x) = \pm \infty$. Then

$$\int_a^b F(x)\,dx$$

is called an *improper integral of the first kind*. It is said to converge and have the finite value A if and only if

$$\lim_{r \to a^+} \int_r^b F(x)\,dx = A.$$

Otherwise it is said to diverge.

If a function F is continuous on $[a, b)$ with $\lim_{x \to b^-} F(x) = \pm \infty$, then an entirely analogous definition is given for the improper integral $\int_a^b F(x)\,dx$. If F is discontinuous at a single point $c \in (a, b)$, where either $\lim_{x \to c^-} F(x)$ or $\lim_{x \to c^+} F(x)$ is infinite, then $\int_a^b F(x)\,dx$ is again an improper integral of the first kind, and is said to converge if and only if $\int_a^c F(x)\,dx$ and $\int_c^b F(x)\,dx$ both converge.

EXAMPLE A

Investigate the convergence of $\int_{-1}^{1} dx/\sqrt[3]{x}$, which is an improper integral of the first kind.

$$\int_{-1}^{1} \frac{dx}{\sqrt[3]{x}} = \lim_{r \to 0^-} \int_{-1}^{r} x^{-1/3} \, dx + \lim_{s \to 0^+} \int_{s}^{1} x^{-1/3} \, dx,$$

$$= \lim_{r \to 0^-} \frac{3}{2} x^{2/3} \Big|_{-1}^{r} + \lim_{s \to 0^+} \frac{3}{2} x^{2/3} \Big|_{s}^{1},$$

$$= \lim_{r \to 0^-} \left[\frac{3}{2} \sqrt[3]{r^2} - \frac{3}{2} \right] + \lim_{s \to 0^+} \left[\frac{3}{2} - \frac{3}{2} \sqrt[3]{s^2} \right],$$

$$= \frac{3}{2} - \frac{3}{2} = 0.$$

This improper integral converges to zero.   ∎

EXAMPLE B

Investigate the convergence of $\int_{-1}^{1} dx/x^3$ which is an improper integral of the first kind.

$$\int_{-1}^{1} \frac{dx}{x^3} = \lim_{r \to 0^-} \int_{-1}^{r} x^{-3} \, dx + \lim_{s \to 0^+} \int_{s}^{1} x^{-3} \, dx$$

$$= \lim_{r \to 0^-} -\frac{1}{2x^2} \Big|_{-1}^{r} + \lim_{s \to 0^+} -\frac{1}{2x^2} \Big|_{s}^{1}.$$

Neither limit is finite. The improper integral diverges. Avoid the temptation to say that the limits

$$\lim_{r \to 0^-} -\frac{1}{2r^2} = -\infty, \qquad \lim_{s \to 0^+} \frac{1}{2s^2} = \infty$$

cancel each other. The definition specifies that the limits must be finite.   ∎

Referring to the integrals discussed thus far as improper integrals of the first kind implies that there must be another kind. We introduce improper integrals of the second kind by considering the area under $y = 1/x^2$ from 1 to an arbitrary number $r > 1$ (Fig. 6.21).

$$\text{Area} = \int_{1}^{r} \frac{dx}{x^2} = -\frac{1}{x} \Big|_{1}^{r} = -\frac{1}{r} + 1.$$

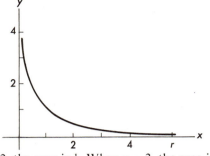

FIG. 6.21

When $r = 2$, the area is $\frac{1}{2}$. When $r = 3$, the area is $\frac{2}{3}$, an increase (naturally). As r increases, the area increases; but the area never exceeds the number 1, which it approaches as $r \to \infty$. We express this by writing

$$\int_1^\infty \frac{dx}{x^2} = 1.$$

This is an example of an improper integral of the second kind.

DEFINITION B

Let F be integrable on $[a, r]$ for every $r > a$. Then we define

$$\int_a^\infty F(x)\, dx = \lim_{r \to \infty} \int_a^r F(x)\, dx.$$

$\int_a^\infty F(x)\, dx$ is called an *improper integral of the second kind*; it is said to converge or diverge according as the indicated limit exists as a finite number or does not.

It is again to be understood that this definition is modified in the obvious way for a function F integrable on $(-\infty, a]$. It may also be used in conjunction with Definition A to discuss integrals improper for two reasons.

EXAMPLE C

Investigate

$$\int_0^\infty \frac{dx}{\sqrt{x}(x + 3)}.$$

Here is an integral that is improper on two counts. We now need to investigate both

$$\lim_{r \to 0^+} \int_r^1 \frac{dx}{\sqrt{x}(x + 3)} \qquad (1)$$

and

$$\lim_{s \to \infty} \int_1^s \frac{dx}{\sqrt{x(x + 3)}}. \tag{2}$$

In both cases we are slowed down by not knowing off hand an antiderivative for $1/[\sqrt{x(x + 3)}]$. We could work at trying to find one, but there is a method for getting around this problem more easily. We first note that on $(0, 1]$, $1/(x + 3) \leq \frac{1}{3}$. Therefore, according to Theorem 6.3E,

$$\int_r^1 \frac{dx}{\sqrt{x(x + 3)}} \leq \int_r^1 \frac{dx}{\sqrt{x}} \cdot \frac{1}{3},$$

$$\leq \frac{1}{3} 2x^{1/2} \Big|_r^1 = \frac{2}{3}(1 - \sqrt{r}),$$

$$\lim_{r \to 0^+} \int_r^1 \frac{dx}{\sqrt{x(x + 3)}} \leq \lim_{r \to 0^+} \left[\frac{2}{3} - \frac{2\sqrt{r}}{3} \right] = \frac{2}{3}.$$

Using the same idea, we note that on $[1, \infty)$, $1/(x + 3) < 1/x$. Therefore,

$$\int_1^s \frac{dx}{\sqrt{x(x + 3)}} < \int_1^s \frac{dx}{x^{3/2}} = -2x^{-1/2} \Big|_1^s = \frac{-2}{\sqrt{s}} + 2,$$

$$\lim_{s \to \infty} \int_1^s \frac{dx}{\sqrt{x(x + 3)}} < \lim_{s \to \infty} \left(-\frac{2}{\sqrt{s}} + 2 \right) = 2.$$

We have thus established that both (1) and (2) are finite; the given improper (on two counts) integral converges. Note that we do not say that it converges to $2 + \frac{2}{3}$ on the basis of our work. We just say it converges. ∎

PROBLEMS

A · Discuss the convergence of the integral $\int_0^\infty dx/\sqrt{x^4 + x^3}$.

B · Discuss the convergence of the integral $\int_0^\infty dx/x^p$ for various values of p.

6·5 EXERCISES

In Exercises 1–6, show that the integrals converge and evaluate them.

1 · $\displaystyle\int_0^1 \frac{dx}{\sqrt{1 - x}}.$

2 · $\displaystyle\int_{-2}^1 \frac{dx}{\sqrt{(x + 2)}}.$

3. $\int_1^\infty \dfrac{dx}{x^3}.$

4. $\int_0^\infty \dfrac{dx}{(x+1)^2}.$

5. $\int_2^3 \dfrac{dx}{\sqrt[3]{x-2}}.$

6. $\int_3^{\sqrt{10}} \dfrac{\times dx}{\sqrt[3]{x^2-9}}.$

Do the following improper integrals converge?

7. $\int_1^\infty \dfrac{dx}{x^2}.$

8. $\int_0^\infty \dfrac{dx}{\sqrt{x}}.$

9. $\int_1^\infty \dfrac{dx}{x^{1/4}}.$

10. $\int_1^\infty \dfrac{dx}{x^{3/4}}.$

11. $\int_0^1 \dfrac{dx}{x^{1/4}}.$

12. $\int_0^1 \dfrac{dx}{x^{3/4}}.$

13. $\int_0^1 \dfrac{dx}{\sqrt{1-x^2}}.$

14. $\int_0^1 \dfrac{dx}{\sqrt{1-x^3}}.$

15. $\int_1^\infty \dfrac{dx}{\sqrt[3]{x^2+1}}.$

16. $\int_1^\infty \dfrac{dx}{\sqrt{x^3+1}}.$

CHAPTER

7

TRANSCENDENTAL FUNCTIONS

We have learned to differentiate polynomials and a large variety of functions obtained by taking sums, products, quotients, powers, and roots of polynomials. The aggregate of all such functions is called the *set of algebraic functions*, and to be able to differentiate any member of this set is an impressive accomplishment. Yet the reader is no doubt familiar with a number of functions that we do not know how to differentiate. These include the trigonometric functions, the logarithm function, functions of the form $T(x) = 2^x$, and others. Such nonalgebraic functions are grouped together under the general heading of *transcendental functions*. We will now study some of these functions.

7·1 Trigonometric Functions

We begin this section with two assumptions. One is that the reader has had a course in trigonometry. The other is that the reader has forgotten much of the trigonometry he learned. Whether or not the second assumption is true, we believe it will be worthwhile to review trigonometry with the purpose of summarizing those aspects of the subject that are most important to us.

We begin by drawing a circle of radius r. We then connect an arbitrary point $P(x, y)$ on this circle with the origin (Fig. 7.1). View the circle as the

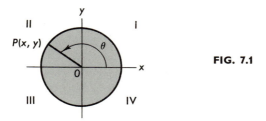

FIG. 7.1

face of a clock, and imagine a second hand that coincides with the positive X axis. Now let the clock run backward, so that the second hand sweeps from its horizontal position to the position of our line segment OP. We say the hand has passed through the positive angle θ. (When the rotation is clockwise, the angle is called *negative*.)

The four regions of the plane formed by the coordinate axes are called quadrants and are numbered as indicated in Fig. 7.1. Angles are classified as first-, second-, third-, or fourth-quadrant angles, according to the quadrant in which the point P falls. (When P falls on one of the axes, we do not speak of the angle as being in a certain quadrant.)

There are two common ways to measure the angle θ; in degrees or in radians. The first method is the best known. The second method is, at least to the mathematician, by far the most important. We will see the reason for this in the next section. We concentrate here on becoming more familiar with the concept of a radian.

Return to the circle we drew earlier. Let Q designate the point $(r, 0)$. Now choose the point P on the circle so that the arc PQ has a length of r and so that the corresponding central angle θ is positive (Fig. 7.2). Then we say

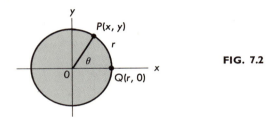

FIG. 7.2

that θ is an angle of 1 radian. We state this formally.

DEFINITION A (Radian)

A radian is the measure of that positive central angle of a circle of radius r that intercepts an arc of length r.

From the formula for the circumference of a circle, we see that

$$\text{one revolution} = 2\pi \text{ radians}$$

EXAMPLE A

Convert the following angles, measured in degrees, to radian measure and sketch each: (a) $30°$; (b) $45°$; (c) $60°$; (d) $90°$; (e) $210°$; (f) $225°$; (g) $-120°$; (h) $-270°$.

We know that $360° = 2\pi$ radians, so $1° = \pi/180$ radians. Then $n° = n(\pi/180)$ radians.

If the first four angles are in mind, it is easier to sketch multiples of them by thinking in terms of radians than by thinking in terms of degrees. For instance, if the reader has a mental picture of $\pi/6$ as one third of a right angle, it is quite easy to visualize $7(\pi/6)$ by counting off seven such portions of a right angle; perhaps easier than visualizing $210°$.

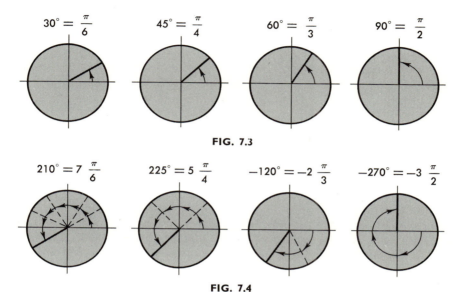

FIG. 7.3

FIG. 7.4

It should be remembered that we shall always use radian measure in our work. With this understanding of how to measure θ, we now turn to defining six functions of the angle θ in Fig. 7.1. ∎

DEFINITION B   (The Trigonometric Functions)

$$\text{sine } \theta = \sin \theta = \frac{y}{r} \qquad\qquad \text{cosecant } \theta = \csc \theta = \frac{r}{y}$$

$$\text{cosine } \theta = \cos \theta = \frac{x}{r} \qquad\qquad \text{secant } \theta = \sec \theta = \frac{r}{x}$$

$$\text{tangent } \theta = \tan \theta = \frac{y}{x} \qquad\qquad \text{cotangent } \theta = \cot \theta = \frac{x}{y}$$

The sine and cosine functions are defined for all possible θ. Each of the other functions is undefined for certain values of θ. (For example, $\tan \pi/2$ is not defined.)

The angles θ, $\theta \pm 2\pi$, $\theta \pm 4\pi$, ... all terminate at the same place. The point (x, y) is therefore the same for all of them, so the functions are sure to be equal for angles whose measures differ by a multiple of 2π. For this reason, the functions are all said to be periodic. A function is said to be periodic with period l if for all x, $f(x) = f(x + l)$. The six trigonometric functions are periodic with period 2π. The tangent and cotangent functions are also periodic with period π.

Certain relations between these functions are obvious from the definitions.

$$\csc \theta = \frac{1}{\sin \theta}, \qquad \cot \theta = \frac{1}{\tan \theta}. \qquad \theta \neq \pm n\pi.$$

$$\sec \theta = \frac{1}{\cos \theta}, \qquad \tan \theta = \frac{\sin \theta}{\cos \theta}, \qquad \theta \neq \frac{\pi}{2} \pm n\pi.$$

Since $r^2 = x^2 + y^2$, a bare minimum of computations gives some further relations.

$$\sin^2 \theta + \cos^2 \theta = 1$$
$$\tan^2 \theta + 1 = \sec^2 \theta \qquad (1)$$
$$1 + \cot^2 \theta = \csc^2 \theta$$

Our knowledge of plane geometry can be used to find the trigonometric functions of certain familiar angles.

EXAMPLE B

Find the value of the sine, cosine, and tangent functions for

(a) $\theta = \dfrac{3\pi}{4}$; (b) $\theta = -\dfrac{\pi}{3}$; (c) $\theta = \dfrac{3\pi}{2}$.

We find it convenient (for purposes of avoiding fractions in the labeling of our pictures) to take the radius to be 2.

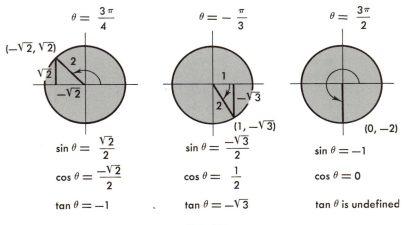

FIG. 7.5

By using such familiar values and plotting points, we may draw graphs of the trigonometric functions. When plotting the values of the angle, note that we are merely plotting real numbers along the horizontal axis. Thus, when plotting π, we place it just to the right of 3. This simple fact is often overlooked. Observe carefully the graphs drawn in Fig. 7.6.

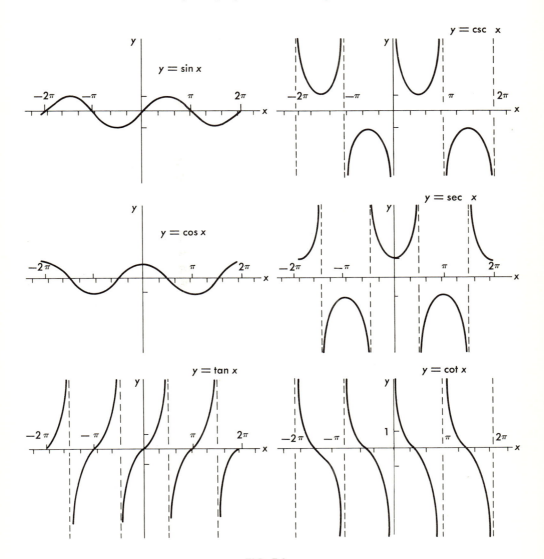

FIG. 7.6

The reader should not have to think too long if asked to sketch a graph of any of these six functions. It is also a good idea to develop some facility in using a knowledge of these graphs to aid in sketching the graphs of other trigonometric functions.

EXAMPLE C
Graph $y = \sin 2x + 2 \cos x$ for $x \in [0, 2\pi]$.

We sketch this graph by "addition of ordinates." First sketch the two curves

$$y = \sin 2x,$$

$$y = 2 \cos x.$$

Then add the two curves to get the desired graph. We simply estimate where the "peaks" and "valleys" of the desired graph actually fall. ∎

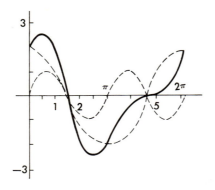

FIG. 7.7

7·1A EXERCISES

1· Express in radian measure the following angles.
(a) 315°.   (b) 135°.
(c) 210°.   (d) −120°.
(e) −45°.   (f) 7°.

2· Express in radian measure the following angles.
(a) 225°.   (b) 150°.
(c) 330°.   (d) −315°.
(e) −60°.   (f) 11°.

3· Find the sine, cosine, and tangent of the following angles.

(a) $\dfrac{5\pi}{6}$.   (b) $-\dfrac{5\pi}{6}$.

(c) $\dfrac{5\pi}{4}$.   (d) $\dfrac{5\pi}{3}$.

(e) $\dfrac{7\pi}{2}$.   (f) $\dfrac{-3\pi}{2}$.

4· Find the sine, cosine, and tangent of the following angles.

(a) $\dfrac{2\pi}{3}$.   (b) $\dfrac{-2\pi}{3}$.

(c) $\dfrac{7\pi}{6}$.   (d) $\dfrac{7\pi}{4}$.

(e) $\dfrac{5\pi}{2}$.   (f) $\dfrac{-\pi}{2}$.

5· Suppose each of the angles below is in the second quadrant. Find in each case the cosine and the tangent of the angle.
(a) $\sin \theta = 3/5$.   (b) $\cot \theta = -2$.
(c) $\sec \theta = -13/5$.   (d) $\csc \theta = 3$.

6· Suppose each of the angles below is in the third quadrant. Find in each case the cosine and tangent of the angle.
(a) $\sin \theta = -12/13$.   (b) $\cot \theta = 3$.
(c) $\sec \theta = -5/3$.   (d) $\csc \theta = -4$.

7· Graph the following for $x \in [0, 2\pi]$.
(a) $y = 2 \cos x$.   (b) $y = \cos 2x$.

(c) $y = \cos \dfrac{x}{2}$.   (d) $y = \cos \dfrac{x}{2} + \cos 2x$.

8· Graph the following for $x \in [0, 2\pi]$.
(a) $y = 3 \sin x$.   (b) $y = \sin 3x$.

(c) $y = \sin \dfrac{x}{3}$.   (d) $y = \sin \dfrac{x}{3} + \sin 3x$.

9· Graph $y = 2 \sin x + \cos 2x$.

10· Graph $y = 3 \sin x + \cos 2x$.

11· Graph $y = \sin^2 x + \cos 2x$.

12· Graph $y = \sin^2 x + \sin 2x$.

We observe from the graphs of Fig. 7.6 (or from the fact that all of the trigonometric functions are periodic) that none of the trigonometric functions have inverses. Now in our discussion of inverses (Section 2.5) we illustrated the technique of restricting the domain of the function so that the restricted function would have an inverse. The student would do well to read Section 2.5 again at this point.

The choice of a restricted domain is not unique. Hence, we shall set up guidelines for choosing the restricted domain for the trigonometric functions. The following characteristics seem to be sensible.

(a)   Let the domain D include those values of x for which the angle is in the first quadrant.

(b)   Choose the domain D so that each possible value of the function is assumed for exactly one $x \in D$.

(c)   If possible, let the domain be a single interval.

Not all the desiderata can be achieved in each case, but with these ideas guiding us, we shall agree to restrict the functions as indicated in Fig. 7.8.

$y = \sin x$ $\qquad\qquad\qquad$ $y = \cos x$ $\qquad\qquad\qquad$ $y = \tan x$

$x \in \left[-\dfrac{\pi}{2}, \dfrac{\pi}{2} \right]$ $\qquad\qquad$ $x \in [0, \pi]$ $\qquad\qquad\qquad$ $x \in \left(-\dfrac{\pi}{2}, \dfrac{\pi}{2} \right)$

$y = \csc x$ $\qquad\qquad\qquad$ $y = \sec x$ $\qquad\qquad\qquad$ $y = \cot x$

$x \in \left[-\dfrac{\pi}{2}, \dfrac{\pi}{2} \right],$ $\quad x \neq 0$ $\qquad$ $x \in [0, \pi],$ $\quad x \neq \dfrac{\pi}{2}$ $\qquad$ $x \in (0, \pi)$

The warning that the restricted domain is arbitrary should be taken seriously. The reader would not have to look through very many calculus texts before he would discover that there is no general agreement on how to restrict either the secant or the cosecant function. For example, a number of authors restrict the secant function to $[-\pi, -\pi/2)$ and $[0, \pi/2)$. The important thing is to be consistent. For this reason, having made our choice, we shall say no more about the other possibilities.

Each of these functions now has an inverse, and we must give names to the respective inverses. We recall that the inverse of a function F was indicated in Section 2.5 by F^{-1}. Hence, the inverse of the sin function is logically indicated by $\sin^{-1}$. The problem with this is that since $(\sin x)^2$ is commonly written $\sin^2 x$, it would be reasonable to suppose $(\sin x)^{-1}$ might similarly be written $\sin^{-1} x$. This then gives two distinct meanings to the symbol $\sin^{-1} x$. Many people run this risk, using $\sin^{-1} x$ to denote the inverse. We prefer to avoid any ambiguity by adopting a second common notation. If $y = \sin x$, then we denote the inverse function by Arcsin and write $x = \text{Arcsin } y$. Similar remarks apply to the other inverse functions:

If $y = \sin x$, $x \in [-\pi/2, \pi/2]$,   then $x = \text{Arcsin } y$, $y \in [-1, 1]$;
If $y = \cos x$, $x \in [0, \pi]$,   then $x = \text{Arccos } y$, $y \in [-1, 1]$;

If $y = \tan x,\ x \in (-\pi/2, \pi/2),$ then $x = \text{Arctan } y,\ y \in (-\infty, \infty);$

If $y = \csc x,\ x \in [-\pi/2, \pi/2],$ then $x = \text{Arccsc } y,\ |y| \geq 1,\ (x \neq 0);$

If $y = \sec x,\ x \in [0, \pi],$ then $x = \text{Arcsec } y,\ |y| \geq 1,\ (x \neq \pi/2);$

If $y = \cot x,\ x \in (0, \pi),$ then $x = \text{Arccot } y,\ y \in (-\infty, \infty).$

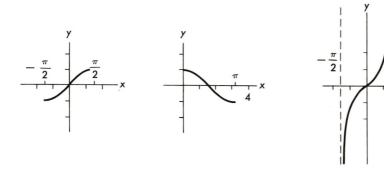

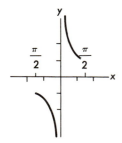
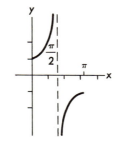
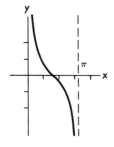

FIG. 7.8

As illustrations, the student should be sure he understands the following:

$$\text{Arcsin}\left(-\frac{1}{2}\right) = -\frac{\pi}{6}; \qquad \text{Arccot}(-1) = \frac{3\pi}{4};$$

$$\text{Arccos}\left(-\frac{1}{2}\right) = \frac{2\pi}{3}; \qquad \text{Arctan}(-1) = -\frac{\pi}{4}$$

If he wishes to write down all the solutions to $\cos\theta = -0.8$, he should write

$$\theta = \text{Arccos}(-0.8) + 2\pi k$$

and

$$\theta = -\text{Arccos}(-0.8) + 2\pi k$$

where k may be any positive or negative integer or zero. The best way to see what to write down is to sketch the angles between 0 and 2π that satisfy the requirement (Fig. 7.9).

$$\alpha = \text{Arccos}(-0.8)$$

$$\beta = -\text{Arccos}(-0.8)$$

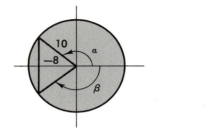

FIG. 7.9

Consider now the following two expressions.

$$x^3 - x^2 - 5x - 3 = 0; \tag{2}$$

$$(x - 1)(x - 5) = (x - 3)^2 - 4. \tag{3}$$

The first is a (conditional) equation; there are at most three distinct numerical values of x for which (2) is a true statement. The second is an identity; it is a true statement for any numerical value of x.

In trigonometry we likewise have both equations such as

$$2 \cos^2 x + \sin x = 1 \tag{4}$$

and identities such as

$$\sin 4x = 8 \sin x \cos^3 x - 2 \sin 2x. \tag{5}$$

If you are not told ahead of time, it is not always easy to recognize identities in trigonometry. Indeed, in a beginning trigonometry class much time is given over to the proving that given statements of equality are actually identities.

Some of the identities commonly proved in the beginning course are of great importance to us in our study of calculus. We list below those identities that make up the bare minimum which should be committed to memory.

$$\sin(\alpha \pm \beta) = \sin \alpha \cos \beta \pm \cos \alpha \sin \beta;$$

$$\cos(\alpha \pm \beta) = \cos \alpha \cos \beta \mp \sin \alpha \sin \beta. \tag{6}$$

$$\cos 2\alpha = \cos^2 \alpha - \sin^2 \alpha,$$

$$\cos 2\alpha = 2\cos^2 \alpha - 1, \qquad \text{so} \qquad \cos^2 \alpha = \frac{1 + \cos 2\alpha}{2} \; ;$$

$$\cos 2\alpha = 1 - 2\sin^2 \alpha, \qquad \text{so} \qquad \sin^2 \alpha = \frac{1 - \cos 2\alpha}{2} \; ; \qquad (7)$$

$$\sin 2\alpha = 2\sin \alpha \cos \alpha.$$

It is assumed that the reader has already memorized those identities listed as (1).

PROBLEMS

A · Given $\beta \in [0, 2\pi]$ such that $\cos \beta = -\frac{4}{5}$ and $\sin \beta = -\frac{3}{5}$, find $\sin \beta/2$. (Hint: Use (7) with $\alpha = \beta/2$.)

B · Show that Arctan $2 +$ Arctan $3 = 3\pi/4$. (Hint: Let $\alpha =$ Arctan 2; $\beta =$ Arctan 3. Find $\cos(\alpha + \beta)$. Then explain why we did not suggest that you find $\sin(\alpha + \beta)$ instead.)

C · Show that Arctan $\frac{1}{2} +$ Arctan $\frac{1}{3} = \pi/4$.

EXAMPLE D

Solve the equation given in (4); $2\cos^2 x + \sin x = 1$.

Express $\cos^2 x$ as $1 - \sin^2 x$. The equation is then

$$2 - 2\sin^2 x + \sin x = 1,$$
$$2\sin^2 x - \sin x - 1 = 0,$$
$$(2\sin x + 1)(\sin x - 1) = 0,$$
$$\sin x = -\tfrac{1}{2} \qquad \sin x = 1,$$
$$x = \frac{7\pi}{6} + 2\pi k, \qquad -\frac{\pi}{6} + 2\pi k, \qquad \frac{\pi}{2} + 2\pi k. \; \blacksquare$$

EXAMPLE E

Show that equation (5) really is an identity; that is, $\sin 4x = 8\sin x \cos^3 x - 2\sin 2x$.

On the left side, use $\sin 2(2x) = 2\sin 2x \cos 2x$.

$$2\sin 2x \cos 2x = 4\sin x \cos x(2\cos^2 x) - 2\sin 2x,$$
$$2\sin 2x \cos 2x = 2(\sin 2x)(2\cos^2 x) - 2\sin 2x,$$
$$2\sin 2x \cos 2x = 2(\sin 2x)(2\cos^2 x - 1).$$

The last statement is clearly an identity. $\blacksquare$

There is one aspect of trigonometry, called *triangle solving*, that we have largely ignored in our quick review. Although this topic is of great practical importance, we shall have use for only one result from this area, the *law of cosines*, already mentioned in Section 1.2. For later reference when we shall need it again, we recall again that the law of cosines relates the three sides a, b, c of an arbitrary triangle to the cosine of the angle A, the vertex opposite side a (Fig. 7.10).

$$a^2 = b^2 + c^2 - 2bc \cos A. \tag{8}$$

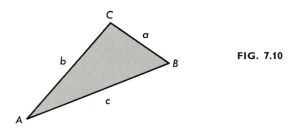

FIG. 7.10

7·1B EXERCISES

1 • Find the sine, cosine, and tangent of the following angles.

(a) $\text{Arctan}(-\sqrt{3})$.

(b) $\text{Arcsin}(-\tfrac{1}{2})$.

(c) $\text{Arccos}\left(-\dfrac{\sqrt{3}}{2}\right)$.

(d) $\text{Arccos}(-1)$.

2 • Find the sine, cosine, and tangent of the following angles.

(a) $\text{Arctan}\dfrac{1}{\sqrt{3}}$.

(b) $\text{Arcsin}\left(-\dfrac{\sqrt{3}}{2}\right)$.

(c) $\text{Arccos}(-\tfrac{1}{2})$.

(d) $\text{Arcsin}(-1)$

3 • Let $\alpha = \text{Arctan}(-\tfrac{1}{2})$ and $\beta = \text{Arcsin }\tfrac{2}{3}$. Find the following.

(a) $\sin(\alpha + \beta)$.

(b) $\cos(\alpha - \beta)$.

(c) $\sin 2\alpha$.

(d) $\cos 2\beta$.

(e) $\sin \dfrac{\alpha}{2}$.

(f) $\cos \dfrac{\beta}{2}$.

4 • Let $\alpha = \text{Arctan}(-\sqrt{3})$ and $\beta = \text{Arccos}(-\tfrac{4}{5})$. Find the values indicated in Exercise 3.

5 • Find $\cos[\text{Arctan}(-1) - \text{Arcsin}(\tfrac{3}{4})]$.

6 · Find $\sin[\text{Arccos}(\frac{2}{3}) + \text{Arctan}(-2)]$.

7 · Find $\sin[2\,\text{Arctan}\,\frac{4}{3}]$ and $\cos[2\,\text{Arctan}\,\frac{4}{3}]$.

8 · Find $\sin[2\,\text{Arccos}(-\frac{3}{5})]$ and $\cos[2\,\text{Arccos}(-\frac{3}{5})]$.

9 · Solve the equation $\cos 2x - \cos x = 0$.

10 · Solve the equation $\tan 2x \sin 2x - \cos 2x = 0$.

11 · Which of the following are identities? Prove your answer in each case.

    (a) $\tan 2x = \dfrac{2 \sin x \sec x}{1 - \tan^2 x}$.

    (b) $\cos 3x = 3 \cos^3 x - \cos x \sin^2 x - 2 \cos x$.
    (c) $2 \sin^2 x = \sin 2x$.

    (d) $\sec x + \tan x = \dfrac{2 \cos^2 x}{2 \cos x - \sin 2x}$.

12 · Which of the following are identities? Prove your answer in each case.
    (a) $8 \sin^2 x = \sec^2 x - \sec^2 x \cos 4x$.
    (b) $\sin 4x - 4 \cos 2x \sin x + 2(\cos 2x)(1 - \cos x) = 0$.
    (c) $\cos x = \sec x - \sin x \sec x + \tan x - \tan x \sin x$.
    (d) $2 \sin x \cos^2 x \sec 2x - \tan 2x = 0$.

7 · 2   Derivatives of the Trigonometric Functions

Let us begin with an attempt to graph $D \sin x$ from what we know about the graph of $y = \sin x$. Our first observation is that the slope of $y = \sin x$ is zero at $x = \pi/2$ and $x = 3\pi/2$. We next observe that the slope at $x = 0$ and the slope at $x = 2\pi$ is the same positive number, say k. Also, the slope at $x = \pi$ will be $-k$. Finally, the value of the slope will be periodic, with with period 2π.

Anyone with a spirit of adventure who looks at Fig. 7.11 would guess that $D \sin x$ would be $\cos x$, or perhaps a multiple of $\cos x$ if $k \neq 1$. With this much feeling for what the answer ought to be, we set about the business of a formal computation. Recall

$$DF(x) = \lim_{h \to 0} \frac{F(x + h) - F(x)}{h}.$$

Therefore,

$$D \sin x = \lim_{h \to 0} \frac{\sin(x + h) - \sin x}{h}.$$

Using (6) of Section 7.1,

$$D \sin x = \lim_{h \to 0} \frac{\sin x \cos h + \cos x \sin h - \sin x}{h}.$$

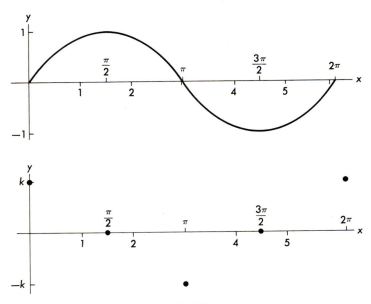

FIG. 7.11

Guided by our expectation of an answer involving cos x, we write this in the form

$$D \sin x = (\cos x) \left[\lim_{h \to 0} \frac{\sin h}{h} \right] + (\sin x) \left[\lim_{h \to 0} \frac{\cos h - 1}{h} \right]. \tag{1}$$

This is justified according to Theorem 2.4B providing that both limits exist and at least one is finite.

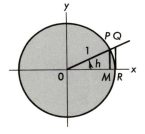

FIG. 7.12

We first direct our attention to evaluating $\lim_{h\to 0}$ (sin h/h). Toward this end, we recall that the length of an arc cut off on the circumference of a circle by a central angle θ is $r\theta$ when θ is measured in radians. (It is the use of this fact at this point that commits us to use radian measure in the study of calculus.)

Now consider a circle of radius $r = 1$ with central angle $h \in (0, \pi/2)$. From Fig. 7.12, comparison of the areas of the triangle OPM, the circular sector OPR, and the triangle ORQ gives us

$$\tfrac{1}{2}OM \cdot MP < \frac{h}{2\pi}\, \pi \cdot 1^2 < \tfrac{1}{2}OR \cdot RQ,$$

$$(\cos h)(\sin h) < h < \tan h,$$

and division by sin h gives

$$\cos h < \frac{h}{\sin h} < \frac{1}{\cos h}.$$

Since $\lim_{h\to 0} \cos h = 1$, it is clear that the terms on the left and right both go to 1. Hence, so does the term caught in the middle,

$$\lim_{h\to 0^+} \frac{h}{\sin h} = 1. \tag{2}$$

We made use in our argument of the fact that h was positive. It remains to show that (2) also holds for $h \to 0^-$.

PROBLEM

A · Show that $\lim_{h\to 0^-} h/\sin h = 1$. (Hint: Let $h = -t$. Then $\lim_{h\to 0^-} h/\sin h = \lim_{t\to 0^+} -t/\sin(-t)$.)

The results of the problem together with (2) enable us to write

$$\lim_{h\to 0} \frac{h}{\sin h} = 1.$$

From this it follows that

$$\lim_{h\to 0} \frac{\sin h}{h} = 1 \qquad (h \text{ measured in radians}). \tag{3}$$

We have now evaluated one of the limits encountered in (1). The second is easily evaluated by virtue of what we know about the first. Multiplication of the numerator and denominator by cos $h + 1$ gives

$$\lim_{h\to 0} \frac{\cos h - 1}{h} = \lim_{h\to 0} \frac{\cos^2 h - 1}{h(\cos h + 1)} = \lim_{h\to 0} \frac{\sin h}{h} \lim_{h\to 0} \frac{-\sin h}{\cos h + 1}.$$

The last statement again depends on both limits existing and being finite (Theorem 2.4C). But the first limit is (3) and the second limit is obviously zero. Hence,

$$\lim_{h \to 0} \frac{\cos h - 1}{h} = 0. \tag{4}$$

From (1) we thus have

$$D \sin x = \cos x.$$

Now if $u = F(x)$, the chain rule immediately gives us

$$\frac{d}{dx} \sin u = (\cos u) \frac{du}{dx}. \tag{5}$$

EXAMPLE A
Find

(a) $\quad D \sin \sqrt{x^2 + 1} \quad$ and $\quad$ (b) $\quad Dx \sin x^2$.

(a) According to the chain rule, where $u = (x^2 + 1)^{1/2}$,

$$D \sin(x^2 + 1)^{1/2} = [\cos(x^2 + 1)^{1/2}]x(x^2 + 1)^{-1/2}.$$

As a practical matter, the reader is cautioned that the answer is *not*

$$\cos(x^2 + 1)^{1/2}x(x^2 + 1)^{-1/2},$$

which someone might be tempted to simplify to $\cos x$. The sure way out of any possible notational confusion is to write

$$D \sin(x^2 + 1)^{1/2} = x(x^2 + 1)^{-1/2} \cos(x^2 + 1)^{1/2}.$$

(b) We recognize that $x \sin x^2$ is a product. Thus,

$$Dx \sin x^2 = x(2x) \cos x^2 + \sin x^2$$
$$= 2x^2 \cos x^2 + \sin x^2. \ \blacksquare$$

We could now similarly compute $D \cos x$ by meeting it head on. We prefer to be coy. From (6) of Section 7.1, with $\alpha = \pi/2$, $\beta = x$, we get

$$\sin\left(\frac{\pi}{2} - x\right) = \cos x.$$

Now if we use the chain rule, we get

$$D \cos x = D \sin\left(\frac{\pi}{2} - x\right) = (-1)\cos\left(\frac{\pi}{2} - x\right).$$

Again using (6) of Section 7.1, we find

$$\cos\left(\frac{\pi}{2} - x\right) = \sin x.$$

Therefore,

$$D \cos x = -\sin x,$$

or, if u is a function of x,

$$\frac{d}{dx} \cos u = (-\sin u)\frac{du}{dx}. \tag{6}$$

PROBLEMS

B · The first two formulas in the following list, which we proved, may be used to develop the others. Do so.

$$D \sin x = \cos x.$$

$$D \cos x = -\sin x.$$

$$D \tan x = \sec^2 x.$$

$$D \sec x = \sec x \tan x.$$

$$D \csc x = -\csc x \cot x.$$

$$D \cot x = -\csc^2 x.$$

(Hint: $\tan x = \sin x/\cos x$. You know how to differentiate the quotient of two functions.)

C · Find $D \tan\sqrt{x^2 + 1}$ by using the chain rule.

When u is a function of x, the formulas developed in **Problem B** along with the chain rule give us

$$D_x \sin u = (\cos u)\, D_x u;$$

$$D_x \cos u = (-\sin u)\, D_x u;$$

$$D_x \tan u = (\sec^2 u)\, D_x u;$$

$$D_x \sec u = (\sec u \tan u)\, D_x u;$$

$$D_x \csc u = (-\csc u \cot u)\, D_x u;$$

$$D_x \cot u = (-\csc^2 u)\, D_x u.$$

The student should memorize and be able to use all of these formulas.

1 • Evaluate the following limits by using (3).

(a) $\lim\limits_{x \to 0} \dfrac{\cos x}{x \cot x}$.

(b) $\lim\limits_{x \to 0} \dfrac{1 - \cos^2 x}{x \sin x}$.

(c) $\lim\limits_{h \to 0+} \dfrac{-\tan^2 h}{h(\sec h + \sec^2 h)}$.

(d) $\lim\limits_{x \to 0+} \dfrac{\sin^2 x \tan x}{x[\sin x - \tan x]}$.

2 • Evaluate the following limits by using (3).

(a) $\lim\limits_{x \to 0} \dfrac{1}{x \csc x}$.

(b) $\lim\limits_{x \to 0} \dfrac{x \sin x \sec^2 x}{\tan^2 x}$.

(c) $\lim\limits_{x \to 0+} \dfrac{x \sec x}{\sec x - 1}$.

(d) $\lim\limits_{x \to 0+} x \dfrac{\cos x - 1}{\sin^2 x}$.

3 • Find the indicated derivatives.

(a) $D \sin(x^2 + 1)^{-1/2}$.

(b) $D(x \tan x)$.

(c) $D \dfrac{\sin x}{\tan x + 1}$.

(d) $D(x^2 \cos x^2)$.

4 • Find the indicated derivatives.

(a) $D \tan(x^3 + x^2)$.

(b) $D(x^2 + 1) \sec x$.

(c) $D \dfrac{\cos 2x}{x + \sin 2x}$.

(d) $D(x \tan x^2)$.

5 • Find the indicated derivatives.

(a) $D \sin^3 x$.

(b) $D \sin x^3$.

(c) $D \sec^3 x$.

(d) $D \sec x \tan x$.

6 • Find the indicated derivatives.

(a) $D \cos^3 x$.

(b) $D \cos x^3$.

(c) $D \csc^3 x$.

(d) $D \csc x \cot x$.

In Exercises 7–12, find the indicated derivative.

7 • $D \dfrac{\tan x}{1 + \sec x}$.

8 • $D \dfrac{\sin x}{1 + \cos x}$.

9 • $D[4 \sin^2 x + (2 \cos^2 x - 1)^2]$.

10 • $D[2 \sin^2 x + \cos 2x - 1]$.

11 • $D \dfrac{\sin x}{x^2 + 1}$.

12 • $D \dfrac{\csc x}{x^2 + 1}$.

We have seen that a knowledge of the derivative is useful in such applications as the drawing of graphs, the solution of maximum-minimum problems, and approximation. We illustrate these three applications in situations that involve trigonometric functions.

EXAMPLE B

Graph $y = F(x) = \sin 2x + 2 \cos x$ for $x \in [0, 2\pi]$.

This is the same function we graphed in Example 7.1C (see Fig. 7.7). We can now find the critical points and identify them.

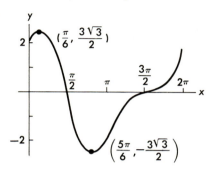

FIG. 7.13

$DF(x) = 2\cos 2x - 2 \sin x;$

$D^2F(x) = -4 \sin 2x - 2 \cos x.$

Critical points are found by solving

$2(1 - 2 \sin^2 x) - 2 \sin x = 0;$

$\sin x = \tfrac{1}{2}; \qquad \sin x = -1;$

$x = \dfrac{\pi}{6}, \dfrac{5\pi}{6}; \qquad x = \dfrac{3\pi}{2}.$

The sketch made in Example 7.1C may be used to classify these critical points as local maximums, minimums, or inflection points. Or, as before, this classification can be made by a study of the second derivative.

$$D^2F\left(\frac{\pi}{6}\right) = -3\sqrt{2} \qquad \text{so a maximum occurs at } \frac{\pi}{6}.$$

$$D^2F\left(\frac{5\pi}{6}\right) = 3\sqrt{2} \qquad \text{so a minimum occurs at } \frac{5\pi}{6}.$$

$$D^2F\left(\frac{3\pi}{2}\right) = 0 \qquad \text{and the sign changes in a neighborhood}$$

of $\dfrac{3\pi}{2}$, so there is a point of inflection at $\dfrac{3\pi}{2}$.

The graph may be further refined by noting that there are inflection points where $\cos x = 0$ and where $\sin x = -\tfrac{1}{4}$. ∎

EXAMPLE C

A 1-ft-wide vent runs perpendicular to a hallway that is 8-ft wide. A certain construction job requires that lengths of pipe be pushed from the hallway through this vent. How long can the lengths of pipe be cut?

We assume in our solution that the pipe is to be kept parallel to the floor. The length of the longest pipe possible corresponding to a given θ (Fig. 7.14) is

$$y = 1 \csc \theta + 8 \sec \theta.$$

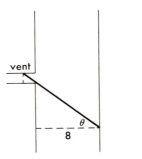

vent

FIG. 7.14

8

The longest pipe that we can use is the minimum value of y obtained for $\theta \in [0, \pi/2]$.

$$\frac{dy}{d\theta} = -\csc \theta \cot \theta + 8 \sec \theta \tan \theta = \frac{-\cos^3 \theta + 8 \sin^3 \theta}{\sin^2 \theta \cos^2 \theta}.$$

Critical points are found by setting the numerator equal to zero. Factoring the numerator as the difference of two cubes, we then have

$$(2 \sin \theta - \cos \theta)(4 \sin^2 \theta + 2 \sin \theta \cos \theta + \cos^2 \theta) = 0.$$

The second factor is nonzero for $\theta \in [0, \pi/2]$. From physical considerations the solution obtained from the first factor must be a minimum.

$$\frac{\sin \theta}{\cos \theta} = \frac{1}{2}.$$

If $\tan \theta = \frac{1}{2}$, then $\csc \theta = \sqrt{5}$ and $\sec \theta = \sqrt{5}/2$.

$$y = \sqrt{5} + 4\sqrt{5} = 5\sqrt{5} \approx 11 \text{ ft.} \quad \blacksquare$$

EXAMPLE D

Write the Taylor polynomial $T_{F(0), 3}$ for $F(x) = \sin x$. Estimate the error that results from using this polynomial to approximate $\sin x$ for $x \in [0 \ \pi/3]$.

$$\sin 0 = 0,$$

$$D \sin x = \cos x, \qquad D \sin 0 = 1,$$

$$D^2 \sin x = -\sin x, \qquad D^2 \sin 0 = 0,$$

$$D^3 \sin x = -\cos x, \qquad D^3 \sin 0 = -1,$$

$$T_3(x) = x - \frac{1}{3!} x^3. \tag{7}$$

We see that if we had been required to write the polynomial T_4 instead of T_3, we would have written exactly the same result; that is, $T_4(x) = T_3(x)$. Hence, we may as well assume that we have written T_4 when we estimate the error, since the larger that n is, the better our estimate is. We therefore compute

$$R_n(x_0, x) = R_4(0, x) = \frac{(x - 0)^5}{5!} D^5 \sin \bar{x}, \qquad \bar{x} \in (0, x).$$

Since $0 < (\cos \bar{x}) < 1$,

$$0 \leq R_4(0, x) \leq x^5 \frac{1}{5!}.$$

The largest possible error for $x \in [0, \pi/3]$ is therefore less than

$$\left(\frac{\pi}{3}\right)^5 \frac{1}{120} < 0.013.$$

Our answer will be correct to within one unit in the first decimal place. This is of course an upper bound on the error; it is the worst possible. We may therefore anticipate that our accuracy might be much better. For example, if we use $\pi \approx 3.14$ and set $x = \pi/6$ in (7) we get

$$T\left(\frac{\pi}{6}\right) = \frac{\pi}{6} - \frac{\pi^3}{6^4} = 0.523 - 0.024 = 0.499.$$

We of course know that $F(\pi/6) = \frac{1}{2}$. ∎

Every time we learn to differentiate a function, we also learn to anti-differentiate a function. For example, $D \sin x = \cos x$ tells us immediately that $D^{-1} \cos x = \sin x$; that is, that $\int \cos x \, dx = \sin x$.

EXAMPLE E

Find the area under one arch of the graph of $y = \sin x$ (Fig. 7.15)

$$\text{Area} = \int_0^\pi y\, dx = \int_0^\pi \sin x\, dx;$$

$$\text{Area} = -\cos x \Big|_0^\pi = -[\cos \pi - \cos 0];$$

$$\text{Area} = 2. \ \blacksquare$$

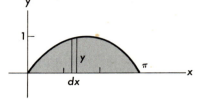

FIG. 7.15

EXAMPLE F

Evaluate

$$\int_0^{\pi/3} \frac{\sin x}{x}\, dx.$$

Our first reaction is to wonder if the integral is improper since $(\sin x)/x$ is not defined at the origin. Because of (3), however, the function

$$F(x) = \begin{cases} \dfrac{\sin x}{x}, & x \neq 0, \\ 1, & x = 0, \end{cases}$$

is continuous on $[0, \pi/3]$. It may therefore be integrated, and this is clearly the value of the integral we seek.

There is nevertheless still a problem; a little experimenting will convince the reader that he does not know of any G for which $DG(x) = (\sin x)/x$. Having just worked Example D, however, we hit upon an alternate procedure. We learned there that

$$\sin x = x - \tfrac{1}{6}x^3 + R_4(0, x)$$

where

$$0 < R_4(0, x) < \frac{x^5}{5!}.$$

Hence,

$$\int_0^{\pi/3} \frac{\sin x}{x}\, dx = \int_0^{\pi/3} \left[1 - \frac{1}{6}x^2 + \frac{1}{x} R_4(0, x)\right],$$

$$= \left(x - \frac{1}{18}x^3\right)\Big|_0^{\pi/3} + \int_0^{\pi/3} \frac{1}{x} R_4(0, x)$$

where

$$0 < \int_0^{\pi/3} \frac{dx}{x} R_4(0, x) < \int_0^{\pi/3} \frac{x^4}{5!}\, dx = \frac{x^5}{5 \cdot 5!}\Big|_0^{\pi/3}.$$

We may therefore write

$$\int_0^{\pi/3} \frac{\sin x}{x}\, dx \approx \frac{\pi}{3} - \frac{\pi^3}{(18)(27)},$$

knowing that our error is less than $\pi^5/(5 \cdot 5! \cdot 3^5)$. ∎

PROBLEMS

D · Find $\int \cos^3 x\, dx$. (Hint: $\cos^3 x = \cos x(1 - \sin^2 x)$.)

E · Find $\int \cos^2 x\, dx$. (Hint: Use (7) of Section 7.1.)

═══════════════════════════════════ 7 · 2B EXERCISES

1 · Write the equation of the line tangent to the graph of $y = \sin x + \cos x$ at the point where $x = 2\pi/3$.

2 · Write the equation of the line tangent to the graph of $y = 2 \sin x + \cos 2x$ at the point where $x = 3\pi/4$.

3 · Find the maximum height of the curve $y = 2 \cos x - 3 \sin x$ above the x axis.

4 · What angle does the curve $y = \sin^3 x$ make with the x axis at $x = 0$?

5 · A light 4 miles from a straight shoreline makes 5 revolutions per minute (rpm). How fast is the light moving along the shore when the beam makes an angle of $\pi/4$ with the shoreline?

6 · A plane flying 2000 ft above the ground is being followed by a searchlight on the ground. Find the angular velocity of the searchlight when the beam makes an angle of $\pi/3$ with the ground if the plane is flying at 360 mph.

7 · A telephone pole is 30 ft high and stands 8 ft from a fence that is $5\sqrt{5}$ ft high. A support wire is to be fastened between the pole and an anchor in the

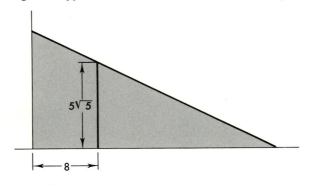

FIG. 7.16

$5\sqrt{5}$

8

ground on the other side of the fence (Fig. 7.16). What is the shortest wire that may be used? How would this problem be changed if the pole were only 18 ft high?

8. Two men carry a $10\sqrt{5}$-ft ladder down an $6\sqrt{3}$-ft-wide corridor. They turn into a second corridor, perpendicular to the first one, while keeping the ladder horizontal. From this find the minimum possible width of the second corridor.

9. Solve Exercise 5 of Section 5.2 again, using as independent variable the angle that the pole makes with the pier.

10. Solve Exercise 6 of Section 5.2 again, using as independent variable the angle that the new sidewalk makes with one of the old sidewalks.

11. Let $S(x) = \sin x$. Find the Taylor polynomials $T_{S(0),1}$, $T_{S(0),2}$, and $T_{S(0),3}$. Graph these three polynomials and the function $S(x)$ on the same coordinate axes for $x \in (-\pi/2, \pi/2]$.

12. Let $C(x) = \cos x$. Solve Exercise 11 using $C(x)$ instead of $S(x)$.

13. The Taylor polynomial for $S(x) = \sin x$ is easy to compute at $x_0 = 0$. Find a general formula for the nth-degree Taylor polynomial $T_{S(0),\,n}$.

14. Solve Exercise 13 for the function $C(x) = \cos x$.

In Exercises 15–24, find the indicated antiderivatives.

15. $\displaystyle\int \sin 3x \, dx.$

16. $\displaystyle\int \frac{dx}{\cos^2 x}.$

17. $\displaystyle\int \sin^2 3x \, dx.$

18. $\displaystyle\int \sin^3 2x \, dx.$

19. $\displaystyle\int \sec^2 x \tan x \, dx.$

20. $\displaystyle\int \csc^2 x \cot x \, dx.$

21. $\displaystyle\int \sin^2 x \cos^2 x \, dx.$

22. $\displaystyle\int (\cos^4 x - \sin^4 x) \, dx.$

23. $\displaystyle\int \cot^2 x \, dx.$

24. $\displaystyle\int \tan^2 x \, dx.$

25. Find the area above the x axis and between the curves
$$y = -\frac{2}{\pi}x + 1,$$
$$y = \sin 2x + \cos x.$$

26. Find the area above the x axis and between the curves
$$y = 1 + \sin x + \cos x,$$
$$y = \frac{2}{\pi}x + 1.$$

In Exercises 27–34, find the value of the integral by finding an antiderivative and using the fundamental theorem.

27. $\displaystyle\int_{3\pi/4}^{\pi} \tan^2 x \, dx.$

28. $\displaystyle\int_{0}^{\pi/6} \sin^2 x \, dx.$

29. $\displaystyle\int_0^{\pi/2} (1 + \sin x)^3 \cos x \, dx.$

30. $\displaystyle\int_{\pi/2}^{3\pi/4} \sqrt{1 - \sin^2 x} \, dx.$

31. $\displaystyle\int_0^{\pi/6} \cos^2 x \, dx.$

32. $\displaystyle\int_0^{\sqrt{\pi}} x \cos x^2 \, dx.$

33. $\displaystyle\int_0^{\sqrt{\pi}} x \sin x^2 \, dx.$

34. $\displaystyle\int_0^{1/3} \frac{\sin x\pi}{\cos^2 x\pi} \, dx.$

In Exercises 35–38, use the appropriate Taylor polynomials of degree three to approximate the value of the integral. Include, in each case, an estimate of the accuracy of the approximation.

35. $\displaystyle\int_0^{1/4} x \sin \pi x \, dx.$

36. $\displaystyle\int_0^{1/4} x \cos \pi x \, dx.$

37. $\displaystyle\int_0^{1/4} \tan \pi x \, dx.$

38. $\displaystyle\int_0^{1/4} \sec \pi x \, dx.$

7·3   Derivatives of the Inverse Trigonometric Functions

Set

$$y = F(x) = \text{Arcsin } x.$$

This means

$$x = F^{-1}(y) = \sin y.$$

It follows that

$$D_y F^{-1}(y) = \cos y.$$

From Theorem 3.4B, we know that if $y = F(x)$, and if the inverse exists and is differentiable, then

$$D_x F(x) = \frac{1}{D_y F^{-1}(y)}.$$

Hence,

$$D_x \text{ Arcsin } x = \frac{1}{\cos y}. \qquad (1)$$

A second way to obtain (1) is to differentiate

$$x = \sin y$$

implicitly with respect to x. Then

$$1 = (\cos y) \frac{dy}{dx}$$

so again,

$$\frac{dy}{dx} = D_x \text{ Arcsin } x = \frac{1}{\cos y}. \tag{1}$$

The answer as we now have it in (1) is in an unsatisfactory form. The derivative D_x Arcsin x ought to be expressed in terms of a function of x. For this purpose we observe that since $y \in [-\pi/2, \pi/2]$, $\cos y \geq 0$; so $\cos y = \sqrt{1 - \sin^2 y}$. But $x = \sin y$; $\cos y = \sqrt{1 - x^2}$. We can now write (1) in the form

$$D \text{ Arcsin } x = \frac{1}{\sqrt{1 - x^2}}.$$

Our choice of the restricted domain for the sine function affects the definition of the inverse function, of course. Hence, it is not surprising that the computation of the derivative of the inverse function requires a knowledge of the particular restricted domain chosen. (We needed to know that $y \in [-\pi/2, \pi/2]$.)

PROBLEMS

A · Using the foregoing work as a pattern, show

$$D \text{ Arccos } x = -\frac{1}{\sqrt{1 - x^2}};$$

$$D \text{ Arctan } x = \frac{1}{1 + x^2};$$

$$D \text{ Arccot } x = -\frac{1}{1 + x^2}.$$

Exactly the same pattern is used in finding D Arcsec x, but a subtlety enters in. Set

$$y = G(x) = \text{Arcsec } x.$$

This means

$$x = G^{-1}(y) = \sec y.$$

We know

$$D_y G^{-1}(y) = D \sec y = \sec y \tan y.$$

Since $y \in [0, \pi]$,

$$\tan y = \begin{cases} \sqrt{\sec^2 y - 1} & \text{if } y \in \left[0, \dfrac{\pi}{2}\right), \\[2ex] -\sqrt{\sec^2 y - 1} & \text{if } y \in \left(\dfrac{\pi}{2}, \pi\right]. \end{cases}$$

$$D_y G^{-1}(y) = \begin{cases} x\sqrt{x^2 - 1} & \text{if } y \in \left[0, \dfrac{\pi}{2}\right) \qquad \text{which means } x \geq 1; \\[2ex] -x\sqrt{x^2 - 1} & \text{if } y \in \left(\dfrac{\pi}{2}, \pi\right] \qquad \text{which means } x \leq -1. \end{cases}$$

Since

$$|x| = \begin{cases} x & \text{if } x \geq 1, \\ -x & \text{if } x \leq -1, \end{cases}$$

$$D_y \, G^{-1}(y) = |x|\sqrt{x^2 - 1}.$$

Therefore, again using Theorem 3.4B, we have

$$D \operatorname{Arcsec} x = \frac{1}{D_y G^{-1}(y)} = \frac{1}{|x|\sqrt{x^2 - 1}}.$$

B · Show

$$D \operatorname{Arccsc} x = \frac{-1}{|x|\sqrt{x^2 - 1}}.$$

If we combine what we have learned about the derivatives of the inverse trigonometric functions with the chain rule, then when u is a function of x, we have

$$D_x \operatorname{Arcsin} u = \frac{D_x u}{\sqrt{1 - u^2}}; \qquad\qquad D_x \operatorname{Arccos} u = \frac{-D_x u}{\sqrt{1 - u^2}},$$

$$D_x \operatorname{Arctan} u = \frac{D_x u}{1 + u^2}; \qquad\qquad D_x \operatorname{Arccot} u = \frac{-D_x u}{1 + u^2};$$

$$D_x \operatorname{Arcsec} u = \frac{D_x u}{|u|\sqrt{u^2 - 1}}; \qquad\qquad D_x \operatorname{Arccsc} u = \frac{-D_x u}{|u|\sqrt{u^2 - 1}}.$$

C · Use the chain rule to find the following

(a) $D \operatorname{Arcsin}(x^2 + 1)^{-1}$.   (b) $D \operatorname{Arctan}(\sin x)$.

EXAMPLE A

Graph

$$\theta = F(t) = \operatorname{Arctan} \frac{t + 1}{t - 1} + \operatorname{Arctan} t.$$

Suppose we begin by finding the critical points. We compute

$$\frac{d\theta}{dt} = \frac{[(t - 1) - (t + 1)]/(t - 1)^2}{1 + [(t + 1)^2/(t - 1)^2]} + \frac{1}{1 + t^2},$$

$$\frac{d\theta}{dt} = \frac{-2}{(t - 1)^2 + (t + 1)^2} + \frac{1}{1 + t^2},$$

$$\frac{d\theta}{dt} = \frac{-2}{2t^2 + 2} + \frac{1}{1 + t^2} = 0.$$

This development catches us by surprise. When we recover, we are tempted to say the function F is constant for all t. (For its derivative is 0 everywhere; see Problem 4.1B.) We find the constant by setting $t = -1$.

$$F(-1) = \operatorname{Arctan} 0 + \operatorname{Arctan}(-1) = -\frac{\pi}{4}.$$

If we are not careful, we will tell the world that for all t,

$$F(t) = \operatorname{Arctan} \frac{t + 1}{t - 1} + \operatorname{Arctan} t = -\frac{\pi}{4}.$$

We will be embarrassed, though, if someone points out that $F(2) = \operatorname{Arctan} 3 + \operatorname{Arctan} 2$, which is, according to Problem 7.1B, equal to $3\pi/4$. We should now go back and read Problem 4.1B more closely. It says "if F is continuous, then ...." A glance at our function shows it to be noncontinuous at $t = 1$. Now it is continuous on any closed interval to the left of $t = 1$; also on any closed interval to the right of $t = 1$. Hence, the graph is as indicated in Fig. 7.17. ∎

The six formulas for differentiation of inverse functions give six formulas for antidifferentiation. Of the six, we need know only two for reasons that will be made clear in Chapter 8. We recommend learning the following, assuming in each case that u is a function of x and that a is a constant.

$$\int \frac{D_x u}{a^2 + u^2} = \frac{1}{a} \operatorname{Arctan} \frac{u}{a} + C;$$

$$\int \frac{D_x u}{\sqrt{a^2 - u^2}} = \operatorname{Arcsin} \frac{u}{a} + C.$$

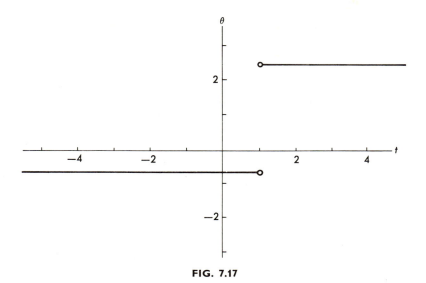

FIG. 7.17

The first follows because

$$D\frac{1}{a}\operatorname{Arctan}\frac{u}{a} = \frac{1}{a}\frac{(1/a)D_x u}{1 + (u/a)^2} = \frac{D_x u}{a^2 + u^2},$$

and the second is verified in the same way.

EXAMPLE B

Evaluate

$$\int_0^1 \frac{dx}{\sqrt{3 - x^2 + 2x}} \quad \text{and} \quad \int_0^1 \frac{dx}{x^2 - 2x + 5}.$$

$$\int_0^1 \frac{dx}{\sqrt{3 - (x^2 - 2x)}} = \int_0^1 \frac{dx}{\sqrt{4 - (x - 1)^2}} = \operatorname{Arcsin}\frac{x - 1}{2}\bigg|_0^1,$$

$$= \operatorname{Arcsin} 0 - \operatorname{Arcsin}\left(-\frac{1}{2}\right) = \frac{\pi}{6}.$$

$$\int_0^1 \frac{dx}{x^2 - 2x + 1 + 4} = \int_0^1 \frac{dx}{(x - 1)^2 + 2^2} = \frac{1}{2}\operatorname{Arctan}\frac{x - 1}{2}\bigg|_0^1,$$

$$= -\tfrac{1}{2}\operatorname{Arctan}(-\tfrac{1}{2}). \ \blacksquare$$

Differentiate the following expressions.

1· Arcsin $\sqrt{x}$.

2· Arctan $\sqrt{x}$.

3· Arcsin $\sqrt{1 - x^2}$.

4· Arctan $\sqrt{x^2 + 1}$.

5· Arccos $\dfrac{3}{x^2 + 2}$.

6· Arccot $\dfrac{4}{3 - x}$.

7· Arctan^2 $4x$.

8· Arcsin^2 $\dfrac{1}{x}$.

9· Arctan $\dfrac{1}{\sqrt{x}}$.

10· x^3 Arcsin $\dfrac{1}{x}$.

11· $F(x) = 2x\sqrt{4 - x^2} + 2 \text{ Arcsin } \dfrac{x}{2} - x(4 - x^2)^{3/2}$.

12· $G(x) = x(4 - x^2)^{3/2} + 6x\sqrt{4 - x^2} + 24 \text{ Arcsin } \dfrac{x}{2}$.

13·* Find $D \cot(\text{Arcsin } \sqrt{1 - x^2})$ by computing it directly. Then compute it a second time by first expressing $\cot(\text{Arcsin } \sqrt{1 - x^2})$ as an algebraic function of x, then finding the derivative of this algebraic expression.

14·* Find $D \sin(\text{Arctan } x/\sqrt{1 - x^2})$ by computing it directly. Then compute it a second time by first expressing $\sin(\text{Arctan } x/\sqrt{1 - x^2})$ as an algebraic function of x, then finding the derivative of this algebraic expression.

15· $\displaystyle\int \frac{2x\,dx}{1 + x^4}$.

16· $\displaystyle\int \frac{x\,dx}{\sqrt{1 - x^2}}$.

17· $\displaystyle\int \frac{-3\,dx}{\sqrt{4 - 9x^2}}$.

18· $\displaystyle\int \frac{-12\,dx}{9 + 16x^2}$.

19· $\displaystyle\int \frac{x\,dx}{\sqrt{9 - x^4}}$.

20· $\displaystyle\int \frac{x\,dx}{4 + x^4}$.

21·* $\displaystyle\int \frac{(2x + 2)\,dx}{x^4 + 4x^3 + 4x^2 + 1}$.

22·* $\displaystyle\int \frac{(2x + 2)\,dx}{\sqrt{4 - x^4 - 4x^3 - 4x^2}}$.

Evaluate the integrals in Exercises 23–26.

23· $\displaystyle\int_{-\sqrt{3}}^{-1} \frac{dx}{\sqrt{4 - x^2}}$.

24· $\displaystyle\int_{1}^{3} \frac{dx}{x^2\sqrt{9 - x^{-2}}}$.

25 · $\displaystyle\int_{-\sqrt{3}}^{-1}\frac{dx}{3+x^2}$

26 · $\displaystyle\int_{0}^{1}\frac{dx}{4x^2+4x+5}$.

27 · An 18-ft ladder leaning against a vertical wall begins to slip. The top of the ladder moves down the wall at 3/2 ft/sec. How fast (in radians per second) is the angle formed by the ladder and the ground changing when the top of the ladder is 6 ft from the ground?

28 · Sand is being poured onto a conical pile in such a way that the radius of the pile grows at the rate of 1 ft/min while the height increases at 3 ft/min. Find the rate (in radians per minute) of change of the base angle of the pile.

29 · A boat moves directly toward shore. Two lights, one located 20 ft from where the boat will hit the beach, the other located 105 ft farther down the shoreline, are trained on the boat when it is 200 ft from shore. The two beams make an angle of θ at the boat. If the lights follow the boat into shore, when will the angle θ be greatest?

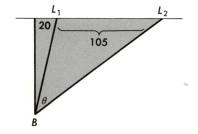

FIG. 7.18

30 · The lower edge of a window is 4 ft above the eye of an observer. The upper edge is 6 ft above his eye. Where should he stand to maximize the angle subtended by the window?

31 ·* A power station on the south shore of a circular lake is to supply power to a factory on the diametrically opposite shore. If it costs twice as much to lay cable under water as it does to run it along the shoreline, how should the cable be put in?

32 · Let $A(x) = $ Arctan x. Find the Taylor polynomial $T_{A(0),5}$. Use it to approximate $A(1)$ and hence to aproximate π.

33 · Let $B(x) = $ Arcsin x. Find the Taylor polynomial $T_{B(0),5}$. Use it to approximate $B(\frac{1}{2})$ and hence to approximate π.

34 ·* The estimates of π obtained in Exercises 32 and 33 are not very good. One way to obtain a better estimate is to use a Taylor polynomial of higher degree. To avoid the forbidding computation of derivatives, proceed as indicated below.

(a) Prove the algebraic identity

$$\frac{1}{1+t^2} = 1 - t^2 + t^4 - t^6 + \ldots + (-1)^n t^{2n} + \frac{(-1)^{n+1}\, t^{2n+2}}{1+t^2}.$$

(b)   Observe Arctan $x = \int_0^x \frac{dt}{1+t^2}$. Set $A(x) = $ Arctan x.

(c)   Substitute the expression for $1/(1+t^2)$ obtained in part (a) into the integral of part (b). From this, show that

$$A(x) = x - \frac{x^3}{3} + \frac{x^5}{5} - \frac{x^7}{7} + \ldots + (-1)^n \frac{x^{2n+1}}{2n+1} + R_{2n+1}(0, x)$$

where

$$|R_{2n+1}(0, x)| = \left| \int_0^x \frac{t^{2n+2}}{1+t^2} \, dt \right|.$$

(d)   In part (c), we found for Arctan x the Taylor polynomial $T_{2n+1}(x)$ and an expression for the remainder term. Use the expression for $R_{2n+1}(0, x)$ to find the choice of n that must be used to be sure that $T_{2n+1}(1)$ will approximate Arctan $1 = \pi/4$ accurate to within one unit in the third decimal place. (Hint: For $t \in [0, 1]$, $t^k/(1+t^2) < t^k$.)

35. *The outline for computing π given in Exercise 34 still involves a lot of work for any kind of accuracy in approximation. The reason is that 1 is so far from the point 0 around which we computed our Taylor polynomial.

(a)   Use the expression for $R_{2n+1}(0, x)$ obtained in part (c) of Exercise 34. Find the value of n necessary to assure accuracy to within one unit in the third decimal place if we restrict x to $[0, \frac{1}{2}]$.

(b)   Using n as chosen in part (a), compute the value of the Taylor polynomial $T_{A(0),n}$ at the points $x = \frac{1}{2}$ and $x = \frac{1}{3}$.

(c)   Find $\pi/4$ from part (b) by using the result of Problem 7.1C.

7·4   The Logarithm and Its Inverse

Logarithms are defined in elementary algebra to be exponents. The logarithm of a positive number N to a base b (any positive number not 1) is said to be that exponent r which raises b to the number N. In symbols

$$\log_b N = r \qquad \text{if and only if} \qquad b^r = N. \qquad (1)$$

After this definition is given, there usually follows a series of exercises in which two of the three numbers are given, and the student is asked to supply the third, as:

$$\log_2 32 = r, \qquad \text{Answer: } r = 5;$$

$$\log_b 27 = \frac{3}{2}, \qquad \text{Answer: } b = 9;$$

$$\log_3 N = -\frac{1}{2}, \qquad \text{Answer: } N = \frac{1}{\sqrt{3}}.$$

From the definition and the properties of exponents, we quickly derive the basic properties of logarithms.

$$\log_b(MN) = \log_b M + \log_b N. \tag{2}$$

$$\log_b \frac{M}{N} = \log_b M + \log_b N. \tag{3}$$

$$\log_b M^p = p \log_b M. \tag{4}$$

Now let us fix $b > 0$ and define a function by

$$y = \mathcal{L}(x) = \log_b x, \qquad x > 0.$$

Since this is a course in calculus, we ask ourselves how to find $D\mathcal{L}(x)$. The "natural" procedure follows. We say "natural" because it seems in line with the way we have always worked with logarithms. We shall see later that our work leaves a number of things to be desired.

$$\frac{\mathcal{L}(x+h) - \mathcal{L}(x)}{h} = \frac{1}{h}[\log_b(x+h) - \log_b x] = \frac{1}{h}\log_b \frac{x+h}{x},$$

$$= \log_b\left(1 + \frac{h}{x}\right)^{1/h}.$$

(Remember that in this discussion, x is some fixed positive number.) Now set $k = h/x$. Note that as $h \to 0$, $k \to 0$. Also note $1/h = 1/kx$.

$$\frac{\mathcal{L}(x+h) - \mathcal{L}(x)}{h} = \log_b[(1+k)^{1/kx}],$$

$$= \log_b[1+k)^{1/k}]^{1/x} = \frac{1}{x}\log_b(1+k)^{1/k}.$$

The last step followed from (4) above.

$$\lim_{h \to 0} \frac{\mathcal{L}(x+h) - \mathcal{L}(x)}{h} = \frac{1}{x}\lim_{k \to 0}\log_b(1+k)^{1/k} \tag{5}$$

Back in Example 2.4A we remarked that $\lim_{k \to 0}(1+k)^{1/k} = e \approx 2.7$. This leads us to suspect that

$$D\mathcal{L}(x) = \frac{1}{x}\log_b e.$$

Now b is arbitrary. Since we have a strong desire to obtain simple answers, it seems desirable at this point to choose $b = e$, because $\log_e e = 1$.

This is, in fact, what is done in calculus. We use a base of e; it is the "natural" thing to do under the circumstances. Logarithms to the base e

are in fact called *natural logarithms*. (Logarithms to the base 10, it may be recalled, are called *common logarithms*.) The definition of $\mathscr{L}$ is thus given by

$$\mathscr{L}(x) = \log_e x.$$

Simple as all this seems, there are several serious difficulties with our work. One criticism centers around the step (5) where we assumed that

$$\lim_{k \to 0} \log_b (1 + k)^{1/k} = \log_b \lim_{k \to 0} (1 + k)^{1/k}.$$

A check with the rules (Theorem 2.4A) shows that this is only legitimate, however, if we know that $\mathscr{L}(x) = \log_b x$ is continuous. We do not know that much about our function $\mathscr{L}$.

But there is a more serious difficulty. Consider the sample exercises at the beginning of this section. They are simple problems, and not difficult to solve. We could easily think of similar questions, however, that are never asked:

$$\log_{10} N = \pi. \tag{6}$$

Such a question is not asked for the obvious reason that students would have no idea of how to find the answer. Nevertheless, students are expected to believe (and all of them seem to do so) that such problems do have answers. In fact, they are taught how to read the answer to (6) from a table of logarithms. When so trained they tell you the answer is 1385.

Now ask yourself what is meant by 10^π. You know that 10^3 means $10 \cdot 10 \cdot 10$. You know that $10^{22/7}$ means the seventh root of 10^{22}. But what does 10^π mean? Chances are that you would have trouble writing down a definition that an eleventh-grade student could understand. Yet that eleventh-grade student will give 1385 as an approximate answer to the problem. And get a gold star too.

The fact is that 10^π is not easy to define in a meaningful way. The best procedure by far is to approach this problem from another direction. The key observation at this point is that if all the difficulties work out, then

$$D\mathscr{L}(x) = \frac{1}{x}.$$

Now we have already observed in our discussion following Theorem 6.4A that the function

$$L(x) = \int_1^x \frac{dt}{t}, \qquad x > 0, \tag{7}$$

has a derivative $DL(x) = 1/x$.

Since $\mathscr{L}$ and L thus have the same derivative, they differ at most by a constant. (The last statement again assumes $\mathscr{L}$ is continuous.) Since $L(1) = 0$ and $\mathscr{L}(1) = \log_e 1 = 0$, the constant difference must be zero. We strongly suspect, therefore, that (7) defines the logarithm function. We have already studied this function somewhat. The next series of problems is designed to verify all of our suspicions about L. Before attempting the problems, the reader may wish to review his solution to Problem 6.3E, where we first introduced the function L defined by

$$L(x) = \int_1^x \frac{dt}{t}. \tag{7}$$

We know

$$DL(x) = \frac{1}{x}.$$

The following problems confirm our suspicion that L will exhibit many properties of the logarithm function.

PROBLEMS

A · Prove that for any two positive numbers a and b, $L(ab) = L(a) - L(1/b)$. (Hint: After setting $x = ab$ in (7), use the result of Problem 6.4G. Finally, use Theorem 6.3C.)

B · For any a and b greater than zero, prove that

(a) $L(b) = -L\left(\dfrac{1}{b}\right)$;   (b) $L(ab) = L(a) + L(b)$;

(c) $L\left(\dfrac{a}{b}\right) = L(a) - L(b)$.

C · Prove that for any rational number p,

$$L(a^p) = pL(a).$$

(Hint: Use the chain rule to compute $DL(x^p)$; compare the answer to $D(pL(x))$.)

We have accomplished quite a bit. Let us review it. Starting with a function

$$\mathscr{L}(x) = \log_e x$$

we ignored a lot of details (continuity, the meaning of $\mathscr{L}(x) = \pi$) and decided that if $\mathscr{L}$ had a derivative, then it must be that $D\mathscr{L}(x) = 1/x$. This reminded us of another function

$$L(x) = \int_1^x \frac{dt}{t},$$

which had the same derivative, $DL(x) = 1/x$. We therefore wondered if L had any properties of the logarithm function. With little difficulty we discovered that L had properties (2), (3), and (4); that is

$$L(ab) = L(a) + L(b),$$

$$L\left(\frac{a}{b}\right) = L(a) - L(b),$$

$$L(a^p) = pL(a), \qquad \text{for any rational } p.$$

We also have more information about the function L. At each point x in the domain $(0, \infty)$ of L, we know that

$$DL(x) = \frac{1}{x}.$$

From this we are certain (Theorem 3.1A) that L is continuous on $(0, \infty)$, and that its derivative is always positive. This in turn means, according to Theorem 5.1A, that L is always increasing. Its graph, which we have already sketched (Problem 6.3E), must be as indicated in Fig. 7.19.

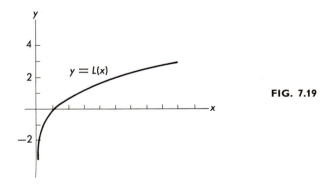

FIG. 7.19

It is thus clear that if x_1 and x_2 are two distinct points in the domain $(0, \infty)$ of L, then $L(x_1) \neq L(x_2)$. This means that L has an inverse defined for every member of its range $(-\infty, \infty)$. Call the inverse E. Then

$$L(x) = y \qquad \text{if and only if} \qquad E(y) = x. \tag{8}$$

That L has many properties of a logarithm is reflected by the fact that its inverse E has many properties of an exponent. Suppose

$$E(r) = a \quad \text{and} \quad E(s) = b$$

so

$$r = L(a) \quad \text{and} \quad s = L(b).$$

Then

$$r + s = L(a) + L(b) = L(ab),$$

so

$$E(r + s) = ab = E(r)E(s).$$

Similarly,

$$r - s = L(a) - L(b) = L\left(\frac{a}{b}\right),$$

so

$$E(r - s) = \frac{a}{b} = \frac{E(r)}{E(s)}.$$

Finally, for any rational number p,

$$pr = pL(a) = L(a^p),$$

so

$$E(pr) = a^p = [E(r)]^p.$$

To complete the identification of our function L with the function $\mathscr{L}$, we need to show that $L(e) = 1$. In this endeavor we are guided by the work we did in guessing at the derivative of $\mathscr{L}$. This time, however, we can fully justify all of our steps. Let us fix $x > 0$. By definition,

$$DL(x) = \lim_{h \to 0} \frac{L(x + h) - L(x)}{h}. \tag{9}$$

If we set $h = kx$, then since x is fixed, it is clear that k will go to zero as h goes to zero. We know that $DL(x) = 1/x$, so (9) becomes

$$\frac{1}{x} = \lim_{k \to 0} \frac{1}{kx} [L(x + kx) - L(x)].$$

Using the logarithm-like properties that have been proved for L, we have

$$\frac{1}{x} = \lim_{k \to 0} \frac{1}{x}\left(\frac{1}{k}\right)L\left(\frac{x + kx}{x}\right),$$

$$\frac{1}{x} = \frac{1}{x}\lim_{k \to 0} L(1 + k)^{1/k}.$$

We know L is continuous. Therefore we have

$$1 = L\left(\lim_{k \to 0}(1 + k)^{1/k}\right) = L(e).$$

From (8),

$$L(e) = 1 \qquad \text{if and only if} \qquad E(1) = e.$$

Let $a > 0$, and suppose p is a rational number. Then a^p is defined as usual, and since E is the inverse of L,

$$E(L(a^p)) = a^p$$

or, since $L(a^p) = pL(a)$, we have

$$E(pL(a)) = a^p.$$

This suggests a natural way to define a^x for irrational values of x, since E is defined for all real numbers.

DEFINITION A

For any real number $a > 0$, and real number $x \in (-\infty, \infty)$,

$$a^x = E(xL(a)).$$

This definition settles the question of what shall be meant by 10^π. It also gives us a neat expression for $E(x)$. If we set $a = e$ in the definition, then $L(e) = 1$. We have the important exponential function (Fig. 7.20)

$$e^x = E(x).$$

The definition itself may be rewritten as

$$a^x = e^{xL(a)}.$$

We may rewrite (8) in a form that more closely parallels (1);

$$L(x) = y \qquad \text{if and only if} \qquad e^y = x.$$

We found in the study of trigonometric functions that for our purposes, angles should always be measured in radians. We have now found out that insofar as we are concerned, the natural base to use for logarithms is e. The numerical value of e, which the reader may compute for himself, is 2.7183, correct to four decimal places. We shall emphasize our use of the natural base e by writing

$$L(x) = \log_e x = \ln x.$$

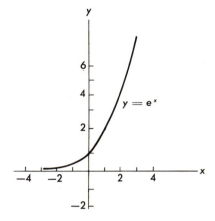

$y = e^x$

FIG. 7.20

We know, using the chain rule as usual, that if u is a function of x, then

$$D_x \ln u = \frac{D_x u}{u}. \qquad (10)$$

Moreover, by the inverse function theorem, if $y = e^x$,

$$D_x e^x = \frac{1}{D_y \ln y} = \frac{1}{1/y} = y = e^x.$$

The exponential function is not altered by the operation of differentiation. If u is a function of x, then of course,

$$D_x e^u = e^u D_x u. \qquad (11)$$

The reader is urged to learn the differentiation formulas (10) and (11). He can practice using them on the following problems.

PROBLEMS

D · Use (10) to show that

(a) $D \ln(x + \sqrt{x^2 - 1}) = \dfrac{1}{\sqrt{x^2 - 1}};$

(b) $D \ln(x + \sqrt{x^2 + 1}) = \dfrac{1}{\sqrt{x^2 + 1}}.$

E · Let

$$C(t) = \frac{e^t + e^{-t}}{2}, \qquad S(t) = \frac{e^t - e^{-t}}{2}.$$

Prove the following:

(a) $DC(t) = S(t)$;   (b) $DS(t) = C(t)$;   (c) $C(t)^2 - S(t)^2 = 1$.

F · Prove that $\lim_{x \to \infty} \ln x = \infty.$

   We could give formulas for the derivative of a logarithm to an arbitrary base and for Da^x where a is an arbitrary constant. We prefer to ask the reader to memorize only the two formulas above. Then by using the chain rule as indicated in the examples that follow, he can quickly figure out any related formula he needs.

EXAMPLE A
   Find (a) $D\, 10^x$;   (b) $D \log_{10} x$;   (c) $Dx^{\sin x}.$

(a)   By definition, $10^x = e^{x \ln 10}$.
Therefore, $D10^x = \ln 10 e^{x \ln 10}$.
(b)   If $y = \log_{10} x$, then $x = 10^y = e^{y \ln 10}$.
Therefore, differentiating implicitly with respect to x, we have

$$1 = \left(\frac{dy}{dx}\right) \ln 10 e^{y \ln 10};$$

$$\left(\frac{dy}{dx}\right) = \frac{1}{\ln 10 e^{y \ln 10}} = \left(\frac{1}{\ln 10}\right)\frac{1}{x}.$$

(c)   Set $y = x^{\sin x}$ so $\ln y = \sin x \ln x$. Differentiate implictly with respect to x;

$$\frac{(dy/dx)}{y} = (\sin x)\left(\frac{1}{x}\right) + (\ln x)\cos x;$$

$$\left(\frac{dy}{dx}\right) = x^{\sin x}\left[\frac{1}{x} \sin x + (\ln x)\cos x\right].$$

The method used here (of taking logarithms, then differentiating) is called *logarithmic differentiation.* ∎

From (10) and (11) we also learn how to find antiderivatives for several more functions.

$$\int \frac{D_x u \, du}{u} = \ln u + C \qquad \text{when } u > 0;$$

(12)

$$\int e^u D_x u \, du = e^u + C.$$

In Problem 5.3D we noticed a gap in our solution in that as far as we could see, $\int x^{-1}$ had no antiderivative. Equation (12) fills this gap in our solution to that problem.

EXAMPLE B
Find

$$\int \frac{\cos x \, dx}{\sin x - 2}.$$

The unwary may say, " Let $u = \sin x - 2$. Then $D_x u = \cos x$."

$$\int \frac{\cos x}{\sin x - 2} \, dx = \int \frac{D_x u}{u} \, du = \ln u = \ln(\sin x - 2) + C.$$

He will be happily on his way unless someone points out to him that $\sin x - 2 < 0$, and that the logarithm of a negative number (and hence his answer) is meaningless.

So we must try again. If we let $u = \sin x - 2$, then $u \not> 0$, so formula (12) will not help. But consider using a little algebra. Then

$$\int \frac{\cos x}{\sin x - 2} \, dx = \int \frac{-\cos x}{2 - \sin x} \, dx.$$

Now let $u = 2 - \sin x$; $u > 0$. $D_x u = -\cos x$. Everything is now on the up and up.

$$\int \frac{\cos x}{\sin x - 2} \, dx = \int \frac{D_x u}{u} \, du = \ln(2 - \sin x) + C.$$

This answer does make sense. ▮

Since the trick given above can always be used if $u < 0$, most people find it more economical simply to memorize (12) in the form

$$\int \frac{D_x u}{u} \, du = \ln |u| + C.$$

(13)

Differentiate the following logarithmic functions.

1· $D \ln(x^2 + x)$.   **2·** $D \ln(x + \sin x)$.

3· $D \ln(\sin x)$.   **4·** $D \ln(\cos x)$.

5· $D \ln(\sec x)$.   **6·** $D \ln(\tan x)$.

7· $D \ln(\sqrt{x^2 + 1} + x)^5$.   **8·** $D \ln(\sqrt{x} - x^3)^7$.

9· $Dx \ln x$.   **10·** $Dx^2 \ln(x^2)$.

Differentiate the following.

11· $De^{x^2 + x}$.   **12·** $De^{\sin x}$.

13· $D \operatorname{Arcsin} \sqrt{x}$.   **14·** $D \operatorname{Arctan} e^x$.

15· $De^{\cos^2 x}$.   **16·** $De^{x + \sqrt{x}}$.

17· $De^x \ln x$.   **18·** $D \sin x \ln(\sin x)$.

19· $De^{x^2} \ln x^2$.   **20·** $De^{x \ln x}$.

In Problem 5.3C we gave a list of functions for which the reader was challenged to find as many antiderivatives as he could. Here is a similar list. Part of the game is to see which functions are easy and which should be left alone for awhile.

21· $\int e^{x^2} \, dx$.   **22·** $\int \sin x e^{\cos x} \, dx$.

23· $\int \tan x \, dx$.   **24·** $\int x e^{x^3} \, dx$.

25· $\int x e^{x^2} \, dx$.   **26·** $\int \dfrac{x \, dx}{1 + x^2}$.

27· $\int \dfrac{e^x \, dx}{1 + e^{2x}}$.   **28·** $\int \dfrac{x^2 \, dx}{1 + x^2}$.

29· $\int \dfrac{e^{2x} \, dx}{1 + e^{2x}}$.   **30·** $\int 3 \sin 2x \sqrt{1 + \cos^2 x} \, dx$.

Graph the following; show critical points where they occur.

31· $y = \ln(x - 1)$.   **32·** $y = x e^{-x}$.

33· $y = x^2 e^{-x}$.   **34·** $y = \ln x^2$.

35· $y = x^2 - 2 \ln x$.   **36·** $y = \dfrac{1}{e^{1/x}}$.

37· $y = \dfrac{4}{1 + e^{(1/x) - 1}}$.   **38·** $y = x^2 - 2e^{-x^2}$.

Use logarithmic differentiation to find dy/dx in Exercises 39–44.

39· $y = x^{\sqrt{x}}$
40· $y = x^x$.

41· $y = (\sin x)^x$.
42· $y = (\cos x)^{\sqrt{x}}$

43· $y = (2x)^x$.
44· $y = (\ln x)^x$.

45· Let $E(x) = e^{-x}$. Write the Taylor polynomial $T_{E(0),5}$.

46· Let $L(x) = \ln x$. Write the Taylor polynomial $T_{L(1),5}$.

47· Let $F(x) = \ln(1 + x)/(1 - x)$. Write the Taylor polynomial $T_{F(0),6}$. Use $T_6(\tfrac{1}{3})$ to approximate ln 2. Show that your approximation is accurate to within one unit in the first decimal place.

48· Let $E(x) = e^x$. Write the Taylor polynomial $T_{E(0),5}$. Use $T_5(1)$ to approximate e. Show that your approximation is accurate to within one unit in the second decimal place.

7·5 The Hyperbolic Functions

Many handbooks contain tables that give the values of $E(x) = e^x$ for different values of x. So far as the author knows, however, no handbooks give a similar table of values of the function $F(x) = x^x$. The reason is that the first function comes up over and over in the applications of mathematics. (The reader will appreciate this statement more after he has finished Chapter 15.) On the other hand, the second function does not often appear. For the same reason, the first function above has a universally recognized name (the *exponential function*), whereas no one has bothered to give the second function a name.

We are about to study some functions that are probably new to the reader. They have names, and we study them for the same reasons that we name and study the function $E(x) = e^x$. These functions do not arise with the frequency of the exponential function, but they are important. (We shall encounter them in Chapter 15.) Moreover, the functions we are now going to study have some surprising properties which make them interesting in and of themselves.

The objects of our attention are the two functions already introduced to the reader in Problem 7.4E. They are

$$C(t) = \frac{e^t + e^{-t}}{2} \quad \text{and} \quad S(t) = \frac{e^t - e^{-t}}{2}.$$

In Problem 7.4E we learned some of their properties.

$$DC(t) = S(t);$$
$$DS(t) = C(t); \tag{1}$$
$$C(t)^2 - S(t)^2 = 1.$$

The functions C and S are called the *hyperbolic cosine* and the *hyperbolic sine*, respectively. The reason for these names is suggested by the following considerations.

Set

$$x = \text{Cosine } t;$$

$$y = \text{Sine } t.$$

Then $x^2 + y^2 = 1$, so the point (x, y) determined by an arbitrary t must be on the circle.

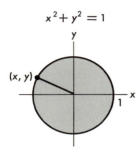

$$x^2 + y^2 = 1$$

FIG. 7.21

Set

$$x = C(t);$$

$$y = S(t).$$

Then $x^2 - y^2 = 1$, so the point (x, y) determined by an arbitrary t must be on the hyperbola.

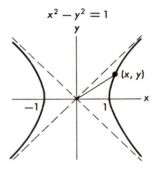

$$x^2 - y^2 = 1$$

FIG. 7.22

The functions $\cos t$ and $\sin t$ are for this reason called *circular functions*. They are designated by

$$\text{Cosine } t = \cos t;$$

$$\text{Sine } t = \sin t.$$

These functions are then used to define other circular functions.

$$\tan t = \frac{\sin t}{\cos t};$$

$$\sec t = \frac{1}{\cos t};$$

$$\csc t = \frac{1}{\sin t};$$

$$\cot t = \frac{\cos t}{\sin t}.$$

The functions $C(t)$ and $S(t)$ are for this reason called *hyperbolic functions*. They are designated by

$$C(t) = \cosh t;$$

$$S(t) = \sinh t.$$

These functions are then used to define other hyperbolic functions.

$$\tanh t = \frac{\sinh t}{\cosh t};$$

$$\text{sech } t = \frac{1}{\cosh t};$$

$$\text{csch } t = \frac{1}{\sinh t};$$

$$\coth t = \frac{\cosh t}{\sinh t}.$$

From these relations, many other relations may be developed. For example,

$$\sin 2t = 2 \sin t \cos t;$$

$$\cos 2t = \cos^2 t - \sin^2 t.$$

From these relations, many other relations may be developed. For example,

$$\sinh 2t = 2 \sinh t \cosh t;$$

$$\cosh 2t = \cosh^2 t + \sinh^2 t.$$

PROBLEMS

A · Prove the formulas given for sinh $2t$ and cosh $2t$.

B · Prove the following formulas for differentiation.

$$D \tanh t = \operatorname{sech}^2 t.$$

$$D \operatorname{sech} t = -\operatorname{sech} t \tanh t.$$

$$D \operatorname{csch} t = -\operatorname{csch} t \coth t.$$

$$D \coth t = -\operatorname{csch}^2 t.$$

We restrict ourselves in the remainder of this section to the two functions $S(x) = \sinh x$ and $C(x) = \cosh x$. Experience suggests that many students confuse the relations

$$\sinh x = \frac{e^x - e^{-x}}{2};$$

$$\cosh x = \frac{e^x + e^{-x}}{2}.$$

The difficulty comes in remembering in which function to subtract and in which one to add. The best aid over this block is to remember that

$$\sinh 0 = \sin 0 = 0,$$

$$\cosh 0 = \cos 0 = 1.$$

Plotting a few points will enable the reader to obtain graphs of the functions as indicated in Fig. 7.23 and 7.24.

It is clear from the graphs of these two functions that $S(x) = \sinh x$ has an inverse, and that $C(x) = \cosh x$ has an inverse if we restrict the domain of C to $x \geq 0$. Exploiting further the analogy with the circular (trigonometric) functions, we adopt similar terminology for the inverses.

If $y = \sinh x$, then $x = \operatorname{Argsinh} y$.

If $y = \cosh x$, then $x = \operatorname{Argcosh} y$, if $x \in [0, \infty)$.

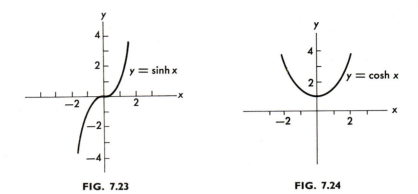

FIG. 7.23 **FIG. 7.24**

Suppose $A(x) = \text{Argcosh } x$. To find $DA(x)$, we proceed as we did in Section 7.3 with the inverse trigonometric functions.

$$y = A(x) = \text{Argcosh } x;$$
$$x = A^{-1}(y) = \cosh y.$$

We know

$$D_y A^{-1}(y) = D_y \cosh y = \sinh y.$$

Since $y \geq 0$, $\sinh y \geq 0$; so from (1),

$$\sinh y = \sqrt{\cosh^2 y - 1},$$
$$D_y A^{-1}(y) = \sqrt{x^2 - 1}.$$

From Theorem 3.4B, we also know that if $y = A(x)$, and if the inverse exists and is differentiable, then

$$D_x A(x) = \frac{1}{D_y A^{-1}(y)}.$$

Hence,

$$D \text{ Argcosh } x = \frac{1}{\sqrt{x^2 - 1}}. \tag{2}$$

PROBLEM

C· Prove

$$D \text{ Argsinh } x = \frac{1}{\sqrt{x^2 + 1}}.$$

Now compare (2) with the answer to Problem 7.4D(a). The logarithm function of Problem 7.4D(a) is continuous for $x \geq 1$, and the inverse hyperbolic sine function is continuous everywhere. We conclude (Problem 4.1B) that for some C

$$\text{Argcosh } x = \ln(x + \sqrt{x^2 - 1}) + C$$

is an identity when $x \geq 1$ (where both are continuous). Setting $x = 1$, we get

$$\text{Argcosh } 1 = \ln(1) + C.$$

This means $C = 0$. It must be that

$$\text{Argcosh } x = \ln(x + \sqrt{x^2 - 1}) \qquad \text{(all } x \geq 1\text{)}. \qquad (3)$$

We can, in fact, prove this directly. Set

$$y = \text{Argcosh } x,$$

$$x = \cosh y = \frac{e^y + e^{-y}}{2}.$$

Algebraic simplification gives

$$e^y - 2x + e^{-y} = 0.$$

Multiplication by e^y gives

$$(e^y)^2 - 2xe^y + 1 = 0,$$

which we recognize as a quadratic equation in e^y.

$$e^y = \frac{2x \pm \sqrt{4x^2 - 4}}{2} = x \pm \sqrt{x^2 - 1}.$$

We know that $x \geq 1$ and that $y \geq 0$. This means $e^y \geq 1$, so we must choose the plus sign, and we have

$$e^y = x + \sqrt{x^2 - 1}.$$

Taking logarithms of each side gives (3).

PROBLEM

D · Compare the result of Problem 7.4D(b) with the result of Problem C. From this, find an expression for Argsinh x in terms of a logarithmic function. Then verify your conclusion by a computation similar to the one above.

It is clear that we can add two more antidifferentiation formulas to our list:

$$\int \frac{dx}{\sqrt{x^2 - 1}} = \text{Argcosh } x + C, \qquad x \geq 1,$$

$$= \ln(x + \sqrt{x^2 - 1}) + C, \qquad x \geq 1;$$

$$\int \frac{dx}{\sqrt{x^2 + 1}} = \text{Argsinh } x + C,$$

$$= \ln(x + \sqrt{x^2 + 1}) + C.$$

These formulas may be memorized. On the other hand, we shall see in the next chapter that if forgotten, the formulas expressed in terms of logarithms may be easily derived by a standard procedure.

7·5 EXERCISES

In Exercises 1–8, prove the indicated identity, either by appealing directly to the definitions in terms of exponentials, or by using identities already proved.

1. $\sinh^2 \dfrac{t}{2} = \dfrac{1}{2}(\cosh t - 1)$.

2. $\tanh(s + t) = \dfrac{\tanh s + \tanh t}{1 + \tanh s \tanh t}$.

3. $\tanh t = \dfrac{\sinh 2t}{1 + \cosh 2t}$.

4. $(\cosh t + \sinh t)^n = \cosh nt + \sinh nt$.

5. $\sinh(s + t) = \sinh s \cosh t + \cosh s \sinh t$.

6. $2 \cosh^2 \dfrac{t}{2} = \cosh t + 1$.

7. $(\cosh t - \sinh t)^n = \cosh nt - \sinh nt$.

8. $\cosh(s + t) = \cosh s \cosh t + \sinh s \sinh t$.

In Exercises 9–16, find the derivative.

9. $y = \text{Arctan } \sinh t^2$.   **10.** $y = \text{Arctan } \cosh t^2$.

11. $y = \ln \tanh t^2$.   **12.** $y = \ln(\cosh t - \sinh t)$.

13. $y = \text{Argsinh } \tan t$.   **14.** $y = \text{Argcosh } \sec t$.

15. $y = \sinh^2 e^{t^2}$.   **16.** $y = \tanh^2 \ln t$.

7 · 6   A Tool for Finding Limits

In Section 2 of this chapter we expended great effort on two problems.

$$\lim_{h \to 0} \frac{\sin h}{h} = ? \tag{1}$$

$$\lim_{h \to 0} \frac{\cos h - 1}{h} = ? \tag{2}$$

The difficulty in each case came from the fact that both the numerator and the denominator approach 0 as h approaches 0. No general conclusion may be drawn in such cases. This is illustrated by the fact (as we found) that the first limit above is 1 while the second limit is 0.

Both problems may be viewed as illustrations of the following general question.

Let F and G be differentiable in a neighborhood of x_0. Suppose that

$$\lim_{x \to x_0} F(x) = 0,$$

$$\lim_{x \to x_0} G(x) = 0.$$

Can we find

$$\lim_{x \to x_0} \frac{F(x)}{G(x)} \, ?$$

Since F is differentiable at x_0, we know that

$$F(x_0 + h) = F(x_0) + F'(x_0)h + |h| R_1(x_0, h),$$

or, setting $h = x - x_0$,

$$F(x) = F(x_0) + F'(x_0)(x - x_0) + |x - x_0| R_1(x_0, x - x_0) \tag{3}$$

where $\lim_{x \to x_0} R_1(x_0, x - x_0) = 0$. Differentiability of F at x also implies that F is continuous at x_0 (Theorem 3.1A), so according to Theorem 2.4A

$$F(x_0) = \lim_{x \to x_0} F(x).$$

This limit, according to our assumption above, is 0, so $F(x_0) = 0$, and (3) says

$$F(x) = (x - x_0)\{F'(x_0) \pm R_1(x_0, x - x_0)\}.$$

Similarly,

$$G(x) = (x - x_0)\{G'(x_0) \pm R_2(x_0, x - x_0)\}.$$

Thus,

$$\lim_{x \to x_0} \frac{F(x)}{G(x)} = \lim_{x \to x_0} \frac{F'(x_0) \pm R_1(x_0, x - x_0)}{G'(x_0) \pm R_2(x_0, x - x_0)}.$$

If $[G'(x_0)] \neq 0$, then Theorem 2.4D enables us to write

$$\lim_{x \to x_0} \frac{F(x)}{G(x)} = \frac{\lim_{x \to x_0} \{F'(x_0) \pm R_1(x_0, x - x_0)\}}{\lim_{x \to x_0} \{G'(x_0) \pm R_2(x_0, x - x_0)\}},$$

$$= \frac{F'(x_0)}{G'(x_0)}.$$

EXAMPLE A
Find

$$\lim_{x \to 2} \frac{e^{x-2} - 1}{\sin \pi x}.$$

Both the numerator and denominator approach zero. The derivative of the denominator, $\pi \cos \pi x$, when evaluated at $x = 2$, is nonzero. Therefore we have

$$\lim_{x \to 2} \frac{e^{x-2} - 1}{\sin \pi x} = \left. \frac{e^{x-2}}{\pi \cos \pi x} \right|_{x=2} = \frac{1}{\pi}. \quad \blacksquare$$

In using this device, the reader must remember that he is not to take the derivative of a quotient. He is to use the derivative of the numerator over the derivative of the denominator; that is,

$$\lim_{x \to x_0} \frac{F(x)}{G(x)} = \frac{F'(x_0)}{G'(x_0)} \qquad \text{if } G'(x_0) \neq 0. \tag{4}$$

A similar but more useful rule can actually be proved, namely,

$$\lim_{x \to x_0} \frac{F(x)}{G(x)} = \lim_{x \to x_0} \frac{F'(x)}{G'(x)} \tag{5}$$

whenever the latter limit exists. The greater utility of the second form is illustrated in the next example.

EXAMPLE B
Find

$$\lim_{x \to 2} \frac{e^{x-2} - x + 1}{\sin^2 \pi x}.$$

Since both numerator and denominator approach 0 as x approaches 2, we may write, according to (5),

$$\lim_{x \to 2} \frac{e^{x-2} - x + 1}{\sin^2 \pi x} = \lim_{x \to 2} \frac{e^{x-2} - 1}{2\pi \sin \pi x \cos \pi x}.$$

The denominator of the expression on the right is 0 at $x = 2$, so we cannot use (4). We can use (5), however, providing

$$\lim_{x \to 2} \frac{e^{x-2} - 1}{2\pi \sin \pi x \cos \pi x}$$

exists. This again is a problem in which the numerator and denominator both approach 0. We therefore use (5) a second time, getting

$$\lim_{x \to 2} \frac{e^{x-2} - 1}{\sin 2\pi x} = \lim_{x \to 2} \frac{e^{x-2}}{2\pi^2 \cos 2\pi x} = \frac{1}{2\pi^2}. \quad \blacksquare$$

Problems of the form considered in (1) and (2), and in Examples A and B, are called *indeterminate forms* 0/0. The rule (5) is often (not always) useful in such cases. (Note, for example, that this rule would not have helped us evaluate (1) when we first encountered it in Section 7.2; why not?)

There is a second kind of problem, also referred to as an indeterminate form. It is typified by the problems

$$\lim_{x \to \infty} \frac{\ln(x + 1)}{x}; \tag{6}$$

$$\lim_{x \to \pi/2^-} \frac{\ln(\pi/2 - x)}{\tan x}. \tag{7}$$

Such expressions are called *indeterminate of the form* ∞/∞. They may be solved by essentially the same procedure as was used for the indeterminate form 0/0. Both procedures are justified by a theorem, which we now state.

THEOREM A (L'Hôpital's Rule)
Suppose either

(a) $\lim_{x \to a} F(x) = 0$ and $\lim_{x \to a} G(x) = 0$

or

(b) $\lim_{x \to a} |F(x)| = \infty$ and $\lim_{x \to a} |G(x)| = \infty.$

Then if

$$\lim_{x \to a} \frac{F'(x)}{G'(x)} = L,$$

it also follows that

$$\lim_{x \to a} \frac{F(x)}{G(x)} = L.$$

It is to be understood that a and L may denote any real number or one of the symbols ∞, $-\infty$.

The reader should be warned that this theorem is one of several variations that go by the name of *L'Hôpital's rule* (sometimes spelled L'Hospital's rule). Proofs are to be found in [2, page 92] and [7, page 121]. By using the theorem, the limits (6) and (7) can be computed. In the first case.

$$\lim_{x \to \infty} \frac{\ln(x + 1)}{x} = \lim_{x \to \infty} \frac{1/(x + 1)}{1} = 0.$$

Computation of the second limit involves more work.

$$\lim_{x \to \pi/2^-} \frac{\ln(\pi/2 - x)}{\tan x} = \lim_{x \to \pi/2^-} \frac{-1/(\pi/2 - x)}{\sec^2 x} = \lim_{x \to \pi/2^-} \frac{\cos^2 x}{x - \pi/2}.$$

This is now indeterminate of the form 0/0. We therefore appeal to the theorem a second time.

$$\lim_{x \to \pi/2^-} \frac{\cos^2 x}{x - (\pi/2)} = \lim_{x \to \pi/2^-} \frac{-2 \cos x \sin x}{1} = 0.$$

There are other situations in which preliminary algebraic manipulation may be combined with L'Hôpital's rule to find certain limits. These are illustrated in our final examples.

EXAMPLE C
Find $\lim_{x \to 0^+} x \ln x$.

This takes the form $0 \cdot (-\infty)$, but can be put into the form $-\infty/\infty$. We write

$$\lim_{x \to 0^+} \frac{\ln x}{x^{-1}} = \lim_{x \to 0^+} \frac{1/x}{-x^{-2}} = \lim_{x \to 0^+} (-x) = 0. \quad \blacksquare$$

EXAMPLE D
Find

$$\lim_{x \to 0^+} \left[\frac{1}{x^2} - \frac{1}{\sin x} \right].$$

This takes the form $\infty - \infty$, but may be put over a common denominator and then seen to take the form $0/0$.

$$\lim_{x \to 0^+} \frac{\sin x - x^2}{x^2 \sin x} = \lim_{x \to 0^+} \frac{\cos x - 2x}{x^2 \cos x + 2x \sin x} = \infty.$$

We observe that since we know $\lim_{x \to 0} (\sin x)/x = 1$, this same result could be obtained quite easily without appeal to L'Hôpital's rule:

$$\lim_{x \to 0^+} \frac{1}{x} \left[\frac{1}{x} - \frac{x}{\sin x} \right] = \infty,$$

since both factors clearly approach ∞. ∎

EXAMPLE E
Find $\lim_{x \to 0^+} x^{1/x}$.

Set $y = x^{1/x}$ so $\ln y = \dfrac{\ln x}{x}$. Then

$$\lim_{x \to 0^+} \ln y = \lim_{x \to 0^+} \frac{\ln x}{x} = -\infty.$$

Now if $\ln y$ approaches $-\infty$, y must approach 0. Thus,

$$\lim_{x \to 0^+} x^{1/x} = 0. ∎$$

7 · 6   E X E R C I S E S

1. $\lim\limits_{x \to \infty} \dfrac{3x^2 + x}{5x^2 - 1}$.

2. $\lim\limits_{x \to \infty} \dfrac{2x^2 + 4}{3x^2 - 5x + 2}$.

3. $\lim\limits_{x \to 1^+} \dfrac{x^4 - 3x^3 + 3x^2 - 3x + 2}{x^6 - 2x^5 + x^4 - 7x^3 + 14x^2 - 7x}$.

4. $\lim\limits_{x \to -1} \dfrac{x^4 - 3x^2 - 2x}{x^5 + 2x^4 + x^3 - x^2 - 2x - 1}$.

5. $\lim\limits_{x \to 2^-} \dfrac{e^{x-2} - \cos \pi x}{\ln(3 - x)}$.

6. $\lim\limits_{x \to 2^-} \dfrac{e^{1/(x-2)}}{\ln(2 - x)}$.

7. $\lim\limits_{x\to 0} \dfrac{\tan x - x}{x^2}$.

8. $\lim\limits_{x\to 0} \dfrac{\cos^2 x - 1}{\sin x}$.

9. $\lim\limits_{x\to 1-} \dfrac{\ln \sin \pi x}{\tan(\pi/2)x}$.

10. $\lim\limits_{x\to 0+} \dfrac{\ln \sin x}{\ln \tan x}$.

11. $\lim\limits_{x\to 0-} \dfrac{\sin x - 2\cos 2x}{x^3 \cos x}$.

12. $\lim\limits_{x\to \infty} \dfrac{\ln x}{\sqrt[3]{1+x^2}}$.

13. $\lim\limits_{x\to 2} \dfrac{2-x}{\ln 2 - \ln x}$.

14. $\lim\limits_{x\to \pi/4} \dfrac{\sec x - \sqrt{2}}{\tan x - 1}$.

15. $\lim\limits_{x\to \pi/3} \dfrac{2\cos x - 1}{2\sin x - \sqrt{3}}$.

16. $\lim\limits_{x\to \infty} \dfrac{x^3}{2^x}$.

17. $\lim\limits_{x\to 0+} (\tan x)^x$.

18. $\lim\limits_{x\to 0+} (\sin x)^x$.

19. $\lim\limits_{x\to 0} \dfrac{e^x - 2^x}{x}$.

20. $\lim\limits_{x\to 0} \dfrac{\sin^2 x - x^2 \cos^2 x}{x^2 \sin^2 x}$.

21. $\lim\limits_{x\to 0} \dfrac{x \csc x - 1}{x^2}$.

22. $\lim\limits_{x\to 0} \dfrac{\text{Arcsin}x}{\text{Arctan } x}$.

23. $\lim\limits_{x\to 0+} x^x$.

24. $\lim\limits_{x\to \infty} x^{1/x}$.

25. $\lim\limits_{x\to 0+} \left(\tan x - \dfrac{1}{x}\right)$.

26. $\lim\limits_{x\to \pi/2-} (\sec^3 x - \tan^3 x)$.

27. $\lim\limits_{x\to 0} x^2(\csc^2 x - \csc x \cot x)$.

28. $\lim\limits_{x\to 0} \sin x \dfrac{\cos x - 1}{x^3}$.

29. $\lim\limits_{x\to \infty} \left(\dfrac{2x}{2x+1}\right)^x$.

30. $\lim\limits_{x\to \infty} \left(\dfrac{x}{x+1}\right)^{x^2}$.

Graph the curves indicated in Exercises 31–40. Be sure to indicate the graph for $x \to \infty$ and $x \to -\infty$, if the function is defined for these values.

31. $y = \ln\dfrac{x+1}{x-1}$.

32. $y = \ln\dfrac{x^2-1}{x^2+1}$.

33. $y = x^2 e^{-x}$.

34. $y = \dfrac{e^x}{x^2}$.

35. $y = \dfrac{1 + \sin x - \cos x - x}{x^2}$.

36. $y = x^2 \sin \dfrac{1}{x}$.

37. $y = \sqrt{x} \ln \dfrac{1}{x^2}$.

38. $y = \dfrac{1}{\sqrt{x}} \ln x^2$.

39. $y = \dfrac{e^{\sin x} - 1}{x}$.

40. $y = \dfrac{\ln \sin x}{x}$.

FINDING ANTIDERIVATIVES

In Section 6.2 we defined the integral of a function F. We referred to work done in Section 6.1 to show that

$$\int_0^2 (x^2 + 2)\, dx = \frac{20}{3}.$$

This was accomplished without any reference to antiderivatives. We later saw, after the fundamental theorem of calculus was proved, that the work involved in finding the value of this integral could have been greatly reduced.

In Example 7.2F we found that

$$\int_0^{\pi/3} \frac{\sin x}{x}\, dx \approx \frac{\pi}{3} - \frac{\pi^3}{(18)(27)}.$$

Here we knew about the fundamental theorem, but we were not able to use it because we did not know of an antiderivative for $(\sin x)/x$.

These two illustrations are cited to emphasize that we do not need to know an antiderivative of F to approximate the integral of F on some interval. Indeed, for some integrable functions, no antiderivative, expressed in terms of familiar functions, can be found. $F(x) = e^{-x^2}$ is such a function. Yet, since it is continuous on $[0, 1]$ (and everywhere else, too), $\int_0^1 e^{-x^2}\, dx$ exists.

Having emphasized that integration does not depend on being able to find an antiderivative, it now seems reasonable to admit that the integration of F is usually made easier if we can find $D^{-1}F$. Even when $D^{-1}F$ exists, however, it is not always easy to find. The reader had his first hint at this fact when he tried in Problem 5.3C(f), to find $D^{-1}(x^2 + 1)^{-3/2}$. The answer was simple enough, but he probably did not guess it. Many such examples could be given, for example,

$$D^{-1} \frac{x + 10}{x^3 + 2x^2 + 4x + 8} = \frac{1}{2} \ln \frac{(x + 2)^2}{x^2 + 4} + \frac{3}{2} \text{Arctan} \frac{x}{2} + C.$$

Try it. It works.

The point of this chapter is to give some methods for finding antiderivatives when they exist. The foregoing remarks warn the reader that he will never be completely effective. But he will improve.

Since the test in finding an antiderivative is to merely differentiate it, we will not greatly concern ourselves with a lot of theory in this chapter. The many examples should be read carefully, for it is here that methods will be observed.

8·1 A Summary

We begin by introducing a little notation. The chain rule tells us that

$$D \sin G(x) = [\cos G(x)]DG(x)$$

so

$$\int [\cos G(x)]DG(x) \, dx = \sin G(x) + C. \tag{1}$$

Now the implications of the chain rule were found to be easier to remember when changing variables in an integral (Theorem 6.4C) if we used the substitution

$$u = G(x), \qquad du = DG(x) \, dx.$$

If we follow this lead, then we are able to write (1) in the form

$$\int \cos u \, du = \sin u + C. \tag{2}$$

We shall now make a list of the formulas we know for certain antiderivatives. We will not list all that we could, but only those we feel the reader should memorize (if he has not done so already). We exploit the notation of the previous paragraph in this list, understanding that u may be a function of x.

$$\int u^r \, du = \begin{cases} \dfrac{1}{r+1} u^{r+1} + C & \text{for } r \neq -1, \\ \ln|u| + C & \text{for } r = -1. \end{cases} \tag{3}$$

$$\int \sin u \, du = -\cos u + C.$$

$$\int \cos u \, du = \sin u + C.$$

$$\int \sec^2 u \, du = \tan u + C.$$

$$\int \csc^2 u \, du = -\cot u + C.$$

$$\int \sec u \tan u \, du = \sec u + C.$$

$$\int \csc u \cot u \, du = -\csc u + C.$$

$$\int \frac{du}{a^2 + u^2} = \frac{1}{a} \operatorname{Arctan} \frac{u}{a} + C.$$

$$\int \frac{du}{\sqrt{a^2 - u^2}} = \text{Arcsin } \frac{u}{a} + C.$$

$$\int e^u \, du = e^u + C.$$

To this list we wish to add two other pieces of information, but they are probably better remembered as techniques than as formulas. We exhibit them as two short but very important examples.

EXAMPLE A
Find $\int \cos^2 3x \, dx$.

We appeal to our trigonometric identities ((7) of Section 7.1) to write

$$\cos^2 3x = \frac{1 + \cos 6x}{2}.$$

Now if we set $u = 6x$,

$$\int \cos^2 3x \, dx = \int (\tfrac{1}{2} + \tfrac{1}{2} \cos 6x) \, dx,$$

$$= \int \tfrac{1}{2} \, dx + \int \tfrac{1}{2} \cos u \, \frac{du}{6},$$

$$= \frac{x}{2} + \frac{1}{12} \sin u + C,$$

$$= \frac{x}{2} + \frac{1}{12} \sin 6x + C. \quad \blacksquare$$

Notice that the preceding example was not finished until we had replaced u by the function of x that it represents. If the function for which we are to find an antiderivative is expressed in terms of x, then our answer should be also.

EXAMPLE B
Find $\int \sec u \, du$.

We solve this one by pulling a trick out of the hat. (The reader may comfort himself with the knowledge that no such inspiration will be expected from him without some kind of hint.)

$$\int \sec u \, du = \int \sec u \, \frac{\sec u + \tan u}{\sec u + \tan u} \, du,$$

$$= \int \frac{\sec^2 u + \sec u \tan u}{\tan u + \sec u} \, du.$$

Now set $v = \tan u + \sec u$; $dv = (\sec^2 u + \sec u \tan u)\, du$.

$$\int \sec u\, du = \int \frac{dv}{v} = \ln|v| + C,$$

$$= \ln|\tan u + \sec u| + C. \ \blacksquare$$

PROBLEMS

A · Find $\int x \sin^2 x^2\, dx$.

B · Find $\int \csc u\, du$. (Hint: Devise a trick similar to the one employed in Example B.)

Most difficulties with problems in antidifferentiation arise because the student fails to get the *du* to match up with his selection of *u* in the integral. In all of the following exercises, write down a selection of *u* and *du*. These exercises are so devised that the reader should be able to express every one of them as a constant multiple of one of the forms indicated in the list above (or in the two examples following the list). The next example illustrates the procedure to follow.

EXAMPLE C
Find $\int \{x \sec^3 (x^2 + 1) \tan(x^2 + 1)\}\, dx$.

After some experimentation with various possibilities that come to mind, we hit upon the substitution

$$u = \sec(x^2 + 1).$$

Then

$$du = 2x \sec(x^2 + 1) \tan(x^2 + 1)\, dx.$$

The problem is now seen to be in the form

$$\tfrac{1}{2}\int u^2\, du = \tfrac{1}{6}u^3 + C.$$

The desired antiderivative is

$$\tfrac{1}{6} \sec^3(x^2 + 1) + C. \ \blacksquare$$

Find the indicated antiderivatives.

1 · $\int \tan x \, dx.$

2 · $\int \cot x \, dx.$

3 · $\int x^2(1 - x^3)^{1/5} \, dx.$

4 · $\int x^{1/2}(4 + x^{3/2})^{1/4} \, dx.$

5 · $\int \cos(3x + 1) \, dx.$

6 · $\int x \cos x^2 \, dx.$

7 · $\int \dfrac{x \, dx}{\sin^2 x^2}.$

8 · $\int \dfrac{3 \, dx}{\cos^2(2x - 1)}.$

9 · $\int \sec^2 x \tan x \, dx.$

10 · $\int \csc^3 x \cot x \, dx.$

11 · $\int x^2 e^{x^3} \, dx.$

12 · $\int e^{\sin x} \cos x \, dx.$

13 · $\int \dfrac{x \, dx}{x^2 + 1}.$

14 · $\int \dfrac{dx}{x^2 + 1}.$

15 · $\int \dfrac{x \, dx}{x^4 + 1}.$

16 · $\int \dfrac{x^3 \, dx}{x^4 + 1}.$

✗ 17 · $\int \dfrac{\sin x \, dx}{\sqrt{1 - \cos^2 x}}.$

18 · $\int \dfrac{e^x \, dx}{\sqrt{1 - e^{2x}}}.$

19 · $\int x e^{x^2} \sin e^{x^2} \, dx.$

20 · $\int \dfrac{dx}{x \ln x}.$

21 · $\int \cos x \sin^3 x \, dx.$

22 · $\int \sin 2x \cos^2 2x \, dx.$

23 · $\int \sin 2x \cos x \, dx.$

24 · $\int \cos 2x \sin x \, dx.$

25 · $\int \sin^3 x \, dx.$

26 · $\int \sec^2 x \tan^3 x \, dx.$

27 · $\int \tan^2 x \, dx.$

28 · $\int \cos^4 x \, dx.$

29 · $\int \sin^4 x \, dx.$

30 · $\int \cos^3 x \, dx.$

31 · $\int \csc^2 2x \cot^2 2x \, dx.$

32 · $\int \dfrac{\sec^2 x \, dx}{4 + \tan x}.$

33 · $\int \dfrac{x \, dx}{(x^2 + 1)^{1/3}}.$

34 · $\int \dfrac{x \, dx}{(2x^2 + 3)^4}.$

35 · $\int \dfrac{\sin^3 2x}{\cos^2 2x} \, dx.$

36 · $\int \dfrac{\sec^2 x}{1 + \tan^2 x} \, dx.$

37 · $\int \dfrac{(x + 2) \, dx}{x^2 + 4x + 5}.$

38 · $\int \dfrac{(2x + 10) \, dx}{x^2 + 10x + 13}.$

39 · $\int (x \cos x + \sin x) e^{x \sin x} \, dx.$ **40** · $\int \dfrac{e^{\tan x}}{\cos^2 x} \, dx.$

8 · 2 Trigonometric Substitutions

Whenever $\sqrt{a^2 - u^2}$, $\sqrt{u^2 - a^2}$, or $\sqrt{a^2 + u^2}$ enter into an expression for which we seek an antiderivative, we should consider using a trigonometric substitution. Each of the three radicals can be thought of as the length of a side of a triangle if the triangle is correctly labeled. This triangle then suggests the proper substitution.

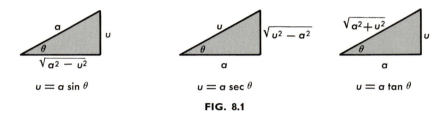

$u = a \sin \theta$ $u = a \sec \theta$ $u = a \tan \theta$

FIG. 8.1

EXAMPLE A

Find $\int (x^2 + 1)^{-3/2} \, dx.$ (This is the problem that caused trouble as Problem 5.3C(f).)

We draw an appropriate triangle (Fig. 8.2).

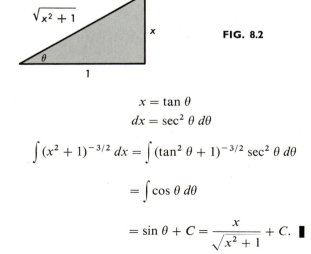

FIG. 8.2

$$x = \tan \theta$$

$$dx = \sec^2 \theta \, d\theta$$

$$\int (x^2 + 1)^{-3/2} \, dx = \int (\tan^2 \theta + 1)^{-3/2} \sec^2 \theta \, d\theta$$

$$= \int \cos \theta \, d\theta$$

$$= \sin \theta + C = \frac{x}{\sqrt{x^2 + 1}} + C. \quad \blacksquare$$

Note that if a substitution is made, the problem is completed by expressing the answer in terms of the original variable. Of course, if we are seeking the antiderivative to evaluate an integral, the substitution can be made directly in the integral, using Theorem 6.4C. Then the limits on the integral will be changed in view of the substitution, and it will not be necessary to change back to the original variable.

EXAMPLE B
Evaluate

$$\int_{\sqrt{2}}^{2} \frac{dx}{x(x^2 - 1)^{1/2}}.$$

The appropriate triangle is sketched in Fig. 8.3.

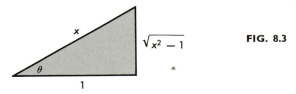

FIG. 8.3

$$x = \sec \theta$$
$$dx = \sec \theta \tan \theta \, d\theta.$$

When $x = \sqrt{2}$, $\theta = \pi/4$; when $x = 2$, $\theta = \pi/3$.

$$\int_{\sqrt{2}}^{2} \frac{dx}{x(x^2 - 1)^{1/2}} = \int_{\pi/4}^{\pi/3} \frac{\sec \theta \tan \theta \, d\theta}{\sec \theta \tan \theta} = \frac{\pi}{3} - \frac{\pi}{4} = \frac{\pi}{12}. \blacksquare$$

Those who have memorized the derivative of the Arcsec function would have done the preceding example more directly, being able to write

$$\int_{\sqrt{2}}^{2} \frac{dx}{x(x^2 - 1)^{1/2}} = \text{Arcsec } x \Big|_{\sqrt{2}}^{2} = \frac{\pi}{12}.$$

There is certainly nothing wrong with this; it is simply a matter of memorization. We suggested memorizing the derivatives (hence, the associated antiderivatives) for only two of the inverse trigonometric functions. This, too, is a matter of preference. Those who do not know even these formulas by memory will not be greatly handicapped if the methods of this section are kept in mind. We work the following example under the assumption that we have forgotten that

$$\int \frac{dx}{4 + x^2} = \frac{1}{2} \text{Arctan } \frac{x}{2} + C.$$

EXAMPLE C
Find

$$\int \frac{dx}{4 + x^2}.$$

We sketch the triangle in Fig. 8.4.

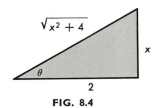

$$x = 2 \tan \theta$$

$$dx = 2 \sec^2 \theta \, d\theta$$

$$\int \frac{dx}{4 + x^2} = \int \frac{2 \sec^2 \theta \, d\theta}{4 + 4 \tan^2 \theta} = \int \frac{d\theta}{2} = \frac{\theta}{2} + C.$$

FIG. 8.4

Since $x = 2 \tan \theta$, it follows that $\theta = \text{Arctan } x/2$.

$$\int \frac{dx}{4 + x^2} = \frac{1}{2} \text{Arctan } \frac{x}{2} + C. \ \blacksquare$$

This last example illustrates the fact that the substitutions of this section can be useful on expressions containing just $a^2 - u^2$, $u^2 - a^2$, or $a^2 + u^2$ without a radical sign.

PROBLEM

A · Find

$$\int \frac{1}{(x^2 + 1)^2} \, dx.$$

The student should also remember that even before he encountered this section, he could handle $\int x\sqrt{x^2 + 1} \, dx$. The moral of this is that we do not need a trigonometric substitution for every expression involving one of the square roots considered in this section.

===   8·2 EXERCISES

Find the antiderivatives indicated in Exercises 1–8.

1 · $\displaystyle\int \frac{x \, dx}{\sqrt{a^2 - x^2}}.$

2 · $\displaystyle\int \frac{x \, dx}{\sqrt{x^2 + a^2}}.$

3 · $\displaystyle\int \frac{x^2 \, dx}{\sqrt{a^2 - x^2}}.$

4 · $\displaystyle\int \frac{x^3 \, dx}{\sqrt{a^2 - x^2}}.$

5. $\int \dfrac{x\,dx}{4 + x^2}.$

6. $\int \dfrac{\sqrt{a^2 - x^2}}{x}\,dx.$

7. $\int \dfrac{\sqrt{x^2 - a^2}}{x}\,dx.$

8. $\int \dfrac{x\,dx}{\sqrt{x^2 - a^2}}.$

Evaluate the integrals below.

9. $\displaystyle\int_3^4 \dfrac{1}{\sqrt{x^2 - 4}}\,dx.$

10. $\displaystyle\int_0^3 \dfrac{1}{\sqrt{9 + x^2}}\,dx.$

11. $\displaystyle\int_0^2 \sqrt{4 - x^2}\,dx.$

12. $\displaystyle\int_0^1 \sqrt{4 - 4x^2}\,dx.$

13. $\displaystyle\int_{1/2}^1 \dfrac{\sqrt{1 - x^2}}{x^2}\,dx.$

14. $\displaystyle\int_{\sqrt{2}}^2 \dfrac{x^2}{x^2 - 1}\,dx.$

8 · 3   Integration by Parts

The formula for the differentiation of a product of two functions says

$$D[F(x)G(x)] = F(x)DG(x) + G(x)DF(x).$$

This means

$$D^{-1}\{F(x)DG(x)\} = F(x)G(x) - D^{-1}\{G(x)DF(x)\}.$$

If we replace D^{-1} by $\int$, we have

$$\int \{F(x)DG(x)\,dx\} = F(x)G(x) - \int \{G(x)DF(x)\,dx\}$$

and this will be easier to remember if we set

$$u = F(x), \qquad\qquad v = G(x),$$
$$du = DF(x)\,dx, \qquad dv = DG(x)\,dx,$$
$$\int u\,dv = uv - \int v\,du. \tag{1}$$

Formula (1) is more useful than it might at first glance appear to be. Its usefulness derives from the fact that it is often possible, after a little experience, to choose u and v so that although $\int u\,dv$ is hard to find, $\int v\,du$ may be easy to find.

EXAMPLE A
Find $\int x^3 e^{x^2}\,dx.$

Our job is to choose u and dv so that

$$u\,dv = x^3 e^{x^2}\,dx.$$

In addition, we want to be able to easily find v (so we would not consider setting $dv = e^{x^2} dx$, for instance) and as a final goal, we hope $\int v\, du$ is easy to find. Guided by these considerations (and a bit of experience) we try setting

$$u = x^2, \qquad dv = xe^{x^2}\, dx,$$
$$du = 2x\, dx, \qquad v = \tfrac{1}{2}e^{x^2} + C_1. \tag{2}$$

Then by the formula

$$\int x^3 e^{x^2}\, dx = (\tfrac{1}{2}e^{x^2} + C_1)x^2 - \int (xe^{x^2} + 2C_1 x)\, dx,$$
$$= \tfrac{1}{2}x^2 e^{x^2} + C_1 x^2 - \tfrac{1}{2}e^{x^2} - C_1 x^2 + C_2,$$
$$= \tfrac{1}{2}x^2 e^{x^2} - \tfrac{1}{2}e^{x^2} + C_2 . \ \blacksquare$$

The term involving the $C_1 x^2$ dropped out through subtraction. This will always happen, so it is customary in step (2) to simply choose $C_1 = 0$.

EXAMPLE B
Find $\int x \sin x\, dx$.

The beginner might try

$$u = \sin x, \qquad dv = x\, dx,$$
$$du = \cos x\, dx, \qquad v = \frac{x^2}{2}.$$

The foregoing would then give

$$\int x \sin x\, dx = \frac{x^2}{2} \sin x - \int \frac{x^2}{2} \cos x\, dx.$$

Now this is correct. But it has increased our trouble, since $\int (x^2/2)\cos x\, dx$ looks more forbidding than the original problem. Hopefully, at this point, our student will have achieved some of the experience we spoke of before, and will try again, this time setting

$$u = x, \qquad dv = \sin x\, dx,$$
$$du = dx, \qquad v = -\cos x,$$
$$\int x \sin x\, dx = -x \cos x - \int (-\cos x)\, dx,$$
$$= -x \cos x + \sin x + C. \ \blacksquare$$

Something can be learned by differentiating the answers we are getting to verify that they work.

EXAMPLE C
Find $\int x^2 e^x\, dx$.

Let

$$u = x^2, \qquad dv = e^x\, dx,$$
$$du = 2x\, dx, \qquad v = e^x,$$

so

$$\int x^2 e^x\, dx = x^2 e^x - 2 \int x e^x\, dx.$$

Now concentrate on $\int x e^x\, dx$, setting

$$U = x, \qquad dV = e^x\, dx,$$
$$dU = dx, \qquad V = e^x.$$

so

$$\int x^2 e^x\, dx = x^2 e^x - 2 \left\{ x e^x - \int e^x\, dx \right\},$$
$$= x^2 e^x - 2x e^x + 2e^x + C. \ \blacksquare$$

EXAMPLE D
Find $\int e^x \sin x\, dx$.

Set

$$u = e^x, \qquad dv = \sin x\, dx,$$
$$du = e^x\, dx, \qquad v = -\cos x,$$
$$\int e^x \sin x\, dx = -e^x \cos x + \int e^x \cos x\, dx.$$

Now in the last term we try again setting

$$U = e^x, \qquad dV = \cos x\, dx,$$
$$dU = e^x\, dx, \qquad V = \sin x,$$
$$\int e^x \sin x\, dx = -e^x \cos x + \left[e^x \sin x - \int e^x \sin x\, dx \right].$$

This simplifies to

$$2 \int e^x \sin x\, dx = e^x(\sin x - \cos x).$$

Therefore,

$$\int e^x \sin x \, dx = \tfrac{1}{2}e^x(\sin x - \cos x) + C. \quad\blacksquare$$

Using the Fundamental Theorem, we can derive a definite integral form of integration by parts. We have

$$\int_a^b u(x) \, dv(x) = u(x)v(x)\Big|_{x=a}^{x=b} - \int_a^b v(x) \, du(x).$$

In this form the result seems geometrically evident from Fig. 8.5.

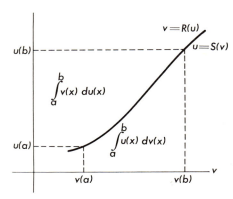

FIG. 8.5

EXAMPLE E
Find

$$\int_0^{2\pi} x \sin x \, dx.$$

Proceeding as in Example B, we have

$$\int_0^{2\pi} x \sin x \, dx = -x \cos x \Big|_{x=0}^{x=2\pi} + \int_0^{2\pi} \cos x \, dx.$$

Since the symmetry of $\cos x$ on $[0, 2\pi]$ makes it evident that $\int_0^{2\pi}\cos x \, dx = 0$, we have

$$\int_0^{2\pi} x \sin x \, dx = (-x \cos x)\Big|_0^{2\pi} = -2\pi - (-0) = -2\pi. \quad\blacksquare$$

PROBLEM

A · Criticize the following proof that $0 = 1$. In the given integral, set $u = (1/t)$ and $dv = dt$. Then since $uv = 1$,

$$\int_1^2 \frac{dt}{t} = 1 + \int_1^2 \frac{dt}{t}.$$

Subtraction of the integral from each side gives the (un)desired conclusion.

8·3   EXERCISES

Find the indicated antiderivatives.

1 · $\displaystyle\int \ln x \, dx.$ **2 ·** $\displaystyle\int x \ln x \, dx.$

3 · $\displaystyle\int x^3(1 + x^2)^{1/3} \, dx.$ **4 ·** $\displaystyle\int x^3 \sin x^2 \, dx.$

5 · $\displaystyle\int \sec^2 x \, dx.$ **6 ·** $\displaystyle\int x \sec x \tan x \, dx.$

7 · $\displaystyle\int \text{Arcsin } x \, dx.$ **8 ·** $\displaystyle\int \text{Arctan } x \, dx.$

9 · $\displaystyle\int \sec^3 x \, dx.$ **10 ·** $\displaystyle\int \sec^2 x \sin x \, dx.$

11 · $\displaystyle\int e^x \cos 2x \, dx.$ **12 ·** $\displaystyle\int e^{3x} \sin x \, dx.$

Evaluate the integrals that follow.

13 · $\displaystyle\int_0^1 x^2 \sin x \, dx.$ **14 ·** $\displaystyle\int_0^1 x^2 e^{2x} \, dx.$

15 · $\displaystyle\int_1^2 x^2 \ln x \, dx.$ **16 ·** $\displaystyle\int_1^2 \ln x \, dx.$

8·4   Rational Functions

In the introduction to this chapter it was pointed out that

$$D^{-1} \frac{x+10}{x^3 + 2x^2 + 4x + 8} = \frac{1}{2} \ln \frac{(x+2)^2}{x^2+4} + \frac{3}{2} \text{Arctan } \frac{x}{2} + C. \tag{1}$$

It was also pointed out that the reader would probably not get this answer by guessing; at least not right away. But suppose he had been supplied with the algebraic identity

$$\frac{x+10}{x^3 + 2x^2 + 4x + 8} = \frac{1}{x+2} - \frac{x}{x^2+4} + \frac{3}{x^2+4}. \tag{2}$$

Each of the three rational functions on the right has an antiderivative that is apparent. The reader who writes them down and combines them in obvious ways will readily arrive at the solution given in (1).

It thus appears that after years of practice in making the right-hand side of (2) look like the left side, we now wish to reverse the process. The development of this algebraic skill is our main business in this section. We begin with a simple observation.

> (a) Whenever the numerator of a rational function is not of degree less than that of the denominator, we may divide in the usual way.

By way of illustration, given $(x^3 + 3x - 1)/(x^2 + 2x)$, we divide.

$$
\begin{array}{r}
x - 2 \\
x^2 + 2x \overline{\smash{\big)}\, x^3 + 0x^2 + 3x - 1} \\
\underline{x^3 + 2x^2} \\
-2x^2 + 3x \\
\underline{-2x^2 - 4x} \\
7x - 1
\end{array}
$$

Then

$$
\frac{x^3 + 3x - 1}{x^2 + 2x} = x - 2 + \frac{7x - 1}{x^2 + 2x}.
$$

A second observation has to do with the factoring of the polynomial in the denominator. We call a polynomial irreducible if it cannot be expressed as a product of other polynomials of lower degree having real coefficients. The polynomial $x^2 + 4$ is irreducible. No polynomial of degree greater than two is irreducible. (A cubic, for example, has at least one real root r because complex roots occur in conjugate pairs. Then $(x - r)$ will be a factor.)

> (b) The polynomial in the denominator may always be expressed as a product of first-degree and possibly second-degree irreducible polynomials.

Because of these observations, we may always assume that the rational function with which we work (a) has a numerator of degree lower than the denominator, and (b) has its denominator factored into a product of irreducible polynomials. Therefore, given an arbitrary rational function for which we are to find an antiderivative, we always begin with two routine steps.

> (1) Perform any long division necessary to reduce the problem to finding an antiderivative for a rational function in which the degree of the denominator exceeds the degree of the numerator.

(2)   Factor the denominator into a product of irreducible polynomials of degree one or two.

The next four examples show the proper procedure to use for four different forms of the factored denominator. When we put together what we learn in these four examples, we will be able to decompose any rational function into its *partial fractions*.

EXAMPLE A
(Denominator Factors into Distinct First-Degree Factors)
Decompose

$$\frac{2x}{(x-3)(x+2)}.$$

In this case we assume a decomposition of the form

$$\frac{2x}{(x-3)(x+2)} = \frac{A}{x-3} + \frac{B}{x+2}. \tag{3}$$

The game is now to choose A and B so that (3) will be an algebraic identity; that is, so that (3) will be true for all values of x. We first note that if (3) is true, then

$$2x = A(x+2) + B(x-3). \tag{4}$$

We may now proceed in either of two ways.

Method 1   Expand the right side of (4), collecting like powers of x. This gives

$$2x = (A+B)x + (2A - 3B).$$

Since this is an identity, the coefficients of like powers of x on each side of the equal sign must be the same. This means

$$2 = A + B,$$

$$0 = 2A - 3B.$$

Solving, we have $A = 6/5$, $B = 4/5$. Substituting these values in (3) gives the desired decomposition.

$$\frac{2x}{(x-3)(x+2)} = \frac{6}{5(x-3)} + \frac{4}{5(x+2)}.$$

Method 2   We again consider (4) and the problem of choosing A and B to make this an identity. Since it is to hold for all x, it certainly holds for

$x = -2$ and $x = 3$. Upon substituting these judicious choices of x, we get for $x = -2$,

$$-4 = B(-5) \qquad \text{so} \qquad B = \tfrac{4}{5}.$$

For $x = 3$, we get

$$6 = A(5) \qquad \text{so} \qquad A = \tfrac{6}{5}. \quad \blacksquare$$

EXAMPLE B
(Denominator Factors into First-Degree Factors, Some Repeated)
Decompose

$$\frac{2x}{(x - 3)^2(x + 2)}.$$

In this case we assume a decomposition of the form

$$\frac{2x}{(x - 3)^2(x + 2)} = \frac{A}{x - 3} + \frac{B}{(x - 3)^2} + \frac{C}{x + 2};$$

$$2x = A(x - 3)(x + 2) + B(x + 2) + C(x - 3)^2.$$

For $x = 3$, we get

$$6 = B(5) \qquad \text{so} \qquad B = \tfrac{6}{5};$$

for $x = -2$,

$$-4 = C(25) \qquad \text{so} \qquad C = -\tfrac{4}{25};$$

for $x = 1$,

$$2 = A(-2)(3) + \tfrac{6}{5}(3) - \tfrac{4}{25}(4) \qquad \text{so} \qquad A = \tfrac{4}{25}.$$

The desired decomposition is

$$\frac{2x}{(x - 3)^2(x + 2)} = \frac{4}{25(x - 3)} + \frac{6}{5(x - 3)^2} - \frac{4}{25(x + 2)}. \quad \blacksquare$$

EXAMPLE C
(Denominator Factors into Distinct Second-Degree Factors)
Decompose

$$\frac{x^3 - 2x}{(x^2 + 1)(x^2 + x + 1)}.$$

We now assume a decomposition of the form

$$\frac{x^3 - 2x}{(x^2 + 1)(x^2 + x + 1)} = \frac{Ax + B}{x^2 + 1} + \frac{Cx + D}{x^2 + x + 1},$$

$$x^3 - 2x = (Ax + B)(x^2 + x + 1) + (Cx + D)(x^2 + 1).$$

Using either of the suggested methods, the reader should verify for himself that

$$A = 0, \qquad B = -3, \qquad C = 1, \qquad D = 3.$$

We get

$$\frac{x^3 - 2x}{(x^2 + 1)(x^2 + x + 1)} = \frac{-3}{x^2 + 1} + \frac{x + 3}{x^2 + x + 1}. \quad \blacksquare$$

Notice the difference in the decomposition assumed for the factor $(x - 3)^2$ in Example B and the factor $x^2 + 1$ in Example C. Corresponding to the first, we introduced two terms:

$$\frac{p(x)}{(x - 3)^2 q(x)} = \frac{A}{x - 3} + \frac{B}{(x - 3)^2} + \cdots.$$

Corresponding to the second, we wrote

$$\frac{p(x)}{(x^2 + 1)q(x)} = \frac{Ax + B}{x^2 + 1} + \cdots.$$

EXAMPLE D

(Denominator Factors with Second-Degree Factors, Some Repeated)

Decompose

$$\frac{3x^3 + x^2 + 10}{(x^2 + 1)^2(x - 3)}.$$

We assume a decomposition

$$\frac{3x^3 + x^2 + 10}{(x^2 + 1)^2(x - 3)} = \frac{Ax + B}{x^2 + 1} + \frac{Cx + D}{(x^2 + 1)^2} + \frac{E}{x - 3},$$

$$3x^3 + x^2 + 10 = (Ax + B)(x^2 + 1)(x - 3) + (Cx + D)(x - 3) + E(x^2 + 1)^2$$

From this identity in x we may determine A, B, C, D, E.

$$A = -1, \qquad B = 0, \qquad C = 0, \qquad D = -3, \qquad E = 1.$$

Thus,

$$\frac{3x^3 + x^2 + 10}{(x^2 + 1)^2(x - 3)} = \frac{-x}{x^2 + 1} + \frac{-3}{(x^2 + 1)^2} + \frac{1}{x - 3}. \blacksquare$$

We have not stated any formal rules, because such a statement would appear quite complicated, whereas the few examples above should make the procedure for a particular case quite clear. Nor have we proved that the indicated methods work. It can be demonstrated that they do, but the method is probably best defended in a particular case by a very simple argument. It works. Any doubter should add up the partial fractions.

Sometimes a problem not involving the quotient of two polynomials will, after an appropriate substitution, depend on the use of partial fractions. We illustrate this by solving Example 8.1B again, this time avoiding the impression of owing our success to unabashed trickery.

EXAMPLE E
Find

$$\int \sec x \, dx.$$

A good rule for any trigonometric antiderivative problem is to express everything in terms of sines and cosines. A second good rule, when stuck, is to use familiar identities to change the form of the problem. Taken together, these pieces of advice lead to

$$\int \sec x \, dx = \int \frac{dx}{\cos x} = \int \frac{\cos x \, dx}{1 - \sin^2 x}.$$

We now see that if we set $u = \sin x$ so $du = \cos x \, dx$, we are led to

$$\int \frac{du}{1 - u^2} = \frac{1}{2} \int \left(\frac{1}{1 + u} + \frac{1}{1 - u} \right) du = \frac{1}{2} \ln \left| \frac{1 + u}{1 - u} \right| + C.$$

We have made obvious use of partial fractions here. It now remains to express our answer in terms of x:

$$\frac{1}{2} \ln \left| \frac{1 + u}{1 - u} \right| = \ln \left| \frac{1 + \sin x}{1 - \sin x} \right|^{1/2} = \ln \left| \frac{(1 + \sin x)^2}{\cos^2 x} \right|^{1/2}$$

$$= \ln \left| \frac{1 + \sin x}{\cos x} \right| = \ln | \sec x + \tan x |. \blacksquare$$

PROBLEMS

A · Beginning with the left-hand side of (2), obtain the right side by the methods of this section.

B · Find

$$\int \frac{3x^3 + x^2 + 10}{(x^2 + 1)^2(x - 3)} dx.$$

(Use the work already done in Example D above. Also refer to Problem 8.2A.)

C · After a rational function has been decomposed by using partial fractions, you are supposed to be able to find antiderivatives of the individual terms. The following problems may well turn up. Find antiderivatives for each.

$$(a) \quad \int \frac{dx}{x^2 + 2x + 5}. \qquad (b) \quad \int \frac{(3x + 5)\, dx}{x^2 + 2x + 5}.$$

D · Using the work already done in Example C, find

$$\int \frac{(x^3 - 2x)\, dx}{(x^2 + 1)(x^2 + x + 1)}.$$

8 · 4 EXERCISES

Find the indicated antiderivatives.

1 · $\int \dfrac{x^4 + 2x^2 + x}{x^2 + 2} dx.$

2 · $\int \dfrac{1 - x}{3x^2 + 9x + 6} dx.$

3 · $\int \dfrac{28 - x}{6x^2 + 6x - 12} dx.$

4 · $\int \dfrac{3x^3 + 2x^2 + 2}{3(x^2 + 1)} dx.$

5 · $\int \dfrac{x^2 + 3x + 1}{x^3 + x^2 + x} dx.$

6 · $\int \dfrac{2x^3 - 5x^2 + 2x - 15}{x^4 + 4x^2 + 3} dx.$

7 · $\int \dfrac{x^3 + 3x^2 + 4x + 3}{x^4 + 5x^2 + 4} dx.$

8 · $\int \dfrac{3x^2 - x + 3}{x^3 - x^2 + x} dx.$

9 · $\int \dfrac{13x^3 + 7x^2 - 4}{2x^4 - 2x^2} dx.$

10 · $\int \dfrac{6x^3 - 17x^2 - 12}{2x^4 - 8x^2} dx.$

11 · $\int \dfrac{2x^3 - 5x^2 - 4x + 10}{2x^2 - 8x + 8} dx.$

12 · $\int \dfrac{x^3 + 2x^2 + 3x - 1}{(x - 1)^3 x^2} dx.$

13 · $\int \dfrac{7x^4 + 56x^2 + 3x + 118}{(x + 2)(x^2 + 4)^2} dx.$

14 · $\int \dfrac{x^6 + 6x^4 + 9x^2 + 3}{x^4 + 6x^2 + 9} dx.$

15 · $\int \dfrac{5x^3 - 9x^2 + 11x - 2}{(x - 1)^2(x^2 - x + 1)} dx.$

16 · $\int \dfrac{6x^3 + 22x^2 + 18x + 9}{3x^4 + 6x^3 + 3x^2} dx.$

8 · 5   Other Methods

We have made a systematic study of three methods of finding antiderivatives. These methods constitute the basic tools needed by anyone who expects to develop facility in finding antiderivatives. We have seen that some problems can be solved by using a combination of our basic tools (Problem 8.4B).

As is true of any job, the skilled craftsman acquires certain specialized tools and techniques. It is not our purpose here to achieve a high level of competence in the business of finding antiderivatives, so we bring to a halt at this point the systematic development of any further methods.

We do wish to emphasize, however, that there is no law against trying something other than what has been suggested here. We do occasionally encounter problems that fairly beg for a certain substitution.

EXAMPLE A
Find

$$\int \frac{x^{1/3}}{1 + x^{2/3}} \, dx.$$

We try setting $x = u^3$; $dx = 3u^2 \, du$. Then

$$\int \frac{x^{1/3}}{1 + x^{2/3}} \, dx = \int \frac{3u^3}{1 + u^2} \, du = \int \left[3\left(u - \frac{u}{u^2 + 1}\right) \right] du,$$

$$= \tfrac{3}{2}u^2 - \tfrac{3}{2} \ln(u^2 + 1) + C,$$

$$= \tfrac{3}{2}x^{2/3} - \tfrac{3}{2} \ln(x^{2/3} + 1) + C. \quad \blacksquare$$

In Section 8.1 we compiled a list of antiderivatives that we knew (3). Every time we find another antiderivative, it can be added to this list. We would of course abandon the idea of memorizing the list, but it would provide a source to which we could turn when confronted with a problem to see if we have already done it. The difficulty with this is that the list would grow, until it might be faster to work the problem from the beginning than to read through the entire list.

Compromise lists have been made. They include the antiderivatives of functions that frequently turn up. They are organized somewhat by the type of functions with which one is working. Such lists are called *tables of integrals* (a definite abuse of language; they are really tables of antiderivatives). Anyone in a situation calling for the evaluating of many integrals (or other business in which it is nice to know a number of antiderivatives) should acquire such a list. Such a list is not, however, a substitute for the techniques of this chapter, since it rarely happens that the tables have the desired antiderivatives in just the form needed. As the man said after trying to borrow money at the bank, "You can't get help unless you really don't need it."

8 · 5   EXERCISES

Find the indicated antiderivatives if you can. Some are easily found directly. For many of them, one or more of the three principal methods of this chapter will work. In a few you will need a substitution that requires a little imagination. A very few have no antiderivative among the elementary functions.

1. $\int x \tan x^2 \, dx.$

2. $\int \dfrac{dx}{\sqrt{8 - 4x - 4x^2}}.$

3. $\int \dfrac{\sec^2 x}{\sqrt{1 + \tan x}} \, dx.$

4. $\int \dfrac{\sin x}{1 + \cos^2 x} \, dx.$

5. $\int x^2 \operatorname{Arctan} x \, dx.$

6. $\int \dfrac{1}{x^2} \ln x \, dx.$

7. $\int \dfrac{1}{x} \ln x \, dx.$

8. $\int \dfrac{\operatorname{Arctan} x}{1 + x^2} \, dx.$

9. $\int \dfrac{e^x \, dx}{1 + e^{2x}}.$

10. $\int \dfrac{\sin 2x}{1 + \sin^4 x} \, dx.$

11. $\int \dfrac{dx}{\sqrt{1 + x + x^2}}.$

12. $\int \dfrac{x}{\cos^2 x^2} \, dx.$

13. $\int \dfrac{dx}{\sqrt{1 - x^2 - x}}.$

14. $\int \dfrac{3x + 1}{x^3 + 1} \, dx.$

15. $\int \dfrac{x}{x^3 - 1} \, dx.$

16. $\int \sin x \sqrt{1 + \cos x} \, dx.$

17. $\int \dfrac{\tan x}{\cos x + \sec x} \, dx.$

18. $\int x \operatorname{Arcsin} x \, dx.$

19. $\int \dfrac{\operatorname{Arctan} x}{x^2} \, dx.$

20. $\int \dfrac{dx}{\sqrt{4x^2 - 4x + 2}}.$

21. $\int \dfrac{dx}{1 + e^{-x}}.$

22. $\int \cos^3 x \sin^5 x \, dx.$

23. $\int \sec^4 x \, dx.$

24. $\int \dfrac{dx}{(x^2 - 2x + 1)(x^2 - x + 1)}.$

25. $\int \dfrac{dx}{\sqrt{6x - x^2}}.$

26. $\int x^2 e^{x/2} \, dx.$

27 · $\displaystyle\int \frac{x^2\,dx}{(x+2)^{1/2}}.$

28 · $\displaystyle\int e^{x^3}\,dx.$

29 · $\displaystyle\int \frac{(x^2-9)^{1/2}}{x^3}\,dx.$

30 · $\displaystyle\int \frac{dx}{(2x-x^2)^{3/2}}.$

31 · $\displaystyle\int \frac{x^3+2x}{x^2-5x+6}\,dx.$

32 · $\displaystyle\int \frac{\sec x}{\cos x e^{\tan x}}\,dx.$

33 · $\displaystyle\int \cos x e^{x^2}\,dx.$

34 · $\displaystyle\int \frac{x^3}{(3-x^2)^{1/3}}\,dx.$

35 · $\displaystyle\int 2x^2 \cos x\,dx.$

36 · $\displaystyle\int \frac{dx}{3x^2-4x+2}.$

37 · $\displaystyle\int \frac{dx}{(x^2+2x+1)(x^2+x+1)}.$

38 · $\displaystyle\int \frac{x}{(10-x)^{1/3}}\,dx.$

39 · $\displaystyle\int \frac{x^3}{(x^2+4)^{1/2}}\,dx.$

40 · $\displaystyle\int \frac{x^3-4}{x\sqrt{x^3-4}}\,dx.$

41 · $\displaystyle\int e^{3x} \sin x\,dx.$

42 · $\displaystyle\int x^4 \ln x\,dx.$

43 · $\displaystyle\int \frac{3x}{(x^2-9)(x+2)}\,dx.$

44 · $\displaystyle\int e^{2x} \cos 3x\,dx.$

45 · $\displaystyle\int \frac{4x^3}{(4+9x^2)^{1/3}}\,dx.$

46 · $\displaystyle\int \frac{dx}{x^2-5x+6}.$

In Exercises 47–66, find (if you can) the value of the integral correct to one decimal place. You may have to rely on more than one method to evaluate an integral; you will also need some tables.

47 · $\displaystyle\int_0^1 \sin x\,dx.$

48 · $\displaystyle\int_0^{\sqrt{3}} \frac{3x+1}{x^2+1}\,dx.$

49 · $\displaystyle\int_0^{\pi/6} x \sin x\,dx.$

50 · $\displaystyle\int_0^2 xe^x\,dx.$

51 · $\displaystyle\int_0^2 \frac{dx}{6x^2+5x+1}.$

52 · $\displaystyle\int_0^{\pi/3} \frac{\sin x}{1+\cos^2 x}\,dx.$

53 · $\displaystyle\int_0^3 \frac{dx}{\sqrt{6x-x^2}}.$

54 · $\displaystyle\int_0^3 \ln x\,dx.$

55 · $\displaystyle\int_0^{\pi/4} \frac{1-\cos x}{x}\,dx.$

56 · $\displaystyle\int_0^1 e^{\sin x} \cos x\,dx.$

57 · $\displaystyle\int_4^{\infty} \frac{dx}{\sqrt{x^2-2}}.$

58 · $\displaystyle\int_{\sqrt{3}}^{\infty} \frac{x\,dx}{1+x^4}.$

59· $\int_0^1 \dfrac{x^3 + x}{x^4 + 1}\,dx.$

60· $\int_0^{1/3} e^{x^2}\,dx.$

61· $\int_0^{\pi/4} \dfrac{\sec x}{1 - \tan^2 x}\,dx.$

62· $\int_0^{\pi/4} \dfrac{\sin x}{x}\,dx.$

63· $\int_0^{1/2} e^{x^3}\,dx.$

64· $\int_0^2 \dfrac{x\,dx}{x^3 + 1}.$

65· $\int_1^{3/2} \ln x\,dx.$

66· $\int_0^{1/2} \dfrac{dx}{\sqrt{x^3 + 2x}}.$

67· Find the centroid of the area enclosed between $y = e^x$ and $y = (e - 1)x + 1$.

68· Find the centroid of the area enclosed by the curves $x = 1(e - 1)y$ and $y = \ln x$.

69· Find the area of the ellipse $(x^2/a^2) + (y^2/b^2) = 1$.

70· Find the area enclosed by the hyperbola $(x^2/a^2) - (y^2/b^2) = 1$ and the line $x = 2a$.

71· Suppose the ellipse of Exercise 69 is cut from a material that has a density that is proportional to the distance from the Y axis. Find the mass of the ellipse.

72· Suppose the area of Exercise 70 is cut from a material that has a density that is proportional to the distance from the Y axis. Find the mass of the area.

CHAPTER

9

CURVES IN THE PLANE

What is a curve? It is the mark left by a pen drawn in a continuous, not necessarily straight, stroke between two points. It is a geometric figure having just one dimension (length but no width). It is a (perhaps bent or twisted) piece of wire. It is a line segment that has been deformed.

All of these descriptions convey the intuitive picture we have in mind when we speak of a curve. None of them proves very helpful, however, to a mathematician trying to work with curves. The fact is that this seemingly elementary concept is very difficult to define in a way that cannot be misunderstood.

We shall see in this chapter that attempts to work with curves come up against two major difficulties.

(1) Obviously different mathematical formulas used to describe curves often turn out to represent the same set of points.

(2) Definitions of "curve" that look reasonable are satisfied by geometrical configurations that look unreasonable.

An adequate treatment of these difficulties is not appropriate to our aims. We hope to acquire some understanding of techniques that prove helpful in the many situations where the curves involved are nicely behaved. In the course of our work we shall point to problems lurking in the corners, and we shall give a few references that will help those readers who are attracted to dark corners. Beyond this we shall not go.

9 · 1 Parametric Equations

Cheerleaders at the homecoming football game are often equipped with gas-filled balloons that they release at the kickoff. Suppose we are told that the height y of such a balloon at time t seconds is $y = 6 + 8t^{3/2}$. (Obviously the balloon cannot rise according to this formula forever, but to avoid a complicated set of equations here, we will take this to be descriptive of the behavior for the time during which we are interested.) Suppose further that a wind of $7\frac{1}{2}$ mph (11 ft/sec) is blowing. Then if the positive x axis is taken in the direction of the wind and the y axis coincides with the position of the cheerleader, the position of the balloon at the end of t sec is given by

$$y = 6 + 8t^{3/2},$$
$$x = 11t. \tag{1}$$

The path that the balloon follows in the air is determined by eliminating t from Eq. (1). This gives

$$y = 6 + 8\left(\frac{x}{11}\right)^{3/2}. \tag{2}$$

We observe that there is more information in Eq. (1) than there is in Eq. (2). From (1) we can tell the position of the balloon at a specific time. When $t = 1$, $x = 11$ and $y = 14$. (Substituting values of t into (1) is, in fact, a very easy way to obtain the graph in Fig. 9.1.) All we can tell from (2) is that the

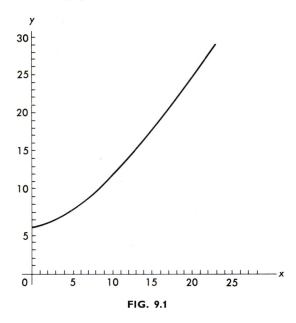

FIG. 9.1

balloon does pass through $(11, 14)$; we cannot tell when it passes through this point. Observe that a balloon traveling according to the equations

$$y = 6 + t^{3/2},$$

$$x = \frac{11}{4} t \tag{3}$$

follows the same path described by (2), but this balloon passes through $(11, 14)$ when $t = 4$.

Equations of the form (1) and (3), which relate x and y by means of a third variable, are called *parametric equations*. The third variable is called the *parameter*. The collection of all points (x, y) obtained as t ranges over some (possibly infinite) interval forms the *path* (or *trace*) of the parametric equations. We have already seen that the two different sets of parametric equations may have the same path.

Parametric equations are useful in a variety of applications. There is no reason to think of the parameter as representing time, though this is often a helpful interpretation. Neither is there any reason to think it desirable (or even possible in all cases) to eliminate the parameter.

EXAMPLE A

A wheel of radius b rolls along a level surface without slipping. Describe the path of a fixed point on the rim of the wheel.

Place the wheel on the x axis with its center on the y axis and the fixed point at the origin. Then roll it to a position so that the radius to the fixed point makes an angle of θ with the radius to the point of contact on the x axis (Fig. 9.2). Since the wheel has rolled without slipping, the arc $\overset{\frown}{CP}$ cut off by

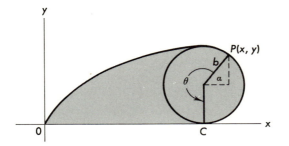

FIG. 9.2

the central angle θ must be equal to the length of OC. Let α be the angle indicated in Fig. 9.2.

$$x = OC + b \cos \alpha,$$

$$x = b\theta + b \cos \alpha,$$

and

$$y = b + b \sin \alpha.$$

Since $\theta + \alpha = 3\pi/2$, we use the trigonometric identities (6) of Section 7.1 to write

$$\cos \alpha = \cos\left(\frac{3\pi}{2} - \theta\right) = -\sin \theta,$$

$$\sin \alpha = \sin\left(\frac{3\pi}{2} - \theta\right) = -\cos \theta,$$

The parametric equations of the desired path are

$$x = b(\theta - \sin \theta),$$

$$y = b(1 - \cos \theta). \quad \blacksquare$$

(4)

Now suppose we were given the equations

$$x = 2(t - \sin t),$$
$$y = 2(1 - \cos t),$$

(5)

with no indication of where they came from. (They are, of course, obtained from (4) by setting $b = 2$, $\theta = t$.) It is unlikely that the inexperienced would know the source of these equations or the meaning of the parameter. It is indeed more likely that the beginner would again think of (5) as equations giving the location (x, y) of a point on a path at time t. We emphasize that there would be nothing wrong with such an interpretation.

Given (5) "out of the blue," suppose we set out to sketch the corresponding path. A little effort convinces us that it is not practical to eliminate t. We decide to utilize the method of plotting the points corresponding to some choices of t.

t	0	$\pi/2$	π	$3\pi/2$	2π	$5\pi/2$	3π
x	0	$\pi - 2$	2π	$3\pi + 2$	4π	$5\pi - 2$	6π
y	0	2	4	2	0	2	4

This curve is called a *cycloid*. The author's experience is that students are reluctant to accept the accuracy of Fig. 9.3 at the point C. They logically

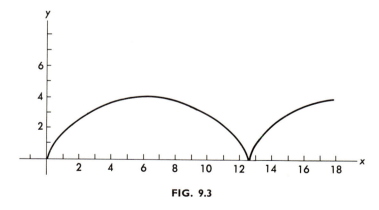

FIG. 9.3

ask, particularly when not thinking about the geometric development of Eq. (5), how we can be sure that the correct picture at C does not look like Fig. 9.4.

One easy way to answer the question is to find the slope of the path and study it near the point C. Since we have not been successful in expressing y in terms of x, we need to appeal to the chain rule to find (dy/dx) from (5).

$$\frac{dy}{dt} = \frac{dy}{dx} \cdot \frac{dx}{dt}.$$

Therefore,

$$\frac{dy}{dx} = \frac{dy/dt}{dx/dt}.$$

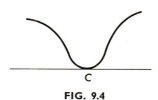

FIG. 9.4

From (5) we get

$$dy = 2 \sin t \, dt,$$

$$dx = (2 - 2 \cos t) \, dt,$$

$$\frac{dy}{dx} = \frac{\sin t}{1 - \cos t}.$$

Now at the point C, $t = 2\pi$, so dy/dx does not exist. Our well-established pattern by now is to creep up on 2π in such cases. With the help of L'Hôpital's rule,

$$\lim_{t \to 2\pi} \frac{\sin t}{1 - \cos t} = \lim_{t \to 2\pi} \frac{\cos t}{\sin t} = \begin{cases} \infty & \text{if} \quad t \to 2\pi^+ \\ -\infty & \text{if} \quad t \to 2\pi^- \end{cases}$$

This means the graph at C as we pictured it in Fig. 9.3 is correct.

While thinking about finding the derivative dy/dx from Eq. (5), let us consider the problem of finding the second derivative. We already know

$$\frac{dy}{dx} = \frac{\sin t}{1 - \cos t}.$$

Set $dy/dx = u$. Our problem is to find

$$\frac{d}{dx}\left(\frac{dy}{dx}\right) = \frac{du}{dx}.$$

We simply proceed as before:

$$du = \frac{(1 - \cos t)\cos t - (\sin t)\sin t}{(1 - \cos t)^2} \, dt,$$

$$dx = (2 - 2 \cos t) \, dt.$$

Hence,

$$\frac{d}{dx}\left(\frac{dy}{dx}\right) = \frac{\cos t - (\cos^2 t + \sin^2 t)}{2(1 - \cos t)^3} = \frac{-1}{2(1 - \cos t)^2}.$$

EXAMPLE B

For the following parametric equations, find dy/dx, d^2y/dx^2 and sketch the graph for $t \in [0, \pi]$.

$$y = 2 \sin t - 2t \cos t,$$

$$x = 2 \cos t + 2t \sin t,$$

$$\frac{dy}{dx} = \frac{2 \cos t - 2(t(-\sin t) + \cos t)}{-2 \sin t + 2(t \cos t + \sin t)} = \tan t,$$

$$\frac{d}{dx}\left(\frac{dy}{dx}\right) = \frac{\sec^2 t}{2t \cos t} = \frac{1}{2t} \sec^3 t. \ \blacksquare$$

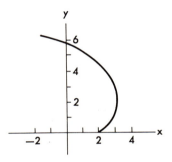

FIG. 9.5

9.1 EXERCISES

Sketch the path described by each set of parametric equations. Then find both the first and second derivatives of y with respect to x. Find the Cartesian equation of the path whenever you can.

1. $x = 2 \cos t,$
   $y = 2 \sin t.$

2. $x = 4 \cos t,$
   $y = \sin t.$

3. $x = 2 \sin t,$
   $y = 2 \cos t.$

4. $x = \sin t,$
   $y = 4 \cos t.$

5. $x = e^t,$
   $y = e^{t^2}.$

6. $x = \sqrt{4 - t},$
   $y = t.$

7. $x = \tan t,$
   $y = \sec t.$

8. $x = \sin 2t,$
   $y = \cos t.$

9. $x = \dfrac{1 + t}{1 - t},$
   $y = \dfrac{t}{1 - t}.$

10. $x = \dfrac{t}{1 + t},$
    $y = \dfrac{1 - t}{1 + t}.$

11. $x = \dfrac{1 + e^t}{1 - e^t}$,

 $y = \dfrac{e^t}{1 - e^t}$.

12. $x = \dfrac{e^t}{1 + e^t}$,

 $y = \dfrac{1 - e^t}{1 + e^t}$.

In Exercises 13–16, C is the circle of radius 4 centered at the origin.

13*. A second circle of radius 1 has a fixed point P on its circumference. This circle is placed with its center at $(5, 0)$ so that the point P touches C, and is then rolled along the outside of C without slipping. Find parametric equations describing the path traced out by the point P. (This curve is called an *epicycloid.*)

14*. A second circle of radius 1 has a fixed point P on its circumference. This circle is placed with its center at $(3, 0)$ so that the point P touches C, and is then rolled along the inside of C without slipping. Find parametric equations describing the path traced out by the point P. (This curve is called a *hypocloid.*)

15.* A string wound tightly about C has its end point P at $(4, 0)$. The string is unwound from C, being pulled taut as it is unwound. Find parametric equations describing the path traced out by the point P. (This curve is called an *involute.*)

16*. A line segment of length 8π is placed in a vertical position with the bottom point, say P, at $(4, 0)$. This segment is then rolled around C without slipping. Write parametric equations describing the path traced out by the point P.

9 · 2 Curves

We have now encountered situations in which, given a real number t, two numbers x and y may be determined from some parametric equations

$$x = F(t), \qquad y = G(t).$$

F and G are real-valued functions of the real variable t.

This could be viewed as a description of a single function that takes real numbers t as raw material and produces points in the plane as a finished product. Call the function R. If R accepts all t in some given interval $[a, b]$, we have in our mind the following picture (Fig. 9.6).

Our function R is described by the equation

$$R(t) = (F(t), G(t)). \tag{1}$$

It is here that we encounter the first of the difficulties mentioned in the introduction to this chapter. Two different functions of the type described in (1) may produce exactly the same set of points in the plane.

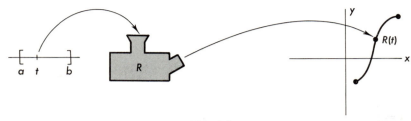

FIG. 9.6

Consider, for example, the two sets of parametric equations described in (1) and (3) of Section 9.1. They give rise to the two functions

$$R(t) = (11t,\ 6 + 8t^{3/2}),\tag{2}$$

$$Q(t) = \left(\frac{11}{4}\,t,\ 6 + t^{3/2}\right).\tag{3}$$

We have come to the heart of the difficulty. Equations (2) and (3) both describe the path pictured in Fig. 9.1. This path has certain geometric properties (length from (0, 6) to (11, 14); slope at an arbitrary point; etc.). If we wish to study these properties of the path, should we use the representation (2), the representation (3), or another, as yet undetermined?

The best way to cope with the difficulty is to adopt a point of view that places R and Q into a so-called class of equivalent functions.†

Our method will be to take one function that represents a curve and use it to study the things in which we are interested. We will then comment on the effect of using a different representation, but the truth of our comments will not be proved.

PROBLEM

A · Write down several representations (in the form of (1)) of the curve that is the upper half of the circle $x^2 + y^2 = a^2$.

Our final remarks in this section have to do with another means of expressing (1). Our purpose is to develop a notation that emphasizes the range of R as the path of a continuously moving point.

Imagine $R(t) = (F(t),\ G(t))$ to be the end-point of a radius vector. As t varies from a to b, the terminal point of this vector follows the path represented

---

† The details of carrying out such a program are described by J. W. T. Youngs, *Curves and surfaces*, Amer. Math. Monthly **51** (1944), 1–11.

by (1). It is with this point of view that one often refers to **R** as a vector valued function, and writes (Fig. 9.7)

$$\mathbf{R}(t) = F(t)\mathbf{i} + G(t)\mathbf{j}.$$

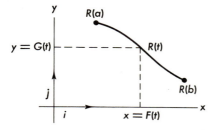

FIG. 9.7

=== 9·2 EXERCISES

In each of the following exercises you are given two sets of parametric equations. Each set describes the position at time t, $t \in [0, 2\pi]$, of a child walking in freshly fallen snow. Do the following for each exercise.

(a) Find, for each child, the Cartesian equation (the equation in the variables x and y) of the tracks left in the snow.
(b) Sketch the tracks left by each child (remember that $t \in [0, 2\pi]$).
(c) Find those places where the tracks intersect.
(d) Find those places, if any, where the children collide.

1.

A: $\begin{aligned} x &= 2\cos t, \\ y &= \sin^2 t. \end{aligned}$

B: $\begin{aligned} x &= 4t, \\ y &= 3t. \end{aligned}$

2.

A: $\begin{aligned} x &= \sqrt{1 + t^2}, \\ y &= t. \end{aligned}$

B: $\begin{aligned} x &= t^3, \\ y &= 3 - t^3. \end{aligned}$

3.

A: $\begin{aligned} x &= 2t^{1/3}, \\ y &= 8t - 6t^{1/3} + 1. \end{aligned}$

B: $\begin{aligned} x &= 3^t - 1, \\ y &= 3^t. \end{aligned}$

4.

A: $\begin{aligned} x &= 1 - 2\cos\frac{t}{2}, \\ y &= 4\cos^2\frac{t}{2} - 4\cos\frac{t}{2}. \end{aligned}$

B: $\begin{aligned} x &= \frac{9}{4}t - 1, \\ y &= \frac{9t}{4}. \end{aligned}$

A:
$$x = e^t,$$
$$y = e^{-t}.$$

A:
$$x = 2 \sin \frac{\pi t}{4},$$
$$y = 2 \cos \frac{\pi t}{4}.$$

5.

6.

B:
$$x = 2^{2\pi - t},$$
$$y = 2^{t - 2\pi}.$$

B:
$$x = \frac{2t}{\sqrt{t^2 + 1}},$$
$$y = \frac{2}{\sqrt{t^2 + 1}}.$$

A:
$$x = e^t - 1,$$
$$y = e^{t^2}.$$

A:
$$x = \ln(t + e^{t-1}),$$
$$y = \ln \frac{e^{-t}}{1 + te^{1-t}}.$$

7.

8.

B:
$$x = \frac{2t + 1}{(t + 1)^2},$$
$$y = \frac{t}{t + 1}.$$

B:
$$x = 2^t - 2,$$
$$y = 4^t - 1.$$

A:
$$x = \tan \frac{\pi t}{2t^2 + 1},$$
$$y = \tan \frac{\pi(2t^2 - 2t + 1)}{4t^2 + 2}.$$

A:
$$x = 2 \sin \frac{\pi t}{2(t + 1)},$$
$$y = 2 \sin \frac{\pi}{2(t + 1)}.$$

9*.

10*.

B:
$$x = \frac{2t - \sqrt{2}}{2},$$
$$y = t^2 - t\sqrt{2} + \tfrac{1}{2}.$$

B:
$$x = \frac{\sqrt{8 + t^2}}{2},$$
$$y = \frac{8 + t^2}{12}.$$

9 · 3   Differentiation of Vector-Valued Functions

In Section 9.2 we discussed a function **R** defined for $t \in [a, b]$ that produced points in the plane. The points in the range of this function seemed to constitute what we intuitively call a curve. To emphasize this, we decided to represent the points in the range of R as the path of the terminal point of the radius vector,

$$\mathbf{R}(t) = F(t)\mathbf{i} + G(t)\mathbf{j}, \qquad t \in [a, b]. \tag{1}$$

We now face up to the second of the difficulties mentioned in the introduction to this chapter. Unless we place some restrictions on R, equations of the form (1) can be used to describe a number of geometric configurations that most people would not call a curve. Some care is needed, for example, to exclude the set of all points in the solid square as a curve between two opposite vertices A and B. (See [1, pages 91–93].) Yet this set is too "fat" (Fig. 9.8) to meet our intuitive idea of a curve.

FIG. 9.8

To avoid such difficulties, we are going to place some restrictions on R. The restrictions we choose are actually more severe than would be necessary to avoid the worst of the problems, but they have the advantage of allowing us to go quickly to the applications we have in mind.†

We have seen that when given a function described by (1), it is often helpful to think of t as representing time. Then $R(t)$ gives the position of a moving particle at time t. It is natural to follow this up by asking for the velocity of the particle at time t. Since velocity is associated in our mind with derivatives, we will need to know what is meant by the derivative of a function $\mathbf{R}$ as defined in (1). We are guided in this matter by our sense of what ought to be.

DEFINITION A

Let $\mathbf{R}(t) = F(t)\mathbf{i} + G(t)\mathbf{j}$ be a vector valued function defined on $[a, b]$. If the functions F and G are differentiable at $t \in [a, b]$, then we define

$$\mathbf{R}'(t) = F'(t)\mathbf{i} + G'(t)\mathbf{j}.$$

It follows from the properties of differentials for functions that $\mathbf{R}'(t)\,dt = F'(t)\,dt\mathbf{i} + G'(t)\,dt\mathbf{j}$ may be used to approximate the difference $\mathbf{R}(t + dt) - \mathbf{R}(t)$ for small dt.

EXAMPLE A

Let $\mathbf{R}(t) = (4 - t^2)\mathbf{i} + t\mathbf{j}$.
(a)   Sketch the curve that $\mathbf{R}$ represents for $t \in [0, 2]$.
(b)   Find an expression for $\mathbf{R}'(t)$ at any t

---

† Those interested in the various types of restrictions that can be used are referred to the article by G. T. Whyburn, *What is a curve?* Amer. Math. Monthly **49** (1942), 493–497.

(c) Find an expression for $\mathbf{R}'(\tfrac{3}{2})\,dt$.
(d) Find $\mathbf{R}(\tfrac{3}{2}+\tfrac{1}{4})-\mathbf{R}(\tfrac{3}{2})$.
(e) Use the result of part (c) to approximate the answer to part (d).
(f) Show the vectors of parts (c) and (d) in the sketch of the curve, putting the initial point of each at $\mathbf{R}(\tfrac{3}{2})$.

Since $x = (4 - t^2)$ and $y = t$, we see that the curve is a portion of the set described by $x = 4 - y^2$. For $t \in [0, 2]$ we have $x \in [0, 4]$ and $y \in [0, 2]$. The curve is drawn in Fig. 9.9.

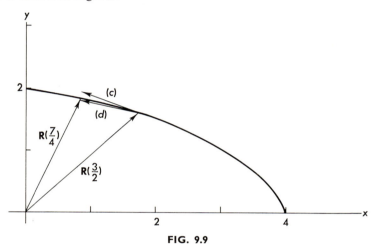

FIG. 9.9

Differentiation gives for (b),

$$\mathbf{R}'(t) = -2t\mathbf{i} + \mathbf{j}.$$

It is then clear that the expression sought in part (c) is

$$\mathbf{R}'(\tfrac{3}{2})\,dt = -3\,dt\mathbf{i} + dt\mathbf{j}.$$

Direct computation gives, as the result for part (d),

$$\mathbf{R}(\tfrac{3}{2} + \tfrac{1}{4}) - \mathbf{R}(\tfrac{3}{2}) = (\tfrac{15}{16}\mathbf{i} + \tfrac{7}{4}\mathbf{j}) - (\tfrac{7}{4}\mathbf{i} + \tfrac{3}{2}\mathbf{j}),$$

$$\mathbf{R}(\tfrac{3}{2} + \tfrac{1}{4}) - \mathbf{R}(\tfrac{3}{2}) = -\tfrac{13}{16}\mathbf{i} + \tfrac{1}{4}\mathbf{j}.$$

Now we know from the definition of the derivative that

$$\mathbf{R}(\tfrac{3}{2} + \tfrac{1}{4}) - \mathbf{R}(\tfrac{3}{2}) \approx \mathbf{R}'(\tfrac{3}{2})(\tfrac{1}{4}).$$

Substitution in (c) gives, for part (e),

$$R'(\tfrac{3}{2})(\tfrac{1}{4}) = -\tfrac{3}{4}\mathbf{i} + \tfrac{1}{4}\mathbf{j},$$

which does indeed approximate the answer to (d). The vectors obtained as solutions to (c) and (d) are so labeled in Fig. 9.9. ∎

PROBLEMS

A · Suppose $R(t) = F(t)\mathbf{i} + G(t)\mathbf{j}$ is differentiable at t_0. Verify that $R(t_0 + h)$ $- R(t_0) = R'(t_0)h + |h|r(t_0, h)$ where $r(t_0, h)$ is a vector-valued function that approaches $\mathbf{0}$ as h approaches 0.

B · Let $R(t) = F(t)\mathbf{i} + G(t)\mathbf{j}$ and $S(t) = H(t)\mathbf{i} + K(t)\mathbf{j}$. Verify that

$$D[R(t) \cdot S(t)] = R(t) \cdot [S'(t)] + [R'(t)] \cdot S(t)$$

where the · stands for the dot product of the respective vectors.

We are now able to concisely state the restrictions we want to place on **R** if it is to represent the kind of a curve with which we wish to work.

DEFINITION B

Suppose the vector-valued function

$$R(t) = F(t)\mathbf{i} + G(t)\mathbf{j}, \qquad t \in [a, b]$$

has a continuous nonzero derivative $R'(t)$ throughout the interval. Then we say **R** represents a smooth curve.

Consider again the motivation for defining a derivative of the function **R**. We wanted to find an expression for the velocity of a particle at time t if its position was given by $R(t)$. As used by physicists, velocity is a technical word to be distinguished from speed. Velocity describes the direction in which a body moves, as well as the instantaneous reading of a speedometer. Thus, when we spoke of the velocity of a falling body we used $+$ and $-$ to designate up or down, as well as a number (the speed) that measured the rate of descent or ascent in absolute units of distance and time. We thus hope that the vector $R'(t_0)$ might give the direction of the moving particle on the curve at $t = t_0$; that is, we hope $R'(t_0)$ is in the direction of a tangent to the curve at $R(t_0)$. The sketch in Fig. 9.9 seems to suggest that this is indeed the case.

THEOREM A

Let $R(t)$ describe a smooth curve in a neighborhood of t_0. Then $R'(t_0)$ is tangent to the curve at $R(t_0)$.

PROOF:

The tangent to the curve at $\mathbf{R}(t_0)$, if it exists, must be the limiting position of the vector (Fig. 9.10)

$$\mathbf{R}(t_0 + h) - \mathbf{R}(t_0) \tag{2}$$

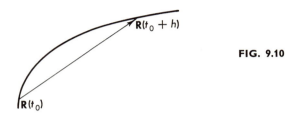

FIG. 9.10

as h approaches 0. Suppose we consider the angle θ made by the vector (2) and $\mathbf{R}'(t_0)$. From properties of the dot product,

$$\cos \theta = \frac{\mathbf{R}'(t_0) \cdot \{\mathbf{R}(t_0 + h) - \mathbf{R}(t_0)\}}{|\mathbf{R}'(t_0)| \, |\mathbf{R}(t_0 + h) - \mathbf{R}(t_0)|}. \tag{3}$$

(We observe here that we depend on the fact that in a smooth curve, $\mathbf{R}'(t_0) \neq 0$.) Since $\mathbf{R}$ is differentiable at t_0, (2) may be written (according to Problem A) as

$$\mathbf{R}'(t_0)h + |h|\mathbf{r}(t_0, h)$$

or, assuming $h > 0$ so that motion along the curve is in the direction of the particle as t increases,

$$h\{\mathbf{R}'(t_0) + \mathbf{r}(t_0, h)\}.$$

Substitution in (3) gives

$$\cos \theta = \frac{\mathbf{R}'(t_0) \cdot \{\mathbf{R}'(t_0) + \mathbf{r}(t_0, h)\}}{|\mathbf{R}'(t_0)| \, |\mathbf{R}'(t_0) + \mathbf{r}(t_0, h)|}.$$

Now as h approaches zero, we have (using $\mathbf{v} \cdot \mathbf{v} = |\mathbf{v}|^2$) the cosine of θ approaching one. This means θ approaches 0. ∎

So far, so good. $\mathbf{R}'(t_0)$ does point in the right direction to give the velocity. Let us try pushing a little further the analogy with falling bodies. The speed was the absolute value of the velocity. Does $|\mathbf{R}'(t_0)|$, the length of the vector $\mathbf{R}'(t_0)$, represent the speed of the particle along its path? Let us investigate an example in which the speed is obvious, and see how the length of the vector $\mathbf{R}'(t_0)$ compares with the obvious answer.

EXAMPLE B

Let $\mathbf{R}(t) = t^2\mathbf{i} + (3t^2 + 2)\mathbf{j}$. Show that R represents a curve that is straight. Then find an expression for the speed of a particle at time t if its position is given by $\mathbf{R}(t)$. Compare this with $|\mathbf{R}'(t)|$.

The graph is that portion of

$$y = 3x + 2$$

for which $x \geq 0$ (Fig. 9.11). The distance s of the point $(x, y) = (t^2, 3t^2 + 2)$ from the starting point $(0, 2)$ is

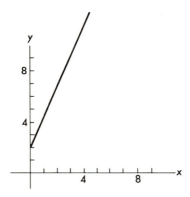

FIG. 9.11

$$s = \{(x - 0)^2 + (y - 2)^2\}^{1/2} = \{t^4 + 9t^4\}^{1/2}.$$

Naturally s depends on t, and we have just shown that

$$s = S(t) = \sqrt{10t^2}.$$

Then the speed of the particle at time t is

$$S'(t) = 2\sqrt{10}t.$$

Now $|\mathbf{R}'(t)| = |2t\mathbf{i} + 6t\mathbf{j}| = (4t^2 + 36t^2)^{1/2},$

$$|\mathbf{R}'(t)| = 2\sqrt{10}t. \quad \blacksquare$$

PROBLEMS

C · The motion of a particle is described by

$$\mathbf{R}(t) = (\cos t + t \sin t)\mathbf{i} + (\sin t - t \cos t)\mathbf{j}.$$

(a)   Sketch the path followed by this particle for $t \in [0, 2\pi]$.
(b)   Sketch the velocity vector $\mathbf{R}'(\pi/2)$ with its initial point at $\mathbf{R}(\pi/2)$.
(c)   Sketch the velocity vector $\mathbf{R}'(\pi)$ with its initial point at $\mathbf{R}(\pi)$.

D · The motion of a particle is described by

$$\mathbf{R}(t) = \frac{5t}{1+t}\mathbf{i} + \frac{1-5t}{1+t}\mathbf{j}.$$

(a)   Show that the path of the particle is a familiar curve.

(b)   Because of the special nature of the path, you should be able to find a formula for $S(t)$, the distance of the particle from the point $(0, 1)$ at time t.

(c)   Find $\mathbf{R}'(t)$. Draw the vector $\mathbf{R}'(2)$ with its initial point at $\mathbf{R}(2)$.

(d)   Find $|\mathbf{R}'(t)|$. Compare this with the derivative $S'(t)$ of the function obtained in part (b).

We are encouraged to define $|\mathbf{R}'(t)|$ to be the speed of a particle at time t if the position of the particle is given by $\mathbf{R}(t)$. We can, of course, make any definition we please, but once made, we must live with it. This sometimes leads to unforeseen complications. Suppose, for example, that we do agree to let $S(t)$ represent the distance a particle travels from some initial point along a given curve in time t, so that $S'(t)$ represents the speed; and suppose we just define

$$S'(t) = |\mathbf{R}'(t)|.$$

Then, according to Theorem 6.4A, we are committed to agree that

$$S(t_1) = \int_{t_0}^{t_1} |\mathbf{R}'(t)|\, dt. \tag{4}$$

This, whether we like it or not, gives us a formula for the length of a curve, or for that portion of a curve traveled by our particle as t ranges from t_0 to t_1. We will like this only if (as a bare minimum requirement) it gives the answers we expect for the length of common curves.

EXAMPLE C

In Example B, it was clear that as t went from 0 to 2, the particle moved from $(0, 2)$ to $(4, 14)$, a distance of $\sqrt{16 + 144} = 4\sqrt{10}$. Evaluate the distance as given by (4)

$$S(2) = \int_0^2 |R'(t)|\, dt = \int_0^2 2\sqrt{10t}\, dx = \sqrt{10t^2}\,\Big|_0^2,$$

$$S(2) = 4\sqrt{10}. \quad \blacksquare$$

Before we congratulate ourselves on a job well done, let us look at one other possible source of trouble. We recall that two different functions R

and Q may be used to represent the same curve. It would indeed be a pity if the length of the curve according to our definition were to change, depending on which representation was used.

EXAMPLE D

Two functions first met in Section 9.1 were seen to describe the curve shown in Fig. 9.12. The functions were

$$R(t) = 11t\mathbf{i} + (6 + 8t^{3/2})\mathbf{j},$$

$$Q(t) = \tfrac{11}{4}t\mathbf{i} + (6 + t^{3/2})\mathbf{j}.$$

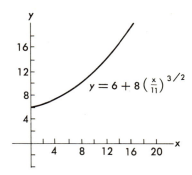

$$y = 6 + 8\left(\tfrac{x}{11}\right)^{3/2}$$

FIG. 9.12

Show the following:

(a) $\mathbf{R}'(t)$ and $\mathbf{Q}'(t)$ have the same direction at $(11, 14)$.
(b)   Either representation may be used to obtain the same length of the curve from $(0, 6)$ to $(11, 14)$.

We note that $\mathbf{R}(0) = \mathbf{Q}(0) = 6\mathbf{j}$. $\mathbf{R}(1) = \mathbf{Q}(4) = 11\mathbf{i} + 14\mathbf{j}$,

$$\mathbf{R}'(1) = 11\mathbf{i} + 12\mathbf{j},$$

$$\mathbf{Q}'(4) = \tfrac{11}{4}\mathbf{i} + 3\mathbf{j}.$$

Thus, $\mathbf{R}'(1) = 4\mathbf{Q}'(4)$, so both have the same direction. $\mathbf{R}'(1) = 4\mathbf{Q}'(4)$ signifies that the particle whose motion is described by $\mathbf{Q}$ moves along the curve more slowly than the particle whose motion is described by $\mathbf{R}$.

If we use the Cartesian equation for the curve, then

$$\frac{dy}{dx} = 12\left(\frac{x}{11}\right)^{1/2}\left(\frac{1}{11}\right)$$

and the slope of the curve at $x = 11$ is $12/11$. This is consistent with the value of the slope of either of the vectors $\mathbf{R}'(1)$ or $\mathbf{Q}'(4)$. Finally, using $\mathbf{R}$,

$$S(t) = \int_0^1 |\mathbf{R}'(t)|\, dt = \int_0^1 (121 + 144t)^{1/2}\, dt$$

and, using $\mathbf{Q}$

$$S(t) = \int_0^4 |\mathbf{Q}'(t)|\, dt = \int_0^4 \left(\frac{121}{16} + \frac{9}{4}t\right)^{1/2} dt.$$

If in the second integral we let $t = 4u$, then $dt = 4\, du$.

$$S(t) = \int_0^1 \left(\frac{121}{16} + 9u\right)^{1/2} 4\, du = \int_0^1 (121 + 144u)^{1/2}\, du.$$

Thus, the two integrals are the same. ∎

PROBLEM

E · In Problem 9.2A, you were asked to write down several representations of the curve that is the upper half of the circle $x^2 + y^2 = a^2$. Use these representations to show that the length of this half circle is just what you expect it to be.

Example D and Problem E illustrate a happy truth. So long as we confine ourselves to well-behaved representations of smooth curves, the length of arc as given by (4) is not dependent on the particular representation chosen. We are trusting the reader at this point to understand that this "proof by two illustrations" appeals more to the charity than the clarity of one's thinking.

Much of this section has been devoted to motivation and illustration. We may summarize the results very quickly.

Let the position of a moving particle in the plane be given by the position vector

$$\mathbf{R}(t) = F(t)\mathbf{i} + G(t)\mathbf{j}, \qquad t \in [a, b],$$

where $\mathbf{R}$ is differentiable in the interval. Then the velocity vector is

$$\mathbf{R}'(t) = F'(t)\mathbf{i} + G'(t)\mathbf{j}.$$

The velocity vector, placed with its initial point at $\mathbf{R}(t)$, will be tangent to the path of the moving particle at that point. The length of the curve as t increases from t_0 is given by

$$S(t) = \int_{t_0}^t |\mathbf{R}'(t)|\, dt$$

so the speed, being ds/dt, is given by

$$\frac{ds}{dt} = |\mathbf{R}'(t)|.$$

We see that the speed at time t is equal to the length of the velocity vector $\mathbf{R}'(t)$.

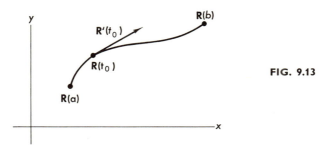

FIG. 9.13

When finding areas under curves, we pointed out several heuristic devices that could be used to help in setting up the integral properly. Such devices also exist as an aid to setting up the integral for arc length. If a curve is described by

$$\mathbf{R}(t) = F(t)\mathbf{i} + G(t)\mathbf{j},$$

then the same curve is described by the parametric equations

$$x = F(t),$$

$$y = G(t).$$

From these we recall that

$$dx = F'(t)\,dt,$$

$$dy = G'(t)\,dt,$$

where dx and dy are the quantities pictured in Fig. 9.14. Suppose we define ds by

$$ds^2 = dx^2 + dy^2,$$

as our picture suggests. Continuing to follow our nose, we write the length s of the arc C from a to b as

$$s = \int_C ds = \int_C \sqrt{dx^2 + dy^2},$$

$$= \int_a^b \sqrt{[F'(t)]^2 + [G'(t)]^2}\,dt. \tag{5}$$

This happens to be the formula for s as we defined it.

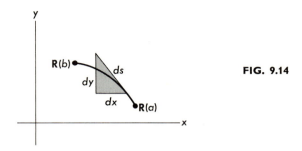

FIG. 9.14

As a final remark on the problem of finding the length of a curve, suppose the curve is described in the first place by $y = H(x)$. We could express the equation in parametric form:

$$x = t,$$

$$y = H(t).$$

Then if $x \in [a, b]$, so is t, and we get

$$s = \int_a^b \sqrt{1 + [H'(t)]^2} \, dt.$$

Now this is the same (see Problem 6.4E) as the integral

$$s = \int_a^b \sqrt{1 + [H'(x)]^2} \, dx.$$

We could have written this down directly from (5) if we simply manipulated algebraically to write

$$s = \int_c \sqrt{\left(1 + \left(\frac{dy}{dx}\right)^2\right) dx^2} = \int_c \sqrt{1 + [H'(x)]^2} \, dx.$$

The length of an arc may be related to the area of a surface. Suppose that an arc is rotated about an axis. It generates a surface. The area of this surface is defined to be the integral of the arc length ds multiplied by 2π times the distance from the arc to the axis.

EXAMPLE E

Find the surface area of a doughnut (known more technically as a *torus*).

Our first problem is to generate a doughnut. This we do by rotating the circle pictured in Fig. 9.15 about the y axis. Equations for the circle may be written in the form

$$x = b + a \cos \theta,$$

$$y = a \sin \theta.$$

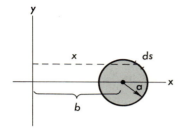

FIG. 9.15

(This convenient parameterization may be verified by noting that $(x - b)^2 + y^2 = a^2(\cos^2 \theta + \sin^2 \theta) = a^2$.) The integral for the surface area of the top half of the doughnut, multiplied by 2, is

$$A = 2\int_0^\pi 2\pi x \sqrt{dx^2 + dy^2},$$

$$= 4\pi \int_0^\pi (b + a \cos \theta)\sqrt{(-a \sin \theta)^2 + (a \cos \theta)^2}\, d\theta,$$

$$= 4\pi a(b\theta + a \sin \theta)\Big|_0^\pi = 4\pi^2\, ab. \quad \blacksquare$$

== 9 · 3   EXERCISES

Find the distance between the indicated points along the indicated curves.

1· From $(0, 0)$ to $(2, 4)$ along $y = x^2$.

2· From $(0, 0)$ to $(4, 8)$ along $y = x^{3/2}$.

3· From $(1, 0)$ to $(\sqrt{3}, \ln \sqrt{3})$ along $y = \ln x$.

4· From $(0, 0)$ to $(\pi/3, \ln \frac{1}{2})$ along $y = \ln \cos x$.

5· From $(8, 0)$ to $(0, 8)$ along $x^{2/3} + y^{2/3} = 4$.

6· From $(0, -1)$ to $(1, 1)$ along $(y + 1)^2 = 4x^3$.

In Exercise 7–10, find the length of the arc described as t varies from $t = 0$ to $t = 2\pi$.

7· $R(t) = e^t \cos t\, i + e^t \sin t\, j$.

8· $R(t) = (4 + 2t)i + (\frac{1}{2}t^2 + 3)j$.

9. $\mathbf{R}(t) = e^t\mathbf{i} + \left(\dfrac{e^{2t}}{2} - \dfrac{t}{4}\right)\mathbf{j}$.

10. $\mathbf{R}(t) = (t + 1)\mathbf{i} + \left(\dfrac{t^2 + 2t}{2}\right)\mathbf{j}$.

Exercises 11–14 refer to Exercises 13–16 of Section 9.1.

11. Refer to Exercise 13. How far has P traveled when it again comes into contact with circle C?

12. Refer to Exercise 14. How far has P traveled when it again comes into contact with Circle C?

13. Refer to Exercise 15. How far has P traveled when it passes through $(-6\pi, -4)$?

14. Refer to Exercise 16. How far has P traveled when it passes through $(4, -8\pi)$?

15. Use the method of Example E to verify that the area of a right circular cone, radius of the base r, height h, is given by $\pi r \sqrt{r^2 + h^2}$.

16. Use the method of Example E to verify that the area of a sphere of radius r is $4\pi r^2$.

17. Two parallel planes are passed through a sphere. Prove that the surface area intercepted between the planes is independent of where they are passed through the sphere.

18. The curve $y = \frac{2}{3}(1 + x^2)^{3/2}$ is rotated about the y axis. Find the area of the surface cut off by the plane $y = \frac{16}{3}$.

9·4 Acceleration and Curvature

Let the position of a point moving in space be described by the radius vector

$$\mathbf{R}(t) = F(t)\mathbf{i} + G(t)\mathbf{j}, \qquad t \in [a, b].$$

The velocity vector at time t is

$$\mathbf{R}'(t) = F'(t)\mathbf{i} + G'(t)\mathbf{j}, \qquad t \in [a, b],$$

which is another vector-valued function. It, too, may be differentiable on the interval. If so, it is natural to define this derivative to be the acceleration vector

$$\mathbf{R}''(t) = F''(t)\mathbf{i} + G''(t)\mathbf{j}.$$

The velocity vector will in general change both its length and its direction as the point moves. Acceleration in this case represents both kinds of change.

To emphasize that acceleration is a change in direction as well as in speed, we consider the following example, in which the speed remains constant.

EXAMPLE A

The vector $\mathbf{R}(t) = (a \cos \omega t)\mathbf{i} + (a \sin \omega t)\mathbf{j}$, where a and ω are constants, describes the motion of a particle that moves around a circle of radius a (Fig. 9.16).

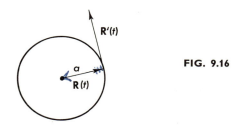

FIG. 9.16

(a) Show that the speed of the particle is constant, and the resulting constant angular velocity of the radius vector drawn to the particle is ω radians per unit time interval.

(b) Show that the acceleration of the particle is always directed toward the center of the circle, and that its magnitude is the product of the radius of the path of motion and the square of the angular velocity ω.

(a) $$\mathbf{R}'(t) = (-a\omega \sin \omega t)\mathbf{i} + (a\omega \cos \omega t)\mathbf{j}.$$

It is clear that this velocity vector is not constant because it continually changes direction. But the speed,

$$|\mathbf{R}'(t)| = \sqrt{(a\omega)^2(\sin^2 \omega t + \cos^2 \omega t)} = a\omega$$

is constant. The distance traveled along the path is a times the central angle subtended by the radius vector, so while the particle travels $a\omega$ in a unit time interval, the radius vector must subtend an angle of ω radians.

(b) $$\mathbf{R}''(t) = (-a\omega^2 \cos \omega t)\mathbf{i} + (-a\omega^2 \sin \omega t)\mathbf{j}.$$

Thus,

$$\mathbf{R}''(t) = -\omega^2 \mathbf{R}(t),$$

so the acceleration vector has the opposite sense of the vector $\mathbf{R}(t)$, which always points from the center of the circle to the particle. Its magnitude is $a\omega^2$. ∎

We have now examined a case in which the speed remained constant. All of the acceleration was expended in changing the direction of the particle. The other extreme is illustrated by the motion of the particle studied in Example 9.3B, where we saw that

$$\mathbf{R}(t) = t^2\mathbf{i} + (3t^2 + 2)\mathbf{j}$$

described a straight line of slope 3. In this case

$$\mathbf{R}'(t) = 2t\mathbf{i} + 6t\mathbf{j},$$
$$\mathbf{R}''(t) = 2\mathbf{i} + 6\mathbf{j}.$$

The acceleration is thus directed along the straight line path of motion. This is what we would expect; if the acceleration were acting in any other way, the particle would be deflected from its straight line course.

Acceleration that acts only to change the direction of a particle must act on the particle in a direction perpendicular to the direction of motion. (Otherwise part of its effect would be to change the speed with which the particle moves along its path.) This acceleration is called *normal acceleration*.

Acceleration that acts only to change the speed of a particle along its path of motion must at any time t be directed along the path of motion. This acceleration is called *tangential acceleration*.

Motion at a constant speed along a circular path illustrates the situation in which all acceleration takes the form of normal acceleration. Motion along a straight line illustrates the situation in which all the acceleration is tangential. Typically, the acceleration of a particle moving in the plane will be a combination of normal and tangential acceleration. The acceleration vector $\mathbf{R}''(t)$ will be directed somewhere between the directions tangent to and normal to the path of motion.

PROBLEM

A · In Problem 9.3C we sketched the path described by

$$\mathbf{R}(t) = (\cos t + t \sin t)\mathbf{i} + (\sin t - t \cos t)\mathbf{j}.$$

We also sketched the velocity vectors $\mathbf{R}'(\pi/2)$ and $\mathbf{R}'(\pi)$.
   (a)   Find $\mathbf{R}''(\pi/2)$. Sketch this vector with its initial point at $\mathbf{R}(\pi/2)$.
   (b)   Find $\mathbf{R}''(\pi)$. Sketch this vector with its initial point at $\mathbf{R}(\pi)$.

Given an arbitrary acceleration vector, it is often useful to know what portion of this acceleration acts normal to the path of motion and what portion is tangential. (It is necessary, for example, for an engineer to know the normal component of acceleration if he is to compute the force tending to throw an accelerating automobile off of the road, or a missile out of its path of motion.)

To accomplish this goal, we begin by finding, for any t, two unit vectors $\mathbf{T}(t)$ and $\mathbf{N}(t)$. The first of these is to be tangent to the path of motion at time t. The second is to be normal to the path (Fig. 9.17). The objective is to

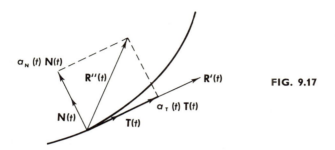

FIG. 9.17

express $\mathbf{R}''(t)$ as some combination of these two vectors. We wish to find two real numbers (which will also change as t does), say $a_T(t)$ and $a_N(t)$ so that

$$\mathbf{R}''(t) = a_T(t)\mathbf{T}(t) + a_N(t)\mathbf{N}(t). \tag{1}$$

It is not hard to find a unit vector tangent to the curve for arbitrary t since we know that $\mathbf{R}'(t)$ is tangent. We merely use the fact that given any vector $\mathbf{u}$, $\mathbf{u}/|\mathbf{u}|$ is a unit vector. Set

$$\mathbf{T}(t) = \frac{\mathbf{R}'(t)}{|\mathbf{R}'(t)|}.$$

Before taking up the problem of how to find a vector $\mathbf{N}(t)$ that is normal to the curve, we wish to investigate $\mathbf{T}(t)$ a little further. As t changes, $\mathbf{T}(t)$ does also; but it changes only in direction, not length. The rate at which $\mathbf{T}(t)$ changes as t increases depends, of course, on how fast the particle is moving along the path. That is, two particles may move on the same path, but at different speeds, and we would expect a derivative of $\mathbf{T}(t)$ with respect to time t (a measure of the rate of change of $\mathbf{T}(t)$) to depend on how fast the particle is moving. But now suppose we let s represent the distance the particle has moved as t has increased from some value $t = a$. For a choice of s, there is a corresponding unit tangent vector $\mathbf{T}(t)$. (See Fig. 9.18.) As s changes, so must $\mathbf{T}(t)$. We may therefore ask how $\mathbf{T}(t)$ changes for a small change in s.

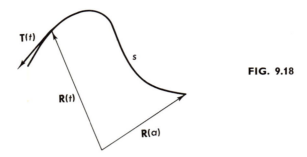

FIG. 9.18

This change depends on the intrinsic properties of the curve and not on the speed of the particle. The sharper the turn in the curve, the greater the change in $\mathbf{T}(t)$ for a small change in s.

DEFINITION A   (Curvature)

Let $\mathbf{T}(t)$ be the unit tangent vector to a curve. Then the curvature is defined to be

$$\kappa(t) = \left| \frac{d}{ds}\, \mathbf{T}(t) \right|.$$

EXAMPLE B

We continue Example A with $\mathbf{R}(t) = (a \cos \omega t)\mathbf{i} + (a \sin \omega t)\mathbf{j}$.
(a)   Find $\mathbf{T}(t)$.
(b)   Find $\kappa(t)$.

By definition,

$$\mathbf{T}(t) = \frac{\mathbf{R}'(t)}{|\mathbf{R}'(t)|} = \frac{(-a\omega \sin \omega t)\mathbf{i} + (a\omega \cos \omega t)\mathbf{j}}{a\omega},$$

$$\mathbf{T}(t) = (-\sin \omega t)\mathbf{i} + (\cos \omega t)\mathbf{j}.$$

Using the chain rule,

$$\frac{d}{ds}\, \mathbf{T}(t) = \frac{d}{dt}\, T(t) \frac{dt}{ds},$$

$$= \{(-\omega \cos \omega t)\mathbf{i} + (-\omega \sin \omega t)\mathbf{j}\}\, \frac{1}{|\mathbf{R}'(t)|}.$$

(Since we are now considering derivatives with respect to s as well as t, we should state explicitly that whenever we use the prime notation, as in $\mathbf{R}'(t)$, we mean $d/dt\ \mathbf{R}(t)$.)

$$\frac{d}{ds}\mathbf{T}(t) = \frac{1}{a}\{(-\cos\omega t)\mathbf{i} + (-\sin\omega t)\mathbf{j}\}.$$

Therefore,

$$\kappa(t) = \frac{1}{a}. \quad \blacksquare$$

A curve that has a curvature of $\kappa(t)$ at a point apparently has a curvature very much like a circle of radius $1/\kappa(t)$. Hence, we define for an arbitrary curve,

$$\text{the radius of curvature} = \frac{1}{\kappa(t)}.$$

PROBLEM   (A Continuation of Problem A)

B · When $\mathbf{R}(t) = (\cos t + t \sin t)\mathbf{i} + (\sin t - t \cos t)\mathbf{j}$,
 (a)   Find $\mathbf{T}(t)$ and $k(t)$.
 (b)   Find $\kappa(\pi/2)$.
 (c)   Find $\kappa(\pi)$.

We now return to the problem of finding the unit normal vector $\mathbf{N}(t)$. The solution is not hard if we recall that for a unit vector $\mathbf{u}$, $\mathbf{u} \cdot \mathbf{u} = 1$. Thus,

$$\mathbf{T}(t) \cdot \mathbf{T}(t) = 1.$$

Differentiating by use of Problem 9.3B, we get

$$\mathbf{T}(t) \cdot \frac{d}{dt}\mathbf{T}(t) + \frac{d}{dt}\mathbf{T}(t) \cdot \mathbf{T}(t) = 0.$$

Since the dot product is commutative, this says

$$2\mathbf{T}(t) \cdot \frac{d}{dt}\mathbf{T}(t) = 0.$$

If $d/dt\ \mathbf{T}(t) \neq \mathbf{0}$, then it must be perpendicular to $\mathbf{T}(t)$. (If $d/dt\,\mathbf{T}(t) = \mathbf{0}$, the second derivative is also $\mathbf{0}$; the particle is moving with constant velocity and we need not worry about expressing $\mathbf{R}''(t)$ in the form (1).) Thus, if $d/dt\ \mathbf{T}(t) \neq \mathbf{0}$, we set

$$\mathbf{N}(t) = \frac{d/dt\ \mathbf{T}(t)}{|d/dt\ \mathbf{T}(t)|}.$$

PROBLEM

C · When $\mathbf{R}(t) = (\cos t + t \sin t)\mathbf{i} + (\sin t - t \cos t)\mathbf{j}$,
   (a)   Find $\mathbf{N}(t)$.
   (b)   Find $\mathbf{N}(\pi/2)$ and sketch it with its initial point at $\mathbf{R}(\pi/2)$.
   (c)   Find $\mathbf{N}(\pi)$ and sketch it with its initial point at $\mathbf{R}(\pi)$.

We have now defined a unit tangent vector $\mathbf{T}(t)$ and a unit normal vector $\mathbf{N}(t)$ for any t. We are ready to try to express $\mathbf{R}''(t)$ in the form suggested in (1).

From the definition of $\mathbf{T}(t)$ and the fact that $ds/dt = |\mathbf{R}'(t)|$, we have that

$$\mathbf{R}'(t) = |\mathbf{R}'(t)|\,\mathbf{T}(t) = \frac{ds}{dt}\,\mathbf{T}(t).$$

Differentiating this product gives

$$\mathbf{R}''(t) = \frac{d^2 s}{dt^2}\,\mathbf{T}(t) + \frac{ds}{dt}\frac{d}{dt}\,\mathbf{T}(t).$$

From the definition of $\mathbf{N}(t)$, $d/dt\,\mathbf{T}(t) = |d/dt\,\mathbf{T}(t)|\,\mathbf{N}(t)$. Hence,

$$\mathbf{R}''(t) = \frac{d^2 s}{dt^2}\,\mathbf{T}(t) + \frac{ds}{dt}\left|\frac{d}{dt}\,\mathbf{T}(t)\right|\mathbf{N}(t).$$

This is in the desired form (1) if we define

$$a_\mathrm{T}(t) = \frac{d^2 s}{dt^2},$$

$$a_\mathrm{N}(t) = \frac{ds}{dt}\left|\frac{d}{dt}\,\mathbf{T}(t)\right|.$$

The expression for $a_\mathrm{N}(t)$ can, with the help of the chain rule, be written in another form.

$$a_\mathrm{N}(t) = \frac{ds}{dt}\left|\frac{d}{ds}\,\mathbf{T}(t)\frac{ds}{dt}\right|,$$

and since $ds/dt = |\mathbf{R}'(t)| \geq 0$,

$$a_\mathrm{N}(t) = \left(\frac{ds}{dt}\right)^2 \kappa(t).$$

We shall state this important result formally.

THEOREM A

Let $\mathbf{R}(t) = F(t)\mathbf{i} + G(t)\mathbf{j}$ give the position of a moving particle at time t. Then the acceleration vector may be written in the form

$$\mathbf{R}''(t) = a_{\mathrm{T}}(t)\mathbf{T}(t) + a_{\mathrm{N}}(t)\mathbf{N}(t)$$

where

$$a_{\mathrm{T}}(t) = \frac{d^2 s}{dt^2} = \frac{d}{dt}|\mathbf{R}'(t)|,$$

$$a_{\mathrm{N}}(t) = \left(\frac{ds}{dt}\right)^2 \kappa(t).$$

PROBLEMS

D · (A continuation of Problem C.) When $\mathbf{R}(t) = (\cos t + t \sin t)\mathbf{i} + (\sin t - t \cos t)\mathbf{j}$,

(a) Find $a_{\mathrm{T}}(t)$ and $a_{\mathrm{N}}(t)$.
(b) Using $a_{\mathrm{T}}(\pi/2)$ and $a_{\mathrm{N}}(\pi/2)$, write $\mathbf{R}''(\pi/2)$ in the form (1).
(c) Using $a_{\mathrm{T}}(\pi)$ and $a_{\mathrm{N}}(\pi)$, write $\mathbf{R}''(\pi)$ in the form (1).

E · Prove: $|\mathbf{R}''(t)|^2 = [a_{\mathrm{T}}(t)]^2 + [a_{\mathrm{N}}(t)]^2$. Hint: Beginning with the expression for $\mathbf{R}''(t)$ given in (1), use $|\mathbf{R}''(t)|^2 = \mathbf{R}''(t) \cdot \mathbf{R}''(t)$.

In a specific problem the hardest part of expresssing the acceleration vector in the form suggested by Theorem A is often finding $\kappa(t)$. We therefore outline a procedure for getting around this computation.

(a) Find $\mathbf{R}''(t)$.
(b) Find $ds/dt = |\mathbf{R}'(t)|$.
(c) Find $a_{\mathrm{T}}(t) = d^2 s/dt^2$.
(d) Find $a_{\mathrm{N}}(t)$ from the relation

$$|a_{\mathrm{N}}(t)|^2 = |\mathbf{R}''(t)|^2 - |a_{\mathrm{T}}(t)|^2.$$

Note in solving this that $a_{\mathrm{N}}(t)$ is always positive. Now if the only purpose were to write the expression for $\mathbf{R}''(t)$ as given in Theorem A, we could now do so, since we have both $a_{\mathrm{T}}(t)$ and $a_{\mathrm{N}}(t)$. If we actually wanted $\kappa(t)$ for some reason,

(e) $\kappa(t) = \dfrac{a_{\mathrm{N}}(t)}{(ds/dt)^2}.$

We shall follow this outline in the proof of the next theorem. The point of this theorem is to enable us to find the curvature when we are given the cartesian equation of the curve.

THEOREM B

Let $y = H(x)$ describe a twice-differentiable curve. The curvature at a point x is then given by

$$\kappa(x) = \frac{H''(x)}{\{1 + [H'(x)]^2\}^{3/2}}.$$

If we set $x = t$, then $y = H(t)$ and the curve $y = H(x)$ is identical to the path traced out by

$$\mathbf{R}(t) = t\mathbf{i} + H(t)\mathbf{j}.$$

Then

$$\mathbf{R}'(t) = \mathbf{i} + H'(t)\mathbf{j} \quad \text{and} \quad \mathbf{R}''(t) = H''(t)\mathbf{j}.$$

$$\frac{ds}{dt} = \{1 + [H'(t)]^2\}^{1/2},$$

$$a_T(t) = \frac{d^2s}{dt^2} = \frac{1}{2}\{1 + [H'(t)]^2\}^{-1/2}2[H'(t)]H''(t),$$

$$a_N(t)^2 = [H''(t)]^2 - \frac{\{H'(t)H''(t)\}^2}{1 + [H'(t)]^2} = \frac{[H''(t)]^2}{1 + [H'(t)]^2},$$

$$a_N(t) = \frac{|H''(t)|}{\{1 + [H'(t)]^2\}^{1/2}},$$

$$\kappa(t) = \frac{|H''(t)|}{\{1 + [H'(t)]^2\}^{3/2}}. \quad \blacksquare$$

9.4 EXERCISES

More is to be learned from much work on one example than will be learned from a little work on many examples. For Exercises 1–8, do each of the following.

(a) Sketch a graph for $t \in [0, 3]$.

(b) Find $\mathbf{R}'(1)$. Sketch it at the point P of the curve corresponding to $t = 1$.

(c) Find $\mathbf{R}''(1)$. Sketch it at P.

(d) Find and draw at P the two vectors $a_T(1)\mathbf{T}(1)$ and $a_N(1)\mathbf{N}(1)$.

(e) Use $S(t) = \int_0^t ds$ to find an expression for the length of the arc corresponding to any t.

(f) Use e to verify that $S'(1) = |\mathbf{R}'(1)|$.

(g) Find the curvature at $t = 1$.

(h)   Draw a line normal to the curve at $t=1$. Write the equation of a circle centered on this line that is tangent to the curve at $t=1$ and has as its radius the radius of curvature of the given curve at $t=1$. Sketch this circle (called the *osculating circle*).

1· $R(t) = \frac{3}{4}t^2\mathbf{i} + \frac{1}{2}t^3\mathbf{j}.$

2· $R(t) = t\mathbf{i} + [\frac{2}{3}(t^2 + 1)]\mathbf{j}.$

3· $R(t) = (3 - t)\mathbf{i} + \dfrac{1}{t}\,\mathbf{j}.$

4· $R(t) = \left(1 - 2\cos\dfrac{\pi t}{3}\right)\mathbf{i} + \left(4\cos^2\dfrac{\pi t}{3} - 4\cos\dfrac{\pi t}{3}\right)\mathbf{j}.$

5· $R(t) = \left(\cos\dfrac{\pi}{4}t + \dfrac{\pi}{4}t\sin\dfrac{\pi}{4}t\right)\mathbf{i} + \left(\sin\dfrac{\pi}{4}t - \dfrac{\pi}{4}t\cos\dfrac{\pi}{4}t\right)\mathbf{j}.$

6· $R(t) = (\pi t - \sin\pi t)\mathbf{i} + (1 - \cos\pi t)\mathbf{j}.$

7· $R(t) = e^t\sin\dfrac{t}{2}\,\mathbf{i} + e^t\cos\dfrac{t}{2}\,\mathbf{j}.$

8· $R(t) = [e^{-2t}\sin\pi t]\mathbf{i} + [e^{-2t}\cos\pi t]\mathbf{j}.$

9 · 5 Polar Coordinates

The radius vector $x\mathbf{i} + y\mathbf{j}$ is completely determined by giving the Cartesian coordinates (x, y) of its terminal point. We can also determine a radius vector by giving its length and the angle it makes with the positive x axis.

Thus, the radius vector $\mathbf{i} + \sqrt{3}\mathbf{j}$ is also determined by the pair $(2, \pi/3)$. It is to be understood in the latter representation that 2 is the length of the vector and $\pi/3$ is the angle made with the positive x axis (Fig. 9.19).

When points are located by giving the length and angle of the corresponding radius vectors, the origin is frequently referred to as the *pole*.

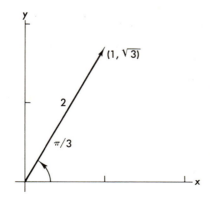

FIG. 9.19

DEFINITION A

Let r be the directed distance of a point P from the origin, θ the angle that vector **OP** makes with the positive x axis. Then (r, θ) are called the *polar coordinates of P*.

If a point has Cartesian coordinates (x, y) and polar coordinates (r, θ), then it is obvious (Fig. 9.20) that

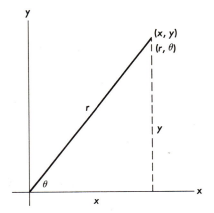

FIG. 9.20

$$x = r \cos \theta,$$
$$y = r \sin \theta,$$
$$x^2 + y^2 = r^2.$$

The polar representation of a point is not unique. In fact, the point $P(r, \theta)$ is equally well represented by $P(r, \theta + 2\pi n)$ or $P(-r, \theta + \pi + 2\pi n)$. Note that the latter representation depends on the notion of r directed negatively.

Relations between polar coordinates can be used to draw a graph just as we have used relations between Cartesian coordinates to draw a graph. Where we have drawn a graph of the function $y = f(x)$ we now draw $r = g(\theta)$. In fact, one of the reasons for using polar coordinates is that it is often easier to write an expression between r and θ to describe a locus than it is to write an expression between x and y. For example, $x^2 + y^2 = 1$ describes the same locus of points as does $r = 1$.

In graphing $r = f(\theta)$, it is best to imagine the radius vector beginning at $\theta = 0$ and either lengthening or shortening as θ increases. Usually, the observation of the value of r at just a few points will enable us to plot the entire curve.

EXAMPLE A

Graph $r = 2 \sin \theta$.

θ	r
0	0
$\pi/6$	1
$\pi/3$	$\sqrt{3}$
$\pi/2$	2

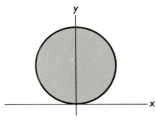

FIG. 9.21

Continue in this fashion to plot points. The graph is indicated in Fig. 9.21. ▮

With just a little practice, the reader will find that he need not plot so many points. It is generally sufficient to examine the equation and choose only those values of θ for which r assumes a local extreme or is zero.

EXAMPLE B

Graph $r = 1 - 2 \sin \theta$.

When $\theta = 0$, $r = 1$. As θ begins to increase, r begins to decrease, taking the value 0 when $\theta = \pi/6$. As θ continues to increase, r becomes negative. By the time $\theta = \pi/2$, we have drawn the curve indicated in Fig. 9.22. As θ con-

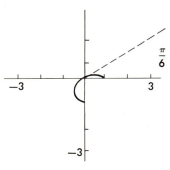

FIG. 9.22

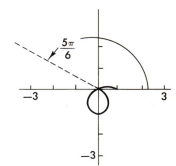

FIG. 9.23

tinues to increase past $\pi/2$, r begins to come back from -1 toward 0; r will be 0 when $\theta = 5\pi/6$ (Fig. 9.23). Moreover, r continues to grow as θ increases to π. At π, $r = 1$. When θ moves into the third quadrant, $\sin \theta$ becomes negative and r exceeds 1 for the first time. It reaches a maximum of 3 at $\theta = 3\pi/2$. The completed graph is indicated in Fig. 9.24. ▮

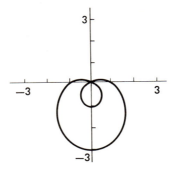

FIG. 9.24

It may happen with certain equations for which we are to draw a graph that a change to another system of coordinates first will make the problem much more simple. Facility in going from polar coordinates to Cartesian coordinates and vice versa is important.

EXAMPLE C

The graph of $r = 2 \sin \theta$ sketched in Example A is oval shaped. Is it in fact a circle?

$$r = 2 \sin \theta,$$
$$r^2 = 2r \sin \theta,$$
$$x^2 + y^2 = 2y,$$
$$x^2 + (y - 1)^2 = 1.$$

We recognize the Cartesian equation as the description of a circle centered at $(0, 1)$ with a radius of 1. ∎

EXAMPLE D

Graph $x^4 + y^4 + 2x^2y^2 - 2xy = 0$.

We rewrite this equation in the form

$$(x^2 + y^2)^2 - 2xy = 0,$$
$$(r^2)^2 - 2r^2 \cos \theta \sin \theta = 0,$$
$$r^2[r^2 - \sin 2\theta] = 0.$$

Setting the second factor equal to zero, we get

$$r = \pm \sqrt{\sin 2\theta}.$$

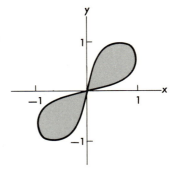

FIG. 9.25

The best tactic here is to graph $r = \sqrt{\sin 2\theta}$. (Note in drawing this graph, we only get values for r when $\sin 2\theta \geq 0$.) Then reflect this picture in the origin to get $r = -\sqrt{\sin 2\theta}$. Since $r = \sqrt{\sin 2\theta}$ passes through the origin, setting the first factor $r^2 = 0$ contributes nothing new to our graph (Fig. 9.25). ▮

PROBLEMS

A · Graph $r = 1 - \sin \theta$.

B · Graph $r = -(1 + \sin \theta)$.

C · Graph $r = a$ (constant) from $\theta = \alpha$ to $\theta = \beta$. What is the area of the region bounded by this curve and the lines $\theta = \alpha$, $\theta = \beta$?

The last problem is a special case of a more general question. Suppose $r = G(\theta)$ is graphed for $\theta \in [\alpha, \beta]$. How do we find the area enclosed by this curve and the lines $\theta = \alpha$, $\theta = \beta$ (Fig. 9.26)? We could change to Cartesian coordinates in some cases, but even when this is algebraically simple, the shape of the region will probably not easily lend itself to being filled up with rectangles. Instead we look at the wedge sketched in Fig. 9.26. The labeling is to suggest that we partition $[\alpha, \beta]$ into subintervals. The area of the wedge is according to Problem C,

$$\tfrac{1}{2}r^2(\theta_k - \theta_{k-1}).$$

If we sum the areas of each wedge obtained in the partitioning of $[\alpha, \beta]$, we get the Riemann sum

$$\sum_{k=1}^{n} \tfrac{1}{2}r^2(\theta_k - \theta_{k-1}) = \sum_{k=1}^{n} \tfrac{1}{2}G(\bar{\theta}_k)^2(\theta_k - \theta_{k-1})$$

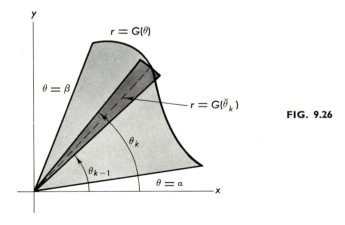

FIG. 9.26

where $\bar{\theta}_k \in [\theta_{k-1}, \theta_k]$. Without supplying the details showing that this sum actually converges, it is hoped that the reader will be convinced from this argument that

$$A = \tfrac{1}{2} \int_\alpha^\beta r^2 d\theta. \tag{1}$$

We shall indicate another proof of this formula in Section 13.3. Note that if we remember the result of Problem C and the picture in Fig. 9.26, then formula (1) follows quite naturally.

EXAMPLE E

Find the area of the region between the inner and outer loops of the curve $r = 1 - 2 \sin \theta$ graphed in Example B.

We will find the area on the right side of the y axis and double it. We must remember the range of the θ that gave rise to the various portions of the curve. We get the area inside the outer loop from

$$A_0 = \tfrac{1}{2} \int_0^{\pi/6} r^2 \, d\theta + \tfrac{1}{2} \int_{3\pi/2}^{2\pi} r^2 \, d\theta.$$

We must then subtract the area of the inner loop

$$A_i = \tfrac{1}{2} \int_{\pi/2}^{5\pi/6} r^2 \, d\theta.$$

Computation of the first integral gives

$$\tfrac{1}{2} \int_0^{\pi/6} 1 - 4 \sin \theta + 4 \sin^2 \theta) = \tfrac{1}{2}[\theta + 4 \cos \theta + 2\theta - \sin 2\theta] \Big|_0^{\pi/6},$$

$$= \frac{\pi + 3\sqrt{3} - 8}{4}.$$

Similarly compute

$$\tfrac{1}{2} \int_{3\pi/2}^{2\pi} r^2 \, d\theta = \frac{3\pi + 8}{4} \qquad \text{and} \qquad \tfrac{1}{2} \int_{\pi/2}^{5\pi/6} r^2 \, d\theta = \frac{2\pi - 3\sqrt{3}}{4}.$$

The desired area is $(\pi + 3\sqrt{3})$. ■

In Exercises 1–10, graph the indicated curve.

1 · $r = 1 + \sin \theta.$ **2 ·** $r = 1 + \cos \theta.$

· 3 · $r = \sin 2\theta.$ **4 ·** $r = \cos 2\theta.$

5 · $r^2 = \cos 3\theta.$ **6 ·** $r^2 = \cos 2\theta.$

7 · $r = 3 - 2 \cos \theta.$ **8 ·** $r = 3 + 2 \sin \theta.$

9 · $r = \sec \theta.$ **10 ·** $r = \csc \theta.$

Graph the curves of Exercises 11–16 and then write the Cartesian equation of the same graph. (Reverse this order if it seems expedient.)

11 · $r(\cos \theta + \sin \theta) = 3.$ **12 ·** $r^2 - 6r \cos \theta + 4r \sin \theta = 48.$

13 · $r = \dfrac{8}{2 + \sin \theta}.$ **14 ·** $r + r \sin \theta = 4.$

15 · $r^2 = \dfrac{144}{9 \cos^2 \theta + 16 \sin^2 \theta}.$ **16 ·** $r = \dfrac{4}{1 + \cos \theta}.$

17–24 ·   Find the areas enclosed by the curves described in Exercises 1–8.

In Exercises 25–36, graph the indicated curve.

25 · $r = \sin 3\theta.$ **26 ·** $r = \cos 3\theta.$

27 · $r = \sin^2 \theta.$ **28 ·** $r = \cos^2 \theta.$

29 · $r = 4 \cos \theta + 3.$ **30 ·** $r = 4 \sin \theta + 3.$

31 · $r^2 = 2 \cos \theta + 1.$ **32 ·** $r^2 = 2 \sin \theta + 1.$

33 · $r^2 = \cos 2\theta.$ **34 ·** $r^2 = \sin 2\theta.$

35 · $r = 2\theta.$ **36 ·** $r = \theta \sin \theta.$

CHAPTER

10

LINEAR ALGEBRA

10 · 1 Solving Linear Equations

Consider the two equations

$$a_{11}x + a_{12}y = 0,$$
$$a_{21}x + a_{22}y = 0; \tag{1}$$

where we assume that neither equation vanishes (has all its coefficients zero). The graphs corresponding to these equations are straight lines through the origin, so we always have the trivial solution $x = y = 0$. We distinguish, however, two possibilities.

Case I. The lines have different slopes, and the trivial solution is the only solution (Fig. 10.1).

Case II. The lines coincide, and there are an infinite number of solutions (Fig. 10.2).

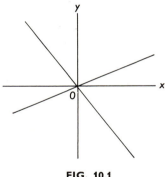

FIG. 10.1

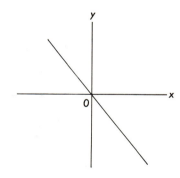

FIG. 10.2

In case II, the system reduces to just one equation since one must be a scalar multiple of the other. If $a_{11} = 0$, all solutions are described by $(x, 0)$ where x is arbitrary. If $a_{11} \neq 0$, all solutions are given by $(-a_{12}y/a_{11}, y)$ where y is arbitrary.

We wish to explore the solution of systems like (1), involving m equations in n unknowns:

$$a_{11}x_1 + a_{12}x_2 + \cdots + a_{1n}x_n = 0$$
$$\vdots \tag{2}$$
$$a_{m1}x_1 + a_{m2}x_2 + \ldots + a_{mn}x_n = 0.$$

Such a system always has the trivial solution $x_1 = x_2 = \cdots = x_n = 0$. The problem of interest is to determine whether or not there are any nontrivial solutions.

Given a system like (2), our goal (roughly stated) is to write down an equivalent system in a form so that all solutions may be simply obtained. In saying that the simplified system should be equivalent to the given system, we mean that both systems should have the same solution. There are three elementary operations which if performed on a given system of equations, will clearly result in an equivalent system:

I. We may change the order in which the equations are listed under one another;

II. Any equation can be multiplied by a nonzero number;

III. To one equation we may add a multiple of another.

In practice, it is a good procedure to write down a complete set of equations with each step. Never omit an equation unless, at some step, all its coefficients vanish (become zero). It is also very useful to keep coefficients of the same unknowns in different equations lined up in vertical columns.

We wish to illustrate these ideas with the system

$$(r_1) \quad 2x_1 + 8x_2 + 4x_3 + 8x_4 = 0,$$
$$(r_2) \quad 3x_1 + 12x_2 \qquad\quad - 6x_4 = 0, \tag{3}$$
$$(r_3) \quad 5x_1 + 20x_2 + x_3 \;\; - 7x_4 = 0.$$

We shall use the elementary operations described above to obtain an equivalent system in which the coefficient of x_1 in the first equation is 1, and the coefficients of x_1 in the remaining equations are 0. Note that to facilitate the discussion, we have labeled the equations row 1, row 2, and row 3.

We begin by copying r_1 and r_3. We replace r_2 by $-r_1 + r_2$, this being elementary operation III:

$$(r_1') \quad 2x_1 + 8x_2 + 4x_3 + 8x_4 = 0,$$
$$(r_2') \quad\;\; x_1 + 4x_2 - 4x_3 - 14x_4 = 0,$$
$$(r_3') \quad 5x_1 + 20x_2 + x_3 - 7x_4 = 0.$$

Using operation I, interchanging r_1' and r_2',

$$(r_1'') \quad\;\; x_1 + 4x_2 - 4x_3 - 14x_4 = 0,$$
$$(r_2'') \quad 2x_1 + 8x_2 + 4x_3 + 8x_4 = 0,$$
$$(r_3'') \quad 5x_1 + 20x_2 + x_3 - 7x_4 = 0.$$

Now copy (r_1'') and use operation III to replace r_2'' by $-2r_1'' + r_2''$; and replace r_3'' by $-5r_1'' + r_3''$.

$$(r_1''')\qquad x_1 + 4x_2 - 4x_3 - 14x_4 = 0,$$

$$(r_2''')\qquad\qquad\qquad 12x_3 + 36x_4 = 0,$$

$$(r_3''')\qquad\qquad\qquad 21x_3 + 63x_4 = 0.$$

We have accomplished our goal. In fact, we have accomplished more since, without planning it, all the x_2 terms have dropped out of r_2''' and r_3'''. We will find it instructive to pursue this illustration further, but it seems appropriate to pause here to introduce a notational convenience that will eliminate a lot of the messy subscripts now dogging us.

If we agree to keep our columns straight, if we assume our reader can see for himself which elementary operation has been used at each step, and if we agree to use $\sim$ to indicate that two systems are equivalent, then the steps above can be indicated more succinctly using just the coefficients arranged in a rectangular array called a *matrix* (plural—*matrices*):

$$\begin{bmatrix} 2 & 8 & 4 & 8 \\ 3 & 12 & 0 & -6 \\ 5 & 20 & 1 & -7 \end{bmatrix} \sim \begin{bmatrix} 2 & 8 & 4 & 8 \\ 1 & 4 & -4 & -14 \\ 5 & 20 & 1 & -7 \end{bmatrix}$$

$$\sim \begin{bmatrix} 1 & 4 & -4 & -14 \\ 2 & 8 & 4 & 8 \\ 5 & 20 & 1 & -7 \end{bmatrix} \sim \begin{bmatrix} 1 & 4 & -4 & -14 \\ 0 & 0 & 12 & 36 \\ 0 & 0 & 21 & 63 \end{bmatrix}$$

We have now completed the first step toward finding all solutions to the given systems (3). An outline of the complete procedure, including the step already taken is given below. For convenience, we refer to the elementary operations as operations on the matrices.

Step 1.   Obtain an equivalent matrix M_1 having a 1 in the upper left corner and zeros in all other positions in the first column.

Step 2.   Consider the rows of M_1 not having a 1 in the first column. If there is such a row having a nonzero entry in column 2, use operation I if necessary to get this element into the second row, and proceed as before to obtain an equivalent matrix M_2 having as the only nonzero entry in the second column a 1 in the second row. If no such row exists, go to step 3.

Notice that after step 2, column 1 still has only one nonzero entry, the 1 in the upper left hand corner.

Step 3.   Consider the rows of M_2 not having a 1 in the first or second columns. If there is such a row having a nonzero entry in the third column, proceed as in step 2. If no such row exists, go to step 4.

This process clearly terminates after a maximum of n steps (less if some row identically vanishes). The resulting matrix is said to be a *row-reduced echelon matrix*. From such a matrix, all solutions to the system are easily determined as indicated below.

We return to our illustration. In attempting step 2, we see that except for row 1, there are no nonzero entries in the second column. So we move on to step 3. We easily obtain

$$\begin{bmatrix} 1 & 4 & -4 & -14 \\ 0 & 0 & 12 & 36 \\ 0 & 0 & 21 & 63 \end{bmatrix} \sim \begin{bmatrix} 1 & 4 & -4 & -14 \\ 0 & 0 & 1 & 3 \\ 0 & 0 & 1 & 3 \end{bmatrix} \qquad \text{(operation II)}$$

$$\sim \begin{bmatrix} 1 & 4 & 0 & -2 \\ 0 & 0 & 1 & 3 \\ 0 & 0 & 0 & 0 \end{bmatrix} \qquad \text{(operation III)}$$

The process terminates after step 3 because the last row has vanished. Introducing the unknowns again, this last (row reduced echelon) matrix corresponds to

$$x_1 + 4x_2 \quad - 2x_4 = 0,$$
$$x_3 + 3x_4 = 0.$$

It is clear that

$$x_1 \quad = -4x_2 + 2x_4,$$
$$x_3 = \quad - 3x_4,$$

so if we choose x_2 and x_4 arbitrarily, and then let x_1 and x_3 be determined accordingly, we will have a solution. All solutions are then given by

$$(-4x_2 + 2x_4, x_2, -3x_4, x_4). \tag{4}$$

For example, choosing $x_2 = 3$, $x_4 = 2$, we get $(-8, 3, -6, 2)$. The reader should verify that these numbers when substituted for (x_1, x_2, x_3, x_4) in (3) do constitute a solution. He should also generate some solutions of his own by choosing x_2 and x_4 to be some of his favorite numbers.

We make one more observation about notation. In order to exhibit the coefficient matrix and at the same time indicate the unknowns being used, the system of equations indicated in (2) is often written in the matrix form

$$\begin{bmatrix} a_{11} & a_{12} & \cdots & a_{1n} \\ \vdots & & & \vdots \\ a_{m1} & a_{m2} & \cdots & a_{mn} \end{bmatrix} \begin{bmatrix} x_1 \\ \vdots \\ x_n \end{bmatrix} = \begin{bmatrix} 0 \\ \vdots \\ 0 \end{bmatrix}$$

The coefficient matrix of m rows and n columns is referred to as an $m \times n$ *matrix*. In the same vein, one might refer to the column of x's as an $n \times 1$ matrix. One then thinks of the system of equations (2) as the recipe (definition) to be followed in multiplying an $m \times n$ matrix by an $n \times 1$ matrix.

EXAMPLE A

Solve the system $\begin{bmatrix} 2 & -1 & -2 \\ -1 & 2 & -2 \\ -2 & -2 & 8 \end{bmatrix} \begin{bmatrix} x \\ y \\ z \end{bmatrix} = \begin{bmatrix} 0 \\ 0 \\ 0 \end{bmatrix}$

We first obtain a 1 in the upper left corner by multiplying the second row by -1 and then interchanging the first two rows:

$$\begin{bmatrix} 2 & -1 & -2 \\ -1 & 2 & -2 \\ -2 & -2 & 8 \end{bmatrix} \sim \begin{bmatrix} 1 & -2 & 2 \\ 2 & -1 & -2 \\ -2 & -2 & 8 \end{bmatrix}$$

We now use the upper left hand 1 to eliminate the other numbers in the first column; multiply row 1 by -2 and add to the second row; multiply row 1 by 2 and add to the third row. Note that the first step here is to copy down the row being used to effect changes in the other rows:

$$\begin{bmatrix} 1 & -2 & 2 \\ 2 & -1 & -2 \\ -2 & -2 & 8 \end{bmatrix} \sim \begin{bmatrix} 1 & -2 & 2 \\ 0 & 3 & -6 \\ 0 & -6 & 12 \end{bmatrix}$$

It is clear that we can now obtain a 1 in the second position on the diagonal by multiplying row 2 by $\frac{1}{3}$. Then this 1 can be used to eliminate the -2 in the first row and the -6 in the third row:

$$\begin{bmatrix} 1 & -2 & 2 \\ 0 & 3 & -6 \\ 0 & -6 & 12 \end{bmatrix} \sim \begin{bmatrix} 1 & -2 & 2 \\ 0 & 1 & -2 \\ 0 & -6 & 12 \end{bmatrix} \sim \begin{bmatrix} 1 & 0 & -2 \\ 0 & 1 & -2 \\ 0 & 0 & 0 \end{bmatrix}$$

Since the last row turned out to consist of all zeros, there is no hope of eliminating the -2's in the last column. We have come to

$$\begin{bmatrix} 1 & 0 & -2 \\ 0 & 1 & -2 \\ 0 & 0 & 0 \end{bmatrix} \begin{bmatrix} x \\ y \\ z \end{bmatrix} = \begin{bmatrix} 0 \\ 0 \\ 0 \end{bmatrix} \quad \text{or} \quad \begin{aligned} x &= 2z, \\ y &= 2z, \\ 0z &= 0. \end{aligned}$$

Clearly, we can choose z arbitrarily, but then x and y are determined. We say the solution is given by

$$(2z, 2z, z) = z(2, 2, 1). \quad \blacksquare$$

EXAMPLE B

Solve the system $\begin{bmatrix} -7 & -1 & -2 \\ -1 & -7 & -2 \\ -2 & -2 & -1 \end{bmatrix} \begin{bmatrix} x \\ y \\ z \end{bmatrix} = \begin{bmatrix} 0 \\ 0 \\ 0 \end{bmatrix}$

$$\begin{bmatrix} -7 & -1 & -2 \\ -1 & -7 & -2 \\ -2 & -2 & -1 \end{bmatrix} \sim \begin{bmatrix} 1 & 7 & 2 \\ -7 & -1 & -2 \\ -2 & -2 & -1 \end{bmatrix} \sim \begin{bmatrix} 1 & 7 & 2 \\ 0 & 48 & 12 \\ 0 & 12 & 3 \end{bmatrix}$$

$$\sim \begin{bmatrix} 1 & 7 & 2 \\ 0 & 4 & 1 \\ 0 & 4 & 1 \end{bmatrix} \sim \begin{bmatrix} 1 & -1 & 0 \\ 0 & 0 & 0 \\ 0 & 4 & 1 \end{bmatrix}$$

We thus have the system

$$x - y = 0,$$
$$0y = 0,$$
$$4y + z = 0,$$

with solution $(y, y, -4y) = y(1, 1, -4)$. ∎

Having learned to solve homogeneous equations, that is, equations of the form (1) in which the right-hand column matrix consists entirely of zeros, it is but a small step to solve systems in which the right hand column matrix has nonzero entries. The principal difference is that in performing any of the three elementary row operations, it is necessary to keep track of the effect on the right-hand side as well as on the left. This is easily accomplished by listing the entries on the right side as one more column in the coefficient matrix. To serve as a reminder of what has been done, we introduce a vertical line in the matrix to set off the coefficient matrix from the added column vector.

EXAMPLE C

We shall solve the system

$$3x - y + z = 2,$$
$$x + 2y + 3z = 4,$$
$$5x - y + 2z = 3.$$

$$\left[\begin{array}{ccc|c} 3 & -1 & 1 & 2 \\ 1 & 2 & 3 & 4 \\ 5 & -1 & 2 & 3 \end{array}\right] \sim \left[\begin{array}{ccc|c} 1 & 2 & 3 & 4 \\ 3 & -1 & 1 & 2 \\ 5 & -1 & 2 & 3 \end{array}\right] \sim \left[\begin{array}{ccc|c} 1 & 2 & 3 & 4 \\ 0 & -7 & -8 & -10 \\ 0 & -11 & -13 & -17 \end{array}\right]$$

$$\sim \left[\begin{array}{ccc|c} 1 & 2 & 3 & 4 \\ 0 & 21 & 24 & 30 \\ 0 & -22 & 26 & -34 \end{array}\right] \sim \left[\begin{array}{ccc|c} 1 & 2 & 3 & 4 \\ 0 & 21 & 24 & 30 \\ 0 & -1 & -2 & -4 \end{array}\right]$$

$$\sim \left[\begin{array}{ccc|c} 1 & 0 & -1 & -4 \\ 0 & 1 & 2 & 4 \\ 0 & 0 & -18 & -54 \end{array}\right] \sim \left[\begin{array}{ccc|c} 1 & 0 & 0 & -1 \\ 0 & 1 & 0 & -2 \\ 0 & 0 & 1 & 3 \end{array}\right]$$

Thus, we have the solution

$$
\begin{aligned}
x &= -1 \\
y &= -2 \\
z &= 3
\end{aligned}
$$ ∎

It is to be noted that while a homogeneous system always has the trivial solution of $x_1 = \cdots = x_n = 0$, a nonhomogeneous system may have no solution at all. For example,

$$
\begin{aligned}
2x - 4y &= 5, \\
-3x + 6y &= 7,
\end{aligned}
$$

has no solution since the graphs of the two equations are distinct, parallel lines (Fig. 10.3). On the other hand, a nonhomogeneous system may have many solutions.

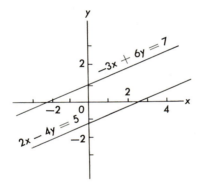

FIG. 10.3

EXAMPLE D

We shall solve the system

$$
\begin{aligned}
2x_1 + 8x_2 + 4x_3 - 4x_4 &= 3, \\
3x_1 + 12x_2 \qquad\quad - 9x_4 &= 5, \\
6x_1 + 24x_2 + 36x_3 \qquad\quad &= 7.
\end{aligned}
$$

$$
\begin{bmatrix}
2 & 8 & 4 & -4 & \bigm| & 3 \\
3 & 12 & 0 & -9 & \bigm| & 5 \\
6 & 24 & 36 & 0 & \bigm| & 7
\end{bmatrix}
\sim
\begin{bmatrix}
2 & 8 & 4 & -4 & \bigm| & 3 \\
1 & 4 & -4 & -5 & \bigm| & 2 \\
0 & 0 & 24 & 12 & \bigm| & -2
\end{bmatrix}
$$

$$
\sim
\begin{bmatrix}
1 & 4 & -4 & -5 & \bigm| & 2 \\
0 & 0 & 12 & 6 & \bigm| & -1 \\
0 & 0 & 24 & 12 & \bigm| & -2
\end{bmatrix}
\sim
\begin{bmatrix}
1 & 4 & -4 & -5 & \bigm| & 2 \\
0 & 0 & 1 & \frac{1}{2} & \bigm| & -\frac{1}{12}
\end{bmatrix}
$$

$$
\sim
\begin{bmatrix}
1 & 4 & 0 & -3 & \bigm| & \frac{5}{3} \\
0 & 0 & 1 & \frac{1}{2} & \bigm| & -\frac{1}{12}
\end{bmatrix}
$$

We have now found the equivalent system

$$x_1 + 4x_2 \quad - 3x_4 = \tfrac{5}{3},$$

$$x_3 + \tfrac{1}{2}x_4 = -\tfrac{1}{12},$$

from which we see that x_2 and x_4 may be chosen arbitrarily so long as we choose $x_1 = \tfrac{5}{3} - 4x_2 + 3x_4$; $x_3 = -\tfrac{1}{12} - \tfrac{1}{2}x_4$. For a specific example of a solution, set $x_2 = \tfrac{2}{3}$ and $x_4 = 0$. We obtain $(-1, \tfrac{2}{3}, -\tfrac{1}{12}, 0)$. ∎

10·1 EXERCISES

Find all solutions to the following systems of homogeneous equations.

1.
$$x + y - z = 0,$$
$$2x - y + z = 0,$$
$$x \quad - 2z = 0.$$

2.
$$2x + y + z = 0,$$
$$x - y + z = 0,$$
$$x \quad + 2z = 0.$$

3.
$$x - y + z = 0,$$
$$x + y + 2z = 0.$$

4.
$$2x + y + 3z = 0,$$
$$x - y + z = 0.$$

5.
$$2x + y + z = 0,$$
$$3x - y + 2z = 0,$$
$$- 5y + z = 0.$$

6.
$$x + 3y - 2z = 0,$$
$$3x + y - z = 0,$$
$$-7x + 3y - z = 0.$$

7.
$$w - x + y + 2z = 0,$$
$$2w + x + 2y - 2z = 0,$$
$$-w + x + 2y - 2z = 0.$$

8.
$$3w - x + 2y + 2z = 0,$$
$$w + x - 2y - 2z = 0,$$
$$2w + x - y - 2z = 0.$$

9.
$$w + x - y + 2z = 0,$$
$$3w - x + 2y + z = 0,$$
$$3w - 5x + 7y - 4z = 0.$$

10.
$$2w - x + y + 3z = 0,$$
$$3w - 2x + 2y + 2z = 0,$$
$$w + x - y + 9z = 0.$$

Find any solutions that exist to the following systems of nonhomogeneous systems of equations.

11.
$$3x - y + z = 4,$$
$$x + 2y + 2z = 10,$$
$$-7x + 7y + z = 8.$$

12.
$$2x + 5y - 3z = 4,$$
$$x + 4y - 5z = -9,$$
$$3x + 6y - z = 17.$$

13.
$$x - y + z = 4,$$
$$2x + y - 3z = -2,$$
$$x - 4y + 6z = 10.$$

14.
$$x + 2y - 3z = -6,$$
$$2x + y - z = 0,$$
$$3y - 5z = 10.$$

15.
$$w + x - y - z = -3,$$
$$w - x - y + 2z = 2,$$
$$2w + x + y + z = 4.$$

16.
$$w - x + y - z = 3,$$
$$2w + x - y - z = -2,$$
$$w - x + 2y - 3z = 3.$$

17.
$$w + x + y - 3z = 4,$$
$$3w + 2x - y + 2z = 5,$$
$$w - x - 2y = -5,$$
$$2w - x - 2y + z = -5.$$

18.
$$2w - x - y + 2z = 0,$$
$$w + 2x - y - 2z = -1,$$
$$3w - x + 2y + 3z = 7,$$
$$w + 2x + 3y - z = 7.$$

10 · 2   More about Vectors

We have already worked with the unit vectors **i** and **j** in the plane. Since our aim in this section is to learn something about vectors in higher dimensions, let us introduce some new terms by first using them in the familiar context of the plane (two dimensional space R^2).

Since any vector $\mathbf{v} \in R^2$ can be written in the form $\mathbf{v} = a\mathbf{i} + b\mathbf{j}$, the vectors **i** and **j** are called *basis vectors*. We say that together they *span* R^2. Any sum of the form $a\mathbf{i} + b\mathbf{j}$ is called a *linear combination* of the vectors **i** and **j**. We said in Chapter 1 that if two nonzero vectors **u** and **v** were such that one was a multiple of the other, that is if there is a real number r such that $\mathbf{u} = r\mathbf{v}$, then **u** and **v** are *linearly dependent*. It is more common, though it amounts to the same thing, to say that two nonzero vectors **u** and **v** are *linearly independent* if and only if $a\mathbf{u} + b\mathbf{v} = 0$ implies that the real numbers a and b must both be 0. Then **u** and **v** are linearly dependent if they are not linearly independent, consistent with the previous notion. Vectors **i** and **j** are linearly independent. Finally, all the vectors obtainable as scalar multiples of a single vector **u** are said to form a one dimensional subspace of R^2, and this subspace is *spanned* by **u**. If we think of **u** as a radius vector, then the subspace spanned by **u** consists of all points on the line through the origin determined by the direction of **u** (Fig. 10.4).

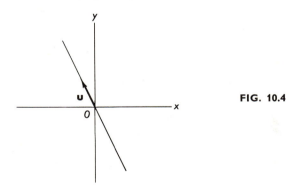

FIG. 10.4

As intimated, we aim at higher things. We now propose to start over again, introducing each of the terms above more formally, and in a way that is useful in spaces other than R^2. But at each step of the way, the reader should verify that our terminology is consistent with his mental picture in the plane. For example, in our definition, we distinguish between the real number 0 and the zero vector **0**. This distinction has already been made in the plane where the zero vector is $\mathbf{0} = (0, 0)$.

DEFINITION A

A real *vector space* is a collection $\mathscr{V} = \{u, v, \ldots\}$ of elements called vectors that satisfy the following:

(1) $u, v \in \mathscr{V}$ implies $u + v \in \mathscr{V}$.
(2) $(u + v) + w = u + (v + w)$.
(3) $0 \in \mathscr{V}$, where $u + 0 = u$ for all u.
(4) $u + v = v + u$.
(5) If r is a real number, then $u \in \mathscr{V}$ implies $ru \in \mathscr{V}$.
(6) $r(u + v) = ru + rv$.
(7) For real numbers r and s, $(r + s)u = ru + su$ and $(rs)u = r(su)$.
(8) $0u = 0$; $1u = u$.

This may be viewed as a list of the properties we have already observed for vectors in the plane. Or it may be viewed much more abstractly. For example the "vectors" may be thought of as polynomials with real coefficients. All the statements remain true. But we shall not be concerned with such interpretations here.

A subset of $\mathscr{V}$ that is itself a vector space is called a *subspace*. Since any subset of $\mathscr{V}$ inherits the properties of the space $\mathscr{V}$, the principal question that needs to be answered when investigating a subset $\mathscr{S} \subset \mathscr{V}$ is the following. If u and v are in $\mathscr{S}$, is it always true that $(u + rv) \in \mathscr{S}$ for any real number r? If the answer is yes for all u and v in $\mathscr{S}$, then $\mathscr{S}$ is a subspace.

Note that for a vector space $\mathscr{V}$, the set consisting just of 0 is a subspace, as is $\mathscr{V}$ itself. These are called the *trivial subspaces*.

EXAMPLE A

Consider, for some fixed n, the collection of all ordered sets of n real numbers, that is, the sets that we may write as $(x_1, x_2, \ldots, x_n)$.

Two such sets will be equal if and only if they are identical in each entry. Define addition and multiplication by a real number by

$$(x_1, \ldots, x_n) + (y_1, \ldots, y_n) = (x_1 + y_1, \ldots, x_n + y_n),$$
$$r(x_1, \ldots, x_n) = (rx_1, \ldots, rx_n).$$

It is an entirely trivial exercise to verify that this system satisfies the eight properties of a vector space, the zero vector being $0 = (0, \ldots, 0)$. We also observe an obvious basis for the space given by

$$\varepsilon_1 = (1, 0, \ldots, 0),$$
$$\varepsilon_2 = (0, 1, \ldots, 0),$$
$$\vdots$$
$$\varepsilon_n = (0, 0, \ldots, 1).$$

To verify that these vectors form a basis, we first note that they are linearly independent. Indeed, if

$$a_1\varepsilon_1 + a_2\varepsilon_2 + \cdots + a_n\varepsilon_n = (a_1, a_2, \ldots, a_n) = (0, 0, \ldots, 0),$$

then by definition of equality, $a_1 = a_2 = \cdots = a_n = 0$. And clearly, for any set $(x_1, \ldots, x_n)$, we have $(x_1, \ldots, x_n) = x_1\varepsilon_1 + \cdots + x_n\varepsilon_n$. Members of this space are called *n*-tuples. The space itself is referred to as *n*-space and is designated by R^n. The basis $\{\varepsilon_1, \ldots, \varepsilon_n\}$ is called the *standard basis*. ∎

EXAMPLE B

Consider the system of equations

$$\begin{bmatrix} a_{11} & a_{12} & \cdots & a_{1n} \\ \vdots & & & \vdots \\ a_{m1} & a_{m2} & \cdots & a_{mn} \end{bmatrix} \begin{bmatrix} x_1 \\ \vdots \\ x_n \end{bmatrix} = \begin{bmatrix} 0 \\ \vdots \\ 0 \end{bmatrix} \tag{1}$$

studied in Section 10.1.

The collection of all solutions constitute a subspace of R^n. It is obvious that the solutions are in R^n, and to show that they form a subspace, we consider any two solutions $(s_1, \ldots, s_n)$ and $(t_1, \ldots, t_n)$. (If there are no solutions other than the trivial solution $(0, \ldots, 0)$, then the solution subspace is the trivial subspace consisting of the one vector $\mathbf{0}$.) Form $(s_1, \ldots, s_n) + r(t_1, \ldots, t_n) = (s_1 + rt_1, \ldots, s_n + rt_n)$. Substitution in any of the m equations gives

$$a_{j1}(s_1 + rt_1) + \cdots + a_{jn}(s_n + rt_n)$$
$$= [a_{j1}s_1 + \cdots + a_{jn}s_n] + r[a_{j1}t_1 + \cdots + a_{jn}t_n] = 0,$$

so we again have a solution. This establishes the assertion, and explains the common terminology of a *solution space* for a system of equations. ∎

We return to terminology defined for an arbitrary vector space.

DEFINITION B

A set of vectors $\mathbf{u}_1, \mathbf{u}_2, \ldots, \mathbf{u}_n$ is said to be linearly independent if $a_1\mathbf{u}_1 + a_2\mathbf{u}_2 + \cdots + a_n\mathbf{u}_n = \mathbf{0}$ implies that $a_1 = a_2 = \cdots = a_n = 0$.

In R^2, the set $\mathbf{u}_1 = 4\mathbf{i} + 2\mathbf{j}$, $\mathbf{u}_2 = -2\mathbf{i} + \mathbf{j}$ is linearly independent. But if we set $\mathbf{u}_3 = 2\mathbf{i} - 5\mathbf{j}$, the set $\mathbf{u}_1, \mathbf{u}_2, \mathbf{u}_3$ is not linearly independent because $3\mathbf{u}_1 + 4\mathbf{u}_2 - 2\mathbf{u}_3 = \mathbf{0}$. (Try it!)

DEFINITION C

A set of vectors $\mathbf{u}_1, \mathbf{u}_2, \ldots, \mathbf{u}_n$ is called a *basis* of vector space $\mathscr{V}$ if (1) the set is linearly independent and (2) given any $\mathbf{v} \in \mathscr{V}$ there is a set of real numbers so that $\mathbf{v} = a_1 \mathbf{u}_1 + a_2 \mathbf{u}_2 + \cdots + a_n \mathbf{u}_n$.

In R^2, the vectors $\mathbf{i}$ and $\mathbf{j}$ form a basis. So do the vectors $\mathbf{u}_1 = 4\mathbf{i} + 2\mathbf{j}$, $\mathbf{u}_2 = -2\mathbf{i} + \mathbf{j}$.

We saw in Example B above that the solutions to a system of homogeneous linear equations in n unknowns constitute a subspace of R^n. The usual process of solving such a system can be viewed as a process for finding a basis for the solution space. Thus, in Example A of Section 10.1, we found that the solution space had a basis consisting of the vector $(2, 2, 1)$; in Example B of that Section, we found $(1, 1, -4)$ to be a basis vector for the solution space. But the system (3) of equations introduced in Section 10.1 had a solution space with a basis of two vectors, since the solutions (4) had the form

$$(-4x_2 + 2x_4, x_2, -3x_4, x_4) = x_2(-4, 1, 0, 0) + x_4(2, 0, -3, 1). \quad \blacksquare$$

PROBLEMS

A · Let $\mathbf{v} = -4\mathbf{i} - 8\mathbf{j}$. Having said that $\mathbf{u}_1 = 4\mathbf{i} + 2\mathbf{j}$ and $\mathbf{u}_2 = -2\mathbf{i} + \mathbf{j}$ constitute a basis for R^2, we should be able to find a_1 and a_2 so that $\mathbf{v} = a_1\mathbf{u}_1 + a_2\mathbf{u}_2$. Do so.

B · Generalize Problem A by supposing $\mathbf{v} = r\mathbf{i} + s\mathbf{j}$. Choose a_1 and a_2 so that $\mathbf{v} = a_1\mathbf{u}_1 + a_2\mathbf{u}_2$.

C · Prove that in R^2, any set of three vectors $\mathbf{u}_1$, $\mathbf{u}_2$, and $\mathbf{u}_3$ must be linearly dependent.

A single vector in R^2 spans a one-dimensional subspace. Three vectors are always linearly dependent (Problem C). Hence, a basis for R^2 must always consist of exactly two vectors.

Recall that $\mathbf{i}$ and $\mathbf{j}$ were defined as the unit vectors directed along the positive coordinate axes in the plane. If we introduce the vector $\mathbf{k}$ as the unit vector along the positive vertical axis in space, then any vector in space may be expressed in the form $\mathbf{v} = a\mathbf{i} + b\mathbf{j} + c\mathbf{k}$. Vectors $\mathbf{i}, \mathbf{j}$, and $\mathbf{k}$ form a basis for R^3. The nontrivial subspaces are more interesting, consisting of spaces spanned by one vector (corresponding to a line through the origin) or by a pair of linearly independent vectors (corresponding to a plane through the origin). All possible linear combinations of a set of vectors form a subspace said to be spanned by the set.

PROBLEMS

D · Show that any $v \in R^3$ can be written in the form $v = a_1 u_1 + a_2 u_2 + a_3 u_3$ where $u_1 = -i$, $u_2 = j + k$, $u_3 = j - k$. (Hint: We know $v = ai + bj + ck$. Also show that u_1, u_2, and u_3 are linearly independent.)

E · Because of Problem D, we know that u_1, u_2, and u_3 form a basis for R^3. Write $v = -4i - j - 5k$ in the form $v = a_1 u_1 + a_2 u_2 + a_3 u_3$.

F · Show that the set $v_1 = 5i + 5j - 3k$, $v_2 = -i + 3j - k$ and $v_3 = 5i - k$ spans the same subspace as does the set $w_1 = i + 2j - k$ and $w_2 = 2i - j$. (Hint: Show that v_1, v_2, and v_3 are each linear combinations of w_1 and w_2; and conversely.)

We saw above that any basis of R^2 has exactly two vectors. We know that i, j, and k form a basis for R^3, and we found in Problem D a second set of three vectors u_1, u_2, and u_3 that also form a basis for R^3. It is natural to suspect that every basis of R^3 has exactly three vectors.

THEOREM A

Any basis of R^3 has exactly three vectors.

PROOF:

Let $v_1, v_2, \ldots, v_n$ be a basis. Then the system $x_1 v_1 + x_2 v_2 + \cdots + x_n v_n = 0$ has only the trivial solution. But let us write each of the basis vectors $v_i \in R^3$ as a column vector:

$$v_1 = \begin{bmatrix} a_{11} \\ a_{21} \\ a_{31} \end{bmatrix}, \quad v_2 = \begin{bmatrix} a_{12} \\ a_{22} \\ a_{32} \end{bmatrix}, \quad \ldots, \quad v_n = \begin{bmatrix} a_{1n} \\ a_{2n} \\ a_{3n} \end{bmatrix}.$$

Then $x_1 v_1 + x_2 v_2 + \cdots + x_n v_n = 0$ becomes

$$\begin{bmatrix} a_{11} x_1 \\ a_{21} x_1 \\ a_{31} x_1 \end{bmatrix} + \begin{bmatrix} a_{12} x_2 \\ a_{22} x_2 \\ a_{32} x_2 \end{bmatrix} + \cdots + \begin{bmatrix} a_{1n} x_n \\ a_{2n} x_n \\ a_{3n} x_n \end{bmatrix} = \begin{bmatrix} 0 \\ 0 \\ 0 \end{bmatrix}$$

or, after adding vectors in the usual way,

$$a_{11} x_1 + a_{12} x_2 + \cdots + a_{1n} x_n = 0,$$

$$a_{21} x_1 + a_{22} x_2 + \cdots + a_{2n} x_n = 0,$$

$$a_{31} x_1 + a_{32} x_2 + \cdots + a_{3n} x_n = 0.$$

It is clear, however, that such a system will always have more than the trivial solution $x_1 = x_2 = \cdots = x_n = 0$ unless $n \leq 3$. Thus, for $v_1, v_2, \ldots, v_n$ to be a basis, $n \leq 3$. Now if $n < 3$, we may suppose $\mathbf{v}_1$ and $\mathbf{v}_2$ form a basis. It will follow that

$$\mathbf{i} = a_{11}\mathbf{v}_1 + a_{12}\mathbf{v}_2,$$

$$\mathbf{j} = a_{21}\mathbf{v}_1 + a_{22}\mathbf{v}_2,$$

$$\mathbf{k} = a_{31}\mathbf{v}_1 + a_{32}\mathbf{v}_2.$$

We now consider, for real numbers x_1, x_2, and x_3 the sum $x_1\mathbf{i} + x_2\mathbf{j} + x_3\mathbf{k}$. Since $\mathbf{i}$, $\mathbf{j}$, and $\mathbf{k}$ are linearly independent, this sum cannot be $\mathbf{0}$ unless $x_1 = x_2 = x_3 = 0$. But we notice that

$$x_1\mathbf{i} + x_2\mathbf{j} + x_3\mathbf{k} = (a_{11}x_1 + a_{21}x_2 + a_{31}x_3)\mathbf{v}_1 + (a_{12}x_1 + a_{22}x_2 + a_{32}x_3)\mathbf{v}_2.$$

Certainly we can find a nontrivial solution to the system

$$a_{11}x_1 + a_{21}x_2 + a_{31}x_3 = 0,$$

$$a_{12}x_1 + a_{22}x_2 + a_{32}x_3 = 0,$$

contradicting the linear independence of $\mathbf{i}$, $\mathbf{j}$, and $\mathbf{k}$. Hence, $n = 3$. ∎

In exactly the same way we can prove that if a vector space $\mathscr{V}$ has a basis $\mathbf{u}_1, \mathbf{u}_2, \ldots, \mathbf{u}_m$ and a second basis $\mathbf{v}_1, \mathbf{v}_2, \ldots, \mathbf{v}_n$, then $m = n$. This fixed number of elements in any basis of the space $\mathscr{V}$ is called the *dimension* of $\mathscr{V}$. In particular, we note that the space R^n (see Example A) has dimension n. Surprise!

There is one more feature of R^n to which we must draw attention. The idea was introduced for R^2 in Section 1.2 where, for two vectors $\mathbf{v} = a\mathbf{i} + b\mathbf{j}$ and $\mathbf{w} = c\mathbf{i} + d\mathbf{j}$, we introduced the dot product $\mathbf{v} \cdot \mathbf{w} = ac + bd$. The natural extension to vectors of R^n is obvious.

DEFINITION D

The dot product of two vectors $\mathbf{v} = (a_1, \ldots, a_n)$ and $\mathbf{w} = (b_1, \ldots, b_n)$ is $\mathbf{v} \cdot \mathbf{w} = a_1b_1 + \cdots + a_nb_n$.

It is the introduction of the notion of dot product that paves the way for a concept of length in R^n. Taking our cue from Problem 1.2D, we define the *length* (also called the *norm*) of a vector $\mathbf{v} \in R^n$ to be

$$|\mathbf{v}| = \sqrt{\mathbf{v} \cdot \mathbf{v}}.$$

It is clear that this definition of length corresponds to the usual notion in the plane, and the reader may easily verify that it is "right" for a vector $\mathbf{v} = a\mathbf{i} + b\mathbf{j} + c\mathbf{k}$ in R^3. A vector of length one is called a unit vector.

The dot product also gives us a way to define the concept of an angle between two vectors in R^n. When $n > 3$, we have no intuitive picture in mind (regard with suspicion people making claims to the contrary), but we can use what we know about the case $n = 2$ to guide us into a definition for higher dimensions. Given two nonzero vectors $\mathbf{p}$ and $\mathbf{q}$ in R^n, we define the angle θ between them to be the solution of

$$\theta = \text{Arccos } \frac{\mathbf{p} \cdot \mathbf{q}}{|\mathbf{p}||\mathbf{q}|}.$$

To be certain that this definition always makes sense, we need to be sure that

$$\frac{|\mathbf{p} \cdot \mathbf{q}|}{|\mathbf{p}||\mathbf{q}|} \leq 1.$$

That is, we need to be sure that $|\mathbf{p} \cdot \mathbf{q}| \leq |\mathbf{p}||\mathbf{q}|$. This verification is the substance of Problem J.

Recall that in R^2, two nonzero vectors $\mathbf{v}$ and $\mathbf{w}$ are perpendicular if and only if $\mathbf{v} \cdot \mathbf{w} = 0$. Such vectors are called *orthogonal*, and again we carry the definition to R^n. In particular a set of basis vectors $\{\mathbf{u}_1, \mathbf{u}_2, \ldots, \mathbf{u}_n\}$ is called orthogonal if for any distinct indices i and j, $\mathbf{u}_i \cdot \mathbf{u}_j = 0$. If in addition, each of the basis vectors is a unit vector, then we say that we have an *orthonormal basis*. The basis $\{\boldsymbol{\varepsilon}_1, \boldsymbol{\varepsilon}_2, \ldots, \boldsymbol{\varepsilon}_n\}$ introduced in Example A is an orthonormal basis for R^n. The basis $\{\mathbf{u}_1, \mathbf{u}_2, \mathbf{u}_3\}$ of R^3 introduced in Problem D is not orthonormal.

PROBLEMS

G · Let $\mathbf{u} = (a_1, \ldots, a_n)$, $\mathbf{v} = (b_1, \ldots, b_n)$, $\mathbf{w} = (c_1, \ldots, c_n)$ be vectors in R^n. Prove (i) $\mathbf{u} \cdot \mathbf{v} = \mathbf{v} \cdot \mathbf{u}$; (ii) $\mathbf{u} \cdot (\mathbf{v} + \mathbf{w}) = \mathbf{u} \cdot \mathbf{v} + \mathbf{u} \cdot \mathbf{w}$; (iii) for a real number r, $\mathbf{u} \cdot r\mathbf{v} = r(\mathbf{u} \cdot \mathbf{v})$.

H · The vectors $\mathbf{u}_1 = (1/\sqrt{2}, -1/\sqrt{2}, 0)$ and $\mathbf{u}_2 = (2/\sqrt{3}, 2/\sqrt{3}, 1/\sqrt{3})$ are orthonormal in R^3. Can you find a third vector $\mathbf{u}_3$ so that $\{\mathbf{u}_1, \mathbf{u}_2, \mathbf{u}_3\}$ will form an orthonormal basis?

I · Vectors $\mathbf{u}_1 = (\frac{3}{5}, -\frac{4}{5})$ and $\mathbf{u}_2 = (\frac{4}{5}, \frac{3}{5})$ form an orthonormal basis for R^2.

(a) Since $\mathbf{v} = (-2, 5)$ is in R^2, there exist two real numbers r and s such that $\mathbf{v} = r\mathbf{u}_1 + s\mathbf{u}_2$. Find r and s.

(b) Consider an arbitrary vector $\mathbf{w} \in R^2$. Give instructions for finding r and s such that $\mathbf{w} = r\mathbf{u}_1 + s\mathbf{u}_2$.

J · Prove that for any two vectors $\mathbf{p}$ and $\mathbf{q}$ in R^n, $|\mathbf{p} \cdot \mathbf{q}| \leq |\mathbf{p}||\mathbf{q}|$. (Hint: $(\mathbf{p} \pm t\mathbf{q}) \cdot (\mathbf{p} \pm t\mathbf{q}) \geq 0$ for any real number t. The inequality is obvious if $\mathbf{q} = 0$. If $\mathbf{q} \neq 0$, set $t = |\mathbf{p}|/|\mathbf{q}|$.) Note that if $\mathbf{q} \neq \mathbf{0}$, then equality holds if and only if $\mathbf{p} = t\mathbf{q}$ for some real t.

If $\{\mathbf{u}_1, \mathbf{u}_2, \ldots, \mathbf{u}_n\}$ is a basis of R^n and $\mathbf{v} \in R^n$, then there are real numbers r_i such that $\mathbf{v} = r_1 \mathbf{u}_1 + \cdots + r_n \mathbf{u}_n$. One great advantage to having an orthonormal basis is that given an arbitrary $\mathbf{v}$, it is very simple to find the corresponding r_i. We merely observe that since the dot product obeys the rules indicated in Problem G,

$$\mathbf{u}_i \cdot \mathbf{v} = \mathbf{u}_i \cdot (r_1 \mathbf{u}_1 + \cdots + r_i \mathbf{u}_i + \cdots + r_n \mathbf{u}_n)$$

$$= r_1 \mathbf{u}_i \cdot \mathbf{u}_1 + \cdots + r_i \mathbf{u}_i \cdot \mathbf{u}_i + \cdots + r_n \mathbf{u}_i \cdot \mathbf{u}_n,$$

so $\mathbf{u}_i \cdot \mathbf{v} = r_i$. Hence, $\mathbf{v} = (\mathbf{u}_1 \cdot \mathbf{v})\mathbf{u}_1 + \cdots + (\mathbf{u}_n \cdot \mathbf{v})\mathbf{u}_n$. Note that Problem I is a special case of this general procedure.

We have now identified one great advantage in having an orthonormal basis. This introduces another problem to think about. Suppose we have a subspace, perhaps the solution subspace of a system of equations for example. Can we always find an orthonormal basis for this subspace? If we find an orthonormal basis to the subspace, can we find a few vectors to add to this basis in order to have an orthonormal basis for the entire space? This second question is the one we raised in a special case in Problem H above. Both questions can be answered if we learn how, given a basis to any subspace (possibly the space itself), to construct another basis that is orthonormal.

Notice first that if $\{\mathbf{u}_1, \ldots, \mathbf{u}_r\}$ is an orthogonal basis to an r-dimensional subspace of R^n, $r \leq n$, then it is simple to construct an orthonormal basis $\{\mathbf{v}_1, \ldots, \mathbf{v}_r\}$ of the same subspace by setting $\mathbf{v}_i = \mathbf{u}_i / |\mathbf{u}_i|$. If the given basis is not orthogonal, then it is more of a challenge to construct an orthonormal basis. But it can always be done using the *Gram–Schmidt orthogonalization process* that we now describe.

Suppose $\{\mathbf{u}_1, \ldots, \mathbf{u}_r\}$ is a given basis. Set $\mathbf{v}_1 = \mathbf{u}_1 / |\mathbf{u}_1|$. Then set $\mathbf{w}_2 = \mathbf{u}_2 - \alpha_1 \mathbf{v}_1$ where α_1 is a real number yet to be chosen. We see that

$$\mathbf{v}_1 \cdot \mathbf{w}_2 = \mathbf{v}_1 \cdot (\mathbf{u}_2 - \alpha_1 \mathbf{v}_1) = \mathbf{v}_1 \cdot \mathbf{u}_2 - \alpha_1 \mathbf{v}_1 \cdot \mathbf{v}_1 = \mathbf{v}_1 \cdot \mathbf{u}_2 - \alpha_1,$$

since $\mathbf{v}_1 \cdot \mathbf{v}_1 = |\mathbf{v}_1|^2 = 1$. Hence $\mathbf{v}_1 \cdot \mathbf{w}_2 = 0$ if we choose $\alpha_1 = \mathbf{v}_1 \cdot \mathbf{u}_2$. Let us then define $\mathbf{v}_2 = \mathbf{w}_2 / |\mathbf{w}_2|$. Summarizing, we have set

$$\mathbf{v}_1 = \mathbf{u}_1 / |\mathbf{u}_1|,$$

$$\mathbf{w}_2 = \mathbf{u}_2 - (\mathbf{v}_1 \cdot \mathbf{u}_2)\mathbf{v}_1,$$

$$\mathbf{v}_2 = \mathbf{w}_2 / |\mathbf{w}_2|.$$

We then proceed according to the same pattern. Set

$$\mathbf{w}_3 = \mathbf{u}_3 - (\mathbf{v}_1 \cdot \mathbf{u}_3)\mathbf{v}_1 - (\mathbf{v}_2 \cdot \mathbf{u}_3)\mathbf{v}_2$$

$$\mathbf{v}_3 = \mathbf{w}_3 / |\mathbf{w}_3|.$$

In this way we generate a new basis $\{v_1, \ldots, v_r\}$ that is orthonormal, and since each v_i is a linear combination of the u_i, the new basis spans the same subspace as did the original basis.

EXAMPLE C

Find an orthonormal basis for the solution space $x + y - 4z = 0$.

Since we have only one equation, it is clear that y and z may be chosen arbitrarily so long as we set $x = -y + 4z$; all solutions are given by

$$[-y + 4z \quad y \quad z] = y[-1 \quad 1 \quad 0] + z[4 \quad 0 \quad 1].$$

The solution space has basis $u_1 = [-1 \quad 1 \quad 0]$, $u_2 = [4 \quad 0 \quad 1]$. But these vectors are not orthogonal, so we have need of the Gram–Schmidt orthogonalization process. We set

$$v_1 = u_1/|u_1| = [-1/\sqrt{2} \quad 1/\sqrt{2} \quad 0],$$

$$w_2 = u_2 - (u_2 \cdot v_1)v_1 = [4 \quad 0 \quad 1] - (-4/\sqrt{2})[-1/\sqrt{2} \quad 1/\sqrt{2} \quad 0]$$

$$w_2 = [2 \quad 2 \quad 1],$$

$$v_2 = \frac{w_2}{|w_2|} = [\tfrac{2}{3} \quad \tfrac{2}{3} \quad \tfrac{1}{3}];$$

v_1 and v_2 form an orthonormal basis. ∎

===   10 · 2   E X E R C I S E S

Do the following sets of three vectors constitute bases for R^3?

1. $u_1 = (1, 0, 1)$
   $u_2 = (-1, 2, 0)$
   $u_3 = (9, -6, -10)$.

2. $u_1 = (4, 0, -2)$
   $u_2 = (3, -1, 2)$
   $u_3 = (-1, 3, 10)$.

3. $u_1 = (4, 1, -1)$
   $u_2 = (1, 0, 2)$
   $u_3 = (3, 1, 1)$.

4. $u_1 = (-1, 2, 3)$
   $u_2 = (2, 1, 4)$
   $u_3 = (0, 1, -1)$.

Find the acute angle between the two given radius vectors u and v in R^3.

5. $u = (3, 0, 1)$,
   $v = (1, -1, 1)$.

6. $u = (-2, 1, 2)$,
   $v = (1, 2, -2)$.

An orthonormal basis for R^3 is formed by the vectors

$$\mathbf{u}_1 = (2/3, \ 1/\sqrt{2}, \ 1/3\sqrt{2}),$$

$$\mathbf{u}_2 = (2/3, \ -1/\sqrt{2}, \ 1/3\sqrt{2}),$$

$$\mathbf{u}_3 = (1/3, \ 0, \ -4/3\sqrt{2}).$$

Express each of the following vectors in the form $a_1\mathbf{u}_1 + a_2\mathbf{u}_2 + a_3\mathbf{u}_3$.

7. $(4, 0, 3)$.   **8.** $(5, -1, 1)$.

9. $\varepsilon_1 = (1, 0, 0)$.   **10.** $\varepsilon_2 = (0, 1, 0)$.

11. The vectors of Exercise 1 span a subspace $\mathscr{S}$ of R^3 (possibly R^3 itself). Taking $\mathbf{v}_1 = \mathbf{u}_1/|\mathbf{u}_1|$, use the Gram–Schmidt orthogonalization process to find an orthonormal basis for the subspace $\mathscr{S}$.

12. Solve Exercise 11, using the vectors of Exercise 2.

13. Solve Exercise 11, using the vectors of Exercise 3.

14. Solve Exercise 11, using the vectors of Exercise 4.

Find an orthonormal basis to the solution space of the following systems of equations.

15.
$$\begin{aligned} w + x - y - z &= 0, \\ 2w + 3x + y - 2z &= 0, \\ x + 3y &= 0. \end{aligned}$$

16.
$$\begin{aligned} 2w - x - y + z &= 0, \\ 3w + x - y - 3z &= 0, \\ -w + 3x + y - 5z &= 0. \end{aligned}$$

17. Suppose that in the process of solving the system

$$a_{11}x_1 + a_{12}x_2 + \cdots + a_{1n}x_n = 0,$$
$$\vdots$$
$$a_{n1}x_1 + a_{n2}x_2 + \cdots + a_{nn}x_n = 0,$$

the coefficient matrix is found to be equivalent to a row-reduced echelon matrix having r nonvanishing rows. What can you say about the dimension of the solution space $\mathscr{S}$?

18. We now consider a non-homogeneous system

$$a_{11}x_1 + a_{12}x_2 + \cdots + a_{1n}x_n = b_1,$$
$$\vdots$$
$$a_{n1}x_1 + a_{n2}x_2 + \cdots + a_{nn}x_n = b_n,$$
(2)

having the same coefficient matrix as the homogeneous system of Exercise 17. Let $\mathbf{p} = (\bar{x}_1, \bar{x}_2, \ldots, \bar{x}_n)$ be a solution to this system. Prove that all solutions of the system (2) are given by $\mathbf{p} + \mathbf{b}$ where $\mathbf{b} \in \mathscr{S}$, the solution space of Exercise 17.

10 · 3 Linear Transformations

Consider the equations

$$r = x^2 - 3xy,$$
$$s = x + y^2.$$

(1)

Given x and y, r and s are determined. These equations may be thought of as describing a function T from the xy plane into the rs plane (Fig. 10.5). Functions from one space of dimension greater than one into a second space are often called *transformations*.

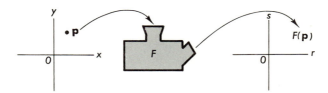

FIG. 10.5

It is our intention in this section to single out for study a particular class of such transformations; the so-called linear transformations. A *linear transformation T* is a transformation satisfying the two conditions

(1) $T(\mathbf{p} + \mathbf{q}) = T(\mathbf{p}) + T(\mathbf{q})$,
(2) $T(r\mathbf{p}) = rT(\mathbf{p})$, r a real number.

The transformation described by the equations (1) is not linear. To see this, let $\mathbf{p} = (2, 1)$ and $\mathbf{q} = (-1, 1)$. Then $\mathbf{p} + \mathbf{q} = (1, 2)$ and we have

$$T(\mathbf{p}) + T(\mathbf{q}) = (-2, 3) + (4, 0) = (2, 3),$$

but

$$T(\mathbf{p} + \mathbf{q}) = (-5, 5).$$

We often refer to $T(\mathbf{p})$ as the *image* of $\mathbf{p}$. And of course $\mathbf{p}$ is the *preimage* of $T(\mathbf{p})$.

PROBLEM

A · Prove that if T is a *linear* transformation with the property that for any unit vector $\mathbf{u}$, $T(\mathbf{u}) = \mathbf{0}$, then it follows that $T(\mathbf{p}) = \mathbf{0}$ for any vector $\mathbf{p}$; that is $T = \mathcal{O}$, the zero transformation.

The fact is that linear transformations are always described by very simple equations. In the case of a transformation from R^2 into R^2, these equations are easily determined. Suppose we agree to a particular basis $\{\mathbf{u}_1, \mathbf{u}_2\}$. Then any $\mathbf{p} \in R^2$ takes the form $\mathbf{p} = x\mathbf{u}_1 + y\mathbf{u}_2$, and the linearity of T enables us to write

$$T(\mathbf{p}) = T(x\mathbf{u}_1 + y\mathbf{u}_2),$$
$$= T(x\mathbf{u}_1) + T(y\mathbf{u}_2), \tag{2}$$
$$T(\mathbf{p}) = xT(\mathbf{u}_1) + yT(\mathbf{u}_2).$$

The key to determining the equations is to know what T does to the basis vectors $\mathbf{u}_1$ and $\mathbf{u}_2$. Suppose we know $T(\mathbf{u}_1) = (a_{11}, a_{21})$ and $T(\mathbf{u}_2) = (a_{12}, a_{22})$ where the coordinates are given with respect to the understood basis $\{\mathbf{u}_1, \mathbf{u}_2\}$. Using (r, s) as the coordinates of $T(\mathbf{p})$, and (for reasons to appear) writing all of these points as column vectors, equation (2) becomes

$$\begin{bmatrix} r \\ s \end{bmatrix} = x \begin{bmatrix} a_{11} \\ a_{21} \end{bmatrix} + y \begin{bmatrix} a_{12} \\ a_{22} \end{bmatrix} = \begin{bmatrix} a_{11}x + a_{12}y \\ a_{21}x + a_{22}y \end{bmatrix}$$

If we now equate the first and second entries of our column vectors, we get in the form of equations (1) above, a description of the general linear transformation $T: R^2 \to R^2$:

$$r = a_{11}x + a_{12}y,$$
$$s = a_{21}x + a_{22}y.$$

The use of column vectors therefore leads in a natural way to the form we were expecting for our answer. Beyond this, it also suggests the use of the matrix notation introduced in Section 10.2, giving us

$$\begin{bmatrix} r \\ s \end{bmatrix} = \begin{bmatrix} a_{11} & a_{12} \\ a_{21} & a_{22} \end{bmatrix} \begin{bmatrix} x \\ y \end{bmatrix}$$

These simple ideas may immediately be extended to linear transformations T from R^m to R^n. We outline the procedure in this more general setting:

(i) Let $\{\mathbf{u}_1, \ldots, \mathbf{u}_m\}$ be a basis for R^m and $\{\mathbf{v}_1, \ldots, \mathbf{v}_n\}$ be a basis for R^n.

(ii) The linear transformation $T: R^m \to R^n$ corresponds to an $n \times m$ (n rows, m columns) matrix.

(iii) To determine the matrix, find the image of each basis vector in R^m. Suppose

$$T(\mathbf{u}_j) = a_{1j}\mathbf{v}_1 + a_{2j}\mathbf{v}_2 + \cdots + a_{nj}\mathbf{v}_n.$$

Then the jth column vector in the matrix is

$$\begin{bmatrix} a_{1j} \\ a_{2j} \\ \vdots \\ a_{nj} \end{bmatrix}$$

It should be clear that for a different basis, a transformation T will correspond to a different matrix. More about this in Section 10.5. For now, wishing to distinguish between a transformation T and its matrix with respect to some basis, we shall consistently write $[T]$ when referring to a matrix corresponding to a transformation T.

There is another convention we wish to adopt. Having already commented upon the advantages of thinking in terms of column vectors, we wish from now on that whenever we talk about $\mathbf{p} \in R^n$, the reader would always understand us to mean the column vector

$$\mathbf{p} = \begin{bmatrix} a_1 \\ \vdots \\ a_n \end{bmatrix}$$

When we do wish to write $\mathbf{p}$ as a row vector (as in cases where we want to avoid a whole page of space consuming vertical columns), we will write

$$\mathbf{p}^t = [a_1 \ \ldots \ a_n].$$

This is called the *transpose* of the vector $\mathbf{p}$.

EXAMPLE A
We shall consider a linear transformation $T: R^3 \to R^2$.

Let $\{\varepsilon_1, \varepsilon_2, \varepsilon_3\}$ be the standard basis (Example 10.2A) for R^3, and let us use as a basis for R^2 the vectors $\mathbf{u}_1{}^t = [2 \ 3]$ and $\mathbf{u}_2{}^t = [-1 \ 1]$. Suppose that T has the property that

$$T(\varepsilon_1) = \ 3\mathbf{u}_1 \ + \mathbf{u}_2,$$
$$T(\varepsilon_2) = -\mathbf{u}_1 + 2\mathbf{u}_2,$$
$$T(\varepsilon_3) = \ \ \ \mathbf{u}_1 \ + \mathbf{u}_2.$$

Then with respect to the bases $\{\varepsilon_1, \varepsilon_2, \varepsilon_3\}$ and $\{\mathbf{u}_1, \mathbf{u}_2\}$,

$$[T] = \begin{bmatrix} 3 & -1 & 1 \\ 1 & 2 & 1 \end{bmatrix} \tag{3}$$

Suppose now that we consider the standard bases $\bar{\varepsilon}_1 = [1, 0]$ and $\bar{\varepsilon}_2 = [0, 1]$ for R^2. We see that

$$T(\varepsilon_1) = 3\mathbf{u}_1 + \mathbf{u}_2 = [5 \quad 10] = 5\bar{\varepsilon}_1 + 10\bar{\varepsilon}_2,$$

$$T(\varepsilon_2) = -\mathbf{u}_1 + 2\mathbf{u}_2 = [-4 \quad -1] = -4\bar{\varepsilon}_1 - \bar{\varepsilon}_2,$$

$$T(\varepsilon_3) = \mathbf{u}_1 + \mathbf{u}_2 = [1 \quad 4] = \bar{\varepsilon}_1 + 4\bar{\varepsilon}_2.$$

With respect to the bases $\{\varepsilon_1, \varepsilon_2, \varepsilon_3\}$ and $\{\bar{\varepsilon}_1, \bar{\varepsilon}_2\}$,

$$[T] = \begin{bmatrix} 5 & -4 & 1 \\ 10 & -1 & 4 \end{bmatrix} \tag{4}$$

Now let $\mathbf{p}^t = [-1 \quad -2 \quad 3]$. Using (3)

$$T(\mathbf{p}) = \begin{bmatrix} 3 & -1 & 1 \\ 1 & 2 & 1 \end{bmatrix} \begin{bmatrix} -1 \\ -2 \\ 3 \end{bmatrix} = \begin{bmatrix} 2 \\ -2 \end{bmatrix}$$

so the image of $\mathbf{p}$ is $2\mathbf{u}_1 - 2\mathbf{u}_2$. But if we use (4), then we have

$$T(\mathbf{p}) = \begin{bmatrix} 5 & -4 & 1 \\ 10 & -1 & 4 \end{bmatrix} \begin{bmatrix} -1 \\ -2 \\ 3 \end{bmatrix} = \begin{bmatrix} 6 \\ 4 \end{bmatrix}$$

so the image of $\mathbf{p}$ is $6\bar{\varepsilon}_1 + 4\bar{\varepsilon}_2$. Now T is the same transformation and $\mathbf{p}$ is the same point, so $T(\mathbf{p})$ ought to be the same no matter which basis we use. Happily,

$$T(\mathbf{p}) = 2\mathbf{u}_1 - 2\mathbf{u}_2 = 2\begin{bmatrix} 2 \\ 3 \end{bmatrix} - 2\begin{bmatrix} -1 \\ 1 \end{bmatrix} = \begin{bmatrix} 6 \\ 4 \end{bmatrix} = 6\begin{bmatrix} 1 \\ 0 \end{bmatrix} + 4\begin{bmatrix} 0 \\ 1 \end{bmatrix}$$

$$= 6\bar{\varepsilon}_1 + 4\bar{\varepsilon}_2. \quad \blacksquare$$

PROBLEMS

B . Let $S = \{\varepsilon_1, \varepsilon_2\}$ be the standard basis for R^2. Suppose $T(\varepsilon_1) = -2\varepsilon_1 + 4\varepsilon_2$; $T(\varepsilon_2) = 6\varepsilon_1 - 2\varepsilon_2$. Find the matrix of T with respect to S. Then find $T(\mathbf{u}_1)$ and $T(\mathbf{u}_2)$ where $\mathbf{u}_1^t = [1 \quad 1]$ and $\mathbf{u}_2^t = [1 \quad -1]$.

C . Note that, using $\mathbf{u}_1$ and $\mathbf{u}_2$ from Problem B, $2\varepsilon_1 = \mathbf{u}_1 + \mathbf{u}_2$ and $2\varepsilon_2 = \mathbf{u}_1 - \mathbf{u}_2$. Also, $B = \{\mathbf{u}_1, \mathbf{u}_2\}$ is a basis. Express $T(\mathbf{u}_1)$ and $T(\mathbf{u}_2)$ in terms of the basis B. Then find the matrix of T with respect to basis B.

D . Let $\mathbf{p} = 3\mathbf{u}_1 - 2\mathbf{u}_2$. Show that $\mathbf{p} = \varepsilon_1 + 5\varepsilon_2$. Using $[T]_S$ as the matrix obtained in Problem B and $[T]_B$ as the matrix obtained in Problem C, verify that

$$[T]_B \begin{bmatrix} 3 \\ -2 \end{bmatrix} = [T]_S \begin{bmatrix} 1 \\ 5 \end{bmatrix}.$$

Always assume, unless it is otherwise stated, that the basis of R^n is taken to be the standard basis $\varepsilon_1, \varepsilon_2, \ldots, \varepsilon_n$.

1. $L: R^3 \to R^2$ is described by $[L] = \begin{bmatrix} 3 & -1 & 4 \\ 2 & 1 & 0 \end{bmatrix}$. Find $L(1, -1, 3); L(4, 2, -1);$ $L(-1, 0, 2)$.

2. $L: R^2 \to R^3$ is described by $[L] = \begin{bmatrix} 4 & -1 \\ -1 & 3 \\ 2 & -3 \end{bmatrix}$. Find $L(1, -3); L(0, 2); L(-1, 4)$.

3. $L: R^3 \to R^4$. $L(\varepsilon_1) = (1, -1, 4, 3); L(\varepsilon_2) = (0, 1, 3, 4); L(\varepsilon_3) = (2, 1, -1, 3)$. Find $[L]$.

4. $L: R^3 \to R^2$. $L(\varepsilon_1) = (3, 5); L(\varepsilon_2) = (1, -1); L(\varepsilon_3) = (-5, 0)$. Find $[L]$.

5. $L: R^2 \to R^2$. $L(2, 1) = (7, 0); L(1, -1) = (-1, 3)$. Find $[L]$. Then find $L(3, 2)$.

6. $L: R^2 \to R^2$. $L(3, 1) = (1, 4); L(-1, 2) = (-5, 1)$. Find $[L]$. Then find $L(4, -3)$.

We now wish to consider the composition of two linear transformations. As a starter, using the standard basis of R^2 throughout, let us define

$$F: \quad \begin{aligned} r &= a_{11}x + a_{12}y, \\ s &= a_{21}x + a_{22}y, \end{aligned}$$

$$G: \quad \begin{aligned} u &= b_{11}r + b_{12}s, \\ v &= b_{21}r + b_{22}s. \end{aligned}$$

The composition $H = G \circ F$ takes points from the xy plane into the uv plane (Fig. 10.6). It is not hard to determine the equations describing H. Substitution for r and s in the equations describing G gives

$$H: \quad \begin{aligned} u &= b_{11}(a_{11}x + a_{12}y) + b_{12}(a_{21}x + a_{22}y), \\ v &= b_{21}(a_{11}x + a_{12}y) + b_{22}(a_{21}x + a_{22}y). \end{aligned}$$

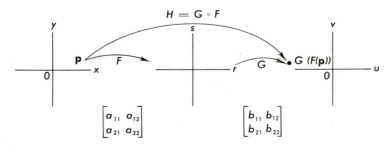

FIG. 10.6

Writing each of these transformations using matrix notation,

$$F: \quad \begin{bmatrix} r \\ s \end{bmatrix} = \begin{bmatrix} a_{11} & a_{12} \\ a_{21} & a_{22} \end{bmatrix} \begin{bmatrix} x \\ y \end{bmatrix}, \qquad G: \quad \begin{bmatrix} u \\ v \end{bmatrix} = \begin{bmatrix} b_{11} & b_{12} \\ b_{21} & b_{22} \end{bmatrix} \begin{bmatrix} r \\ s \end{bmatrix}$$

$$H: \quad \begin{bmatrix} u \\ v \end{bmatrix} = \begin{bmatrix} b_{11}a_{11} + b_{12}a_{21} & b_{11}a_{12} + b_{12}a_{22} \\ b_{21}a_{11} + b_{22}a_{21} & b_{21}a_{12} + b_{22}a_{22} \end{bmatrix} \begin{bmatrix} x \\ y \end{bmatrix}$$

(5)

If we substitute in the matrix equation for G the expression for $\begin{bmatrix} r \\ s \end{bmatrix}$ as given in the matrix equation for F, we are led in a purely formal way to

$$\begin{bmatrix} u \\ v \end{bmatrix} = \begin{bmatrix} b_{11} & b_{12} \\ b_{21} & b_{22} \end{bmatrix} \begin{bmatrix} a_{11} & a_{12} \\ a_{21} & a_{22} \end{bmatrix} \begin{bmatrix} x \\ y \end{bmatrix}$$

(6)

This motivates our definition of the multiplication of two matrices. We want to define

$$\begin{bmatrix} b_{11} & b_{12} \\ b_{21} & b_{22} \end{bmatrix} \begin{bmatrix} a_{11} & a_{12} \\ a_{21} & a_{22} \end{bmatrix} = \begin{bmatrix} c_{11} & c_{12} \\ c_{21} & c_{22} \end{bmatrix}$$

so that (6) will give, after multiplication, the answer indicated in (5). This means we must define

$$c_{11} = b_{11}a_{11} + b_{12}a_{21}, \qquad c_{12} = b_{11}a_{12} + b_{12}a_{22},$$

$$c_{21} = b_{21}a_{11} + b_{22}a_{21}, \qquad c_{22} = b_{21}a_{12} + b_{22}a_{22}.$$

These formulas are more easily remembered if we state them in terms of inner products. Let B and A be matrices. Then we define

$$BA = C$$

where the entry in the ith row, the jth column of C (that is the entry c_{ij}) is obtained as the dot product of the ith row of B and the jth column of A. Thus c_{11} above is the dot product of the first row $[b_{11} \; b_{12}]$ of $B = [G]$ and the first column $\begin{bmatrix} a_{11} \\ a_{21} \end{bmatrix}$ of $A = [F]$.

EXAMPLE B

$$\overset{\displaystyle M}{\begin{bmatrix} 3 & -1 & 2 \\ 4 & 0 & 5 \end{bmatrix}} \overset{\displaystyle N}{\begin{bmatrix} 1 & -1 & 0 \\ 3 & 2 & 1 \\ 0 & 4 & -2 \end{bmatrix}} = \overset{\displaystyle P}{\begin{bmatrix} 0 & 3 & -5 \\ 4 & 16 & -10 \end{bmatrix}}$$

Some remarks are in order. First, we review how some of the entries in the product matrix P were obtained. The 3 in the first row, second column is the dot product of the first row of M and the second column of N:

$$[3 \quad -1 \quad 2]\begin{bmatrix} -1 \\ 2 \\ 4 \end{bmatrix} = 3(-1) + (-1)2 + 2(4) = 3. \tag{7}$$

To obtain the -10 in the second row, third column,

$$[4 \quad 0 \quad 5]\begin{bmatrix} 0 \\ 1 \\ -2 \end{bmatrix} = 4(0) + 0(1) + 5(-2) = -10.$$

Note that in order to multiply two matrices $M \cdot N$, the number of columns in M must equal the number of rows in N. We could *not* form the product $N \cdot M$ in the present example. The product matrix P has as many rows as there are rows in M, and as many columns as there are columns in N. ∎

The dimensions of matrix products correspond directly to what we would expect from consideration of the dimension of the spaces involved in a composition transformation. Suppose we have

$$F: \quad R^m \to R^n \qquad G: \quad R^n \to R^p$$

$$\underset{n}{[F]} \qquad\qquad \underset{p}{[G]}$$

Then $H = G \circ F$, $H: R^m \to R^p$, corresponds to

$$\underset{p}{[G]} \ \underset{n}{[F]} = \underset{p}{[H]}.$$

It is sometimes helpful to think of a vector in R^n as an $n \times 1$ matrix. The dot product of two vectors $\mathbf{p} = [a_1, \ldots, a_n]^t$ and $\mathbf{q} = [b_1, \ldots, b_n]^t$ then appears as ordinary multiplication of two matrices if we write $\mathbf{p} \cdot \mathbf{q} = \mathbf{p}^t \mathbf{q}$ (see (7) above).

The transpose now is a concept defined for an $n \times 1$ matrix. We extend the definition to an arbitrary matrix in an obvious way. The *transpose* of a matrix A is obtained by interchanging the rows and columns of A; so the transpose of an $n \times m$ matrix is an $m \times n$ matrix.

If

$$A = \begin{bmatrix} a_{11} & a_{12} \\ a_{21} & a_{22} \\ a_{31} & a_{32} \end{bmatrix}$$

then

$$A^t = \begin{bmatrix} a_{11} & a_{21} & a_{31} \\ a_{12} & a_{22} & a_{32} \end{bmatrix}.$$

It is immediately clear that the transpose of a sum is the sum of the transposes; that is $(A + B)^t = A^t + B^t$. The transpose of a product is not quite so straightforward as we shall see in Problem N below.

PROBLEMS

E · Let $F: R^m \to R^n$ and $G: R^n \to R^p$ be linear transformations. Prove directly from the definition of a linear transformation that $H = G \circ F$ is linear.

F · Let $A = \begin{bmatrix} 1 & 3 \\ 0 & 2 \end{bmatrix}$, $B = \begin{bmatrix} 4 & -3 \\ 1 & -1 \end{bmatrix}$, $C = \begin{bmatrix} 5 & 7 \\ 2 & 3 \end{bmatrix}$.
Find $(A \cdot B)C$ and $A(B \cdot C)$.

G · Using matrices A and B of Problem F, find AB and BA.

H · Using $I = \begin{bmatrix} 1 & 0 \\ 0 & 1 \end{bmatrix}$ and matrix A of Problem F, find AI and IA.

I · For two matrices having the same dimension, we define addition in the most obvious way; we add the corresponding entries. Thus for the matrices of Problem F,

$$B + C = \begin{bmatrix} 4 & -3 \\ 1 & -1 \end{bmatrix} + \begin{bmatrix} 5 & 7 \\ 2 & 3 \end{bmatrix} = \begin{bmatrix} 9 & 4 \\ 3 & 2 \end{bmatrix}.$$

Find $A(B + C)$. Does this equal $AB + AC$?
Find $(B + C)A$. Does this equal $BA + CA$?

J · Using A and B from Problem F and the matrices $R = \begin{bmatrix} 1 & -\frac{3}{2} \\ 0 & \frac{1}{2} \end{bmatrix}$,
$S = \begin{bmatrix} 1 & -3 \\ 1 & -4 \end{bmatrix}$, form the products AR, RA, BS, and SB.

K · Let J and K be $n \times n$ matrices, and let I be the $n \times n$ matrix consisting entirely of 0 entries except for 1's on the diagonal (Problem H shows the special case where $n = 2$). We call J the inverse of K if $JK = KJ = I$. Find the inverse of matrix C in Problem F.

L · Can you find an inverse for the matrix $\begin{bmatrix} 4 & 6 \\ 2 & 3 \end{bmatrix}$?

M · Try to discover an easy "recipe" for finding the inverse of a 2×2 matrix. Look carefully at your answer to Problem K. It might help also to find and reflect upon the inverses to the matrices

$$R = \begin{bmatrix} 10 & 5 \\ 3 & 2 \end{bmatrix} \quad \text{and} \quad S = \begin{bmatrix} 5 & 7 \\ 3 & 4 \end{bmatrix}.$$

N · For the matrices A and B of Problem F, find the products $A^t B^t$ and $B^t A^t$. Compare these products with $(AB)^t$ and $(BA)^t$.

Besides giving practice in the multiplication of matrices, the preceding problems were intended to illustrate the following properties of matrix multiplication. These properties hold for matrices of arbitrary dimensions. Proofs are not hard, but when given for arbitrary dimensions, they present problems in notation. The reader should have no difficulty in writing out proofs in the case of 2×2 matrices.

A   Matrix multiplication is associative   This means that for any three matrices A, B, C with dimensions such that the indicated products can be formed, $(AB)C = A(BC)$.

B   Matrix multiplication is not commutative   Although some matrices may commute, generally speaking, given two $n \times n$ matrices A and B, $AB \neq BA$.

C   In the set of all $n \times n$ matrices, there is an identity matrix I   The identity matrix, having the property that $AI = IA = A$ for all matrices A, is the matrix

$$I = \begin{bmatrix} 1 & & & 0 \\ & 1 & & \\ & & \ddots & \\ 0 & & & 1 \end{bmatrix}$$

having 1's on the diagonal as its only nonzero entries.

D   Matrix multiplication obeys the distributive laws   This means $A(B + C) = AB + AC$ and $(B + C)A = BA + CA$. Notice that since multiplication is not commutative, it is necessary to be careful of the order in which products are written down.

E   Some, but not all matrices A have an inverse, denoted by A^{-1}   An inverse to A, by definition, has the property that $AA^{-1} = A^{-1}A = I$. It is a fact (that we shall later prove) that if $AB = I$, then $BA = I$ so $B = A^{-1}$.

F   The transpose of a product of matrices is the product of the transposes taken in reverse order.   In symbols, $(AB)^t = B^t A^t$.

We close this section with the observation that the solution of n equations in n unknowns is directly related to the existence of an inverse for the coefficient matrix of the system. The equations

$$a_{11}x_1 + a_{12}x_2 + \cdots + a_{1n}x_n = b_1,$$
$$\vdots$$
$$a_{n1}x_1 + a_{n2}x_2 + \cdots + a_{nn}x_n = b_n,$$

can be written in the form

$$A \begin{bmatrix} x_1 \\ \vdots \\ x_n \end{bmatrix} = \begin{bmatrix} b_1 \\ \vdots \\ b_n \end{bmatrix}$$

Then if we can find an inverse matrix A^{-1}, we have the solution

$$\begin{bmatrix} x_1 \\ \vdots \\ x_n \end{bmatrix} = A^{-1} \begin{bmatrix} b_1 \\ \vdots \\ b_n \end{bmatrix}$$

EXAMPLE C

Consider the system of equations

$$ax + by = u$$

$$cx + dy = v$$

Now if we agree to write

$$A = \begin{bmatrix} a & b \\ c & d \end{bmatrix}, \quad \mathbf{p} = \begin{bmatrix} x \\ y \end{bmatrix}, \quad \mathbf{q} = \begin{bmatrix} u \\ v \end{bmatrix}$$

then the system may be written in the matrix form $A\mathbf{p} = \mathbf{q}$. Solution depends upon finding A^{-1}. It is hoped that Problems K, L, and M above suggested to the student the formula, whenever $ad - bc = D \neq 0$,

$$A^{-1} = \begin{bmatrix} d/D & -b/D \\ -c/D & a/D \end{bmatrix}$$

Whether or not this occurred to the reader, he can easily verify that it works by multiplying $A^{-1}A$, and AA^{-1}. If we now multiply both sides of $A\mathbf{p} = \mathbf{q}$ on the left by A^{-1}, we get $A^{-1}A\mathbf{p} = A^{-1}\mathbf{q}$, or since $I\mathbf{p} = \mathbf{p}$,

$$\begin{bmatrix} x \\ y \end{bmatrix} = \begin{bmatrix} \dfrac{d}{D} & \dfrac{-b}{D} \\ \dfrac{-c}{D} & \dfrac{a}{D} \end{bmatrix} \begin{bmatrix} u \\ v \end{bmatrix}$$

If $D = 0$, then of course $ad = bc$ and the graphs of the two equations are either parallel or coincident lines, so we do not expect unique solutions. ∎

PROBLEMS

O · If $A = \begin{bmatrix} 7 & 8 \\ 4 & 3 \end{bmatrix}$, use the formula above to find A^{-1}. Verify that it works.

P · Use your answer to Problem O to help you solve

$$7x + 8y = 10,$$
$$4x + 3y = 1.$$

1 · Let $R: R^2 \to R^2$ and $S: R^2 \to R^2$ have as their respective matrices

$$R \leftrightarrow \begin{bmatrix} 1 & -1 \\ 2 & 3 \end{bmatrix}, \qquad S \leftrightarrow \begin{bmatrix} 0 & 2 \\ -3 & 1 \end{bmatrix}.$$

Find the matrix of $T = R \circ S$. Find $S(3, 4)$. Then find $R(S(3, 4))$. Finally, use the matrix of T to find $T(3, 4)$.

2 · Let R and S be defined as in Exercise 1. Find the matrix of $W = S \circ R$. Find $R(3, 4)$. Then find $S(R(3, 4))$. Finally, use the matrix of W to find $W(3, 4)$.

The following exercises all refer to the matrices

$$A = \begin{bmatrix} 1 & -7 & -2 \\ -1 & -3 & -1 \\ 2 & -1 & 0 \end{bmatrix} \qquad B = \begin{bmatrix} 1 & 0 & -1 \\ 0 & 2 & 3 \\ 1 & -1 & -3 \end{bmatrix} \qquad C = \begin{bmatrix} 2 & 1 & -1 \\ 4 & 0 & 1 \\ 0 & 1 & 3 \end{bmatrix}$$

3 · Find $(AB)C$. Show by actually multiplying that this equals $A(BC)$.

4 · Find $(AC)B$. Show by actually multiplying that this equals $A(CB)$.

5 · Find BC and CB.

6 · Find AB and BA.

7 · Verify that $A(B + C) = AB + AC$.

8 · Verify that $(B + C)A = BA + CA$.

9 · Verify that $A^{-1} = \begin{bmatrix} 1 & -2 & -1 \\ 2 & -4 & -3 \\ -7 & 13 & 10 \end{bmatrix}$

10 · Verify that $B^{-1} = \begin{bmatrix} 3 & -1 & -2 \\ -3 & 2 & 3 \\ 2 & -1 & -2 \end{bmatrix}$

11 · Verify that $(AB)^t = B^t A^t$.

12 · Verify that $(AC)^t = C^t A^t$.

13 • Using A^{-1} as identified in Exercise 9, solve by matrix methods

$$x - 7y - 2z = 3,$$
$$-x - 3y - z = 8,$$
$$2x - y \qquad = -8.$$

14• Using B^{-1} as identified in Exercise 10, solve by matrix methods

$$x \qquad -z = -8,$$
$$2y + 3z = 5,$$
$$x - y - 3z = -12.$$

15• In Exercise 9, we produced a matrix that acted as an inverse for A. Are there any other 3×3 matrices that could be used instead?

10 · 4 Determinants

In example C of the last section, we saw that given the 2×2 matrix

$$A = \begin{bmatrix} a & b \\ c & d \end{bmatrix} \tag{1}$$

the inverse A^{-1} exists if and only if the number

$$D = ad - bc$$

is nonzero. This number is called the *determinant* of the matrix A. It is often written

$$\det A = \begin{vmatrix} a & b \\ c & d \end{vmatrix} = ad - bc,$$

where the vertical lines indicate the determinant of the matrix (1).

THEOREM A
Let $[R]$ and $[S]$ be matrices. Then

$$\det([R][S]) = \det[R] \det[S].$$

PROOF:

$$\text{Let } [R] = \begin{bmatrix} r_{11} & r_{12} \\ r_{21} & r_{22} \end{bmatrix}, \quad [S] = \begin{bmatrix} s_{11} & s_{12} \\ s_{21} & s_{22} \end{bmatrix}.$$

Then

$$\det[R][S] = \begin{vmatrix} r_{11}s_{11} + r_{12}s_{21} & r_{11}s_{12} + r_{12}s_{22} \\ r_{21}s_{11} + r_{22}s_{21} & r_{21}s_{12} + r_{22}s_{22} \end{vmatrix}$$

$$= (r_{11}s_{11} + r_{12}s_{21})(r_{21}s_{12} + r_{22}s_{22})$$
$$\quad - (r_{21}s_{11} + r_{22}s_{21})(r_{11}s_{12} + r_{12}s_{22})$$

$$= r_{11}r_{21}s_{11}s_{12} + r_{11}r_{22}s_{11}s_{22} + r_{12}r_{21}s_{12}s_{21} + r_{12}r_{22}s_{21}s_{22}$$
$$\quad - r_{11}r_{21}s_{11}s_{12} - r_{11}r_{22}s_{12}s_{21} - r_{12}r_{21}s_{11}s_{22} - r_{12}r_{22}s_{21}s_{22}$$

$$= r_{11}r_{22}(s_{11}s_{22} - s_{12}s_{21}) - r_{12}r_{21}(s_{11}s_{22} - s_{12}s_{21})$$

$$= (r_{11}r_{22} - r_{12}r_{21})(s_{11}s_{22} - s_{12}s_{21}) = \det[R]\,\det[S]. \quad \blacksquare$$

We now recall from Section 10.1 the three elementary operations used in solving equations. They were, phrased in terms of the coefficient matrix:

I.   The interchange of two rows;
II.   Multiplication of a row by a number $r \neq 0$;
III. Addition to one row of a multiple of another row.

The interesting thing is that each of these operations can be brought about by multiplication on the left by a so-called *elementary matrix*. Consider the products

I. $$\begin{bmatrix} 0 & 1 \\ 1 & 0 \end{bmatrix} \begin{bmatrix} a & b \\ c & d \end{bmatrix} = \begin{bmatrix} c & d \\ a & b \end{bmatrix}$$

II. $$\begin{bmatrix} 1 & 0 \\ 0 & r \end{bmatrix} \begin{bmatrix} a & b \\ c & d \end{bmatrix} = \begin{bmatrix} a & b \\ rc & rd \end{bmatrix}$$

III. $$\begin{bmatrix} 1 & 0 \\ r & 1 \end{bmatrix} \begin{bmatrix} a & b \\ c & d \end{bmatrix} = \begin{bmatrix} a & b \\ ra + c & rb + d \end{bmatrix}$$

An *elementary matrix* is obtained by performing an elementary operation on the identity matrix. Thus:

I. $$\begin{bmatrix} 0 & 1 \\ 1 & 0 \end{bmatrix}$$ is obtained by interchanging the first and second rows of I;

II. $$\begin{bmatrix} 1 & 0 \\ 0 & r \end{bmatrix}$$ is obtained by multiplying the second row of I by r.

III. $$\begin{bmatrix} 1 & 0 \\ r & 1 \end{bmatrix}$$ is obtained by adding to the second row of I r times the first row.

The determinants of the elementary matrices are easily computed:

$$\begin{vmatrix} 0 & 1 \\ 1 & 0 \end{vmatrix} = -1, \qquad \begin{vmatrix} 1 & 0 \\ 0 & r \end{vmatrix} = r, \qquad \begin{vmatrix} 1 & 0 \\ r & 1 \end{vmatrix} = 1.$$

Then, according to Theorem A, we have

$$\text{I.} \quad (-1)\begin{vmatrix} a & b \\ c & d \end{vmatrix} = \begin{vmatrix} c & d \\ a & b \end{vmatrix}$$

$$\text{II.} \quad r\begin{vmatrix} a & b \\ c & d \end{vmatrix} = \begin{vmatrix} a & b \\ rc & rd \end{vmatrix}$$

$$\text{III.} \quad \begin{vmatrix} a & b \\ c & d \end{vmatrix} = \begin{vmatrix} a & b \\ ra + c & rb + d \end{vmatrix}$$

Stated in words, we have three properties of determinants:

Property I *If two rows of a matrix are interchanged, the sign of the determinant changes.*

Property II *If a row of a matrix is multiplied by r, the value of the determinant is multiplied by r.*

Property III *If a multiple of one row of a matrix is added to a second row, the determinant is unchanged.*

The determinant of a 3×3 matrix is also defined. Though the definition is somewhat complicated, it is designed so that Theorem A remains valid for the product of two 3×3 matrices. Using the same notation as before for a matrix A, we define det A to be

$$\begin{vmatrix} a_{11} & a_{12} & a_{13} \\ a_{21} & a_{22} & a_{23} \\ a_{31} & a_{32} & a_{33} \end{vmatrix} = a_{11}\begin{vmatrix} a_{22} & a_{23} \\ a_{32} & a_{33} \end{vmatrix} - a_{12}\begin{vmatrix} a_{21} & a_{23} \\ a_{31} & a_{33} \end{vmatrix} + a_{13}\begin{vmatrix} a_{21} & a_{22} \\ a_{31} & a_{32} \end{vmatrix}$$

The validity of Theorem A is established as before by direct computation; but this time the notation makes the proof much longer. We therefore omit the proof, asking instead that the reader at least try a few examples.

PROBLEMS

A · Given $A = \begin{bmatrix} 3 & 0 & -1 \\ 4 & 2 & 1 \\ 5 & -1 & 2 \end{bmatrix}$ and $B = \begin{bmatrix} 1 & -1 & 5 \\ 0 & 4 & -2 \\ 3 & 2 & 1 \end{bmatrix}$, verify that

$\det(AB) = (\det A)(\det B)$.

B · Find $\det(BA)$ in Problem A.

Again it is possible to bring about the elementary operations by multiplication on the left by an elementary matrix:

I. $$\begin{bmatrix} 1 & 0 & 0 \\ 0 & 0 & 1 \\ 0 & 1 & 0 \end{bmatrix} \begin{bmatrix} a_{11} & a_{12} & a_{13} \\ a_{21} & a_{22} & a_{23} \\ a_{31} & a_{32} & a_{33} \end{bmatrix} = \begin{bmatrix} a_{11} & a_{12} & a_{13} \\ a_{31} & a_{32} & a_{33} \\ a_{21} & a_{22} & a_{23} \end{bmatrix}$$

II. $$\begin{bmatrix} 1 & 0 & 0 \\ 0 & r & 0 \\ 0 & 0 & 1 \end{bmatrix} \begin{bmatrix} a_{11} & a_{12} & a_{13} \\ a_{21} & a_{22} & a_{23} \\ a_{31} & a_{32} & a_{33} \end{bmatrix} = \begin{bmatrix} a_{11} & a_{12} & a_{13} \\ ra_{21} & ra_{22} & ra_{23} \\ a_{31} & a_{32} & a_{33} \end{bmatrix}$$

III. $$\begin{bmatrix} 1 & 0 & 0 \\ 0 & 1 & 0 \\ r & 0 & 1 \end{bmatrix} \begin{bmatrix} a_{11} & a_{12} & a_{13} \\ a_{21} & a_{22} & a_{23} \\ a_{31} & a_{32} & a_{33} \end{bmatrix}$$

$$= \begin{bmatrix} a_{11} & a_{12} & a_{13} \\ a_{21} & a_{22} & a_{23} \\ ra_{11} + a_{31} & ra_{12} + a_{32} & ra_{13} + a_{33} \end{bmatrix}$$

Again the elementary matrices are obtained from the (3×3) identity I by elementary operations. Thus,

I. $E_1 = \begin{bmatrix} 1 & 0 & 1 \\ 0 & 0 & 1 \\ 0 & 1 & 0 \end{bmatrix}$ is obtained by interchanging rows 2 and 3 of I.

II. $E_2 = \begin{bmatrix} 1 & 0 & 0 \\ 0 & r & 0 \\ 0 & 0 & 1 \end{bmatrix}$ is obtained by multiplying the second row of I by r.

III. $E_3 = \begin{bmatrix} 1 & 0 & 0 \\ 0 & 1 & 0 \\ r & 0 & 1 \end{bmatrix}$ is obtained by adding to the third row of I r times the first row.

Direct computations give det $E_1 = -1$, det $E_2 = r$, and det $E_3 = 1$. Thus, Properties I, II, and III remain valid for 3×3 matrices.

So far, all the operations we have considered are on the rows of a determinant, but one of the nice properties of determinants is that all one learns about rows can be applied just as well to columns. To begin with, note that the number

$$\begin{vmatrix} a_1 & b_1 & c_1 \\ a_2 & b_2 & c_2 \\ a_3 & b_3 & c_3 \end{vmatrix} = a_1 \begin{vmatrix} b_2 & c_2 \\ b_3 & c_3 \end{vmatrix} - a_2 \begin{vmatrix} b_1 & c_1 \\ b_3 & c_3 \end{vmatrix} + a_3 \begin{vmatrix} b_1 & c_1 \\ b_2 & c_2 \end{vmatrix}$$

is exactly the same as the numbers previously defined to be the determinant.

Next, consider the elementary 3×3 matrices E_1, E_2, and E_3 above. We could just as well think of these matrices as having been obtained from the identity I by performing elementary operations on the columns. Thus,

I. E_1 is obtained by interchanging columns 2 and 3 of I.

II. E_2 is obtained by multiplying the second column of I by r.

III. E_3 is obtained by adding to the first column of I r times the third column.

Multiplication of any 3×3 matrix A on the right by the elementary matrices accomplishes the corresponding operations on the columns of A.

We therefore see that Properties I, II, and III remain true if we substitute the word column for the word row in each statement.

The properties of determinants can be used to simplify the problem of evaluation. We illustrate this with a series of examples.

EXAMPLE A

$$\begin{vmatrix} 2 & 4 & -2 \\ 1 & 3 & 2 \\ 0 & 1 & 2 \end{vmatrix} = \begin{vmatrix} 2 & 0 & 0 \\ 1 & 1 & 3 \\ 0 & 1 & 2 \end{vmatrix}$$

We have added -2 times the first column to the second, and we have added the first column to the third. With two zeros in the first row, it is now an easy calculation, working directly from the definition to obtain the answer

$$2 \begin{vmatrix} 1 & 3 \\ 1 & 2 \end{vmatrix} = 2(2 - 3) = -2.$$

EXAMPLE B

$$\begin{vmatrix} 4 & 0 & 2 \\ 1 & 5 & -7 \\ 3 & 0 & 2 \end{vmatrix} = -\begin{vmatrix} 0 & 4 & 2 \\ 5 & 1 & -7 \\ 0 & 3 & 2 \end{vmatrix} = -(-5)\begin{vmatrix} 4 & 2 \\ 3 & 2 \end{vmatrix}$$
$$= 5(8 - 6) = 10.$$

EXAMPLE C

$$\begin{vmatrix} a_1 & b_1 & c_2 \\ 0 & b_2 & c_2 \\ 0 & 0 & c_3 \end{vmatrix} = a_1 \begin{vmatrix} b_2 & c_2 \\ 0 & c_3 \end{vmatrix} = a_1 b_2 c_3.$$

Example C actually proves another useful fact about determinants.

Property IV   *If all entries below the diagonal (or above the diagonal) are 0, then the determinant equals the product of the entries on the diagonal.*

We have now defined the determinant of a 3×3 matrix, and we have established four properties of such determinants. They in turn can be used to establish other properties. The reader should, for example, find it easy to see that the determinant of a matrix A will equal the determinant of its transpose A^t. (Simply note that $A = E_n \cdots E_1 R$ where the E_i are elementary matrices and the row-reduced echelon matrix R has all entries below the diagonal equal to 0.)

It is possible to define the determinant of an $n \times n$ matrix; in fact Exercises 17–20 at the end of this section suggest how to proceed inductively by giving the definition for the case $n = 4$. The comforting fact is that the properties we have developed remain valid for all n. This is also true of the procedures now to be developed.

Let us turn our attention once again to systems of equations. By inserting zero coefficients as needed, we can assume that we are dealing with a " square " system, n equations in n unknowns.

$$
\begin{aligned}
a_{11}x_1 + \cdots + a_{1n}x_n &= b_1, \\
&\vdots \\
a_{n1}x_1 + \cdots + a_{nn}x_n &= b_n.
\end{aligned}
\tag{2}
$$

Call the coefficient matrix A. Solution of the system involves using the three elementary operations to bring matrix A to row-reduced echelon form. We have seen that these operations each correspond to multiplication on the left by an elementary matrix. Thus, we obtain

$$
E_k \cdots E_2 E_1 A = R,
\tag{3}
$$

where R is in row-reduced echelon form. This means, of course, that all entries in R under the diagonal are 0. So by Property 4, det R is the product of entries on the diagonal. If the system (2) has a unique solution, then of course $R = I$ and det $R = 1$. And if the system does not have a unique solution, then at least one entry on the diagonal is 0, so det $R = 0$. None of the elementary matrices has determinant 0, so it follows that det $A = 0$ if and only if det $R = 0$.

THEOREM B

The system (2) has a unique solution if and only if det $A \neq 0$.

COROLLARY A

If a system (2) of homogeneous equations (all $b_i = 0$) has a solution other than $x_1 = \cdots = x_n = 0$, then the determinant of the coefficient matrix must be zero.

COROLLARY B

The column vectors of the matrix

$$A = \begin{bmatrix} a_{11} & a_{12} \cdots a_{1n} \\ a_{21} & a_{22} & a_{2n} \\ \vdots & \vdots & \vdots \\ a_{n1} & a_{n2} & a_{nn} \end{bmatrix}$$

will be linearly independent if and only if det $A \neq 0$.
Corollary B follows from consideration of

$$\begin{bmatrix} a_{11} \\ a_{21} \\ \vdots \\ a_{n1} \end{bmatrix} x_1 + \begin{bmatrix} a_{12} \\ a_{22} \\ \vdots \\ a_{n2} \end{bmatrix} x_2 + \cdots + \begin{bmatrix} a_{1n} \\ a_{2n} \\ \vdots \\ a_{nn} \end{bmatrix} x_n = \begin{bmatrix} 0 \\ 0 \\ \vdots \\ 0 \end{bmatrix}$$

The vectors are linearly independent if the only solution of this equation is $x_1 = \cdots = x_n = 0$, and this will be the only solution if and only if det $A \neq 0$.

The procedure indicated by (3) could, of course, be applied to any matrix. If $R = I$, then we have demonstrated that A has an inverse. And if $R \neq I$, then as before det $A = 0$ and A^{-1} cannot exist (since det $A \cdot$ det $A^{-1} = 1$).

THEOREM C

Matrix A has an inverse if and only if det $A \neq 0$.

We have more here than a theoretic proof of conditions under which A^{-1} exists. If A^{-1} exists, then

$$E_k \cdots E_2 E_1 A = I,$$

and multiplication on the right by A^{-1} gives us $E_k \cdots E_2 E_1 I = A^{-1}$. So if we perform the same operations on I as we do on A, we will obtain A^{-1}.

EXAMPLE D

We shall find A^{-1} for $A = \begin{bmatrix} 2 & -1 & -1 \\ 1 & 4 & 2 \\ 3 & -2 & -2 \end{bmatrix}$

We begin by writing I to the left of A. Then we perform on I exactly those operations performed in bringing A to row-reduced echelon form:

$$\begin{bmatrix} 1 & 0 & 0 & 2 & -1 & -1 \\ 0 & 1 & 0 & 1 & 4 & 2 \\ 0 & 0 & 1 & 3 & -2 & -2 \end{bmatrix} \sim \begin{bmatrix} 1 & 0 & 0 & 2 & -1 & -1 \\ 2 & 1 & 0 & 5 & 2 & 0 \\ -2 & 0 & 1 & -1 & 0 & 0 \end{bmatrix}$$

$$\sim \begin{bmatrix} -2 & 0 & 1 & -1 & 0 & 0 \\ -8 & 1 & 5 & 0 & 2 & 0 \\ -3 & 0 & 2 & 0 & -1 & -1 \end{bmatrix}$$

$$\sim \begin{bmatrix} 2 & 0 & -1 & 1 & 0 & 0 \\ -4 & \frac{1}{2} & \frac{5}{2} & 0 & 1 & 0 \\ -7 & \frac{1}{2} & \frac{9}{2} & 0 & 0 & -1 \end{bmatrix}$$

Thus,

$$A^{-1} = \begin{bmatrix} 2 & 0 & -1 \\ -4 & \frac{1}{2} & \frac{5}{2} \\ 7 & -\frac{1}{2} & -\frac{9}{2} \end{bmatrix}. \quad \blacksquare$$

If a matrix A has no inverse (that is, if det $A = 0$), then we say that A is *singular*. Thus, a *nonsingular* matrix A is one for which A^{-1} exists (that is, one for which det $A \neq 0$).

We know that, in general, matrix multiplication is not commutative. Yet in talking about an inverse for a matrix A, we have not specified whether our inverse is a left inverse (so $A^{-1}A = I$) or a right inverse (so $AA^{-1} = I$). The fact is that if there is a matrix B so that $BA = I$, then $AB = I$. This makes it possible to simply refer to an inverse without specifying right or left. Moreover, we may refer to *the* inverse because a nonsingular matrix has only one inverse.

PROBLEMS

C · Show that if A is an $n \times n$ matrix, and if there exist two other such matrices B and C such that $BA = I$ and $AC = I$, then $B = C$.

D · Show that if $AB = I$ and $AC = I$, then $B = C$. (Thus, the inverse is unique.)

E · For two nonsingular matrices A and B, $(AB)^{-1} = B^{-1}A^{-1}$.

The elementary matrices can be used to prove many properties of matrices. We shall illustrate by proving

Property V   *Suppose matrix A and matrix B are identical except for their i^{th} row, and suppose C is identical to A and B except that the ith row of C is the sum of the ith rows of A and B. Then det $C = $det $A + $det B.*

PROOF:

For the case $i = 1$, this says that

$$\det \begin{bmatrix} a_1 & b_1 & c_1 \\ a_2 & b_2 & c_2 \\ a_3 & b_3 & c_3 \end{bmatrix} + \det \begin{bmatrix} a_1' & b_1' & c_1' \\ a_2 & b_2 & c_2 \\ a_3 & b_3 & c_3 \end{bmatrix} = \det \begin{bmatrix} a_1 + a_1' & b_1 + b' & c_1 + c_1' \\ a_2 & b_2 & c_2 \\ a_3 & b_3 & c_3 \end{bmatrix}$$

And this follows directly from the definition:

$$a_1 \begin{vmatrix} b_2 & c_2 \\ b_3 & c_3 \end{vmatrix} + b_1 \begin{vmatrix} a_2 & c_2 \\ a_3 & c_3 \end{vmatrix} + c_1 \begin{vmatrix} a_2 & b_2 \\ a_3 & b_3 \end{vmatrix}$$

$$+ a_1' \begin{vmatrix} b_2 & c_2 \\ b_3 & c_3 \end{vmatrix} + b_1' \begin{vmatrix} a_2 & c_2 \\ a_3 & c_3 \end{vmatrix} + c_1' \begin{vmatrix} a_2 & b_2 \\ a_3 & c_3 \end{vmatrix}$$

$$= (a_1 + a_1') \begin{vmatrix} b_2 & c_2 \\ b_3 & c_3 \end{vmatrix} + (b_1 + b_1') \begin{vmatrix} a_2 & c_2 \\ a_3 & c_3 \end{vmatrix} + (c_1 + c_1') \begin{vmatrix} a_2 & b_2 \\ a_3 & b_3 \end{vmatrix}$$

In the case where $i \neq 1$, we first note that by interchanging rows 1 and i with an elementary matrix E, of type I, we have

$$\det(E\,A) + \det(E\,B) = \det(E\,C).$$

But this says, since $\det E = -1$, that

$$(-\det A) + (-\det B) = -\det C.$$

Multiplication by -1 gives the result. ∎

We shall have need later for one more fact about linear transformations that is related to the notion of determinants. Consider four points $\mathbf{p}_1$, $\mathbf{p}_2$, $\mathbf{p}_3$, $\mathbf{p}_4$ that are consecutive vertices of a parallelogram in the plane. This is equivalent to asserting that the vectors $\mathbf{p}_3 - \mathbf{p}_4 = \mathbf{p}_2 - \mathbf{p}_1$ (Fig. 10.6). If $T: R^2 \to R^2$ is linear, then

$$T(\mathbf{p}_3) - T(\mathbf{p}_4) = T(\mathbf{p}_3 - \mathbf{p}_4) = T(\mathbf{p}_2 - \mathbf{p}_1) = T(\mathbf{p}_2) - T(\mathbf{p}_1),$$

so the image points are again vertices of a parallelogram. Our interest here is based on a fact illustrated by Problem F.

PROBLEM

F • Let $T: R^2 \to R^2$ be described by $[T] = \begin{bmatrix} 2 & 1 \\ -1 & -2 \end{bmatrix}$.

(a) Find the images of $\mathbf{p}_1' = (0, 0)$, $\mathbf{p}_2' = (0, 2)$, $\mathbf{p}_3' = (3, 2)$, $\mathbf{p}_4' = (3, 0)$.

(b) Find the area of the parallelogram with vertices $T(\mathbf{p}_1)$, $T(\mathbf{p}_2)$, $T(\mathbf{p}_3)$, and $T(\mathbf{p}_4)$.

(c) Find det[T]. Find the area of the parallelogram (rectangle in this case) with vertices $\mathbf{p}_1$, $\mathbf{p}_2$, $\mathbf{p}_3$, and $\mathbf{p}_4$.

(d) Multiply the two numbers obtained in (c). Compare with your answer to (b).

The preceding problem illustrates the fact that the area of the image parallelogram equals the area of the original parallelogram multiplied by det[T]. In fact, a theorem from linear algebra states this for n dimensions.

THEOREM D

Let $T: R^n \to R^n$ be linear. Let P_1 be a parallelepiped in R^n. The image of P_1 is a parallelepiped P_2, and

$$\text{vol } P_2 = |\det[T]| \cdot (\text{vol } P_1).$$

=== 10 · 4   EXERCISES

Evaluate the following determinants.

1. $\begin{vmatrix} 2 & 4 & -1 \\ 3 & 1 & 7 \\ 1 & -1 & 2 \end{vmatrix}$

2. $\begin{vmatrix} 1 & 5 & 6 \\ 3 & -1 & 8 \\ 7 & 2 & -3 \end{vmatrix}$

3. $\begin{vmatrix} 3 & -5 & 1 \\ 6 & -7 & 3 \\ 3 & -5 & 1 \end{vmatrix}$

4. $\begin{vmatrix} 5 & 0 & 5 \\ -1 & 3 & -1 \\ 2 & 4 & 2 \end{vmatrix}$

Verify by computing all three determinants in each exercise.

5. $\begin{vmatrix} 3 & 1 & 2 \\ -5 & 2 & 8 \\ 4 & 1 & 3 \end{vmatrix} + \begin{vmatrix} 3 & 1 & 2 \\ -5 & 2 & 8 \\ -2 & 2 & 1 \end{vmatrix} = \begin{vmatrix} 3 & 1 & 2 \\ -5 & 2 & 8 \\ 2 & 3 & 4 \end{vmatrix}$

6. $\begin{vmatrix} 1 & -1 & 4 \\ 0 & 5 & -2 \\ 2 & 1 & 1 \end{vmatrix} + \begin{vmatrix} 1 & -1 & -2 \\ 0 & 5 & 1 \\ 2 & 1 & 2 \end{vmatrix} = \begin{vmatrix} 1 & -1 & 2 \\ 0 & 5 & -1 \\ 2 & 1 & 3 \end{vmatrix}$

7. See number 9 of Exercise 10.3B. Obtain A^{-1} yourself.

8. See number 10 of Exercise 10.3B. Obtain B^{-1} yourself.

9. Vectors $\mathbf{v}_1{}^t = (4, 0, 0)$, $\mathbf{v}_2{}^t = (3, 4, 0)$ and $\mathbf{v}_3{}^t = (0, 0, 2)$ are radius vectors that form edges of a parallelepiped P_1 in R^3. This parallelepiped is carried into a second parallelepiped P_2 by

$$[T] = \begin{bmatrix} 1 & 3 & 0 \\ 0 & 1 & 4 \\ 0 & 3 & 0 \end{bmatrix}.$$

Find the volume of P_2.

10 · Solve Exercise 9 where $[T] = \begin{bmatrix} -1 & 4 & 2 \\ 0 & 1 & 3 \\ 1 & 0 & -2 \end{bmatrix}$.

11 · Prove that a determinant with two equal rows must be 0.

12 · Prove that a determinant having an entire row of zeros must be zero.

13 · Without evaluating, prove

$$\begin{vmatrix} b-a & c-b & a-c \\ a & b & c \\ d & d & d \end{vmatrix} = \begin{vmatrix} a & b & c \\ d & d & d \\ b & c & a \end{vmatrix}$$

14 · Without evaluating, prove

$$\begin{vmatrix} 1 & 1 & 1 \\ a & b & c \\ a^2 & b^2 & c^2 \end{vmatrix} = (c-b)(c-a)(b-a).$$

15 · Suppose that $T: R^3 \to R^3$ is linear, and that the entire range of T is contained in a two-dimensional subspace $\mathscr{S} \subset R^3$. Then the image of a parallelepiped must be a two dimensional parallelogram. Is Theorem C still correct in this situation? Why?

16 · $T: R^3 \to R^3$ is linear, and there are two distinct points p and q such that $T(p) = T(q)$. Show that $\det[T] = 0$.

The determinant of a 4×4 matrix is defined by

$$\begin{vmatrix} a_1 & b_1 & c_1 & d_1 \\ a_2 & b_2 & c_2 & d_2 \\ a_3 & b_3 & c_3 & d_3 \\ a_4 & b_4 & c_4 & d_4 \end{vmatrix} = a_1 \begin{vmatrix} b_2 & c_2 & d_2 \\ b_3 & c_3 & d_3 \\ b_4 & c_4 & d_4 \end{vmatrix} - b_1 \begin{vmatrix} a_2 & c_2 & d_2 \\ a_3 & c_3 & d_3 \\ a_4 & c_4 & d_4 \end{vmatrix}$$

$$+ c_1 \begin{vmatrix} a_2 & b_2 & d_2 \\ a_3 & b_3 & d_3 \\ a_4 & b_4 & d_4 \end{vmatrix} - d_1 \begin{vmatrix} a_2 & b_2 & c_2 \\ a_3 & b_3 & c_3 \\ a_4 & b_4 & c_4 \end{vmatrix}$$

·All the properties of determinants remain valid, and they should be used in evaluating 4×4 determinants. Evaluate the following.

17 · $\begin{vmatrix} 0 & 1 & 0 & -1 \\ 3 & -7 & 1 & 4 \\ 0 & 5 & 1 & -2 \\ 0 & -4 & 0 & 1 \end{vmatrix}$

18 · $\begin{vmatrix} 3 & 0 & 1 & -1 \\ 5 & 4 & -2 & 1 \\ -1 & 0 & 2 & 0 \\ 2 & 0 & 5 & 0 \end{vmatrix}$

19 · Find the inverse of the matrix corresponding to the determinant of Exercise 17.

20 · Find the inverse of the matrix corresponding to the determinant of Exercise 18.

10 · 5 Change of Basis

Suppose that $B = \{\mathbf{u}_1, \ldots, \mathbf{u}_n\}$ is a basis of R^n, and that $T: R^n \to R^n$ is a linear transformation. We know that

$$[T] = [T(\mathbf{u}_1), \ldots, T(\mathbf{u}_n)].$$

(It is important in reading this to remember our convention: $T(\mathbf{u}_i)$ is a column vector, and so $[T(\mathbf{u}_1), \ldots, T(\mathbf{u}_n)]$ is indeed an $n \times n$ matrix.)

Clearly the matrix $[T]$ depends on the basis. Thus, if $C = [\mathbf{v}_1, \ldots, \mathbf{v}_n]$ is a second basis,

$$[T] = [T(\mathbf{v}_1), \ldots, T(\mathbf{v}_n)].$$

These matrices look very different. For purposes of a discussion in which two bases are under consideration, it therefore behooves us to distinguish between them, writing

$$[T]_B \quad \text{or} \quad [T]_C.$$

In this section we address ourselves to the following problem. Suppose $[T]_B$ is known. How can we determine $[T]_C$? This is equivalent to asking, how does a change of basis from B to C affect the matrix of a transformation T?

Since the new set C is a basis, each $\mathbf{u}_i$ may be expressed as a linear combination of the $\mathbf{v}_i$. Thus,

$$\mathbf{u}_1 = r_{11}\mathbf{v}_1 + \cdots + r_{n1}\mathbf{v}_n = [\mathbf{v}_1, \ldots, \mathbf{v}_n]\begin{bmatrix} r_{11} \\ \vdots \\ r_{n1} \end{bmatrix}$$
$$\vdots$$
$$\mathbf{u}_n = r_{1n}\mathbf{v}_1 + \cdots + r_{nn}\mathbf{v}_n = [\mathbf{v}_1, \ldots, \mathbf{v}_n]\begin{bmatrix} r_{1n} \\ \vdots \\ r_{nn} \end{bmatrix}$$

These equations may be summarized by a matrix equation:

$$[\mathbf{u}_1, \ldots, \mathbf{u}_n] = [\mathbf{v}_1, \ldots, \mathbf{v}_n]\begin{bmatrix} r_{11} & \cdots & r_{1n} \\ \vdots & & \vdots \\ r_{n1} & \cdots & r_{nn} \end{bmatrix}$$

Adopting obvious notation, this is further abbreviated to

$$U = VR.$$

The column vectors of U constitute a basis of R^n, so they are surely linearly independent and (Corollary 10.4B) det $U \neq 0$. Similarly det $V \neq 0$, and it must be (Theorem 10.4A) that det $R \neq 0$. We conclude (Theorem 10.4C) that U, V, and R all have inverses.

Let the coordinates of a point $\mathbf{p}^t$ be $[\bar{p}_1, \ldots, \bar{p}_n]$ with respect to basis C and $[p_1, \ldots, p_n]$ with respect to B. That is,

$$[\mathbf{v}_1, \ldots, \mathbf{v}_n] \begin{bmatrix} \bar{p}_1 \\ \vdots \\ \bar{p}_n \end{bmatrix} = [\mathbf{u}_1, \ldots, \mathbf{u}_n] \begin{bmatrix} p_1 \\ \vdots \\ p_n \end{bmatrix}$$

or, using the notation adopted above

$$V \begin{bmatrix} \bar{p}_1 \\ \vdots \\ \bar{p}_n \end{bmatrix} = U \begin{bmatrix} p_1 \\ \vdots \\ p_n \end{bmatrix} = VR \begin{bmatrix} p_1 \\ \vdots \\ p_n \end{bmatrix}$$

Since V^{-1} exists, the expressions on the left and right of this equality enable us to write

$$\begin{bmatrix} \bar{p}_1 \\ \vdots \\ \bar{p}_n \end{bmatrix} = R \begin{bmatrix} p_1 \\ \vdots \\ p_n \end{bmatrix}$$

If $\mathbf{q} = T(\mathbf{p})$, then $\mathbf{q}^t$ similarly has coordinates $[\bar{q}_1, \ldots, \bar{q}_n]$ with respect to C and $[q_1, \ldots, q_n]$ with respect to B. And again

$$\begin{bmatrix} \bar{q}_1 \\ \vdots \\ \bar{q}_n \end{bmatrix} = R \begin{bmatrix} q_1 \\ \vdots \\ q_n \end{bmatrix}$$

Now we are assuming $[T]_B$ known, so that we have

$$\begin{bmatrix} q_1 \\ \vdots \\ q_n \end{bmatrix} = [T]_B \begin{bmatrix} p_1 \\ \vdots \\ p_n \end{bmatrix}$$

The expression for $[T]_C$ is now obtained from

$$\begin{bmatrix} \bar{q}_1 \\ \vdots \\ \bar{q}_n \end{bmatrix} = R \begin{bmatrix} q_1 \\ \vdots \\ q_n \end{bmatrix} = R[T]_B \begin{bmatrix} p_1 \\ \vdots \\ p_n \end{bmatrix} = R[T]_B R^{-1} \begin{bmatrix} \bar{p}_1 \\ \vdots \\ \bar{p}_n \end{bmatrix}$$

Thus, $[T]_C = R[T]_B R^{-1}$.

In summary, let two bases $B = \{\mathbf{u}_1, \ldots, \mathbf{u}_n\}$ and $C = \{\mathbf{v}_1, \ldots, \mathbf{v}_n\}$ be related by

$$[\mathbf{u}_1, \ldots, \mathbf{u}_n] = [\mathbf{v}_1, \ldots, \mathbf{v}_n] \, R.$$

Then $[T]_C = R[T]_B R^{-1}$.

EXAMPLE A

With respect to basis $B = \{\mathbf{u}_1, \mathbf{u}_2\}$, the transformation $T: R^2 \to R^2$ has matrix

$$[T]_B = \begin{bmatrix} -3 & 7 \\ -1 & 4 \end{bmatrix}$$

We shall find the matrix of T with respect to the basis $C = \{\mathbf{v}_1, \mathbf{v}_2\}$ where

$$\mathbf{u}_1 = \mathbf{v}_1 - \mathbf{v}_2,$$
$$\mathbf{u}_2 = \mathbf{v}_1 + 2\mathbf{v}_2.$$

In this case, $R = \begin{bmatrix} 1 & 1 \\ -1 & 2 \end{bmatrix}$. (Note the coefficients of $\mathbf{v}_1$ and $\mathbf{v}_2$ in the expression for $\mathbf{u}_1$ form the first *column* vector of R.) It is easy to compute $R^{-1} = \begin{bmatrix} \frac{2}{3} & -\frac{1}{3} \\ \frac{1}{3} & \frac{1}{3} \end{bmatrix}$. Thus,

$$[T]_C = \begin{bmatrix} 1 & 1 \\ -1 & 2 \end{bmatrix} \begin{bmatrix} -3 & 7 \\ -1 & 4 \end{bmatrix} \begin{bmatrix} \frac{2}{3} & -\frac{1}{3} \\ \frac{1}{3} & \frac{1}{3} \end{bmatrix}$$

$$[T]_C = \begin{bmatrix} 1 & 5 \\ 1 & 0 \end{bmatrix}$$

As a check of sorts, let us consider the point $\mathbf{p} = 3\mathbf{u}_1 - 2\mathbf{u}_2 = 3(\mathbf{v}_1 - \mathbf{v}_2) - 2(\mathbf{v}_1 + 2\mathbf{v}_2)$, or $\mathbf{p} = \mathbf{v}_1 - 7\mathbf{v}_2$:

$$[T]_B \begin{bmatrix} 3 \\ -2 \end{bmatrix} = \begin{bmatrix} -3 & 7 \\ -1 & 4 \end{bmatrix} \begin{bmatrix} 3 \\ -2 \end{bmatrix} = \begin{bmatrix} -23 \\ -11 \end{bmatrix}$$

$$[T]_C \begin{bmatrix} 1 \\ -7 \end{bmatrix} = \begin{bmatrix} 1 & 5 \\ 1 & 0 \end{bmatrix} \begin{bmatrix} 1 \\ -7 \end{bmatrix} = \begin{bmatrix} -34 \\ 1 \end{bmatrix}$$

To check, note that

$$-23\mathbf{u}_1 - 11\mathbf{u}_2 = -23(\mathbf{v}_1 - \mathbf{v}_2) - 11(\mathbf{v}_1 + 2\mathbf{v}_2)$$
$$= -34\mathbf{v}_1 + \mathbf{v}_2. \ \blacksquare$$

=== 10 · 5 EXERCISES

1. $B = \{\mathbf{u}_1 = [4 \quad -2]^t; \ \mathbf{u}_2 = [1 \quad -1]^t\}$
   $C = \{\varepsilon_1, \varepsilon_2 ; \text{ the standard basis}\}$

   $[T]_B = \begin{bmatrix} 3 & 1 \\ 2 & 4 \end{bmatrix}$. Find $[T]_C$.

2. $B = \{\mathbf{u}_1 = [-2 \quad 1]^t; \mathbf{u}_2 = [1 \quad 3]^t\}$

$C = \{\varepsilon_1, \varepsilon_2; \text{ the standard basis}\}$

$[T]_B = \begin{bmatrix} -1 & 1 \\ 2 & 3 \end{bmatrix}$. Find $[T]_C$.

3. Using the same bases as in Exercise 1, suppose

$[T]_C = \begin{bmatrix} 4 & 1 \\ 2 & 0 \end{bmatrix}$ is given. Find $[T]_B$.

4. Using the same bases as in Exercise 2, suppose

$[T]_C = \begin{bmatrix} -1 & 2 \\ 0 & -2 \end{bmatrix}$. Find $[T]_B$.

5. $T: R^3 \to R^3$ is described by $[T]_C = \begin{bmatrix} 2 & 1 & 0 \\ 0 & -1 & 1 \\ 4 & -1 & 3 \end{bmatrix}$

where C is the standard basis. Find $[T]_B$ where

$$B = \{\mathbf{u}_1 = [1 \quad 2 \quad 0]^t, \mathbf{u}_2 = [-1 \quad 0 \quad -1]^t, \mathbf{u}_3 = [0 \quad 4 \quad 1]^t\}.$$

6. Solve Exercise 5 if $[T]_C = \begin{bmatrix} 3 & -1 & 2 \\ 2 & 0 & 1 \\ 0 & 2 & -1 \end{bmatrix}$

Two $n \times n$ matrices A and B are *similar*, written $A \overset{s}{=} B$, if there is a nonsingular matrix R such that $A = RBR^{-1}$.

7. Prove that if $A \overset{s}{=} B$, then $\det A = \det B$.

8. Prove that if $A \overset{s}{=} B$, then $[A - I] \overset{s}{=} [B - I]$ where I is the $n \times n$ identity matrix.

9. Suppose A is an $n \times n$ matrix for which A^{-1} exists. Prove that for any $n \times n$ matrix B, $AB \overset{s}{=} BA$.

10. Suppose A has an inverse A^{-1}, and that $A \overset{s}{=} B$. Prove that B has an inverse and that $B^{-1} \overset{s}{=} A^{-1}$.

11*. Suppose that $T: R^2 \to R^3$, and that with respect to the standard basis in each space,

$$[T]_C = \begin{bmatrix} 1 & -1 \\ -2 & 0 \\ -1 & 2 \end{bmatrix}$$

Now if we make a change of basis in each space, using the basis B from Exercise 1 for R^2 and the basis B from Exercise 5 for R^3, find $[T]_B$.

10 · 6   Bilinear Transformations and Quadratic Forms

Let $\mathbf{u} = [a \quad b]^t$ and $\mathbf{v} = [c \quad d]^t$ be arbitrary vectors of R^2, and suppose that for any two such vectors, we define by means of a determinant the function

$$B(\mathbf{u}, \mathbf{v}) = \begin{vmatrix} a & c \\ b & d \end{vmatrix}$$

From Properties V and II of determinants, we see immediately that

$$B(\mathbf{u}_1 + \mathbf{u}_2, \mathbf{v}) = B(\mathbf{u}_1, \mathbf{v}) + B(\mathbf{u}_2, \mathbf{v})$$
$$B(\alpha\mathbf{u}, \mathbf{v}) = \alpha B(\mathbf{u}, \mathbf{v}). \tag{1}$$

Similarly,

$$B(\mathbf{u}, \mathbf{v}_1 + \mathbf{v}_2) = B(\mathbf{u}, \mathbf{v}_1) + B(\mathbf{u}, \mathbf{v}_2),$$
$$B(\mathbf{u}, \alpha\mathbf{v}) = \alpha B(\mathbf{u}, \mathbf{v}). \tag{2}$$

Thus, $B(\mathbf{u}, \mathbf{v})$ is linear in $\mathbf{u}$ and linear in $\mathbf{v}$. Any function $B(\mathbf{u}, \mathbf{v})$ that is linear in $\mathbf{u}$ and $\mathbf{v}$ (that is, which satisfies (1) and (2)) is said to be a *bilinear transformation*.

We have given an example of a real-valued bilinear transformation. Just as there is a correspondence between all linear transformations of $R^2 \to R^2$ and the set of 2×2 matrices, so there is a correspondence between all bilinear transformations of $R^2 \to R$ and the 2×2 matrices.† It is determined as follows. Let $\mathbf{u} = [x_1 \quad y_1]^t$ and $\mathbf{v} = [x_2 \quad y_2]^t$. Then for an arbitrary bilinear transformation B, we have (using ε_1 and ε_2 as the standard basis vectors)

$$\begin{aligned} B(\mathbf{u}, \mathbf{v}) &= B(x_1\varepsilon_1 + y_1\varepsilon_2, \mathbf{v}) \\ &= x_1 B(\varepsilon_1, \mathbf{v}) + y_1 B(\varepsilon_2, \mathbf{v}) \\ &= x_1 B(\varepsilon_1, x_2\varepsilon_1 + y_2\varepsilon_2) + y_1 B(\varepsilon_2, x_2\varepsilon_1 + y_2\varepsilon_2) \\ &= x_1[x_2 B(\varepsilon_1, \varepsilon_1) + y_2 B(\varepsilon_1, \varepsilon_2)] \\ &\quad + y_1[x_2 B(\varepsilon_2, \varepsilon_1) + y_2 B(\varepsilon_2, \varepsilon_2)]. \end{aligned}$$

Now if we set $B(\varepsilon_i, \varepsilon_j) = a_{ij}$ for $i = 1, 2, j = 1, 2$, then

$$B(\mathbf{u}, \mathbf{v}) = x_1 x_2 a_{11} + x_1 y_2 a_{12} + x_2 y_1 a_{21} + y_1 y_2 a_{22}$$

$$= [x_1 \quad y_1] \begin{bmatrix} a_{11} & a_{12} \\ a_{21} & a_{22} \end{bmatrix} \begin{bmatrix} x_2 \\ y_2 \end{bmatrix} \tag{3}$$

---

† Since a bilinear transformation is defined for pairs of vectors in R^2, it is more proper to speak of such transformations as defined on the so-called *product space* defined by $R^2 \times R^2 = \{(\mathbf{u}, \mathbf{v}): \mathbf{u}, \mathbf{v} \in R^2\}$. To avoid digressing on product spaces, however, we will continue to speak of bilinear transformations defined on R^2 (or R^n).

Thus, $B \leftrightarrow [a_{ij}]$. This same procedure allows us to put bilinear transformations of $R^n \to R$ into correspondence with $n \times n$ matrices.

PROBLEMS

A · For vectors $\mathbf{u} = [x_1 \quad y_1]^t$ and $\mathbf{v} = [x_2 \quad y_2]^t$ and the bilinear transformation $B = [a_{ij}]$, we wrote out in (3) above an expression for $B(\mathbf{u}, \mathbf{v})$. Using the same notation, write out a similar expression for $B(\mathbf{v}, \mathbf{u})$.

B · Specify conditions on the matrix $\begin{bmatrix} a_{11} & a_{12} \\ a_{21} & a_{22} \end{bmatrix}$ for which the corresponding bilinear transformation B will have the property that $B(\mathbf{u}, \mathbf{v}) = B(\mathbf{v}, \mathbf{u})$ for all $\mathbf{u}$ and $\mathbf{v}$.

Bilinear transformations B are called *symmetric* if for all $\mathbf{u}$ and $\mathbf{v}$, $B(\mathbf{u}, \mathbf{v}) = B(\mathbf{v}, \mathbf{u})$. The matrix corresponding to a symmetric bilinear form is easily recognized. Since it must be true for all i and j that $B(\varepsilon_i, \varepsilon_j) = B(\varepsilon_j, \varepsilon_i)$, it follows that for all i and j, $a_{ij} = a_{ji}$. This means that the first row of the matrix is identical to the first column, the second row to the second column, etc. (Compare this to your answer to Problem B above.) Such matrices are called symmetric. The matrix

$$\begin{bmatrix} 2 & -1 & 4 \\ -1 & -3 & 2 \\ 4 & 2 & 1 \end{bmatrix}$$

is a symmetric 3×3 matrix.

If B is a bilinear symmetric transformation on R^n and $\mathbf{u} \in R^n$, then $B(\mathbf{u}, \mathbf{u})$ is called a *quadratic form*. A quadratic form on R^2 where $\mathbf{u} = [x \quad y]^t$ takes the form

$$[x \quad y] \begin{bmatrix} a & b \\ b & c \end{bmatrix} \begin{bmatrix} x \\ y \end{bmatrix} = ax^2 + 2bxy + cy^2. \tag{4}$$

It is most instructive to consider in this light the problem of removing the product term of the general second-degree equation by rotation of axes.

For convenience we summarize at this point our work in Example 1.4F where the problem was to graph

$$4xy - 3y^2 = 8. \tag{5}$$

We began with the substitution (Fig. 10.7)

$$x = (\cos \theta)u - (\sin \theta)v,$$
$$y = (\sin \theta)u + (\cos \theta)v. \tag{6}$$

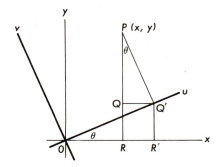

FIG. 10.7

We showed that if we pick θ so that $\sin \theta = 1/\sqrt{5}$, $\cos \theta = 2/\sqrt{5}$, then substitution of

$$x = 2/\sqrt{5}\,u - 1/\sqrt{5}\,v,$$
$$y = 1/\sqrt{5}\,u + 2/\sqrt{5}\,v, \tag{7}$$

into (5) gave, after simplification

$$u^2 - 4v^2 = 8. \tag{8}$$

We now use matrix notation to state the same result. Writing (5) in the form (4),

$$[x \ \ y] \begin{bmatrix} 0 & 2 \\ 2 & -3 \end{bmatrix} \begin{bmatrix} x \\ y \end{bmatrix} = 8. \tag{5'}$$

Setting

$$\begin{bmatrix} x \\ y \end{bmatrix} = \begin{bmatrix} \cos \theta & -\sin \theta \\ \sin \theta & \cos \theta \end{bmatrix} \begin{bmatrix} u \\ v \end{bmatrix} \tag{6'}$$

and choosing $\theta = \operatorname{Arc\,cos} 2/\sqrt{5}$ so that

$$\begin{bmatrix} x \\ y \end{bmatrix} = \begin{bmatrix} 2/\sqrt{5} & -1/\sqrt{5} \\ 1/\sqrt{5} & 2/\sqrt{5} \end{bmatrix} \begin{bmatrix} u \\ v \end{bmatrix} \tag{7'}$$

and

$$[x \ \ y] = [u \ \ v] \begin{bmatrix} 2/\sqrt{5} & 1/\sqrt{5} \\ -1/\sqrt{5} & 2/\sqrt{5} \end{bmatrix}$$

we have by substitution into (5')

$$[u \ \ v] \begin{bmatrix} 2/\sqrt{5} & 1/\sqrt{5} \\ -1/\sqrt{5} & 2/\sqrt{5} \end{bmatrix} \begin{bmatrix} 0 & 2 \\ 2 & -3 \end{bmatrix} \begin{bmatrix} 2/\sqrt{5} & -1/\sqrt{5} \\ 1/\sqrt{5} & 2/\sqrt{5} \end{bmatrix} \begin{bmatrix} u \\ v \end{bmatrix} = 8.$$

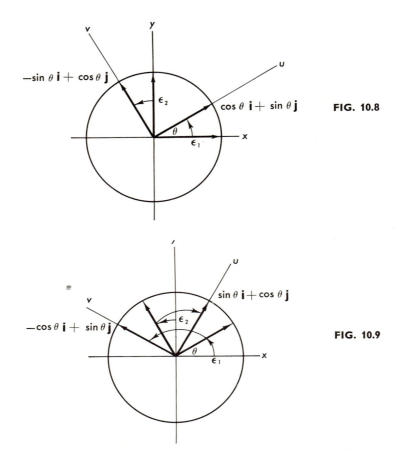

FIG. 10.8

FIG. 10.9

After simplification (multiplying together of the three matrices), we have the quadratic form

$$[u \quad v] \begin{bmatrix} 1 & 0 \\ 0 & 4 \end{bmatrix} \begin{bmatrix} u \\ v \end{bmatrix} = 8. \tag{8'}$$

The substitution (6) was originally suggested by a geometric argument about the effect of rotating the coordinate axes (Section 1.4). Viewed in our present context, the equations are more easily derived. It is obvious (Fig. 10.8) that a rotation is a linear transformation that takes the basis vectors $\varepsilon_1 = [1 \quad 0]^t$ and $\varepsilon_2 = [0 \quad 1]^t$ into the vectors $[\cos \theta \quad \sin \theta]^t$ and $[-\sin \theta \quad \cos \theta]^t$. Then by using these image vectors as the column vectors of our transformation matrix, we have (6') immediately.

There is a related transformation we should mention here. Consider the transformation that rotates points through an angle of θ, then reflects them in the y axis. This transformation (Fig. 10.9) takes $\varepsilon_1 = [1 \quad 0]^t$ and $\varepsilon_2 = [0 \quad 1]^t$ into $[-\cos \theta \quad \sin \theta]^t$ and $[\sin \theta \quad \cos \theta]^t$. The transformation is therefore described by

$$\begin{bmatrix} x \\ y \end{bmatrix} = \begin{bmatrix} -\cos \theta & \sin \theta \\ \sin \theta & \cos \theta \end{bmatrix} \begin{bmatrix} u \\ v \end{bmatrix} \tag{9}$$

The matrices of the transformations described in (6′) and (9) share a common and important property. Let us call them

$$B_1 = \begin{bmatrix} \cos \theta & -\sin \theta \\ \sin \theta & \cos \theta \end{bmatrix}, \qquad B_2 = \begin{bmatrix} -\cos \theta & \sin \theta \\ \sin \theta & \cos \theta \end{bmatrix}$$

Then using our usual technique for finding the inverse of a 2×2 matrix, we have

$$B_1^{-1} = \begin{bmatrix} \cos \theta & \sin \theta \\ -\sin \theta & \cos \theta \end{bmatrix}, \qquad B_2^{-1} = \begin{bmatrix} -\cos \theta & \sin \theta \\ \sin \theta & \cos \theta \end{bmatrix}$$

In each case we notice that

$$B^{-1} = B^t \qquad \text{so that} \quad BB^t = I. \tag{10}$$

An $n \times n$ matrix satisfying (10) is called *orthogonal*.

PROBLEMS

C · In each of the matrices B_1 and B_2, let $\theta = \text{Arcsin } \frac{4}{5}$. Then find the image of $\mathbf{u} = [2 \quad 1]^t$ and $\mathbf{v} = [1 \quad 3]^t$ under these transformations. Draw pictures in each case, verifying that B_1 rotates $\mathbf{u}$ and $\mathbf{v}$, while B_2 rotates them and then reflects them in the y axis.

D · Prove that if B is orthogonal, then $\det(B) = \pm 1$.

E · Let the columns of an $n \times n$ orthogonal matrix be considered as vectors $\mathbf{v}_1, \ldots, \mathbf{v}_n$ in R^n. Show that these vectors form an orthonormal basis.

F · Suppose that the matrix of $T: R^n \to R^n$ is orthogonal. Prove that for any $\mathbf{v} \in R^n$, $|T\mathbf{v}| = |\mathbf{v}|$.

G · Show that the transformation T of Problem F preserves angles between arbitrary vectors $\mathbf{u}$ and $\mathbf{v}$ of R^n.

Because of the properties of orthogonal matrices established in Problems F and G, the corresponding transformations are called *rigid motions*. Consider Problem C in which the angle from $\mathbf{u}$ to $\mathbf{v}$ was positive. The transformations B_1 and B_2 both preserved the angles (Check this!), but with B_1 the orientation of the angle was preserved, while with B_2 it was reversed; that is the angle from $B_2\mathbf{u}$ to $B_2\mathbf{v}$ is negative. It can be shown that this corresponds to the fact that $\det(B_1) = 1$ while $\det(B_2) = -1$. Orthogonal matrices with determinant 1 define *rotations*; orthogonal matrices with determinant -1 define *reflections* (more properly rotation–reflections).

10 · 6 EXERCISES

1 · Let $\mathbf{u}_1 = [1 \quad -1]^t$, $\mathbf{u}_2 = [-1 \quad 2]^t$. The bilinear transformation B takes the following values: $B(\mathbf{u}_1, \mathbf{u}_1) = 0$; $B(\mathbf{u}_1, \mathbf{u}_2) = 3$; $B(\mathbf{u}_2, \mathbf{u}_1) = 4$; $B(\mathbf{u}_2, \mathbf{u}_2) = -2$. Find $[B]$.

2 · Using $\mathbf{u}_1$ and $\mathbf{u}_2$ above, suppose $B(\mathbf{u}_1, \mathbf{u}_1) = -1$; $B(\mathbf{u}_1, \mathbf{u}_2) = 1$; $B(\mathbf{u}_2, \mathbf{u}_1) = 4$; $B(\mathbf{u}_2, \mathbf{u}_2) = 2$. Find $[B]$.

3 · If in Exercise 1, we change $B(\mathbf{u}_2, \mathbf{u}_1)$ to be 3, then B is symmetric. Find $[B]$.

4 · If in Exercise 2, we change $B(\mathbf{u}_2, \mathbf{u}_1)$ to be 1, then B is symmetric. Find $[B]$.

5 · Verify that $A = \begin{bmatrix} \frac{2}{3} & -\frac{2}{3} & -\frac{1}{3} \\ \frac{1}{3} & \frac{2}{3} & -\frac{2}{3} \\ \frac{2}{3} & \frac{1}{3} & \frac{2}{3} \end{bmatrix}$ is orthogonal.

If $\mathbf{p} = [1 \quad 2 \quad -1]^t$ and $\mathbf{q} = [3 \quad 0 \quad -3]^t$, verify that the angle formed by $\mathbf{p}$ and $\mathbf{q}$ equals the angle formed by $A\mathbf{p}$ and $A\mathbf{q}$.

6 · Verify that $B = \begin{bmatrix} \frac{3}{5} & -\frac{4}{5} & 0 \\ \frac{4}{5} & \frac{3}{5} & 0 \\ 0 & 0 & -1 \end{bmatrix}$ is orthogonal. Then proceed as in Exercise 5, using B instead of A.

For each of the following statements, either provide a proof or a counterexample.

7 · If A and B are symmetric $n \times n$ matrices, then AB is also symmetric.

8 · If A and B are orthogonal $n \times n$ matrices, then AB is also orthogonal.

9 · If B is any $n \times n$ matrix, then $A = \frac{1}{2}[B + B^t]$ is symmetric.

10 · If A and B are symmetric, then $A + B$ is symmetric.

11 · If A is orthogonal and $A \stackrel{s}{=} B$, then B is orthogonal.

12 · If A is symmetric and $A \stackrel{s}{=} B$, then B is symmetric.

13 · If $T: R^2 \to R^2$ and $|T\mathbf{v}| = |\mathbf{v}|$ for every $\mathbf{v} \in R^2$, then $[T]$ is orthogonal.

14 · If $T: R^2 \to R^2$ and the angle between $\mathbf{u}$ and $\mathbf{v}$ is always equal to the angle between $T(\mathbf{u})$ and $T(\mathbf{v})$, then $[T]$ is orthogonal.

10 · 7   Symmetric Matrices

Let $[S] = \begin{bmatrix} a & b \\ b & c \end{bmatrix}$ be an arbitrary symmetric matrix. It corresponds to a second degree equation in x and y according to

$$[x \quad y] \begin{bmatrix} a & b \\ b & c \end{bmatrix} \begin{bmatrix} x \\ y \end{bmatrix} = ax^2 + 2bxy + cy^2.$$

Following the procedure illustrated by (6′) of the preceding section, it is clear that we can always find a linear transformation B so that the substitution.

$$\begin{bmatrix} x \\ y \end{bmatrix} = [B] \begin{bmatrix} u \\ v \end{bmatrix}$$

leads to the transformed equation

$$[u \quad v][B]^t[S][B] \begin{bmatrix} u \\ v \end{bmatrix} = \lambda_1 u^2 + \lambda_2 v^2,$$

in which the product term uv is missing. Or, phrased in terms of matrices, we can always find an orthogonal matrix $[B]$ so that the product

$$[B]^t[S][B] = \begin{bmatrix} \lambda_1 & 0 \\ 0 & \lambda_2 \end{bmatrix}$$

is a diagonal matrix as in (8′) of Section 10.6.

We wish now to address ourselves to the problem of removing product terms from the general second degree equation in three variables. Again such equations correspond in a natural way to symmetric matrices;

$$[x \quad y \quad z] \begin{bmatrix} a & b & c \\ b & d & e \\ c & e & f \end{bmatrix} \begin{bmatrix} x \\ y \\ z \end{bmatrix} = ax^2 + dy^2 + fz^2 + 2bxy + 2cxz + 2eyz.$$

Our problem is to determine an orthogonal matrix $[B]$ so that the substitution $[x \ y \ z]^t = [B][u \ v \ w]^t$ will again eliminate the product terms. Or, equivalently, we wish to determine an orthogonal $[B]$ so that for a given symmetric matrix $[S]$, $[B]^t[S][B]$ will be a diagonal matrix. The procedure used to eliminate the product term in a quadratic expression of two variables suggests rotating the three coordinate axes about the origin. But it is very difficult (we are tempted to say impossible) to determine the correct transformation from a picture. We therefore seek another method.

In order to pave the way for solving our new problem, we shall solve our problem in two variables by another method. We note that when $[B]$ has been determined,

$$[B]^t[S][B] \begin{bmatrix} 1 \\ 0 \end{bmatrix} = \begin{bmatrix} \lambda_1 \\ 0 \end{bmatrix} \qquad [B]^t[S][B] \begin{bmatrix} 0 \\ 1 \end{bmatrix} = \begin{bmatrix} 0 \\ \lambda_2 \end{bmatrix}$$

Or, since $[B]^t = [B]^{-1}$ (and it is here that we are making use of the orthogonality of $[B]$), we have after multiplication on the left by $[B]$,

$$[S][B] \begin{bmatrix} 1 \\ 0 \end{bmatrix} = \lambda_1[B] \begin{bmatrix} 1 \\ 0 \end{bmatrix} \qquad [S][B] \begin{bmatrix} 0 \\ 1 \end{bmatrix} = \lambda_2[B] \begin{bmatrix} 0 \\ 1 \end{bmatrix}$$

Now the transformation B is determined if we know the two column vectors

$$\mathbf{v}_1 = [B] \begin{bmatrix} 1 \\ 0 \end{bmatrix} \qquad \text{and} \qquad \mathbf{v}_2 = [B] \begin{bmatrix} 0 \\ 1 \end{bmatrix}$$

And these vectors satisfy $[S]\mathbf{v}_1 = \lambda_1\mathbf{v}_1$: $[S]\mathbf{v}_2 = \lambda_2\mathbf{v}_2$. This suggests solving the matrix equation $[S]\mathbf{v} = \lambda\mathbf{v}$, or equivalently, $[S - \lambda I]\mathbf{v} = 0$. In order that this equation should have the desired nontrivial solutions (we are not interested in the obvious solution $\mathbf{v} = \mathbf{0}$), it is clear (Corollary 10.4A) that the values of λ must be such that

$$\det[S - \lambda I] = 0. \tag{1}$$

The equation (1) is called the *characteristic equation* for the matrix $[S]$. The roots λ are called the *characteristic values* of $[S]$.

Returning to the two variable example used in Section 1.4 and again in Section 10.6, we had

$$[S] = \begin{bmatrix} 0 & 2 \\ 2 & -3 \end{bmatrix}$$

The characteristic equation is

$$\det[S - \lambda I] = \begin{vmatrix} 0 - \lambda & 2 \\ 2 & -3 - \lambda \end{vmatrix} = \lambda^2 + 3\lambda - 4 = 0.$$

The characteristic roots are $\lambda_1 = 1$ and $\lambda_2 = -4$. The vector corresponding to $\lambda_1 = 1$ is obtained by solving

$$\begin{bmatrix} -1 & 2 \\ 2 & -4 \end{bmatrix} \begin{bmatrix} x \\ y \end{bmatrix} = \begin{bmatrix} 0 \\ 0 \end{bmatrix}$$

which leads to one equation in two unknowns. The solution space is spanned by $[2 \quad 1]^t$ and since we want a unit vector, we may use either $\mathbf{v}_1 = [-2/\sqrt{5} \ -1/\sqrt{5}]^t$ or $[2/\sqrt{5} \ 1/\sqrt{5}]^t$. Similarly, solving

$$\begin{bmatrix} 4 & 2 \\ 2 & 1 \end{bmatrix} \begin{bmatrix} x \\ y \end{bmatrix} = \begin{bmatrix} 0 \\ 0 \end{bmatrix}$$

leads us to $2x + y = 0$ and we may use either $\mathbf{v}_2 = [1/\sqrt{5} \ -2/\sqrt{5}]^t$ or $[-1/\sqrt{5} \ \ 2/\sqrt{5}]^t$. There is, therefore, a bit of freedom in our choice of B, and the fact is that any of the indicated choices of $\mathbf{v}_1$ and $\mathbf{v}_2$ as column vectors of $[B]$ will work equally well. (This corresponds to the several choices of θ available in the rotation of axes until we arbitrarily decided to choose θ in the first quadrant.) If we choose $\mathbf{v}_1 = [2/\sqrt{5} \ \ 1/\sqrt{5}]$ and $\mathbf{v}_2 = [-1/\sqrt{5} \ \ 2/\sqrt{5}]$, we obtain

$$[B] = \begin{bmatrix} 2/\sqrt{5} & -1/\sqrt{5} \\ 1/\sqrt{5} & 2/\sqrt{5} \end{bmatrix}$$

as before.

The vectors corresponding to particular characteristic values are called the *characteristic vectors*. In an interesting combination of languages, some authors refer to these numbers and their corresponding vectors as *eigenvalues* and *eigenvectors*.

We are now ready to try the same procedure to remove the product terms from a second degree equation in three variables.

EXAMPLE A

By a rotation of axes, transform the equation

$$2x^2 + 2y^2 + 8z^2 - 2xy - 4yz - 4xz = 12 \tag{2}$$

to an equation in u, v, w in which there are no terms uv, vw, or uw.

The corresponding symmetric matrix is

$$S = \begin{bmatrix} 2 & -1 & -2 \\ -1 & 2 & -2 \\ -2 & -2 & 8 \end{bmatrix}$$

and the characteristic equation is

$$\begin{vmatrix} 2-\lambda & -1 & -2 \\ -1 & 2-\lambda & -2 \\ -2 & -2 & 8-\lambda \end{vmatrix} = (2-\lambda)[(2-\lambda)(8-\lambda)-4] \\ + 1[-(8-\lambda)-4]-2[2+2(2-\lambda)] = 0.$$

Simplification gives $\lambda^3 - 12\lambda^2 + 27\lambda = 0$ or $\lambda(\lambda - 3)(\lambda - 9) = 0$. The characteristic roots are 0, 3, 9. Characteristic vectors corresponding to each root are found from solving the systems below, two of which we have already solved.

For $\lambda = 0$,

$$\begin{bmatrix} 2 & -1 & -2 \\ -1 & 2 & -2 \\ -2 & -2 & 8 \end{bmatrix} \begin{bmatrix} x \\ y \\ z \end{bmatrix} = \begin{bmatrix} 0 \\ 0 \\ 0 \end{bmatrix}$$

with solution space (Example 10.1A) spanned by $[2 \quad 2 \quad 1]^t$.

For $\lambda = 9$,

$$\begin{bmatrix} -7 & -1 & -2 \\ -1 & -7 & -2 \\ -2 & -2 & -1 \end{bmatrix} \begin{bmatrix} x \\ y \\ z \end{bmatrix} = \begin{bmatrix} 0 \\ 0 \\ 0 \end{bmatrix}$$

with solution space (Example 10.1B) spanned by $[1 \quad 1 \; -4]^t$.

For $\lambda = 3$, the reader may verify that

$$\begin{bmatrix} -1 & -1 & -2 \\ -1 & -1 & -2 \\ -2 & -2 & 5 \end{bmatrix} \begin{bmatrix} x \\ y \\ z \end{bmatrix} = \begin{bmatrix} 0 \\ 0 \\ 0 \end{bmatrix}$$

has its solution space spanned by $[-1 \quad 1 \quad 0]^t$.

The vectors spanning the three solution spaces are the desired characteristic vectors to be used in defining matrix B. Since we want the column vectors of B to be unit vectors, we can use

$$\pm \begin{bmatrix} 2/3 \\ 2/3 \\ 1/3 \end{bmatrix}, \qquad \pm \begin{bmatrix} -1/\sqrt{2} \\ 1/\sqrt{2} \\ 0 \end{bmatrix}, \qquad \pm \begin{bmatrix} 1/3\sqrt{2} \\ 1/3\sqrt{2} \\ -4/3\sqrt{2} \end{bmatrix}$$

Though we had no guarantee (other than that examples in books seem to work out somehow) that our vectors would be orthogonal, a quick check establishes the happy fact that they are, no matter how we choose the respective ambiguous signs. Suppose we decide to try the transformation determined by

$$\begin{bmatrix} 2/3 & -1/\sqrt{2} & 1/3\sqrt{2} \\ 2/3 & 1/\sqrt{2} & 1/3\sqrt{2} \\ 1/3 & 0 & -4/3\sqrt{2} \end{bmatrix}$$

We know this matrix is orthogonal, so it defines a rigid motion. To adhere to the letter of the law, we were to find a rotation, so we need to check the determinant. And alas, this computation (left to the student) shows the determinant to be -1. So we shall arbitrarily choose one of the ambiguous signs above to be negative, thereby giving us

$$\begin{bmatrix} x \\ y \\ z \end{bmatrix} = \begin{bmatrix} 2/3 & 1/\sqrt{2} & 1/3\sqrt{2} \\ 2/3 & -1/\sqrt{2} & 1/3\sqrt{2} \\ 1/3 & 0 & -4/3\sqrt{2} \end{bmatrix} \begin{bmatrix} u \\ v \\ w \end{bmatrix}$$

Then equation (2) is transformed from

$$[x \quad y \quad z] \begin{bmatrix} 2 & -1 & -2 \\ -1 & 2 & -2 \\ -2 & -2 & 8 \end{bmatrix} \begin{bmatrix} x \\ y \\ z \end{bmatrix} = 12,$$

to the equation

$$[u \quad v \quad w] \begin{bmatrix} 2/3 & 2/3 & 1/3 \\ 1/\sqrt{2} & -1/\sqrt{2} & 0 \\ 1/3\sqrt{2} & 1/3\sqrt{2} & -4/3\sqrt{2} \end{bmatrix} \begin{bmatrix} 2 & -1 & -2 \\ -1 & 2 & -2 \\ -2 & -2 & 8 \end{bmatrix}$$

$$\begin{bmatrix} 2/3 & 1/\sqrt{2} & 1/3\sqrt{2} \\ 2/3 & -1/\sqrt{2} & 1/3\sqrt{2} \\ 1/3 & 0 & -4/3\sqrt{2} \end{bmatrix} \begin{bmatrix} u \\ v \\ w \end{bmatrix} = 12.$$

Simplified, this gives

$$[u \quad v \quad w] \begin{bmatrix} 0 & 0 & 0 \\ 0 & 3 & 0 \\ 0 & 0 & 9 \end{bmatrix} \begin{bmatrix} u \\ v \\ w \end{bmatrix} = 3v^2 + 9w^2 = 12.$$

Now the happy truth is that we knew the product of these three matrices would give a diagonal matrix, and that the entries would be the characteristic roots. (This foreknowledge is no small help in getting the correct answer.) Strictly speaking then, our problem was solved as soon as the characteristic roots were found. An equation in u, v, and w equivalent to (2) is $v^2 + 3w^2 = 4$; so (2) has as its graph an elliptic cylinder (Fig. 10.10). ∎

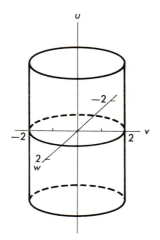

FIG. 10.10

To be certain that the procedure just used will always work, we really need to know two things. First, it is essential that the characteristic equation should have real (not complex) roots. Theorem A assures us that this will always be the case. Secondly, it is important that the characteristic vectors should be mutually orthogonal, a fact that happened above for no better reason than that "examples in books seem to somehow work out." Theorem B below gives a more satisfying reason.

THEOREM A

Let A be a real $n \times n$ symmetric matrix. Then all the characteristic roots are real.

PROOF:

The characteristic equation of A is an nth degree polynomial equation with real coefficients. Let $a + bi$ be a characteristic root, meaning of course that $\det[A - (a + bi)I] = 0$. Since roots of polynomials occur in conjugate pairs, $a - bi$ is also a root. We form

$$[B] = [A - (a + bi)I][A - (a - bi)I] = [A - aI]^2 + b^2I, \tag{3}$$

the right hand expression showing that $[B]$ is a real matrix. It follows from the product rule for determinants that $\det B = 0$. Therefore we can find a nonzero vector $\mathbf{v}$ such that $[B]\mathbf{v} = \mathbf{0}$. Thus, $\mathbf{v}^t[B]\mathbf{v} = 0$, but using (3), this gives

$$\mathbf{v}^t[B]\mathbf{v} = \mathbf{v}^t([A - aI]^2 + b^2I)\mathbf{v}$$
$$= \mathbf{v}^t[A - aI][A - aI]\mathbf{v} + b^2\mathbf{v}^t I\mathbf{v} = 0.$$

If we set $\mathbf{w} = [A - aI]\mathbf{v}$, then $\mathbf{w}^t = \mathbf{v}^t[A^t - aI] = \mathbf{v}^t[A - aI]$, the last equality due to the symmetry of A. Using $\mathbf{w}^t\mathbf{w} = |\mathbf{w}|^2$, we can now write

$$\mathbf{v}^t[B]\mathbf{v} = |\mathbf{w}|^2 + b^2|\mathbf{v}|^2 = 0.$$

Clearly $|\mathbf{w}|^2 \geq 0$ and we know $|\mathbf{v}|^2 > 0$, so the only way this sum can be 0 is for $b = 0$. And this means the characteristic root $a + bi = a$ is real. ∎

THEOREM B

Let A be a real $n \times n$ symmetric matrix. Then the characteristic vectors form an orthogonal basis of R^n.

PROOF:

Let the distinct characteristic values of A be $r_1, \ldots, r_k$. Since $\det[A - r_1 I] = 0$, there is a nontrivial solution, hence a nonzero vector $\mathbf{v}$ such that $[A - r_1 I]\mathbf{v} = \mathbf{0}$. It is easy to verify that the vectors satisfying $A\mathbf{v} = r_1\mathbf{v}$ form a subspace, and so we shall associate with each r_i a subspace S_i. Choose an orthogonal basis for each S_i. We now show that if $\mathbf{u}$ and $\mathbf{v}$ are basis vectors chosen from distinct subspaces S_i and S_j, then $\mathbf{u}$ and $\mathbf{v}$ are themselves orthogonal:

$$r_i\mathbf{u}^t\mathbf{v} = (A\mathbf{u})^t\mathbf{v} = \mathbf{u}^t[A]^t\mathbf{v} = \mathbf{u}^t[A]\mathbf{v} = \mathbf{u}^t r_j\mathbf{v}.$$

Thus, $(r_i - r_j)\mathbf{u}^t\mathbf{v} = 0$, and since $r_i \neq r_j$, $\mathbf{u}$ and $\mathbf{v}$ are orthogonal as stated. This means that the collection of the basis vectors of each S_i is itself an orthogonal set in R^n. Let us indicate this collection by $\mathbf{u}_1, \mathbf{u}_2, \ldots, \mathbf{u}_k$, spanning $S \subset R^n$. If S is not all of R^n, extend $\mathbf{u}_1, \ldots, \mathbf{u}_k, \ldots, \mathbf{u}_n$ to an orthogonal basis of R^n. Choose $\mathbf{v}$ in the space $S^\perp$ spanned by $\mathbf{u}_{k+1}, \ldots, \mathbf{u}_n$. Then for any $\mathbf{u}_i \in S$,

$$\mathbf{u}_i{}^t[A]\mathbf{v} = \mathbf{u}_i{}^t[A]^t\mathbf{v} = (A\mathbf{u}_i)^t\mathbf{v} = r_i\mathbf{u}_i{}^t\mathbf{v} = 0,$$

where r_i is the characteristic value corresponding to u_i. Hence $S^\perp \to S^\perp$ under A. A is symmetric on $S^\perp$, hence has a real characteristic root and a nontrivial characteristic vector in $S^\perp$. But such a vector would already have been included in S. Hence $S^\perp$ is empty, $S = R^n$, and the characteristic vectors $\mathbf{u}_1, \mathbf{u}_2, \ldots, \mathbf{u}_k = \mathbf{u}_n$ span R^n. ∎

The proof of Theorem B allows for the possibility of less than n distinct real roots to the nth degree characteristic equation of an $n \times n$ symmetric matrix. As the reader might guess, a more careful analysis would show that if r is a root of multiplicity m, then the subspace S_m of characteristic vectors corresponding to r has dimension m. It is in this way that the total number of characteristic vectors is n. Multiple roots of the characteristic equation present no difficulty in actual computation.

EXAMPLE B

Find an orthonormal matrix P such that $P^t S P$ is a diagonal matrix, given the symmetric matrix

$$S = \begin{bmatrix} \frac{3}{2} & -\frac{1}{2} & 2 \\ -\frac{1}{2} & \frac{3}{2} & 2 \\ 2 & 2 & -6 \end{bmatrix}$$

$$|S - \lambda I| = \begin{vmatrix} \frac{3}{2} - \lambda & -\frac{1}{2} & 2 \\ -\frac{1}{2} & \frac{3}{2} - \lambda & 2 \\ 2 & 2 & -6 - \lambda \end{vmatrix} = -\lambda^3 - 3\lambda^2 + 24\lambda - 28 = 0.$$

The roots of this cubic equation are 2, 2, and -7. Solving for $\lambda = 2$,

$$\begin{bmatrix} -\frac{1}{2} & -\frac{1}{2} & 2 \\ -\frac{1}{2} & -\frac{1}{2} & 2 \\ 2 & 2 & -8 \end{bmatrix} \begin{bmatrix} x \\ y \\ z \end{bmatrix} = \begin{bmatrix} 0 \\ 0 \\ 0 \end{bmatrix}$$

with solution space (Example 10.2C) spanned by the orthonormal vectors

$$\mathbf{u}_1{}^t = [-1/\sqrt{2} \quad 1/\sqrt{2} \quad 0],$$
$$\mathbf{u}_2{}^t = [2/3 \qquad 2/3 \qquad 1/3],$$

and for $\lambda = -7$,

$$\begin{bmatrix} -\frac{11}{2} & -\frac{1}{2} & 2 \\ -\frac{1}{2} & -\frac{11}{2} & 2 \\ 2 & 2 & 1 \end{bmatrix} \begin{bmatrix} x \\ y \\ z \end{bmatrix} = \begin{bmatrix} 0 \\ 0 \\ 0 \end{bmatrix}$$

with solution space spanned by
$$\mathbf{u}_3{}^t = [1/3\sqrt{2} \quad 1/3\sqrt{2} \quad -4/3\sqrt{2}].$$

We observe with pleasure that $\mathbf{u}_3$ is orthogonal to both $\mathbf{u}_1$ and $\mathbf{u}_2$, so we can obtain the desired orthonormal matrix $[P]$ by setting $[P] = [\mathbf{u}_1\ \mathbf{u}_2\ \mathbf{u}_3]$. Moreover, without doing the computation, we can add that

$$[P]^t[S][P] = \begin{bmatrix} 2 & 0 & 0 \\ 0 & 2 & 0 \\ 0 & 0 & -7 \end{bmatrix} \quad \blacksquare$$

We know that a real $n \times n$ symmetric matrix S has n real, not necessarily distinct characteristic values $r_1, \ldots, r_n$. If each $r_i \geq 0$, we say that S is a *positive symmetric matrix*, and if each $r_i > 0$, we call it *positive definite*. If each $r_i \leq 0$ (or $r_i < 0$), we call S a *negative* (or *negative definite*) matrix.

PROBLEM

A · Let $[S]$ be a real $n \times n$ symmetric matrix. Then $[S]$ is positive if and only if for every nonzero vector $\mathbf{v} \in R^n$, $\mathbf{v}^t[S]\mathbf{v} \geq 0$. Similarly, $[S]$ is positive definite if and only if for every nonzero vector $\mathbf{v} \in R^n$, $\mathbf{v}^t[S]\mathbf{v} > 0$.

10·7 EXERCISES

In Exercises 1–4 you are given an equation in x, y, z that involves product terms. Find a transformation P such that $[x \quad y \quad z]^t = P[u \quad v \quad w]^t$ will result in an equation in u, v, w having no product terms. Determine this equation.

1· $3x^2 + y^2 + 6yz + z^2 = 0$.

2· $z^2 - x^2 - y^2 - 6xy = 0$.

3· $2x^2 + 2y^2 - 4z^2 - 5xy - 2yz - 2xz = 3$.

4. $3x^2 + 2y^2 + 3z^2 - 2xy - 2yz = 4$.

5. Construct an equation in x, y, z as in Exercises 1–4 so that the transformation determined by

$$P = \begin{bmatrix} 2/\sqrt{6} & 0 & 1/\sqrt{3} \\ -1/\sqrt{6} & 1/\sqrt{2} & 1/\sqrt{3} \\ -1/\sqrt{6} & -1/\sqrt{2} & 1/\sqrt{3} \end{bmatrix}$$

results in the equation $2v^2 + 3w^2 = 18$.

6. Solve Exercise 5 if

$$P = \begin{bmatrix} 2/3 & 2/3 & 1/3 \\ -1/\sqrt{2} & 1/\sqrt{2} & 0 \\ 1/\sqrt{18} & 1/\sqrt{18} & -4/\sqrt{18} \end{bmatrix}$$

and the transformed equation is $9u^2 - 9v^2 - 9w^2 = 0$.

7. Prove that if $A \overset{s}{=} B$, then A and B have the same characteristic values.

8. Prove that A and A^t have the same characteristic values.

9. If A is symmetric, then there is orthogonal matrix P such that $P^t A P$ is diagonal. Is the converse true?

10. Suppose $T: R^n \to R^n$, and that with respect to the standard basis C, $[T]_C$ is symmetric. Prove that there is a basis B such that $[T]_B$ is diagonal.

11*. The graph of $z = ax^2 + bxy + cy^2$ is a surface in R^3. What would you expect to be true of this surface in the case where

$$\begin{bmatrix} a & b/2 \\ b/2 & c \end{bmatrix}$$

is positive definite. Prove your conclusion.

12*. Set $F(x, y, z) = 3x^2 + 2y^2 + 3z^2 - 2xy - 2yz$.

(a) Choose values for x, y, z at random. Verify that in each case, $F(x, y, z) \geq 0$.
(b) Using the notions of this section, explain why $F(x, y, z) \geq 0$. (You have done some of the work already if you solved Exercise 4.)
(c) Without using the ideas of this section, prove that $F(x, y, z) \geq 0$ for all choices of the variables.

FUNCTIONS OF SEVERAL VARIABLES

In Chapter 1, we studied vectors in the plane and then used them to study straight lines. We also studied some analytic geometry in the plane, giving special attention to the conic sections. In Chapter 2, we studied functions, comparing them to manufacturing machines, snow blowers, and cannon. But we quickly narrowed our interest to real-valued functions of a single real variable. In this context we studied graphs, measures of good behavior such as continuity, and approximation.

The present chapter is correctly viewed as a simple generalization of Chapters 1 and 2. We begin with a discussion of vectors in R^3 (really a review since we have studied them in R^n). Vectors are used to study lines and planes in three-dimensional space. We also study some analytic geometry in space, giving special attention to certain simple surfaces like spheres, cones, etc.

With this preparation we can discuss functions of several variables; that is, functions taking their raw material (or ammunition) from R^n. Again we come to notions of graphs and continuity. And, as in Chapter 2, we close with a problem in approximation.

11 · 1   Vectors in R^3

Recall that the standard basis vectors in R^3 are

$$\mathbf{i} = \boldsymbol{\varepsilon}_1 = [1 \quad 0 \quad 0]^t,$$

$$\mathbf{j} = \boldsymbol{\varepsilon}_2 = [0 \quad 1 \quad 0]^t,$$

$$\mathbf{k} = \boldsymbol{\varepsilon}_3 = [0 \quad 0 \quad 1]^t.$$

Definitions for equality of vectors, addition of vectors, and multiplication of vectors by a real number have been given (Example 10.2A). We also defined the dot product, the length of a vector, and the angle between two vectors.

The geometrical interpretation of these definitions is exactly as one would expect, following the pattern already established for the plane. Thus, given two points $P_1(x_1, y_1, z_1)$ and $P_2(x_2, y_2, z_2)$ located with respect to the familiar coordinate system, let us identify a directed line segment from P_1 to P_2 with a vector according to the rule

$$\mathbf{P}_1\mathbf{P}_2 = [x_2 - x_1 \quad y_2 - y_1 \quad z_2 - z_1]. \tag{1}$$

(This vector is designated by $\overrightarrow{P_1P_2}$ on blackboards, homework, or wherever boldface type is not available or convenient.)

With this interpretation of a vector, the definition of the equality of two vectors says that two directed line segments are to be thought of as representing the same vector if they

(a)   are parallel;
(b)   have the same sense (point the same way);
(c)   have the same length.

Geometrically speaking, we think of two vectors as being equal if by sliding one of them through a parallel displacement (keeping it parallel to its original position), it can be made to coincide with the other. Because of this, we may never be sure where the initial or terminal point of a given vector is in a pictorial representation. If, however, we know that the initial point is at the origin, then the terminal point is fixed. In fact, it is clear that if the three-dimensional vector $[a \quad b \quad c]$ is drawn with its initial point at the origin, then its terminal point will be at the point usually designated by (a, b, c). Vectors having their initial points at the origin are called *radius vectors*.

According to our definition, vector $\mathbf{P_1 P_2}$ described in (1) above has length

$$|\mathbf{P_1 P_2}| = \sqrt{(x_2 - x_1)^2 + (y_2 - y_1)^2 + (z_2 - z_1)^2},$$

which corresponds to our usual idea of the length between two points in space.

As was the case in the plane, so too in space we may use vectors to solve a wide variety of problems. Again we warn the student of the common error of writing down the negative of the vector he really wants. Thus, the vector from $A(a_1, a_2, a_3)$ to $B(b_1, b_2, b_3)$ is

$$\mathbf{AB} = [b_1 - a_1 \quad b_2 - a_2 \quad b_3 - a_3].$$

EXAMPLE A

The points $A(4, -1, 2)$, $B(0, 1, 4)$, and $C(2, 5, -2)$ determine a triangle in R^3. Find the point where the medians intersect.

We have sketched Fig. 11.1 as a guide. From it we see that the midpoint M of the segment joining A and C is the terminal point of the radius vector $\mathbf{OM} = \mathbf{OA} + \frac{1}{2}\mathbf{AC}$. From this we can find the coordinates of M. Now the medians intersect at a point two-thirds of the way from any vertex to the opposite midpoint. Hence, calling the desired point K, we have

$$\mathbf{OK} = \mathbf{OB} + \tfrac{2}{3}\mathbf{BM}.$$

Now using the given points,

$$\mathbf{OM} = [4 \quad -1 \quad 2] + \tfrac{1}{2}[-2 \quad 6 \quad -4] = [3 \quad 2 \quad 0].$$

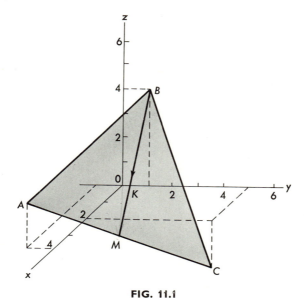

FIG. 11.1

(This same method could be used to prove in general that, as happened here, the midpoint of the segment joining A and C has coordinates that are the averages of the coordinates of A and C.)

$$M = (3, 2, 0),$$

$$\mathbf{OK} = \begin{bmatrix} 0 & 1 & 4 \end{bmatrix} + \frac{2}{3} \begin{bmatrix} 3 & 1 & -4 \end{bmatrix} = \begin{bmatrix} 2 & \frac{5}{3} & \frac{4}{3} \end{bmatrix},$$

$$K = \left(2, \frac{5}{3}, \frac{4}{3} \right). \quad \blacksquare$$

11 · 1 EXERCISES

Are the three points given in each of the Exercises 1–6 in a straight line ?

1· $(2, 1, -3, 4), (1, -1, 0, 2), (4, -1, -3, 8)$
2· $(1, 3, -1, 2), (2, 0, 1, 3), (8, 6, 1, 4)$
3· $(4, 0, 1, 2), (-1, 3, -1, 2), (-13, 3, -4, -4)$
4· $(5, 1, -2, 1), (4, -7, 3, 6), (-6, -9, 7, 4)$
5· $(-1, -2, 3, 1), (2, 0, 4, -2), (-5, -2, -4, 5)$

6· $(0, 1, 0, -1), (1, 3, 7, -2), (-1, -1, -7, 0)$

7· All the points equidistant from $(3, -1, 2)$ and $(1, -4, 3)$ lie on a plane. Find the equation of the plane.

8· All the points equidistant from $(1, 4, -1)$ and $(5, 2, 7)$ lie on a plane. Find the equation of the plane.

9· The points $A(1, 1, 3)$ $B(2, -1, -4)$, and $C(3, 1, -1)$ determine a triangle in R^3. Find the angle formed at A. Find $\sin A$.

10· The points $A(2, 1, 4)$, $B(3, -1, -2)$, and $C(5, 1, -2)$ determine a triangle in R^3. Find the angle formed at A. Find $\sin A$.

11· Find the area of the triangle described in Problem 9.

12· Find the area of the triangle described in Problem 10.

13· Find a point D such that the points A, B, C of Problem 9, together with D form a parallelogram (with vertex D opposite vertex B).

14· Find a point D such that the points A, B, C of Problem 10, together with D form a parallelogram (with vertex D opposite vertex B).

15*· Suppose the points $A(2, 1, -3)$ and $B(5, 1, 1)$ are two vertices of an equilateral triangle. The triangle is in the plane determined by A, B, and $D(3, 4, -2)$. Find the point C such that ABC form the triangle. Devise a way to verify your answer.

11 · 2 Lines and Planes in R^3

Two points P_0 and P_1 determine a line in space. For any point P on this line, we have the vector equation

$$\mathbf{OP} = \mathbf{OP_0} + r(\mathbf{OP_1} - \mathbf{OP_0})$$

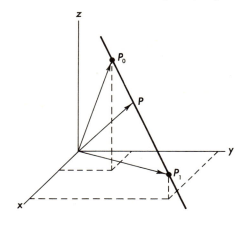

FIG. 11.2

where r is real (Fig. 11.2). This is clearly one way to define a line. The beginner sometimes has a different idea of how to get at a line, reasoning that the relation $Ax + By + C = 0$ defines a line in two dimensions, so he wishes to consider $Ax + By + Cz + D = 0$ in the case of three dimensions. Further consideration will (or should) convince him, however, that this equation in three variables describes not a line, but rather a plane in space. (Sketch a graph showing all points that satisfy $x + y + z = 1$. Note that all the points $x = 0$, $y + z = 1$ are on the graph, as are all the points $y = 0$, $x + z = 1$ and the points $z = 0$, $x + y = 1$. See Fig. 11.3.) Thus, the correct point of view with respect to equations of this

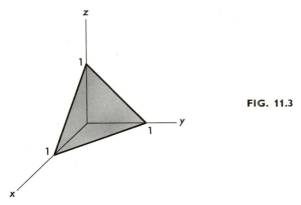

FIG. 11.3

type is that they describe a plane. Then think of a line in two space as the analog of a plane in three space. (It is a flat surface of dimension one less than the dimension of the space.)

Let us summarize;

Lines   The set of all points $P(x, y, z)$ satisfying the vector equation

$$\mathbf{OP} = \mathbf{OP}_0 + r(\mathbf{OP}_1 - \mathbf{OP}_0), \qquad r \text{ real,}$$

lie on the straight line determined by $P_0(x_0, y_0, z_0)$ and $P_1(x_1, y_1, z_1)$.

Planes   The set of all points $P(x, y, z)$ satisfying an equation of the form

$$Ax + By + Cz + D = 0,$$

lie on a plane.

We wish to explore each of these notions a bit further. Beginning with the line, let us set $\mathbf{v} = \mathbf{OP}_1 - \mathbf{OP}_0 = [a \quad b \quad c]$. Then the vector equation can be written

$$[x \quad y \quad z] = [x_0 \quad y_0 \quad z_0] + r[a \quad b \quad c], \qquad r \text{ real.}$$

Using the definitions for multiplication by a real number, addition of vectors, and then equality of vectors, we have

$$x = x_0 + ra,$$
$$y = y_0 + rb,$$
$$z = z_0 + rc.$$

This may be written in the form

$$\frac{x - x_0}{a} = \frac{y - y_0}{b} = \frac{z - z_0}{c} \tag{1}$$

if a, b, and c are nonzero. If $a = 0$, then we see from the expression above for x that $x - x_0 = 0$. Hence if we interpret a zero in the denominator of (1) as meaning that the numerator is zero, we can use (1) without exception to represent a straight line through (x_0, y_0, z_0) and parallel to $[a \quad b \quad c]$.

Caution We have not assigned a meaning to division by zero. We do not mean to imply that a zero in the denominator implies in all cases that the numerator is zero. We are merely agreeing to a convention here: that when an expression of the form (1) is used to represent a line in R^3, then a zero in the denominator of a fraction is understood to mean that the corresponding numerator is zero.)

We now turn our attention to planes. A plane is determined if we know a point $P_0(x_0, y_0, z_0)$ through which it passes and a vector $\mathbf{v} = [a, b, c]$ to which it is perpendicular. Let $P(x, y, z)$ be a point in such a plane. Then the vector

$$\mathbf{PP_0} = [x - x_0 \quad y - y_0 \quad z - z_0],$$

is perpendicular (orthogonal) to $\mathbf{v}$, so $\mathbf{PP_0} \cdot \mathbf{v} = 0$;

$$a(x - x_0) + b(y - y_0) + c(z - z_0) = 0.$$

We now have two pieces of information that together form the basis for solving most problems involving lines and planes in space. We state them as a theorem.

THEOREM A (See Fig. 11.4)

Let $P_0 = (x_0, y_0, z_0)$ be a point of R^3, and let $\mathbf{v} = [a \quad b \quad c]$ be an arbitrary vector. The line through P_0 that is parallel to $\mathbf{v}$ is

$$\frac{x - x_0}{a} = \frac{y - y_0}{b} = \frac{z - z_0}{c}.$$

The plane through P_0 and perpendicular to $\mathbf{v}$ is

$$a(x - x_0) + b(y - y_0) + c(z - z_0) = 0.$$

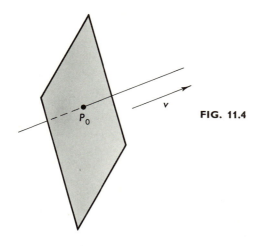

FIG. 11.4

EXAMPLE A

A line is perpendicular to $3x - y + 2z = 4$ and passes through $(1, -1, -2)$. Where does it pierce the plane $x + y + z = 3$?

The vector $[3 \quad -1 \quad 2]$ is perpendicular to the given plane. The equation of the line through $(1, -1, -2)$ that is parallel to this vector (hence perpendicular to the plane) is

$$\frac{x - 1}{3} = \frac{y + 1}{-1} = \frac{z + 2}{2}.$$

Thus

$$x = 1 + 3r.$$
$$y = -1 - r,$$
$$z = -2 + 2r.$$

Substituting in the equation of the plane gives

$$4r = 5.$$

The solution is therefore $(\frac{19}{4}, -\frac{9}{4}, \frac{1}{2})$. ∎

EXAMPLE B

The three points $A(1, 2, -1)$, $B(3, -2, 4)$, and $C(-2, 1, 3)$ determine a plane. Find the equation.

In solving this, we may use any one of the points as P_0 in Theorem A. If we use $(1, 2, -1)$, then the equation we seek is seen to be

$$a(x - 1) + b(y - 2) + c(z + 1) = 0. \tag{2}$$

Now the other two points satisfy this equation. Hence

$$2a - 4b + 5c = 0,$$

$$-3a - b + 4c = 0.$$

We thus have two equations in three unknowns. Proceeding as in Section 10.1, we obtain the equivalent system

$$a \quad -\tfrac{11}{14}c = 0$$

$$-b + \tfrac{23}{14}c = 0$$

which enables us to substitute for a and b in (2);

$$c\{\tfrac{11}{14}(x - 1) + \tfrac{23}{14}(y - 2) + (z + 1)\} = 0.$$

The equation of the plane is therefore

$$11(x - 1) + 23(y - 2) + 14(z + 1) = 0. \quad \blacksquare$$

We can introduce another idea here that simplifies the solution of our last example. Consider Eq. (2). If at this stage we could have obtained a vector perpendicular to the desired plane, we would have been through. (Why?) A vector perpendicular to the plane would be perpendicular to

$$\mathbf{AC} = \begin{bmatrix} -3 & -1 & 4 \end{bmatrix} \quad \text{and} \quad \mathbf{AB} = \begin{bmatrix} 2 & -4 & 5 \end{bmatrix}.$$

There is a way, given two vectors in R^3, to find a vector perpendicular to each of them.

THEOREM B

Let $\mathbf{p} = \begin{bmatrix} a_1 & a_2 & a_3 \end{bmatrix}$ and $\mathbf{q} = \begin{bmatrix} b_1 & b_2 & b_3 \end{bmatrix}$ be arbitrary vectors in R^3. Define $\mathbf{p} \times \mathbf{q} = \begin{bmatrix} a_2 b_3 - a_3 b_2 & a_3 b_1 - a_1 b_3 & a_1 b_2 - a_2 b_1 \end{bmatrix}$. (This vector is called the *cross product of* $\mathbf{p}$ *and* $\mathbf{q}$.) The vector $\mathbf{p} \times \mathbf{q}$ is perpendicular to both $\mathbf{p}$ and $\mathbf{q}$.

PROOF:

The reader is left merely to check that this is so by computing $\mathbf{p} \cdot (\mathbf{p} \times \mathbf{q})$ and $\mathbf{q} \cdot (\mathbf{p} \times \mathbf{q})$.

A student trying to memorize the formula for $\mathbf{p} \times \mathbf{q}$ might give up in despair, so we come to his aid with the following observation:

$$\mathbf{p} \times \mathbf{q} = \begin{vmatrix} \mathbf{i} & \mathbf{j} & \mathbf{k} \\ a_1 & a_2 & a_3 \\ b_1 & b_2 & b_3 \end{vmatrix}$$

EXAMPLE B′

We saw that having once written down (2), we had need of a vector perpendicular to $[-3 \quad -1 \quad 4]$ and $[2 \quad -4 \quad 5]$:

$$\begin{vmatrix} \mathbf{i} & \mathbf{j} & \mathbf{k} \\ -3 & -1 & 4 \\ 2 & -4 & 5 \end{vmatrix} = \mathbf{i}(-5 + 16) - \mathbf{j}(-15 - 8) + \mathbf{k}(12 + 2),$$

$$= [11 \quad 23 \quad 14].$$

Using these numbers in (2) gives

$$11(x - 1) + 23(y - 2) + 14(z + 1) = 0. \quad \blacksquare$$

---

11 · 2 EXERCISES

1. A line passes through $(1, -1, 2)$. It is parallel to the line of intersection of the planes $3x + y - z = 4$ and $2x + y - z = 1$. Write the equations of the line in the form (1).

2. A line passes through $(4, 2, -1)$. It is parallel to the line of intersection of the planes $x - y + z = 1$ and $2x + 3y - z = 4$. Write the equations of the line in the form (1).

3. A line passes through $(2, 1, 4)$ and $(3, -1, -1)$. Where does it pierce the plane $x + y - 3z = 4$?

4. A line passes through $(1, 2, -1)$ and $(4, 1, 2)$. Where does it pierce the plane $2x - y + z = 7$?

5. Find the acute angle formed by the intersection of the plane $x - y + z = 4$ and $3x - y - 2z = 1$.

6. Find the acute angle formed by the intersection of the planes $2x - 2y + z = 6$ and $3x + 2y + 4z = 7$.

7. A line passes through $(1, 0, -1)$ and intersects the line

$$\frac{x - 1}{2} = \frac{y + 2}{3} = \frac{z - 3}{-1}$$

at right angles. Find the point of intersection.

8. A line passes through $(4, 2, 3)$ and intersects the line

$$\frac{x+2}{3} = \frac{y-1}{2} = \frac{z+1}{4}$$

at right angles. Find the point of intersection.

9. Write the equation of the plane determined by the points $(1, -1, 2)$, $(0, 3, 1)$, and $(2, -1, 2)$.

10. Write the equation of the plane determined by the points $(2, 1, 3)$, $(-1, 2, 0)$, and $(1, 0, 1)$.

11. The following lines intersect. Where ?

$$\frac{x-1}{2} = \frac{y+2}{3} = \frac{z-1}{4},$$

$$\frac{x-1}{1} = \frac{y+11}{-3} = \frac{z-1}{2}.$$

12. The following lines intersect. Where ?

$$\frac{x+3}{6} = \frac{y-1}{-1} = \frac{z+1}{2},$$

$$\frac{x}{1} = \frac{3y+5}{6} = \frac{3z-7}{-6}.$$

13. The lines of Exercise 11 determine a plane. Write the equation of the plane.

14. The lines of Exercise 12 determine a plane. Write the equation of the plane.

15*. Let A, B, C be points in R^3. They determine a triangle. Show that the area of this triangle equals $\frac{1}{2}|\mathbf{AB} \times \mathbf{AC}|$.

16*. Use the result of Exercise 15 to solve Exercise 15 in Section 11.1.

11 · 3 Surfaces in R^3

We have seen that a statement of equality among three variables x, y, z of the form $Ax + By + Cz + D = 0$ describes a set of points in three space, all of which lie in a plane. Other statements of equality between x, y, z similarly determine specific sets in three space, though they are usually not so easily described. One means of getting some idea of the kind of set described is to look at various cross sections obtained by setting one variable equal to a constant. The technique is illustrated in the following examples.

EXAMPLE A
Graph $z^2 = x + y - 4$.

It is usually helpful to look at the trace in each coordinate plane; that is, at the points obtained by setting one of the variables equal to zero.

$$x = 0; \qquad z^2 = y - 4;$$
$$y = 0; \qquad z^2 = x - 4;$$
$$z = 0; \qquad x + y = 4.$$

We also notice that when $z = c$, a constant, the plane determined by $z = c$ cuts our set in a line $x + y = 4 + c^2$. Thus, the set could be generated by moving a straight line, keeping it parallel to $x + y = 4$, $z = 0$, and guiding it by the parabolas that are the traces in the xz and yz planes. See Fig. 11.5. ∎

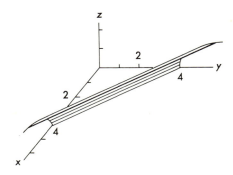

FIG. 11.5

EXAMPLE B
Graph $36x^2 - 9y^2 + 18y - 4z^2 + 40z = 145$.

The appearance of the same variable to the first and second power suggests the same procedure as was used for an equation involving only x and y. We complete the square.

$$36x^2 - 9(y^2 - 2y + 1) - 4(z^2 - 10z + 25) = 145 - 9 - 100,$$

$$\frac{x^2}{1} - \frac{(y-1)^2}{4} - \frac{(z-5)^2}{9} = 1.$$

When $z = 5$, we recognize the cross section as a hyperbola. This is also true in the plane $y = 1$. When $x = 0$, we get no points; and we will not get any until $|x|$ is big enough so that $1 - x^2 < 0$. From these considerations we sketch the graph in Fig. 11.6. Notice that in drawing this graph, we have

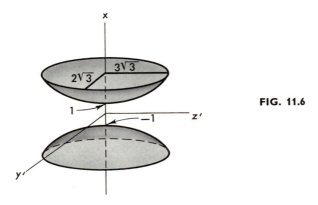

FIG. 11.6

labeled the axes differently than is usual. This is often done if it facilitates the drawing of an object in space. Also note that the primes on the indicated axes of the figure mean that the axis is parallel to the corresponding coordinate axis. The axes shown, x, y', z', intersect at $(0, 1, 5)$. ∎

Certain second-degree equations in x and y described sets in the plane having familiar names. The same is true for certain second-degree equations in x, y, and z. The surface in Fig. 11.6. is called a *hyperboloid of two sheets*. The forms of other equations that describe common surfaces are indicated in Figs. 11.7–11.11.

$$\frac{(x - x_1)}{a^2} + \frac{(y - y_1)}{b^2} - \frac{(z - z_1)^2}{c^2} = 1$$

$$\frac{(x - x_1)^2}{a^2} + \frac{(y - y_1)^2}{b^2} + \frac{(z - z_1)^2}{c^2} = 1$$

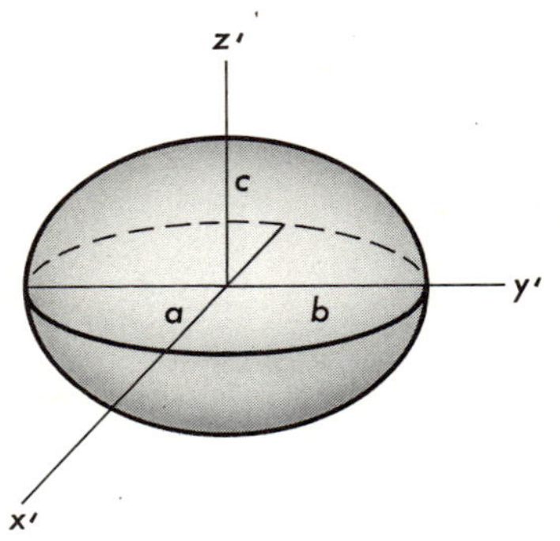

FIG. 11.7 Ellipsoid with center at (x_1, y_1, z_1)

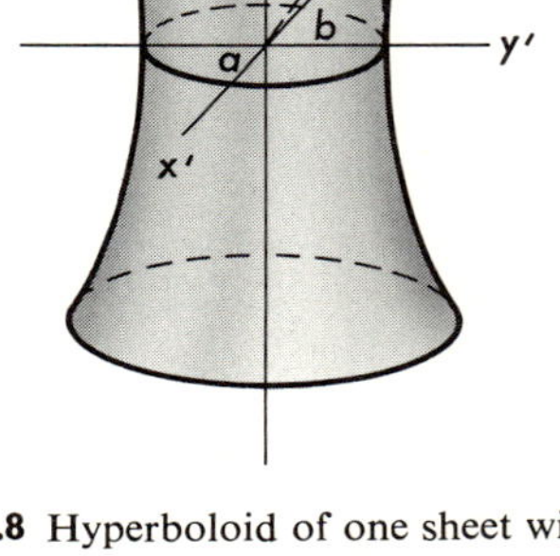

FIG. 11.8 Hyperboloid of one sheet with center at (x_1, y_1, z_1)

$$\frac{(x-x_1)^2}{a^2} + \frac{(y-y_1)^2}{b^2} = \frac{(z-z_1)^2}{c^2}$$

$$\frac{(x-x_1)^2}{a^2} + \frac{(y-y_1)^2}{b^2} = \frac{z-z_1}{c}$$

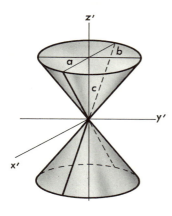

FIG. 11.9 Elliptic cone with apex at $(x_1, \ y_1, \ z_1)$

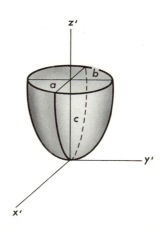

FIG. 11.10 Elliptic paraboloid with vertex at $(x_1, \ y_1, \ z_1)$

$$\frac{(x-x_1)^2}{a^2} - \frac{(y-y_1)^2}{b^2} + \frac{z-z_1}{c} = 0$$

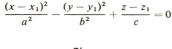

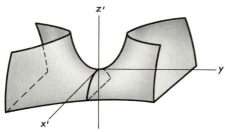

FIG. 11.11 Hyperbolic paraboloid with saddle point at $(x_1, \ y_1, \ z_1)$

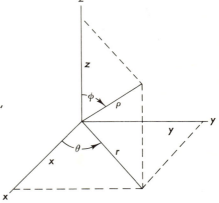

FIG. 11.12

All but the last of these (Fig. 11.11) are easily drawn by considering the obvious cross sections. There is, therefore, no real point to memorizing these various forms. Once they have been sketched, the reader should, however, be able to name them.

We close this section with the observation that a point in R^3 is sometimes located by other means than the familiar cartesian coordinates. Two other systems are commonly used: the cylindrical coordinates $(r, \ \theta, \ z)$ and the spherical coordinates $(\rho, \ \phi, \ \theta)$. The relations to the coordinates (x, y, z) are given below. Figure 11.12 indicates the geometrical meaning.

Cylindrical coordinates:

$$x = r \cos \theta, \qquad r = \sqrt{x^2 + y^2},$$
$$y = r \sin \theta, \qquad \tan \theta = y/x.$$
$$z = z.$$

Spherical coordinates:

$$x = \rho \sin \phi \cos \theta, \qquad \rho = \sqrt{x^2 + y^2 + z^2},$$
$$y = \rho \sin \phi \sin \theta, \qquad \cos \phi = z/\rho,$$
$$z = \rho \cos \phi, \qquad \tan \theta = y/x.$$

The sphere $x^2 + y^2 + z^2 = 25$ has the spherical equation $\rho = 5$. The cylinder $x^2 + y^2 = 25$ has the cylindrical equation $r = 5$. An equation given in terms of cartesian coordinates may be expressed in terms of the other coordinate systems by means of the indicated equations.

Cylindrical coordinates:

$$x = r \cos \theta;$$
$$y = r \sin \theta;$$
$$z = z.$$

Spherical coordinates:

$$x = \rho \sin \phi \cos \theta;$$
$$y = \rho \sin \phi \sin \theta;$$
$$z = \rho \cos \phi.$$

11 · 3 EXERCISES

1. Write the equation of a plane tangent to the sphere $x^2 + y^2 + z^2 = 6$ at the point $(1, -1, 2)$. (Hint: The tangent plane is perpendicular to the radius at the point of tangency.)

2. Write the equation of a plane tangent to the sphere $x^2 + y^2 + z^2 = 17$ at the point $(3, -2, 2)$. See the hint to Exercise 1.

In Exercises 3–16, put the given equations into one of the standard forms, name the curve, and sketch.

3. $9x^2 + 18x + 4z^2 - 8z - y^2 - 6y = 32$.

4. $y^2 - 4x^2 + 8x - z^2 = 20$.

5. $x^2 - 2x + 3z^2 + 6z - 12y + 4 = 0$.

6. $7x^2 + 7x + y^2 - 3y = 0$.

7. $x^2 - 2x + 4y^2 - 8y + 9z^2 = 7$.

8. $3x^2 - 6x + y^2 - 3z^2 + 3 = 0$.

9. $x^2 - y^2 + 2y - 4z^2 - 8z = 14$.

10. $x^2 - 2x - 3z^2 + 6z - 12y + 4 = 0$.

11. $y^2 + 6y - 4z^2 - 8z - 4x + 13 = 0$.

12. $3x^2 + 6x + y^2 + 12z^2 + 72z + 99 = 0$.

13. $2x^2 - 4x - 3y^2 + 4z^2 + 16z + 18 = 0$.

14. $y^2 + 6y + 4z^2 - 8z - 4x + 13 = 0$.

15. $3x^2 + 3x + z^2 - 3z = 1$.

16. $y^2 - 2y + 4z^2 + 16z - x^2 + 6x = 4$.

Sketch the set of points in the first quadrant satisfying the equations given in Exercises 17–22.

17. $(y - z)^2 = 1 - x$.

18. $z = \dfrac{8 - y}{2 + x + y}$.

19. $z = \dfrac{4 - \sqrt{x^2 + y^2}}{1 + x}$.

20. $y^3 - 2y + x + z = 0$.

21. $z = \dfrac{8 - x^2}{1 + x^2 + y^2}$.

22. $z^2 = 4xy$.

23. A surface is described in cylindrical coordinates by $r = 2 \sin \theta$. Write the equation in (a) cartesian coordinates, (b) spherical coordinates.

24. A surface is described in cylindrical coordinates by $r = 2 \cos \theta$. Write the equation in (a) cartesian coordinates, (b) spherical coordinates.

25. A surface is described in cartesian coordinates by $x^2 + y^2 = z$. Write the equation in (a) cylindrical coordinates, (b) spherical coordinates.

26. A surface is described in cartesian coordinates by $xy = z^2$. Write the equation in (a) cylindrical coordinates, (b) spherical coordinates.

27. A surface is described in spherical coordinates by $\rho = \cos \phi$. Write the equation in (a) cartesian coordinates, (b) cylindrical coordinates.

28. A surface is described in spherical coordinates by $\rho = \sin \phi(\cos \theta + \sin \theta)$. write the equation in (a) cartesian coordinates, (b) cylindrical coordinates.

11 · 4 Functions

Suppose $\mathbf{p} = [x_1 \quad x_2 \quad \cdots \quad x_n]$ is a vector. If drawn as a radius vector (that is, if drawn with its initial point at the origin), then $\mathbf{p}$ terminates at the point $p = (x_1, x_2, \ldots, x_n)$. In the work that follows, we shall often use terms in talking about p that really apply to the associated vector $\mathbf{p}$, and we use bold face $\mathbf{p}$ even when thinking of p. This should not cause any difficulty.

The functions studied in Chapter 1 had both domain and range among the real numbers. Identifying the real numbers with points on the line, these functions are seen as machines that take in points from $R^1 = R$ and produce other points in R. The generalization of this idea is obvious. We now imagine a machine built to accommodate points from some subset U of R^n. It produces points in R^m. $F(\mathbf{p})$ is called the *image of* $\mathbf{p}$ *under* F. See Fig. 11.13.

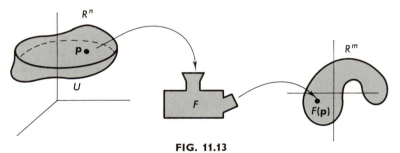

FIG. 11.13

Certain functions are often given other names, though there is no general agreement. When m and n are both greater than 1, F is sometimes called a *mapping* or a *transformation*. When $n = 1$ and $m > 1$, F is called a *vector-valued function*, or a *curve* if F has certain other properties (Chapter 9). When $m = 1$ and $n > 1$, F is called a (real-valued) *function of several variables*.

To form the composition $G \circ F$, the domain of G must contain the range of F. Thus, G takes points from R^m. In particular, it appears that the inverse of F, if it exists, must have its domain in R^m, its range in R^n.

In talking about the continuity of functions where $m = n = 1$, we spoke about neighborhoods centered at a point x_0. A neighborhood of radius r centered at x_0 was described as the set of all x for which $|x - x_0| < r$. We can consider the same kind of neighborhood of a point $\mathbf{p}_0 \in R^n$. It consists of all points $\mathbf{p}$ for which the distance from $\mathbf{p}$ to $\mathbf{p}_0$ is less than r. We call such a neighborhood $N_r(\mathbf{p}_0)$.

$$N_r(\mathbf{p}_0) = \{\mathbf{p} : |\mathbf{p} - \mathbf{p}_0| < r\}.$$

Our concept of a continuous function is exactly the same as in the case where $m = n = 1$. If F is continuous at $\mathbf{p}_0$, then points close to $\mathbf{p}_0$ must be "fired" into points close to $F(\mathbf{p}_0)$. Restated, if someone gives us an arbitrary target neighborhood $N_\varepsilon(F(\mathbf{p}_0))$, we must be able to produce a neighborhood $N_\delta(\mathbf{p}_0)$ of $\mathbf{p}_0$ so that if $\mathbf{p} \in N_\delta(\mathbf{p}_0)$, then $F(\mathbf{p}) \in N_\varepsilon(F(\mathbf{p}_0))$. See Fig. 11.14.

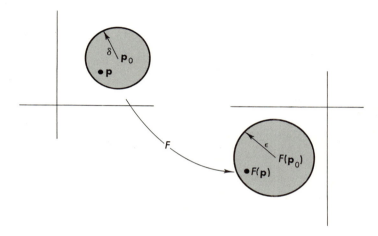

FIG. 11.14

We reap practical benefits from using the same definition of continuity for arbitrary m and n. It means that once the properties of continuous functions (see Theorems A–F of Section 2.6) are proved, they remain valid in the present context.

For many theorems in mathematics, it is important to know that we have some room around a point in which to work. Continuity is one idea in which we need to think of what happens to points near $\mathbf{p}_0$. When we wish to indicate that about each point of a set, there is some neighborhood of the point still in the set, we say the set is *open*. The set of points (x, y) in the plane for which $x^2 + y^2 < 1$ is open. If we add to this set just one point on the "rim," then the set is no longer open, since any neighborhood about the rim point must contain points not inside the circle. Since this idea is important, we shall state it formally.

D E F I N I T I O N A (Open Set)

       The set U is open in R^n if for each $\mathbf{p} \in U$, there is some neighborhood $N_r(\mathbf{p})$ contained in U.

In R the open interval (a, b) is an open set.

Suppose now that F is defined on U, a set (open or not)in R^n, and suppose further that the range of F is in R^m. We abbreviate this long sentence as follows.

$$F: U \to R^m \qquad \text{where } U \text{ is contained in } R^n.$$

Let the points of the domain be $\mathbf{p} = (x_1, x_2, \ldots, x_n)$, and denote those of the range by $\mathbf{q} = (y_1, y_2, \ldots, y_m)$. There are m numbers, $y_1, y_2, \ldots, y_m$ to be determined for each point $\mathbf{p}$. The function F is thus described by m equations.

$$\begin{aligned}
y_1 &= F^1(x_1, x_2, \ldots, x_n), \\
y_2 &= F^2(x_1, x_2, \ldots, x_n), \\
&\vdots \\
y_m &= F^m(x_1, x_2, \ldots, x_n).
\end{aligned} \tag{1}$$

The functions $F^1, F^2, \ldots, F^m$, all real-valued functions of n variables, are called the *coordinate functions*.

The only times we find it convenient to sketch a graph of a function are when $m = n = 1$, the case already discussed; occasionally when $m = 1, n = 2$; or when $n = 1, m = 2$. When $m = n = 2$, a common way to get some "picture" of the function is to indicate certain subsets of the domain and the corresponding image sets in the range.

EXAMPLE A

Let $F: R^2 \to R^2$ be described by the coordinate functions

$$y_1 = x_1^2 - x_2^2,$$

$$y_2 = 2x_1 x_2.$$

For what sets in the $x_1 x_2$ plane are the image sets lines parallel to the axes y_1 and y_2 ?

When $x_1^2 - x_2^2 = c^2$, then $y_1 = c^2$ and y_2 ranges over all real values. If $x_1^2 - x_2^2 = -c^2$, then $y_1 = -c^2$. When $2x_1 x_2 = c^2, y_2 = c^2$; if $2x_1 x_2 = -c^2$, then $y_2 = -c^2$. These sets and their images are indicated in Fig. 11.15. ∎

In studying functions of a single variable, we saw the importance of the theory of limits. Besides being important, we also saw that it was tricky business. Not only were we surprised by such results as $\lim_{x \to \infty}(1 + 1/x)^x = e$, but we came to understand the necessity of considering both

$$\lim_{x \to x_0^-} f(x) \qquad \text{and} \qquad \lim_{x \to x_0^+} f(x). \tag{2}$$

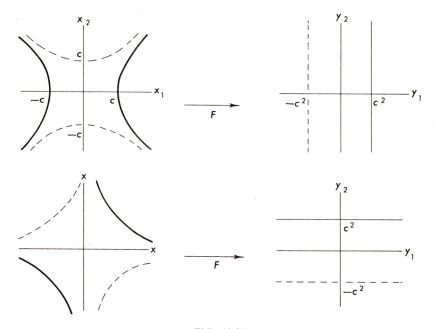

FIG. 11.15

Only when both limits existed and were equal could we talk with meaning about $\lim_{x \to x_0} f(x)$.

For functions of several variables, the study is again both important and tricky. Leaving the importance to emerge in a natural way in Chapter 12, we pause here to point out that the dual considerations (2) encountered in the case of one variable are in a literal way infinitely more complicated as soon as we move to consideration of a function of two variables. The reason is apparent from Fig. 11.16. The single variable x can approach x_0 from one of two directions. The point $p(x, y)$, depending on two variables, can approach $p_0(x_0, y_0)$ from an infinite number of directions along an infinite number of paths.

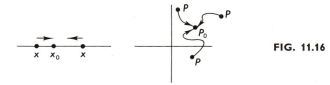

FIG. 11.16

EXAMPLE B

Let $p_0 = (0, 0)$ and $f(x, y) = y/(x + y^2)$. Consider $\lim_{p \to p_0} f(p)$ where $p = (x, y)$.

Suppose we approach p_0 along the line $y = mx$. Then

$$\lim_{p \to p_0} f(p) = \lim_{(x,y) \to (0,0)} f(x, y) = \lim_{(x,y) \to (0,0)} \frac{y}{x + y^2}$$

$$= \lim_{(x,mx) \to 0} \frac{mx}{x + (mx)^2}$$

$$= \lim_{x \to 0} \frac{mx}{x(1 + m^2 x)} = m.$$

Clearly, depending upon the line (hence the value of m) chosen, this limit can be made to take on any finite value. For good measure, we note that by approaching on the curve $y = \sqrt{x}$,

$$\lim_{p \to p_0} f(p) = \lim_{x \to 0^+} \frac{\sqrt{x}}{x + x} = \infty. \quad \blacksquare$$

11 · 5 The Approximation Problem

We now revive the approximation problem introduced in Section 2.7. We state it in the context of a function of two variables.

Suppose F is defined and well behaved in some open neighborhood containing $\mathbf{p}_0 = (x_0, y_0)$, and suppose $z_0 = F(\mathbf{p}_0)$ is known. Can we devise a method for easily approximating $F(\mathbf{p}_0 + \mathbf{h})$ in the event that $\mathbf{h}$ is a small vector in the plane and that the exact value of $F(\mathbf{p}_0 + \mathbf{h})$ is either very difficult or impossible to find ?

The method of approximation that we shall pursue is the one suggested by the procedure followed for functions of one variable. There we used the function having as its graph a line tangent to the graph of F at $(x_0, F(x_0))$. Now we shall use the function having as its graph a plane tangent to the graph of F at $(\mathbf{p}_0, F(\mathbf{p}_0))$.

If they are to be of any practical use, of course, then the geometric notions of tangent lines and tangent planes must be translated into analytic descriptions; that is, we must obtain the equations corresponding to the desired line or plane. In the case of a single variable, the line (Fig. 11.17) through $(x_0, F(x_0))$ took the form

$$y = T(x) = F(x_0) + m(x - x_0). \tag{1}$$

The problem was to determine m so as to obtain the tangent line.

In the case of two variables, the plane (Fig. 11.18) through $(\mathbf{p}_0, F(\mathbf{p}_0))$ is

$$z = T(x, y) = F(x_0, y_0) + m_1(x - x_0) + m_2(y - y_0).$$

Here the problem is to choose m_1 and m_2 so that the plane will be tangent to the surface.

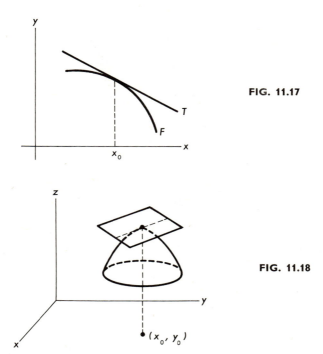

FIG. 11.17

FIG. 11.18

EXAMPLE A

Let $F(x, y) = \dfrac{4 - x}{1 + x + y^2}$.

Then $F(2, 1) = \frac{1}{2}$. Suppose we wish to approximate the values of F at the points (1.9, 0.9), (1.9, 1.1), (2.1, 0.9) and (2.1, 1.1). We are proposing to estimate F at these points by determining a function

$$T(x, y) = \tfrac{1}{2} + m_1(x - 2) + m_2(y - 1),$$

having as its graph a plane tangent to the graph of F at $(2, 1, \frac{1}{2})$ (Fig. 11.19).

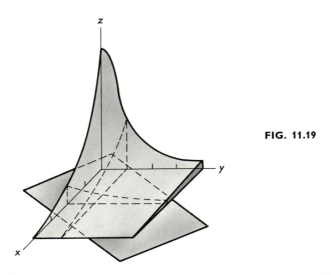

FIG. 11.19

Here is a key observation. The equation that we desire could be written in the form

$$T(x, y) = \tfrac{1}{2} + [m_1 \quad m_2]\begin{bmatrix} x - 2 \\ y - 1 \end{bmatrix}. \tag{2}$$

This hardly seems to be a stroke of genius, but it focuses our attention on the right question. It shows us that our real problem is to determine a linear transformation $L: R^2 \to R$ having matrix

$$[L] = [m_1 \quad m_2].$$

Moreover, it prepares the way for approximating a function of n variables.
The problem stated in the context of n variables is as follows.

The Approximation Problem

Suppose F is defined and well behaved in some open neighborhood containing $\mathbf{p}_0$, and suppose $F(\mathbf{p}_0)$ is known. Can we devise a method for easily approximating $F(\mathbf{p}_0 + \mathbf{h})$ in the event that $\mathbf{h}$ is a small vector in R^n and that the exact value of $F(\mathbf{p}_0 + \mathbf{h})$ is either very difficult or impossible to find ?

The method of approximation again will be to determine a linear transformation $L: R^n \to R$ so that

$$T(\mathbf{p}) = F(\mathbf{p}_0) + L(\mathbf{p} - \mathbf{p}_0) \tag{3}$$

will give us a reasonable approximation to $F(\mathbf{p})$ when $\mathbf{p}$ is close to $\mathbf{p}_0$.

Before closing this section we make several observations about (3). First, by setting $\mathbf{p} = \mathbf{p}_0 + \mathbf{h}$, we see that (3) is equivalent to

$$T(\mathbf{p}_0 + \mathbf{h}) = F(\mathbf{p}_0) + L(\mathbf{h}),$$

where $T(\mathbf{p}_0 + \mathbf{h})$ approximates $F(\mathbf{p}_0 + \mathbf{h})$ as required in the statement of the problem.

Secondly, we note that (2) is the two variable analog of (3). In fact, the linear transformation L in (3) corresponds to a $1 \times n$ matrix:

$$[L] = [m_1 \quad m_2 \ldots m_n].$$

Finally we note that in the one-variable case where we actually know how to complete the solution, (1) is again of the form (3) where we have the trivial correspondence of $[L] = [m]$, a 1×1 matrix. Geometrically, the graph of any real-valued linear transformation of a single variable is a line through the origin. Such lines are in one-to-one correspondence with the real number that is their slope. This is why, in the case of one variable, we were able to solve our problem while thinking in terms of finding a number $F'(x_0)$ and not a linear transformation.

THE DERIVATIVE OF FUNCTIONS OF SEVERAL VARIABLES

[OUR] PRESENTATION, WHICH THROUGHOUT ADHERES STRICTLY TO OUR GENERAL 'GEOMETRIC' OUTLOOK ON ANALYSIS, AIMS AT KEEPING AS CLOSE AS POSSIBLE TO THE FUNDAMENTAL IDEA OF CALCULUS, NAMELY THE 'LOCAL' APPROXIMATION OF FUNCTIONS BY LINEAR FUNCTIONS. IN THE CLASSICAL TEACHING OF CALCULUS, THIS IDEA IS IMMEDIATELY OBSCURED BY THE ACCIDENTAL FACT THAT, ON A ONE-DIMENSIONAL VECTOR SPACE, THERE IS A ONE-TO-ONE CORRESPONDENCE BETWEEN LINEAR FORMS AND NUMBERS, AND THEREFORE THE DERIVATIVE AT A POINT IS DEFINED AS A NUMBER INSTEAD OF A LINEAR FORM. THIS SLAVISH SUBSERVIENCE TO THE SHIBBOLETH OF NUMERICAL INTERPRETATION AT ANY COST BECOMES MUCH WORSE WHEN DEALING WITH FUNCTIONS OF SEVERAL VARIABLES.

—J. Dieudonne
  Foundations of Modern Analysis,
  Academic Press, 1960

In Chapter 3, we defined the derivative of a real-valued function of a real variable to be the limit of a difference quotient;

$$F'(x) = \lim_{h \to 0} \frac{F(x + h) - F(x)}{h}. \tag{1}$$

In Section 3.5 we made the observation that the numbers $F'(x)$, being slopes, are in one-to-one correspondence with the linear functions of one variable, all of which take the form $L(x) = mx$. This fact was used to motivate a second definition of the derivative.

Let F be a real-valued function defined in some interval containing x_0, and suppose there exists a *linear* function L for which, when h is small,

$$F(x_0 + h) = F(x_0) + L(h) + |h| R(x_0, h) \tag{2}$$

where $R(x_0, h)$ approaches zero whenever h approaches zero. Then we say F is differentiable at x_0 and that its derivative is the *linear* function L.

So long as we restrict our attention to real-valued functions of a real variable, the obvious correspondence between real numbers and *linear* transformations obscures the distinction between the two definitions. It is the second definition, however, that makes sense when we consider transformations from R^n to R^m. For this reason, the reader is likely to find that his understanding of Section 3.5 is greatly increased by studying the derivative in the setting of higher dimensions.

Chapter 11 was introduced as a simple generalization of Chapters 1 and 2. This chapter corresponds to Chapters 3 and 4. Taking definition (2) as our notion of the derivative, the other definitions and theorems of those chapters can be carried into our present context with no conceptual changes. The fact is that this same generalization can be carried off again in the context of abstract normed linear spaces [3, Chapter VIII]. This development is clearly beyond the scope of our work; but that it can be done in a natural way makes irresistible the conclusion that the derivative is correctly understood to be a *linear* transformation.

12·1 The Definition

We closed the last chapter with an example in which we sought a *linear* transformation L to be used in defining

$$T(\mathbf{p}) = F(\mathbf{p}_0) + L(\mathbf{p} - \mathbf{p}_0).$$

The function T, in turn, was to have a graph tangent to the graph of F at $(\mathbf{p}_0, F(\mathbf{p}_0))$. In Chapter 3 we used such a situation to motivate our definition of the derivative. We could do the same here so long as F is a function of two variables and our geometric intuition is helpful. Such motivation fails for functions of more than two variables, however, so we alter our procedure. As suggested in the introduction above, we model our definition after (2). In this way we get a *linear* transformation for use in approximating functions near a point $\mathbf{p}_0$, independent of any intuitive notions of tangency. Then we show that in the case of a function of two variables, the graph of T does indeed fulfill our expectations for a tangent to the graph of F at $\mathbf{p}_0$.

DEFINITION A (The Derivative)

Let F be defined in a neighborhood of $\mathbf{p}_0 \in R^n$, taking values in R^m. Suppose there exists a *linear* transformation $L: R^n \to R^m$ for which, when $\mathbf{h}$ is small, $\mathbf{h} \in R^n$,

$$F(\mathbf{p}_0 + \mathbf{h}) - F(\mathbf{p}_0) = L(\mathbf{h}) + |\mathbf{h}| R(\mathbf{p}_0, \mathbf{h}) \tag{1}$$

where $R(\mathbf{p}_0, \mathbf{h})$ is a vector of R^m that approaches zero as $\mathbf{h}$ approaches zero. Then we say F is differentiable at $\mathbf{p}_0$, and that its derivative is the *linear* function L.

It may be that such an L does not exist, in which case F is not differentiable at $\mathbf{p}$. There cannot be two such *linear* transformations, however. Suppose there were two, say L_1 and L_2. Proceed as we did in Section 3.1. Lines (2) and (3) of that computation remain valid if we replace $m_1 h$ and $m_2 h$ by $L_1(\mathbf{h})$ and $L_2(\mathbf{h})$. Subtraction gives

$$L_2(\mathbf{h}) - L_1(\mathbf{h}) = |\mathbf{h}| [R_1(\mathbf{p}_0, \mathbf{h}) - R_2(\mathbf{p}_0, \mathbf{h})].$$

Multiplying both sides by $1/|\mathbf{h}|$ and using the *linearity* of L_1 and L_2, we get

$$L_2\left(\frac{\mathbf{h}}{|\mathbf{h}|}\right) - L_1\left(\frac{\mathbf{h}}{|\mathbf{h}|}\right) = R_1(\mathbf{p}_0, \mathbf{h}) - R_2(\mathbf{p}_0, \mathbf{h}).$$

The function $L = L_2 - L_1$, defined by $L(x) = L_2(x) - L_1(x)$ is *linear*; $\mathbf{h}/|\mathbf{h}|$ is a unit vector, no matter what the length of $\mathbf{h}$ is. If we decrease the length of $\mathbf{h}$, then the left side of

$$L\left(\frac{\mathbf{h}}{|\mathbf{h}|}\right) = R_1(\mathbf{p}_0, \mathbf{h}) - R_2(\mathbf{p}_0, \mathbf{h})$$

remains constant while the right side goes to zero. Hence, for the arbitrary unit vector $\mathbf{h}/|\mathbf{h}|$,

$$L\left(\frac{\mathbf{h}}{|\mathbf{h}|}\right) = 0.$$

According to Problem 10.3A, $L = \mathcal{O}$. Therefore, $L_1 = L_2$.

When F is differentiable, the *linear* transformation L is designated by $F'(\mathbf{p}_0)$. $F'(\mathbf{p}_0)$ is the derivative of F at $\mathbf{p}_0$.

From (1), it is clear that when $\mathbf{p} = \mathbf{p}_0 + \mathbf{h}$, $F(\mathbf{p})$ is approximated by

$$T(\mathbf{p}) = F(\mathbf{p}_0) + F'(\mathbf{p}_0)(\mathbf{p} - \mathbf{p}_0). \tag{2}$$

Now suppose F is a real-valued function of two variables, differentiable at $\mathbf{p}_0$. The equation $z = F(\mathbf{p}) = F(x, y)$ describes a surface in R^3. The graph of T is a plane A passing through $(\mathbf{p}_0, F(\mathbf{p}_0))$. The picture we have in mind is indicated in Fig. 12.1. Our intention is to prove that the plane A really is tangent to the surface at $\mathbf{p}_0$.

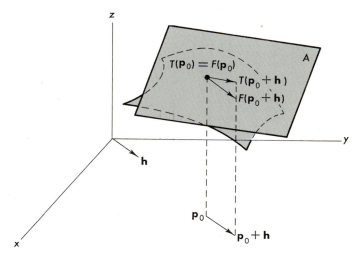

FIG. 12.1

We do this by looking at the plane determined by $\mathbf{p}_0$, $F(\mathbf{p}_0)$, $\mathbf{p}_0 + \mathbf{h}$, and $F(\mathbf{p}_0 + \mathbf{h})$ where $\mathbf{h}$ is arbitrary. This plane intersects the surface in some curve C and intersects plane A in a line k (Fig. 12.2). The reasonable way to define what we mean by saying the A is tangent to the surface is to require that the line k be tangent to C at $(\mathbf{p}_0, F(\mathbf{p}_0))$. This will be the case if, as we shorten $\mathbf{h}$, the slope of the line through $F(\mathbf{p}_0)$ and $F(\mathbf{p}_0 + \mathbf{h})$ approaches the slope of the line k. To avoid any question of how we shorten $\mathbf{h}$ (that is, to be sure we

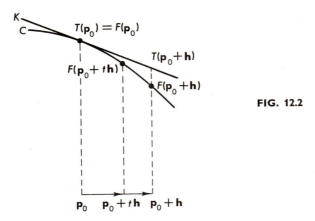

FIG. 12.2

remain parallel to the original $\mathbf{h}$), we consider vectors $\mathbf{p}_0 + t\mathbf{h}$ where the real number t is allowed to approach zero. Then the requirement of tangency is that

$$\lim_{t \to 0} \frac{F(\mathbf{p}_0 + t\mathbf{h}) - F(\mathbf{p}_0)}{t|\mathbf{h}|} = \frac{T(\mathbf{p}_0 + \mathbf{h}) - T(\mathbf{p}_0)}{|\mathbf{h}|} \qquad (3)$$

Since $T(\mathbf{p}_0) = F(\mathbf{p}_0)$, we see from Eq. (2) that

$$T(\mathbf{p}_0 + \mathbf{h}) - T(\mathbf{p}_0) = F'(\mathbf{p}_0)\mathbf{h}.$$

Thus, (3) says the plane is tangent to the graph of F at $(\mathbf{p}_0, F(\mathbf{p}_0))$ if and only if

$$\lim_{t \to 0} \frac{F(\mathbf{p}_0 + t\mathbf{h}) - F(\mathbf{p}_0)}{t} = F'(\mathbf{p}_0)\mathbf{h}. \qquad (4)$$

We know, however, from (1) that

$$\lim_{t \to 0} [F(\mathbf{p}_0 + t\mathbf{h}) - F(\mathbf{p}_0)] = \lim_{t \to 0} [F'(\mathbf{p}_0)t\mathbf{h} + |t\mathbf{h}| R(\mathbf{p}_0, t\mathbf{h})]. \qquad (5)$$

$F'(\mathbf{p}_0)$ is *linear*, so $F'(\mathbf{p}_0)t\mathbf{h} = tF'(\mathbf{p}_0)\mathbf{h}$, and (4) is equivalent to (5). The graph of T is the desired tangent plane.

Suppose a transformation F of the form

$$y_1 = F^1(x_1, x_2, \ldots, x_n),$$

$$y_2 = F^2(x_1, x_2, \ldots, x_n), \qquad (6)$$

$$\vdots$$

$$y_m = F^m(x_1, x_2, \ldots, x_n),$$

is known to be differentiable at $\mathbf{p}_0$. We learned in Section 10.3 that the derivative, a *linear* transformation, is determined by a matrix.

$$\begin{bmatrix} a_{11} & a_{12} & \cdots & a_{1n} \\ a_{21} & a_{22} & \cdots & a_{2n} \\ \vdots & & & \\ a_{m1} & a_{m2} & \cdots & a_{mn} \end{bmatrix}.$$

We need a method for determining the numbers a_{ij}. For this purpose we let $\mathbf{h} = t\boldsymbol{\varepsilon}_j$ where t is a real number and $\boldsymbol{\varepsilon}_j$ is the basis vector of Example 10.2A. We may write (1) in the form

$$\begin{bmatrix} F^1(\mathbf{p}_0 + t\boldsymbol{\varepsilon}_j) \\ F^2(\mathbf{p}_0 + t\boldsymbol{\varepsilon}_j) \\ \vdots \\ F^m(\mathbf{p}_0 + t\boldsymbol{\varepsilon}_j) \end{bmatrix} - \begin{bmatrix} F^1(\mathbf{p}_0) \\ F^2(\mathbf{p}_0) \\ \vdots \\ F^m(\mathbf{p}_0) \end{bmatrix} = \begin{bmatrix} a_{11} & a_{12} & \cdots & a_{1n} \\ a_{21} & a_{22} & \cdots & a_{2n} \\ \vdots & & & \\ a_{m1} & a_{m2} & \cdots & a_{mn} \end{bmatrix} \begin{bmatrix} 0 \\ \vdots \\ t \\ \vdots \\ 0 \end{bmatrix} + |t| \begin{bmatrix} R^1(\mathbf{p}_0, t\boldsymbol{\varepsilon}_j) \\ R^2(\mathbf{p}_0, t\boldsymbol{\varepsilon}_j) \\ \vdots \\ R^m(\mathbf{p}_0, t\boldsymbol{\varepsilon}_j) \end{bmatrix}$$

Vectors are equal, according to the definition, if and only if their entries are all equal. Hence, for any i, $i = 1, 2, \ldots, m$,

$$F^i(\mathbf{p}_0 + t\boldsymbol{\varepsilon}_j) - F^i(\mathbf{p}_0) = ta_{ij} + |t|\, R^i(\mathbf{p}_0, t\boldsymbol{\varepsilon}_j).$$

If we set $\mathbf{p}_0 = (x_1{}^0, x_2{}^0, \ldots, x_n{}^0)$, then this last expression may be written in the form

$$\frac{F^i(x_1{}^0, \ldots, x_j{}^0 + t, \ldots, x_n{}^0) - F^i(x_1{}^0, \ldots, x_j{}^0 \ldots, x_n{}^0)}{t} = a_{ij} \pm R^i(\mathbf{p}_0, t\boldsymbol{\varepsilon}_j).$$

$$(7)$$

As t approaches 0, $\mathbf{h} = t\boldsymbol{\varepsilon}_j$ also approaches 0. The right side of (7) approaches a_{ij}. To see how the left side is computed, consider the function of one variable obtained by setting

$$y = G(x_j) = F^i(x_1{}^0, \ldots, x_{j-1}^0, x_j, x_{j+1}^0, \ldots, x_n{}^0).$$

If you were now asked to find the number dy/dx at $x_j{}^0$, you would say it is the limit of (7) as t approaches 0. Thus, a_{ij} is obtained by treating all the variables except x_j as "frozen" while we find the derivative with respect to x_j according to the usual rules. The number obtained in this way is called *the partial derivative of F^i with respect to x_j*. The partial derivative is a number. It is denoted by various symbols:

$$\frac{\partial}{\partial x_j} F^i(\mathbf{p}_0) = F_j{}^i(\mathbf{p}_0) = D_j F^i(\mathbf{p}_0).$$

The real-valued function that gives the number $\partial/\partial x_j F^i(\mathbf{p}_0)$ at each $\mathbf{p}_0$ is called

$$\frac{\partial}{\partial x_j} F^i = F_j{}^i = D_j F^i$$

We have proved a useful result.

THEOREM A

Let U be an open set of R^n. We suppose

$$F: U \to R^m.$$

is described by the coordinate functions (6), and that F is differentiable at $\mathbf{p} \in U$. Then the *linear* transformation $F'(\mathbf{p})$ has as its matrix

$$\begin{bmatrix} F_1^1(\mathbf{p}) & F_2^1(\mathbf{p}) & \cdots & F_n^1(\mathbf{p}) \\ F_1^2(\mathbf{p}) & F_2^2(\mathbf{p}) & \cdots & F_n^2(\mathbf{p}) \\ \vdots & \vdots & & \vdots \\ F_1^m(\mathbf{p}) & F_2^m(\mathbf{p}) & \cdots & F_n^m(\mathbf{p}) \end{bmatrix}.$$

EXAMPLE A

In Example 11.5A we were given

$$F(x, y) = \frac{4 - x}{1 + x + y^2}.$$

We sought a function T having as its graph a plane that would be tangent to the graph of F at $(2, 1)$. We got as far as writing

$$T(x, y) = \tfrac{1}{2} + L(x - 2, y - 1).$$

It remained for us to determine L. We were also to use T to approximate F at $(1.9, 0.9)$, $(1.9, 1.1)$, $(2.1, 0.9)$, and $(2.1, 1.1)$. We are now able to find the matrix of L. It is simply the matrix of $F'(x, y)$ (which, in accord with our agreement in Section 10.3, is to be designated by $[F'(x, y)]$).

$$[F'(x, y)] = [F_1(x, y) \qquad F_2(x, y)].$$

$F_1(x, y)$ is computed by treating y as a constant while we differentiate $F(x, y)$ with respect to x.

$$F_1(x, y) = \frac{(1 + x + y^2)(-1) - (4 - x)}{(1 + x + y^2)^2} = \frac{-5 - y^2}{(1 + x + y^2)^2}.$$

Similarly,

$$F_2(x, y) = \frac{-(4 - x)2y}{(1 + x + y^2)^2} = \frac{2xy - 8y}{(1 + x + y^2)^2}.$$

At (2, 1),

$$F_1(2, 1) = \frac{-6}{16} = -0.375,$$

$$F_2(2, 1) = \frac{-4}{16} = 0.250.$$

From these computations,

$$T(x, y) = \frac{1}{2} + \begin{bmatrix} -\dfrac{3}{8} & -\dfrac{1}{4} \end{bmatrix} \begin{bmatrix} x - 2 \\ y - 1 \end{bmatrix}.$$

At the points (1.9, 0.9), (1.9, 1.1), (2.1, 0.9) and (2.1, 1.1) where $F(x, y)$ is to be approximated,

$$T(1.9, 0.9) = 0.5 + [-0.375 \quad -0.250] \begin{bmatrix} -0.1 \\ -0.1 \end{bmatrix},$$

$$= 0.5 + 0.0375 + 0.0250 = 0.5625;$$

$$T(1.9, 1.1) = 0.5 + [-0.375 \quad -0.250] \begin{bmatrix} -0.1 \\ 0.1 \end{bmatrix},$$

$$= 0.5 + 0.0375 - 0.0250 = 0.5125.$$

Similarly, compute $T(2.1, 0.9)$ and $T(2.1, 1.1)$. For comparison, we have also computed $F(x, y)$ at these points and tabulated the results.

(x, y)	(1.9, 0.9)	(1.9, 1.1)	(2.1, 0.9)	(2.1, 1.1)
$T(x, y)$	0.5625	0.5125	0.4875	0.4375
$F(x, y)$	0.5660	0.5109	0.4859	0.4408

We shall return to this example again. ∎

A comment about notation is in order. In Section 3.5, considering a real-valued function of a real variable, we saw that the *linear* function $F'(x_0)$ corresponded to a real number, the correspondence being

$$F'(x_0) \leftrightarrow m,$$

where

$$F'(x_0)x = mx.$$

To avoid the cumbersome phrase, "the number m that corresponds to $F'(x_0)$," we adopt the notational device of using brackets about $F'(x_0)$ to designate the number

$$[F'(x_0)] = m.$$

Then

$$F'(x_0)(x) = [F'(x_0)]x. \tag{8}$$

In Section 10.3, we agreed to let $[L]$ denote the matrix corresponding to the *linear* transformation $L: R^n \to R^m$. Then we could write, for $\mathbf{q} \in R^n$

$$L(\mathbf{q}) = [L]\mathbf{q}.$$

where $\mathbf{q}$ is to be thought of as a column vector. In particular, since $F'(\mathbf{p}_0)$ is a *linear* transformation,

$$F'(\mathbf{p}_0)(\mathbf{q}) = [F'(\mathbf{p}_0)]\mathbf{q}. \tag{9}$$

For a real number x, (8) is a special case of (9). This latter formula is clearly the analog of the notation introduced in Section 3.5.

EXAMPLE B

In Example 11.4A we considered F defined by

$$y_1 = F^1(x_1, x_2) = x_1{}^2 - x_2{}^2,$$

$$y_2 = F^2(x_1, x_2) = 2x_1 x_2.$$

Find the derivative of this transformation from R^2 into R^2.

To find $F_1^1(x_1, x_2)$ we treat x_2 as if it were a constant, finding the derivative with respect to x_1 in the usual way.

$$F_1^1(x_1, x_2) = 2x_1.$$

Performing the necessary four computations of this type, we get

$$[F'(\mathbf{p})] = \begin{bmatrix} 2x_1 & -2x_2 \\ 2x_2 & 2x_1 \end{bmatrix}. \ \blacksquare$$

EXAMPLE C

Let $F: R^2 \to R$ be described by

$$F(x, y) = \begin{cases} \dfrac{2x^2 y}{x^2 + y^2}, & (x, y) \neq (0, 0), \\ 0, & (x, y) = (0, 0) \end{cases}$$

Find $F_1(x, y)$ and $F_2(x, y)$.

When $(x, y) \neq (0, 0)$, the denominator is nonzero and we can compute $F_1(x, y)$ in the usual way. We get

$$F_1(x, y) = \frac{(x^2 + y^2)4xy - 2x^2y(2x)}{(x^2 + y^2)^2}$$

$$F_1(x, y) = \frac{4xy^3}{(x^2 + y^2)^2}, \quad (x, y) \neq (0, 0).$$

When $(x, y) = (0, 0)$, our rules do not apply. Going back to the definition,

$$F_1(0, 0) = \lim_{t \to 0} \frac{F(0 + t, 0) - F(0, 0)}{t},$$

$$= \lim_{t \to 0} \frac{0 - 0}{t} = 0.$$

Similarly, compute $F_2(x, y)$. We get

$$F_1(x, y) = \begin{cases} \dfrac{4xy^3}{(x^2 + y^2)^2} \\ 0 \end{cases}$$

$$F_2(x, y) = \begin{cases} \dfrac{2x^2(x^2 - y^2)}{(x^2 + y^2)^2}, & (x, y) \neq (0, 0), \\ 0, & (x, y) = (0, 0). \end{cases}$$

Hence, the partial derivatives exist everywhere. ∎

We know (Theorem A) that if a function F is differentiable, then the coordinate functions of F all have partial derivatives. Example C can be used to show that the converse does not hold. F itself (there are no separate coordinate functions here because $m = 1$; F is real valued) has partial derivatives everywhere. Yet we claim that F does not have a derivative at $(0, 0)$.

To verify that $F'(0, 0)$ does not exist, we suppose that it does. Then, according to Theorem A and the computations of Example C,

$$F'(0, 0) = [0 \quad 0].$$

The approximation property of a derivative then says that for (x, y) close to $(0, 0)$,

$$F(x + 0, y + 0) = F(0, 0) + F'(0, 0)(x, y) + |(x, y)| R((0, 0), (x, y)).$$

In our case this becomes

$$\frac{2x^2 y}{x^2 + y^2} = \sqrt{x^2 + y^2}\, R((0,\, 0),\, (x,\, y)).$$

The limit of R as $|(x, y)|$ approaches zero is to be zero. Hence, we should have

$$\lim_{(x,y)\to(0,0)} \frac{2x^2 y}{(x^2 + y^2)^{3/2}} = 0.$$

If (as in Example 11.4B) we approach along the line $y = mx$, however, we get

$$\lim_{(x,mx)\to(0,0)} \frac{2mx^3}{(1 + m^2)^{3/2} x^3} = \frac{2m}{(1 + m^2)^{3/2}}.$$

This is not zero unless m is. Therefore $F'(0, 0)$ does not exist. ∎

It is significant that in our example, the partial derivatives are not continuous at $(0, 0)$. It is known that if the partials not only exist but are continuous at a point, then the function is differentiable [2, page 243].

Most of the familiar properties of the derivative come along to the new situation as part of the baggage. This must be understood in context, of course.

Rule 1 of Section 3.4 remains unchanged. Rule 2 does not generally make sense, since we have attached no meaning to multiplication if $F(\mathbf{p})$ and $G(\mathbf{p})$ are vectors. In special cases, however, Rule 2 is still of use. When $F(\mathbf{p})$ and $G(\mathbf{p})$ are vectors in the same space, the dot product has meaning and Rule 2 then gives instructions for differentiating (Example 9.3B). Also, if $F(\mathbf{p})$ is real valued while $G(\mathbf{p})$ is a vector in some space R^m, then $F(\mathbf{p})G(\mathbf{p})$ makes sense and Rule 2 is once again meaningful. Similar remarks apply to Rule 3, which makes sense if and only if G is a real-valued function.

Rule 4 calls for special comment. If the composition of $F{\circ}G$ is to be meaningful, then the range of G must be contained in the domain of F. Beyond this, there are certain computational techniques that need to be learned. As an illustration of the general procedure, suppose $H = F{\circ}G$ where

$$G: R^2 \to R^3 \qquad \text{is described by} \qquad \begin{aligned} u &= G^1(x,\, y), \\ v &= G^2(x,\, y), \\ w &= G^3(x,\, y), \end{aligned}$$

and

$$F: R^3 \to R^2 \qquad \text{is described by} \qquad \begin{aligned} r &= F^1(u,\, v,\, w), \\ s &= F^2(u,\, v,\, w). \end{aligned}$$

Then

$$H(\mathbf{p}) = F(G(\mathbf{p})),$$

$$H'(\mathbf{p}) = F'(G(\mathbf{p}))G'(\mathbf{p}). \tag{10}$$

Now $H: R^2 \to R^2$, so we may assume it is described by the coordinate functions

$$r = H^1(x, y),$$

$$s = H^2(x, y).$$

According to (10), $H'(\mathbf{p})$ is the composition of the two *linear* transformations $F'(G(\mathbf{p}))$ and $G'(\mathbf{p})$, so the matrix of $H'(\mathbf{p})$ is the product of the matrices $[F'(G(\mathbf{p}))]$ and $[G'(\mathbf{p})]$.

$$\begin{bmatrix} H_1^1(\mathbf{p}) & H_2^1(\mathbf{p}) \\ H_1^2(\mathbf{p}) & H_2^2(\mathbf{p}) \end{bmatrix} = \begin{bmatrix} F_1^1(G(\mathbf{p})) & F_2^1(G(\mathbf{p})) & F_3^1(G(\mathbf{p})) \\ F_1^2(G(\mathbf{p})) & F_2^2(G(\mathbf{p})) & F_3^2(G(\mathbf{p})) \end{bmatrix} \begin{bmatrix} G_1^1(\mathbf{p}) & G_2^1(\mathbf{p}) \\ G_1^2(\mathbf{p}) & G_2^2(\mathbf{p}) \\ G_1^3(\mathbf{p}) & G_2^3(\mathbf{p}) \end{bmatrix}.$$

After multiplication as indicated on the right, we see that (as one of four examples we could write down),

$$H_1^1(\mathbf{p}) = F_1^1(G(\mathbf{p}))G_1^1(\mathbf{p}) + F_2^1(G(\mathbf{p}))G_1^2(\mathbf{p}) + F_3^1(G(\mathbf{p}))G_1^3(\mathbf{p}).$$

In the case of one variable, the chain rule was more easily remembered by utilizing the quotient notation dy/dx for the derivative. The situation is no different here. The entry $H_1^1(\mathbf{p})$ given above is more easily remembered if we write

$$\frac{\partial r}{\partial x} = \frac{\partial r}{\partial u} \cdot \frac{\partial u}{\partial x} + \frac{\partial r}{\partial v} \cdot \frac{\partial v}{\partial x} + \frac{\partial r}{\partial w} \cdot \frac{\partial w}{\partial x}. \tag{11}$$

This formula is also suggested by drawing a schematic representation of the dependencies involved here (Fig. 12.3). Then each path from x to r represents a term of (11).

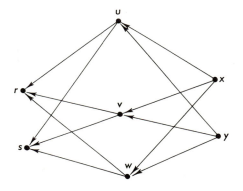

FIG. 12.3

EXAMPLE D

Let $x = t^2$, $y = \sin t$, $z = \sqrt{t^2 + 1}$ and suppose $F(x, y, z) = xy + yz$. Then F can be thought of as a function of t. Find $F'(0)$.

We begin by setting $w = xy + yz$. Then

$$\frac{dw}{dt} = \frac{\partial w}{\partial x} \cdot \frac{dx}{dt} + \frac{\partial w}{\partial y} \cdot \frac{dy}{dt} + \frac{\partial w}{\partial z} \cdot \frac{dz}{dt},$$

$$= y(2t) + (x + z)\cos t + yt(t^2 + 1)^{-1/2}.$$

When $t = 0$, $x = 0$, $y = 0$, $z = 1$;

$$F'(0) = 1. \quad \blacksquare$$

The example above could be given a physical interpretation. Suppose the temperature at a point (x, y, z) in space was known to be $xy + yz$, and suppose the location of a moving body at time t was known to be $(t^2, \sin t, \sqrt{t^2 + 1})$. Then dF/dt gives the rate of change of the temperature of the body at time t.

══ **12·1 EXERCISES**

In each of Exercises 1–6, the three equations describe a function F from R^3 to R^3. Find $F'(p)$ in each case.

1· $u = xy + yz$,
$v = xyz$,
$w = x^2 + y^2 + z^2$.

2· $u = x^2y + y^2z + z^2x$,
$v = xy - yz$,
$w = x^2y^2 + y^2z^2$.

3· $u = xe^{yz}$,
$v = ye^{yz}$,
$w = ze^x + xe^y$.

4· $u = x^2e^y + y^2e^z$,
$v = e^{xyz}$,
$w = xye^z + xze^y$.

5· $u = x \sin yz$,
$v = y \cos \dfrac{y}{z}$,
$w = z \tan x + y \sec x$.

6· $u = x \sin(y + z)$,
$v = \dfrac{y}{z} \tan x$,
$w = x \sin yz + y \cos xz$.

7· Let F be the function defined by the equations of Exercise 1. (a) Find $F(1.1, 1.85, -1.15)$. (b) Using $p_0 = (1, 2, -1)$, approximate the answer to part (a) with $T(1.1, 1.85, -1.15)$ where T is the function described in (2).

8· Let F be the function defined by the equations of Exercise 2. (a) Find $F(1.1, 1.85, -1.15)$. (b) Using $p_0 = (1, 2, -1)$, approximate the answer to part (a) with $T(1.1, 1.85, -1.15)$ where T is the function described in (2).

9· Let $z = F(x, y) = \sqrt{6 - (x^2 + y^2)}$. Use (2) to write the equation of the plane tangent to the graph of F at $(1, -1)$. Sketch a picture. Then compare with Exercise 1 of Section 11.3.

10 · Let $z = F(x, y) = \sqrt{17 - (x^2 + y^2)}$. Use (1.2) to write the equation of the plane tangent to the graph of F at $(3, -2)$. Sketch a picture. Then compare with Exercise 2 of Section 11.3.

11 · $r = uv + vw + uw,$ and $u = e^{xy},$
$s = u^2 + v^2 + w^2,$ $v = \ln(x + y),$
    $w = \sin xy.$

Find $\dfrac{\partial r}{\partial x}, \dfrac{\partial r}{\partial y}, \dfrac{\partial s}{\partial x}, \dfrac{\partial s}{\partial y}$.

12 · $r = \dfrac{u + v + w}{uvw},$ $u = (x^2 + y^2)^3,$
$s = \ln(u^2 + v^2 + w^2),$ $v = \ln xy,$
    $w = e^{xy}.$

Find $\dfrac{\partial r}{\partial x}, \dfrac{\partial r}{\partial y}, \dfrac{\partial s}{\partial x}, \dfrac{\partial s}{\partial y}$.

13 · $r = e^{uv},$
$s = u^2 + v^2,$ $u = x^2 y^2,$
$t = \dfrac{u}{v},$ $v = \sin xy.$

Find $\dfrac{\partial r}{\partial x}, \dfrac{\partial s}{\partial y}, \dfrac{\partial t}{\partial x}$ when $x = 1, y = \pi/2$.

14 · $r = \ln(u^2 + v^2),$ $u = (x^3 + y^3)^{1/2},$
$s = \tan uv,$ $v = \dfrac{x}{y} \sin xy.$
$t = uve^{uv},$

Find $\dfrac{\partial r}{\partial x}, \dfrac{\partial s}{\partial y}, \dfrac{\partial t}{\partial x}$ when $x = 0, y = \pi/2$.

15 · The temperature on a plate of diameter 14 in. varies directly with the square of the distance from the center. The temperature on the outer edge is 588°F. The position of a point moving out from the center is given by $x = t^{1/2}$, $y = 7t/(t + 3)$. How fast is the temperature changing when $t = 4$?

16 · A child's sailboat slips from his grasp at a river's edge. The stream carries it along at 4 ft/sec. A crosswind blows it toward the opposite shore at 3 ft/sec. If the child runs along the shore at 3 ft/sec following his boat, how fast is the boat moving away from him when $t = 5$?

17 · Electric power, measured in watts, obeys

$$P = I^2 R$$

where I is the current in amperes and R is the resistance in ohms. If the current and resistance at time t vary according to

$$I = \frac{3t^2}{t^2 + 4}, \qquad R = \frac{50t}{t + 3},$$

find the rate of change of P when $t = 5$.

18. The resistance R of a parallel combination of two resistances S and T is given by

$$R = \frac{ST}{S+T}.$$

If T and S vary with time according to

$$S = \frac{27t+1}{t+2}, \qquad T = \frac{4t^2}{t^2+1},$$

find the rate of change of R when $t = 3$.

19. A cylindrical membrane is increased in length by 7 units/sec, while it decreases its diameter by 4 units/sec. How fast is the volume increasing when the radius is 3 units and the length is 10 units?

20. Sand is running onto a conical pile. At the instant when the height of the pile is 4 in. and the radius is 6 in., the height is increasing 3 in./min and the radius is increasing at 4 in./min. How fast is the volume increasing?

12·2 Real-Valued Functions of Several Variables

In the remainder of this chapter we restrict ourselves to the case $m = 1$. Our transformation is thus described by

$$y = F(\mathbf{p}) = F(x_1, x_2, \ldots, x_n)$$

The derivative is determined by the matrix

$$[F'(\mathbf{p})] = [F_1(\mathbf{p}) \quad F_2(\mathbf{p}) \quad \ldots, F_n(\mathbf{p})].$$

$F'(\mathbf{p})$ is therefore a member of the space of *linear* transformations from R^n into R, designated by $\mathscr{L}(R^n, R)$. Members of this space, being $1 \times n$ matrices, appear to be row vectors. In order to distinguish them from vectors in the original space, members of $\mathscr{L}(R^n, R)$ are called *covectors*. We recall that if $\mathbf{h} \in R^n$, the linear transformation $F'(\mathbf{p})$ evaluated at $\mathbf{h}$ is

$$F'(\mathbf{p})(\mathbf{h}) = [F'(\mathbf{p})]\mathbf{h}.$$

This is just the dot product of the covector $F'(\mathbf{p})$ and the vector $\mathbf{h}$, so we may in this case also write

$$F'(\mathbf{p})\mathbf{h} = [F'(\mathbf{p})] \cdot \mathbf{h}. \tag{1}$$

The covector $[F'(\mathbf{p})]$ is sometimes called the *gradient vector*, designated ∇y (and read del y). It has several properties to which we shall call attention.

When $y = F(\mathbf{p})$, then for small $\mathbf{h}$, $F'(\mathbf{p})\mathbf{h}$ approximates the change in y that occurs as we move from $\mathbf{p}$ to $\mathbf{p} + \mathbf{h}$. According to (1) and the definition of angle (see paragraph following Definition D of Section 10.2),

$$F'(\mathbf{p})\mathbf{h} = |[F'(\mathbf{p})]| \, |\mathbf{h}| \, \cos \theta.$$

Since $F'(\mathbf{p})$ is linear, we get, in setting $u = \mathbf{h}/|\mathbf{h}|$,

$$F'(\mathbf{p})\left(\frac{\mathbf{h}}{|\mathbf{h}|}\right) = F'(\mathbf{p})\mathbf{u} = |[F'(\mathbf{p})]|\cos\theta. \tag{2}$$

Of course, $\mathbf{u}$ is a unit vector. Now if we seek to find a unit vector for which $|F'(\mathbf{p})\mathbf{u}|$ is maximum, then it is clear that $\mathbf{u}$ must be chosen so that $\cos\theta = 1$; that is, $\mathbf{u}$ must be parallel to the gradient vector. We state this as the first property of the gradient vector.

Property 1 Let $y = F(\mathbf{p})$. The direction in which we move from $\mathbf{p}$ in order that y change the fastest is the direction of the gradient vector $\nabla y = [F'(\mathbf{p})]$. (The change is an increase if motion is in the direction of the gradient; a decrease if motion is opposite to the gradient.)

Another way of stating Property 1 is to say that if $\mathbf{u}$ is a unit vector, then the maximum value of $|F'(\mathbf{p})\mathbf{u}|$ is achieved by choosing $\mathbf{u} = \nabla y/|\nabla y|$; that is, by choosing $\mathbf{u}$ in the direction of the gradient vector. Moreover, (2) makes clear a second property.

Property 2 The maximum value of $|F'(\mathbf{p})\mathbf{u}|$ for a unit vector $\mathbf{u}$ is $|[F'(\mathbf{p})]|$, the length of the gradient vector.

EXAMPLE A
Suppose $z = T(x, y) = x^2 + y^2$ gives the temperature in suitable units at the point (x, y). If a bug at $(1, 2)$ finds the temperature too cool, in which direction should it move to correct the problem most quickly?

The solution is shorter than the problem. The bug should move in the direction of the gradient vector:

$$\nabla z = [2x \quad 2y]|_{(1,2)} = [2 \quad 4].$$

Continuing the same example, suppose the bug finds the temperature to his liking, but must keep moving for some reason. In which direction should he go?

To answer this we refer to (2) again. Here it is clear that to keep $F'(\mathbf{p})\mathbf{u} = 0$ (so that F remains constant), motion should be such that $\cos\theta = 0$; that is, $\mathbf{u}$ should be chosen perpendicular to the gradient vector.

It is also clear, however, that $F(1, 2) = 5$, and that if the bug wishes to keep the temperature constant, it should move along the path described by $F(x, y) = x^2 + y^2 = 5$. Evidently the tangent to $x^2 + y^2 = 5$ at $(1, 2)$ (which is the direction in which the bug begins to move) must be perpendicular to the gradient. It is, of course, geometrically evident that this is the case (Fig. 12.4). ∎

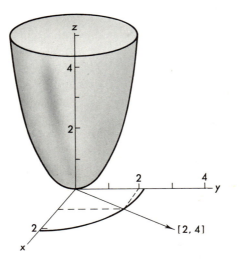

FIG. 12.4

The last part of Example A suggests a final property of the gradient to which we direct attention. We state the result for $n = 3$; the last example illustrates it for $n = 2$. Let $w = F(x, y, z)$. For fixed w_0, $F(x, y, z) - w_0 = 0$ is a statement of equality between three variables; it therefore determines a surface in R^3 (as discussed in Section 11.3). Now suppose R is a curve contained in this surface which passes through (x_0, y_0, z_0). We know (Chapter 9) the curve R is a mapping from an interval $[a, b]$ into R^3 having the form

$$R(t) = [x(t) \quad y(t) \quad z(t)].$$

Since R is contained in the surface $F(x, y, z) = w_0$, it must be that for all $t \in [a, b]$,

$$F(x(t), y(t), z(t)) = w_0.$$

The chain rule then gives, at $R(t_0) = \mathbf{p}_0$,

$$F_1(\mathbf{p}_0)D_t x(t_0) + F_2(\mathbf{p}_0)D_t y(t_0) + F_3(\mathbf{p}_0)D_t z(t_0) = 0;$$

that is,

$$[F'(\mathbf{p}_0)] \cdot [D_t x(t_0) \quad D_t y(t_0) \quad D_t z(t_0)] = 0.$$

The second vector, however, is the tangent to the curve R at t_0. Hence, the gradient vector $[F'(\mathbf{p}_0)]$ is perpendicular to any curve in the surface $F(x, y, z) = w_0$ that passes through $\mathbf{p}_0$. (We have tacitly assumed in making this assertion that $F'(\mathbf{p}_0)$ is not the zero vector.) This establishes our last property.

Property 3 Let F be a real-valued function of two or three variables that has a nonvanishing derivative at $\mathbf{p}_0$. Then the gradient vector $F'(\mathbf{p}_0)$ is perpendicular at $\mathbf{p}_0$ to the surface $F(\mathbf{p}) = F(\mathbf{p}_0)$.

EXAMPLE B

We closed Section 11.5 with an example in which we sought the equation of the plane tangent to the graph of

$$F(x, y) = \frac{4 - x}{1 + x + y^2}$$

at $(2, 1)$. We found the desired equation in Example 12.1A. We now find it again by an alternate method.

Set

$$G(x, y, z) = \frac{4 - x}{1 + x + y^2} - z.$$

Now $G(x, y, z) = 0$ determines a surface; the very one sketched in Fig. 11.18. The point in which we are interested is $(2, 1, \frac{1}{2})$. The gradient here is

$$G'(\mathbf{p}) = \left[\frac{-5 - y^2}{(1 + x + y^2)^2} \quad \frac{2xy - 8y}{(1 + x + y^2)^2} \quad -1 \right]$$

$$G'(2, 1, \tfrac{1}{2}) = [-\tfrac{3}{8} \quad -\tfrac{1}{4} \quad -1]$$

Thus, $[-\tfrac{3}{8} \quad -\tfrac{1}{4} \quad -1]$ is perpendicular to the desired plane at $(2, 1, \frac{1}{2})$. By Theorem 11.2A, the equation of the plane is

$$-\tfrac{3}{8}(x - 2) - \tfrac{1}{4}(y - 1) - (z - \tfrac{1}{2}) = 0.$$

This agrees with the answer obtained in Example 12.1A. ▌

PROBLEM

A · Let $z = F(x, y) = 9 - (x^2 + y^2)$.
(a) Sketch a graph.
(b) In what direction should one move away from $(2, 1)$ in order to achieve the fastest possible growth in F?
(c) Find a vector perpendicular to the graph sketched in part (a) at $(2, 1, 4)$.
(d) Use (2) of 12.1 to write the equation of the tangent plane at $(2, 1, 4)$.
(e) Use the vector obtained in (c) and Theorem 11.2A to write the equation of the tangent plane at $(2, 1, 4)$.

The mean value theorem was of great importance to us in our study of real-valued functions of a single variable. Its n-dimensional form is also very useful. To state the theorem in this form, we introduce an abbreviation for the line segment joining $\mathbf{p}_1$ and $\mathbf{p}_2$.

$$\text{Seg}[\mathbf{p}_1, \mathbf{p}_2) = \{\mathbf{p} \in R^n : \mathbf{p} = \mathbf{p}_1 + t(\mathbf{p}_2 - \mathbf{p}_1), \quad t \in [0, 1)\}.$$

With obvious modifications to allow for various combinations of including or excluding end points, we may now state our theorem.

THEOREM A   (Mean Value)
Let U be an open set in R^n that contains $\text{Seg}[\mathbf{p}_1, \mathbf{p}_2]$.
Suppose $F: U \to R$ has the following properties:

(a) F is continuous on $\text{Seg}[\mathbf{p}_1, \mathbf{p}_2]$;
(b) F is differentiable on $\text{Seg}(\mathbf{p}_1, \mathbf{p}_2)$.

Then there exists a $t_1 \in (0, 1)$ such that

$$F(\mathbf{p}_2) - F(\mathbf{p}_1) = F'(\mathbf{p}_1 + t_1(\mathbf{p}_2 - \mathbf{p}_1))(\mathbf{p}_2 - \mathbf{p}_1).$$

FIG. 12.5

(Figure 12.5 indicates a region U and two points $\mathbf{p}_1$ and $\mathbf{p}_2$ to which the theorem does not apply.)

PROOF:

Define

$$G: \quad \begin{aligned} & [0, 1] \to R \\ & t \to F(\mathbf{p}_1 + t(\mathbf{p}_2 - \mathbf{p}_1)). \end{aligned}$$

G is now a real-valued function of the single variable t. Because of the conditions on F, G satisfies the hypothesis of the mean value theorem as stated in Theorem 4.1A. Hence, there is a t_1 such that

$$G(1) - G(0) = [G'(t_1)].$$

This is just the conclusion we desire. (Don't forget to use the chain rule when differentiating $F(\mathbf{p}_1 + t(\mathbf{p}_2 - \mathbf{p}_1))$.) ∎

PROBLEM

B · Let $F(x, y) = xy + 3x^2$. Set $\mathbf{p}_1 = (2, 1)$ and $\mathbf{p}_2 = (1, 3)$. Find the t_1 guaranteed by Theorem A.

An immediate consequence of the mean value theorem in Chapter 4 was that if a derivative was known to be zero on an interval, then F was constant on the interval (Problem 4.1B). In our present context, this is still true, pro-viding the set U has certain nice properties. The difficulty can be appreciated by looking at Fig. 12.5. Since $\text{Seg}[\mathbf{p}_1, \mathbf{p}_2]$ is not contained in U, the mean value theorem cannot be applied directly. If you think of joining $\mathbf{p}_1$ to $\mathbf{p}_2$ by a series of straight segments staying inside of U, you are thinking along the right lines (pun). But you might have trouble describing such a construction in R^3. One way out of the difficulty is to restrict attention to only those sets where, for any two points $\mathbf{p}_1$, $\mathbf{p}_2$ in the set, $\text{Seg}[\mathbf{p}_1, \mathbf{p}_2]$ is also in the set. Such a set is called *convex*, and such sets have been the object of much attention. Another way out is to give a special name to sets where any two points can be joined by a path of segments (a polygonal path.) Properties of such sets (called *connected sets*) are developed in topology, and they may be used to clinch the argument we suggested above.

The point of the preceding paragraph is that much interesting mathematics remains after this course has been completed. A full discussion of the particular result under consideration is to be found in [2, page 247].

$F'(\mathbf{p})$ is a *linear* transformation; $F'(\mathbf{p})\mathbf{u}$ is a number. If $\mathbf{u}$ is chosen to be a unit vector, then this number depends only on the direction of $\mathbf{u}$. This number is called the *directional derivative* in many texts.

12 · 2   EXERCISES

In Exercises 1–10, find the directional derivative of the given function in the direction of the given vector $\mathbf{h}$. (Note carefully that the definition of the directional derivative involves a unit vector.)

1. $F(x, y, z) = \dfrac{x + y}{z^2}$; $\mathbf{h} = [1 \quad -1 \quad 2.]$

2. $F(x, y, z) = \dfrac{x^2 - y^2}{z}$; $\mathbf{h} = [1 \quad 2 \quad 1].$

3. $F(x, y, z) = xe^{yz}$; $\mathbf{h} = [-3 \quad 0 \quad 1].$

4. $F(x, y, z) = xye^z$; $\mathbf{h} = [0 \quad 1 \quad 0].$

5. $F(x, y) = \sin xy$; $\mathbf{h} = [1 \quad 2].$

6. $F(x, y) = \cos \dfrac{x}{y}$; $\mathbf{h} = [-1 \quad 3].$

7. $F(x, y) = \dfrac{x + y}{x^2 + y^2}$; $\mathbf{h} = [3 \quad 4].$

8. $F(x, y) = (x + y) \ln xy$; $\mathbf{h} = [4 \quad -3].$

9 • $F(x, y, z, w) = \dfrac{x + y}{z + w}$; $\mathbf{h} = [1 \quad -1 \quad 2 \quad 3]$.

10 • $F(x, y, z, w) = \dfrac{xy}{z + w}$; $\mathbf{h} = [0 \quad 1 \quad -1 \quad 2]$.

11 • A line passes through $(2, 3, 6)$ and is normal to the surface $z = xy$, at that point. Write the equation of the line.

12 • A line passes through $(1, -1, 1)$ and is normal to the surface $z = x^2 + xy + y^2$ at that point. Write the equation of the line.

13 • At what point does $z = x^2 + 3xy + y^2 - 4x - y$ have a tangent plane parallel to the xy plane ? Is this point a high or low point?

14 • At what point does $z = x^2 + 5xy - 3y^2 + x - 16y$ have a tangent plane parallel to the xy plane ? Is this point a high or low point?

15 • The electric potential, given in suitable units, at a point in space is $E = (x^2 + y^2 + 4z^2)^{1/2}$. In what direction should you move from $(2, 1, 3)$ in order to decrease the potential most rapidly?

16 • The electric potential, given in suitable units, at a point in space is $E = xy/(x + y + z)$. In what direction should you move from $(1, -2, 3)$ in order to decrease the potential most rapidly?

17 • The profits of a certain company are found to depend on four variables according to the rule $P(x, y, u, v) = x^2 + y^2 - (u + v)/(u^2 + v^2)$. In what ratio should these variables be altered to maximize the increase in profit, starting from $(3, 1, 4, 2)$?

18 • Same problem as Exercise 17, using $P(x, y, u, v) = xy + \sqrt{u^2 + v^2 + x^2}$, starting from $(1, 3, 3, 2)$.

12 · 3   Higher Derivatives

Consider the function $F(x, y, z) = xe^{yz} + \cos xy$.

$$\frac{\partial F}{\partial x} = e^{yz} - y \sin xy.$$

Now there is nothing to prevent us from taking partial derivatives again. With self-explanatory notation

$$\frac{\partial}{\partial x} \left(\frac{\partial F}{\partial x} \right) = -y^2 \cos xy.$$

$$\frac{\partial}{\partial y} \left(\frac{\partial F}{\partial x} \right) = ze^{yz} - yx \cos xy - \sin xy, \qquad (1)$$

$$\frac{\partial}{\partial z} \left(\frac{\partial F}{\partial x} \right) = ye^{yz}.$$

Alternate notations are used.

$$\frac{\partial}{\partial x}\left(\frac{\partial F}{\partial x}\right) = F_{11}(x, y, z),$$

$$\frac{\partial}{\partial y}\left(\frac{\partial F}{\partial x}\right) = F_{12}(x, y, z),$$

$$\frac{\partial}{\partial z}\left(\frac{\partial F}{\partial x}\right) = F_{13}(x, y, z).$$

Similarly,

$$\frac{\partial F}{\partial y} = F_2(x, y, z) = xze^{zy} - x \sin xy,$$

(2)

$$\frac{\partial}{\partial x}\left(\frac{\partial F}{\partial y}\right) = F_{21}(x, y, z) = ze^{yz} - xy \cos xy - \sin xy,$$

and so on.

EXAMPLE A

Find all second partial derivatives of

$$F(x, y) = \frac{4 - x}{1 + x + y^2}.$$

We have already found (Example 12.1A) that

$$F_1(x, y) = \frac{-5 - y^2}{(1 + x + y^2)^2}, \qquad F_2(x, y) = \frac{2xy - 8y}{(1 + x + y^2)^2}.$$

Then

$$F_{11}(x, y) = \frac{(5 + y^2)(2)}{(1 + x + y^2)^3},$$

$$F_{12}(x, y) = \frac{(1 + x + y^2)^2(-2y) + (5 + y^2)2(1 + x + y^2)(2y)}{(1 + x + y^2)^4},$$

$$F_{12}(x, y) = \frac{18y - 2xy + 2y^3}{(1 + x + y^2)^3},$$

$$F_{21}(x, y) = \frac{(1 + x + y^2)^2 2y - (2xy - 8y)2(1 + x + y^2)}{(1 + x + y^2)^4}.$$

$$F_{21}(x, y) = \frac{18y - 2xy + 2y^3}{(1 + x + y^2)^3}.$$

Similarly, compute

$$F_{22}(x, y) = \frac{(2x - 8)(1 + x - 3y^2)}{(1 + x + y^2)^3}. \quad \blacksquare$$

It is now our intention to define the second derivative of a real-valued function of several variables. To do this we need to review some properties of the first derivative.

Suppose that F, a real-valued function, is defined and has a continuous derivative on an open set U in R^n. Now $\mathscr{L}(R^n, R)$ has the same structure as does R^n. (Covectors can be added; they have length; dot products are defined, and so forth). We remind ourselves of this special structure for $\mathscr{L}(R^n, R)$ by writing

$$\mathscr{L}(R^n, R) = R^{n*}$$

For each $\mathbf{p} \in U$, we get a different covector $F'(\mathbf{p}) \in R^{n*}$. F' thus appears as a function that takes points of U into members of R^{n*}.

$$F: U \to R,$$
$$F'(\mathbf{p}): R^n \to R \quad (linearly), \tag{3}$$
$$F': U \to R^{n*}.$$

Since F' takes points of one space R^n into points of a second space R^{n*}, it may itself be differentiable at a point $\mathbf{p}_0$, according to Definition 12.1A. If it is, the resulting *linear* transformation is called $F''(\mathbf{p}_0)$, the second derivative of F at $\mathbf{p}_0$.

EXAMPLE B
The function

$$F(x, y) = \frac{4 - x}{1 + x + y^2}$$

was shown in Example 12.1A to have the derivative

$$F'(x, y) = \left[\frac{-5 - y^2}{(1 + x + y^2)^2} \quad \frac{2xy - 8y}{(1 + x + y^2)^2} \right].$$

Find $F''(2, 1)$.

As a transformation from R^2 to R^{2*}, F' is described by the coordinate functions

$$F': \quad \begin{aligned} u = F_1(x, y) &= \frac{-5 - y^2}{(1 + x + y^2)^2}, \\[2mm] v = F_2(x, y) &= \frac{2xy - 8y}{(1 + x + y^2)^2}. \end{aligned}$$

The derivative, a *linear* transformation from R^2 to R^{2*}, will then be described by the matrix

$$\begin{bmatrix} \dfrac{\partial u}{\partial x} & \dfrac{\partial u}{\partial y} \\[2ex] \dfrac{\partial v}{\partial x} & \dfrac{\partial v}{\partial y} \end{bmatrix} = \begin{bmatrix} F_{11}(x, y) & F_{12}(x, y) \\ F_{21}(x, y) & F_{22}(x, y) \end{bmatrix}.$$

Having computed all these functions in Example A, we merely substitute (2, 1) to get

$$[F''(2, 1)] = \begin{bmatrix} \frac{3}{16} & \frac{1}{4} \\ \frac{1}{4} & 0 \end{bmatrix}.$$

If $\mathbf{h} = [h_1 \quad h_2]$, then

$$F''(2, 1)\mathbf{h} = \begin{bmatrix} \frac{3}{16} & \frac{1}{4} \\ \frac{1}{4} & 0 \end{bmatrix} \begin{bmatrix} h_1 \\ h_2 \end{bmatrix} = \begin{bmatrix} \frac{3}{16}h_1 + \frac{1}{4}h_2 \\ \frac{1}{4}h_1 \end{bmatrix}. \tag{4}$$

$F''(2, 1)\mathbf{h}$ is a member of R^{2*}. It therefore makes sense to talk about $\{F''(2, 1)\mathbf{h}\}\mathbf{k}$ where $\mathbf{h}$ and $\mathbf{k}$ are arbitrary members of R^2. Since $[F''(2, 1)\mathbf{h}]$ comes out as a column vector, we evidently should write, for $\mathbf{h} = [h_1 \quad h_2]$, $\mathbf{k} = [k_1 \quad k_2]$,

$$[F''(2, 1)\mathbf{h}]\mathbf{k} = [k_1 \quad k_2] \begin{bmatrix} \frac{3}{16} & \frac{1}{4} \\ \frac{1}{4} & 0 \end{bmatrix} \begin{bmatrix} h_1 \\ h_2 \end{bmatrix}. \quad \blacksquare \tag{5}$$

When the second derivative $F''(\mathbf{p})$ exists, then according to the definition, we have for small $\mathbf{h}$,

$$F'(\mathbf{p} + \mathbf{h}) - F'(\mathbf{p}) = F''(\mathbf{p})\mathbf{h} + |\mathbf{h}| R(\mathbf{p}, \mathbf{h})$$

where $\lim_{\mathbf{h} \to 0} R(\mathbf{p}, \mathbf{h}) = 0$. We emphasize that $F'(\mathbf{p} + \mathbf{h}) - F'(\mathbf{p})$ is the difference of two covectors. If $\mathbf{k} \in R^n$, then $[F'(\mathbf{p} + \mathbf{h}) - F'(\mathbf{p})]\mathbf{k}$ is a real number. $[F''(\mathbf{p})\mathbf{h}]$, which approximates $F'(\mathbf{p} + \mathbf{h}) - F'(\mathbf{p})$, is likewise a covector. (See (4) in the example above.) $[F''(\mathbf{p})\mathbf{h}]\mathbf{k}$ is a number. Continuing the chart begun in (3),

$$F''(\mathbf{p}): R^n \to R^{n*} \qquad (linearly),$$
$$F''(\mathbf{p})\mathbf{h}: R^n \to R \qquad (linearly).$$

$[F''(\mathbf{p})\mathbf{h}]\mathbf{k}$ is thus *linear* in $\mathbf{h}$ and $\mathbf{k}$. Such a transformation is called *bilinear*. $F''(\mathbf{p})$ is a bilinear transformation. We emphasize this by writing

$$\{F''(\mathbf{p})\mathbf{h}\}\mathbf{k} = F''(\mathbf{p})(\mathbf{h}, \mathbf{k}).$$

$F''(\mathbf{p})$ maps R^n into R^{n*}. It is therefore represented by an $n \times n$ matrix. In fact, using the pattern of (5) in Example B above, we see that to compute $F''(\mathbf{p})(\mathbf{h}, \mathbf{k})$,

$$F''(\mathbf{p})(\mathbf{h}, \mathbf{k}) = [k_1 \quad k_2 \quad \cdots \quad k_n] \begin{bmatrix} F_{11}(\mathbf{p}) & F_{12}(\mathbf{p}) & \cdots & F_{1n}(\mathbf{p}) \\ \vdots & & & \vdots \\ F_{n1}(\mathbf{p}) & F_{n2}(\mathbf{p}) & \cdots & F_{nn}(\mathbf{p}) \end{bmatrix} \begin{bmatrix} h_1 \\ \vdots \\ h_n \end{bmatrix}. \tag{6}$$

We have, of course, already met bilinear transformations in Section 10.6. Readers who remembered this work recognized at line (5) that $F''(\mathbf{p})$ was bilinear, and the general form (6) comes as no surprise.

PROBLEMS

A · Suppose B is a bilinear transformation defined on R^3 by

$$B(\mathbf{h}, \mathbf{k}) = [k_1 \quad k_2 \quad k_3] \begin{bmatrix} 2 & 0 & -1 \\ 3 & 1 & 2 \\ -2 & 1 & -1 \end{bmatrix} \begin{bmatrix} h_1 \\ h_2 \\ h^3 \end{bmatrix}$$

Evaluate this for $\mathbf{h} = (1, -1, 2)$ and $\mathbf{k} = (3, 0, 4)$. For the same choices of $\mathbf{k}$ and $\mathbf{h}$, evaluate $B(\mathbf{k}, \mathbf{h})$.

B · Let $F(x, y, z) = x^2 y + xy^2 z$. Find the 3×3 matrix corresponding to $F''(1, -1, 2)$.

Problem A reminds us that for an arbitrary bilinear transformation B, it is generally the case that $B(\mathbf{h}, \mathbf{k}) \neq B(\mathbf{k}, \mathbf{h})$. It therefore seems important to note the order of $\mathbf{h}$ and $\mathbf{k}$ in writing $F''(\mathbf{p})(\mathbf{h}, \mathbf{k})$. We also recall from Section 10.6 that $B(\mathbf{h}, \mathbf{k}) = B(\mathbf{k}, \mathbf{h})$ if B is symmetric, and we note from Problem B a very happy fact. The matrix $F''(\mathbf{p})$ is symmetric. Will this always be true? Our next theorem assures us that it will be true most of the time in the problems encountered in applications.

THEOREM A

Let U be an open set in R^n, $n \geq 2$. Suppose $F: U \to R$ is twice differentiable throughout U, and that F'' is continuous on U. Then for any two members $\mathbf{h}$ and $\mathbf{k}$ of R^n,

$$F''(\mathbf{p})(\mathbf{h}, \mathbf{k}) = F''(\mathbf{p})(\mathbf{k}, \mathbf{h}).$$

The proof of this theorem is rather involved. The important application of the theorem is simple. We therefore skip the proof and try to emphasize its meaning.

As is indicated in (6), $F''(\mathbf{p})$ corresponds to an $n \times n$ matrix. The entry in the ith row, jth column is $F_{ij}(\mathbf{p})$. The entry in the jth row, ith column is $F_{ji}(\mathbf{p})$. We know that $F''(\mathbf{p})(\mathbf{h}, \mathbf{k}) = F''(\mathbf{p})(\mathbf{k}, \mathbf{h})$ if and only if the entry in the ith row, jth column equals the entry in the jth row, ith column.

We thus conclude, on the basis of Theorem A that $F_{ij}(\mathbf{p}) = F_{ji}(\mathbf{p})$. We state this important fact as a corollary.

COROLLARY

Let F satisfy the hypothesis of Theorem A above. Then

$$F_{ij}(\mathbf{p}) = F_{ji}(\mathbf{p}).$$

We have already made some computations that illustrate this corollary. Compare lines (1) and (2). Also observe that we found in Example A that $F_{21}(x, y) = F_{12}(x, y)$.

Both Theorem A and its corollary can be proved without the assumption that F'' be continuous. In fact, all we need is the continuity of F' and the existence of F'' at $(\mathbf{p})$ [3, page 175].

For real-valued functions of one variable that were n times differentiable, we defined a Taylor polynomial of degree n. This same kind of development can be worked out for functions of n variables. In particular, the Taylor polynominals of degrees 1 and 2 at a point $\mathbf{p_0}$ are:

$$T_{F(\mathbf{p_0}),1}(\mathbf{p}) = F(\mathbf{p_0}) + F'(\mathbf{p_0})(\mathbf{p} - \mathbf{p_0}),$$

$$T_{F(\mathbf{p_0}),2}(\mathbf{p}) = F(\mathbf{p_0}) + F'(\mathbf{p_0})(\mathbf{p} - \mathbf{p_0}) + \tfrac{1}{2}F''(\mathbf{p_0})(\mathbf{p} - \mathbf{p_0}, \mathbf{p} - \mathbf{p_0}).$$

EXAMPLE C

Write the Taylor polynomial of degree 2 at $\mathbf{p_0} = (2, 1)$ for the function

$$F(x, y) = \frac{4 - x}{1 + x + y^2}.$$

Use this polynomial to approximate $F(x, y)$ at the points $(1.9, 0.9)$, $(1.9, 1.1)$, $(2.1, 0.9)$, $(2.1, 1.1)$.

We see that this is a continuation of Example 12.1A. We hope for better approximations than we obtained by using the Taylor polynomial of degree 1.

$$F(2, 1) = \tfrac{1}{2}.$$

Set $\mathbf{p} = (x, y)$, $\mathbf{p_0} = (2, 1)$ so that $\mathbf{p} - \mathbf{p_0} = (x - 2, y - 1)$. From Example 12.1A we have

$$F'(2, 1)(\mathbf{p} - \mathbf{p_0}) = [-\tfrac{3}{8} \quad -\tfrac{1}{4}]\begin{bmatrix} x - 2 \\ y - 1 \end{bmatrix},$$

$$F'(2, 1)(\mathbf{p} - \mathbf{p_0}) = -\tfrac{3}{8}(x - 2) - \tfrac{1}{4}(y - 1).$$

The matrix corresponding to $F''(2, 1)$ was computed in Example B. According to (5),

$$F''(2, 1)(\mathbf{p} - \mathbf{p}_0, \mathbf{p} - \mathbf{p}_0) = [x - 2 \quad y - 1]\begin{bmatrix} \frac{3}{16} & \frac{1}{4} \\ \frac{1}{4} & 0 \end{bmatrix}\begin{bmatrix} x - 2 \\ y - 1 \end{bmatrix},$$

$$\tfrac{1}{2}F''(2, 1)(\mathbf{p} - \mathbf{p}_0, \mathbf{p} - \mathbf{p}_0) = \tfrac{3}{32}(x - 2)^2 + \tfrac{1}{4}(x - 2)(y - 1).$$

The Taylor polynomial of degree 2 is therefore

$$T_{F(\mathbf{p}_0),2}(\mathbf{p}) = \tfrac{1}{2} - \tfrac{3}{8}(x - 2) - \tfrac{1}{4}(y - 1) + \tfrac{3}{32}(x - 2)^2 + \tfrac{1}{4}(x - 2)(y - 1).$$

This expression may be used to compute values that approximate the values of F in a neighborhood of $(2, 1)$. In this way, we add a row to the table of values computed in Example 12.1A. For convenience in comparison, we reproduce the entire table.

(x, y)	$(1.9, 0.9)$	$(1.9, 1.1)$	$(2.1, 0.9)$	$(2.1, 1.1)$
$T_1(x, y)$	0.5625	0.5125	0.4875	0.4375
$T_2(x, y)$	0.5659	0.5109	0.4859	0.4409
$F(x, y)$	0.5660	0.5109	0.4859	0.4408

For real-valued functions of one variable, Taylor's theorem provided an estimate of the error made in using a Taylor polynomial to estimate the value of a function at a given point. Such a result can also be proved for functions of several variables. (See [3, pages 185–186] and [4, pages 91–95].) This development naturally requires a study of higher derivatives, a study we will not pursue here. Example C stands as an illustration of what can be proved. The Taylor polynomial of degree 2 in many cases affords a better approximation to the value of F at a point than does the Taylor polynomial of degree 1.

PROBLEM

C · Let $F(x, y, z) = x^2y + xy^2z$. Set $\mathbf{p}_0 = (1, -1, 2)$. Write the Taylor polynomials $T_{F(\mathbf{p}_0),1}$ and $T_{F(\mathbf{p}_0),2}$. Use each of them to approximate the value of F at $\mathbf{p} = (1.13, -0.92, 1.88)$. Then find the value of F at $\mathbf{p}$. (You have done some of the necessary work in Problem B.)

The final topic that we shall take up in this section has to do with finding relative maxima or minima for functions of several variables. We recall the first step in the solution of such problems for functions of a single variable. It was to find the points at which the first derivative was zero. The idea behind this was that if a function had a relative maximum or minimum, then the slope of its tangent line at that point would be zero.

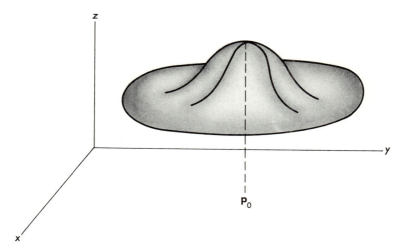

FIG. 12.6

Our geometric intuition leads us to the same kind of conclusion in the case of a function of two variables. Figure 12.6 indicates the graph of a function $z = F(x, y)$ having a relative maximum at $\mathbf{p}_0 = (x_0, y_0)$. In this case the tangent plane at $\mathbf{p}_0$ should be parallel to the xy plane. Now this tangent plane is perpendicular to the gradient vector of

$$G(x, y, z) = F(x, y) - z$$

at $\mathbf{p}_0$. That is, the tangent plane is perpendicular to

$$[F_1(x_0, y_0) \quad F_2(x_0, y_0) \quad -1].$$

Therefore, it is parallel to the xy plane if and only if $F_1(x_0, y_0) = F_2(x_0, y_0) = 0$. This means $F'(x_0, y_0) = \mathbf{0}$ and we are led to the following definition in analogy with the case of a single variable.

DEFINITION A

Let U be an open set in R^n, and suppose $F: U \rightarrow R$ is differentiable in U. The critical points of F are those points for which $F'(\mathbf{p}) = 0$.

It is useful to know that when F is differentiable, any relative minimum or maximum that occurs at a point interior to the domain is always a critical point (proof in Problem D).

DEFINITION B

Suppose $F: U \to R$ in an open set U of R^n. F is said to have a local maximum at $\mathbf{p}_0 \in U$ if there is a neighborhood N of $\mathbf{p}_0$, N contained in U, so that if $\mathbf{p} \in N$, then $F(\mathbf{p}) \leqq F(\mathbf{p}_0)$.

PROBLEMS

D · Suppose F is differentiable in U and that F has a local maximum at $\mathbf{p}_0 \in U$. Show that $\mathbf{p}_0$ is a critical point. Hint: Pick an arbitrary unit vector $\mathbf{h}$. For any real t,

$$F(\mathbf{p}_0 + t\mathbf{h}) - F(\mathbf{p}_0) = tF'(\mathbf{p}_0)(\mathbf{h}) + |t| R(\mathbf{p}_0, t\mathbf{h}).$$

Show $F'(\mathbf{p}_0)\mathbf{h} = 0$ by an argument similar to the one in Problem 3.1B. Then use Problem 10.3A.

E · Define a relative minimum for $F: U \to R$. Show that if F has a relative minimum at $\mathbf{p}_0$ in an open set in which F is differentiable, then $\mathbf{p}_0$ is a critical point.

We need to remember that a critical point may be a local maximum, a local minimum, or neither. For functions of a single variable, the classification of critical points was facilitated by a study of the second derivative. A similar situation prevails for several variables.

We have already seen that for $\mathbf{p}$ close to $\mathbf{p}_0$, the Taylor polynomial of degree two gives us the approximation

$$F(\mathbf{p}) \approx F(\mathbf{p}_0) + F'(\mathbf{p}_0)(\mathbf{p} - \mathbf{p}_0) + \tfrac{1}{2}F''(\mathbf{p}_0)(\mathbf{p} - \mathbf{p}_0, \mathbf{p} - \mathbf{p}_0).$$

Suppose that $\mathbf{p}_0$ is a critical point, so that $F'(\mathbf{p}_0) = \mathcal{O}$. Then setting $\mathbf{h} = \mathbf{p} - \mathbf{p}_0$, the above approximation becomes

$$F(\mathbf{p}_0 + \mathbf{h}) - F(\mathbf{p}_0) \approx \tfrac{1}{2}F''(\mathbf{p}_0)(\mathbf{h}, \mathbf{h}).$$

(See Problem 10.7A, remembering that $F''(\mathbf{p}_0)(\mathbf{h}, \mathbf{h}) = \mathbf{h}^t[F''(\mathbf{p}_0)]\mathbf{h}$.) If the matrix $[F''(\mathbf{p}_0)]$ is positive definite the right side, hence the left side, is positive, meaning that $\mathbf{p}_0$ is a local minimum.

Critical points at which the matrix $[F''(\mathbf{p}_0)]$ is negative are, by a similar argument, local maximums. Finally, as already anticipated in the one variable case, no conclusion can be drawn if $F''(\mathbf{p}_0) = \mathcal{O}$.

We recall that even for the one variable case, use of the second derivative was sometimes more work than the situation justified. That is more likely to be the case where several variables are concerned. Anytime a critical point is to be classified, several techniques should be considered;

(a) Hope that the problem giving rise to the function makes the classification of a critical point "obvious."

(b) Try to determine for small $\mathbf{h}$ the sign of $F(\mathbf{p}_0 + \mathbf{h}) - F(\mathbf{p}_0)$ if $\mathbf{p}_0$ is a critical point.

(c) Investigate the characteristic values of the matrix $[F''(\mathbf{p}_0)]$. If they are all positive, the critical point $\mathbf{p}_0$ is a local minimum. If they are all negative, the critical point $\mathbf{p}_0$ is a local maximum. In all other cases, further analysis is indicated.

EXAMPLE D
Find the point on the plane $x - y + 2z = 6$ that is closest to the origin.

We seek the point (x, y, z) on the plane for which

$$F(x, y, z) = x^2 + y^2 + z^2$$

is a minimum. Since $x = y - 2z + 6$, this is equivalent to seeking the minimum of the function

$$G(y, z) = (y - 2z + 6)^2 + y^2 + z^2.$$

The minimum is a critical point. Hence we find

$$G_1(y, z) = 2(y - 2z + 6) + 2y,$$
$$G_2(y, z) = 2(y - 2z + 6)(-2) + 2z.$$

Setting these two equations equal to zero and solving the resulting system of two equations in two unknowns, we get $(y, z) = (-1, 2)$. Then $x = 1$. It is geometrically evident that the point we have found, $(1, -1, 2)$, must be a minimum. ∎

PROBLEM

F · Use the second derivative test to show that the function $G(y, z)$ of Example D above has $(-1, 2)$ as a local minimum.

In our next example we shall derive a practical procedure used in curve fitting. It is common for students (and research workers on all levels) in a laboratory to obtain experimentally a certain number of points in R^2 that are supposed to fall in a straight line. It is also common for such points to fall into a pattern that only approximates a straight line. If this happens, and if we are trying to find the equation of the line from the points, how should we proceed? (We assume the circumstances are such as to prevent us from employing the time-honored custom of moving a few points.)

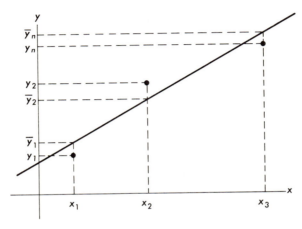

FIG. 12.7

Let the points be $(x_1, y_1), (x_2, y_2), \ldots, (x_n, y_n)$. The straight line we draw will have the equation $y = mx + b$. Corresponding to each x_i, we then have two values of y_i. We have y_i and we have $\bar{y}_i = mx_i + b$ (see Fig. 12.7). Set

$$d_i = y_i - \bar{y}_i.$$

If the line is a perfect fit (that is, passes through each (x_i, y_i)), then of course each d_i is zero. In some sense,

$$M = d_1 + d_2 + \cdots + d_n \tag{7}$$

gives a measure of the "fit" of the line to the points.

PROBLEM

G • Let $(2, 7), (7, 12),$ and $(12, 15)$ be given. Find M as given by (7) for each of the lines:

$$\text{(a) } y = 3x - 10; \quad \text{(b) } y = x + 5.$$

Draw a picture showing the points and the two lines.

The problem illustrates the difficulty with using M as a measure of the "goodness of fit." Large positive values of some d_i are offset by negative values of others. One way to get around this difficulty is to square each of the d_i's. Then differences in signs do not make "bad fits" look good. This procedure of fitting a line to some observed points is called the *method of least squares*. We try to minimize

$$S = d_1{}^2 + d_2{}^2 + \cdots + d_n{}^2.$$

EXAMPLE E
Find the line $y = mx + b$ that best fits the points of Problem G above in the sense of least squares.

$$d_1{}^2 = (2m + b - 7)^2;$$
$$d_2{}^2 = (7m + b - 12)^2;$$
$$d_3{}^2 = (12m + b - 15)^2.$$

S is a function of the two variables m and b.

$$S(m, b) = (2m + b - 7)^2 + (7m + b - 12)^2 + (12m + b - 15)^2.$$

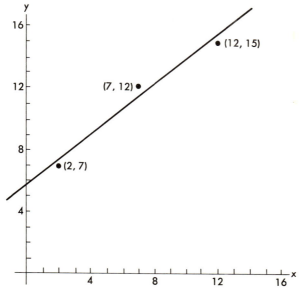

FIG. 12.8

We wish to choose m and b so as to minimize $S(m, b)$.

$$S_1(m, b) = 2(2m + b - 7)(2) + 2(7m + b - 12)(7) + 2(12m + b - 15)(12),$$
$$= 2[197m + 21b - 278];$$

$$S_2(m, b) = 2(2m + b - 7) + 2(7m + b - 12) + 12) + 2(12m + b - 15),$$
$$= 2[21m + 3b - 34].$$

Setting these partial derivatives equal to zero (to find the critical points) and solving the resulting set of two equations in two unknowns, we get $(m, b) = (\frac{4}{5}, \frac{86}{15})$. See Fig. 12.8, where we have graphed

$$y = \tfrac{4}{5}x + \tfrac{86}{15}. \quad \blacksquare$$

12·3 EXERCISES

In Exercises 1–8, write the Taylor polynomial $T_{F(P_0),2}$ at the indicated point.

1. $F(x, y) = x^2y + xy^2$ $\mathbf{p}_0 = (3, -2)$.

2. $F(x, y) = (x^2 + y^2)^{3/2}$ $\mathbf{p}_0 = (-3, 4)$.

3. $F(x, y, z) = xz - 3xy$ $\mathbf{p}_0 = (2, 5, 1)$.

4. $F(x, y, z) = 3xy + 3yz$ $\mathbf{p}_0 = (-1, 3, 0)$.

5. $F(x, y, z) = z \sin xy$ $\mathbf{p}_0 = (\pi, 1, -1)$.

6. $F(x, y, z) = z^2 e^{xy}$ $\mathbf{p}_0 = (2, 0, 3)$.

7. $F(x, y) = \dfrac{3x + 4y}{x^2 + 1}$ $\mathbf{p}_0 = (-4, 3)$.

8. $F(x, y) = \dfrac{2x - 3y}{y^2 + 4}$ $\mathbf{p}_0 = (3, 2)$.

9. Use the second-degree Taylor polynomial corresponding to $F(x, y, z) = (x^2 + y^2 + z^2)^{1/2}$ to approximate $\sqrt{(0.893)^2 + (2.135)^2 + (1.932)^2}$.

10. Use the second-degree Taylor polynomial corresponding to $F(x, y) = \sqrt{25x^2 + 9y^2}$ to approximate $\sqrt{25(0.87)^2 + 9(4.13)^2}$.

In Exercises 11–20, locate the critical points and classify them whenever you can.

11. $z = (x + 3)^2 + (y - 2)^2$.

12. $z = (x + 1)^2 - (y - 1)^2$.

13. $z = y^3 - 6x^2y + x^3$.

14. $z = x^2 + y^2 - 3xy$.

15. $z = xye^{x^2 + y^2}$.

16. $z = (x^2 + y^2)e^{-xy}$.

17. $z = x^3 + y^3 - 12xy + 15$.

18. $z = x^4 - y^4 + 4xy - 2y^2$.

19. $w = xy^2 + 2yz - xz^2 + y$.

20. $w = xz^2 + 3xy - x + y^2z^2$.

21 · Find the maximum value of $F(x, y) = xy - x^2 + 10$ for (x, y) in the rectangular set described by $x \in [0, 4]$, $y \in [0, 3]$.

22 · Find the maximum value of $F(x, y) = xy - 2y^2 + 8$ for (x, y) in the rectangular set described by $x \in [0, 4]$, $y \in [0, 2]$.

23 · Find the maximum value of $F(x, y) = (4 + xy - x)/(1 + y^2)$ for (x, y) in the rectangular set described by $x \in [0, 6]$, $y \in [0, 3]$.

24 · Find the maximum value of $F(x, y) = (6 + 2xy - y)/(3 + x^2)$ for (x, y) in the rectangular set described by $x \in [0, 3]$, $y \in [0, 6]$.

25 · Find the point(s) on the curve $z = 1/(xy^2)$ that is closest to the origin.

26 · Find the point(s) on the curve $z = 4/(1 + x^2 + y^2)^{1/2}$ that is closest to the origin.

In Exercises 27–30, find the line that best fits the given points in the sense of the least squares method. Draw the graph in each case.

27 · $(-3, -1)$, $(2, 1)$, $(5, 3)$.

28 · $(2, -2)$, $(1, -1)$, $(5, -3)$.

29 · $(-3, 1)$, $(3, 2)$, $(6, 4)$, $(9, 5)$.

30 · $(-3, 3)$, $(1, -1)$, $(5, -2)$, $(8, -3)$.

12 · 4   More About Extreme Values of Functions

Let $F(x, y, z) = 0$ describe a "mountain-like" surface in R^3 (Fig. 12.9) and let **S** be a "trail" over the mountain. More technically, **S** is a curve in R^3 that might be described by $\mathbf{S}(t) = (x(t), y(t), z(t))$. Furthermore, to say that the trail is on the mountain means that $F(\mathbf{S}(t)) = 0$.

Suppose we wish to find the highest point on the trail. This is equivalent to maximizing $h(t) = F(\mathbf{S}(t))$, so we find

$$h'(t) = F'(\mathbf{S}(t)) \cdot \mathbf{S}'(t).$$

Now $F'(\mathbf{S}(t))$ is a vector, as is $\mathbf{S}'(t)$. To require that $h'(t) = 0$ is to demand that $F'(\mathbf{S}(t))$ and $\mathbf{S}'(t)$ be orthogonal. Rephrased, the vector $\mathbf{S}'(t)$ tangent to the trail and the vector $F'(\mathbf{S}(t))$ must be orthogonal at the highest point on the trail. As usual, the vanishing of $h'(t)$ is only a necessary condition for determining high (or low) points, but we can ordinarily rely on the circumstances of the problem to aid us in interpreting results.

A more troublesome matter is likely to be the form in which information about the trail is given. A curve in space is often described as the intersection of two surfaces. Thus, the intersection of the plane $y + z = 6$ and the cylinder $x^2 + (y - 3)^2 = 1$ describes a curve (Fig. 12.10). Our problem then takes the

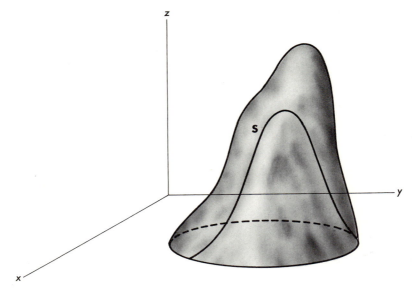

FIG. 12.9

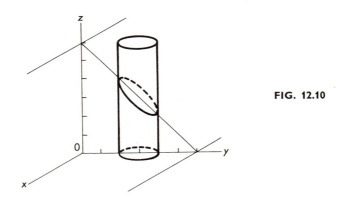

FIG. 12.10

form of finding the extreme points for a function $F(x, y, z)$ with the so-called *side conditions* that (x, y, z) be on the curve determined by the surfaces $g_1(x, y, z) = 0$ and $g_2(x, y, z) = 0$.

The gradient vectors $g_1'(x, y, z)$ and $g_2'(x, y, z)$, being perpendicular to the surfaces determined by g_1 and g_2, must be orthogonal to the vector $\mathbf{S}'(t)$ tangent to their curve of intersection (Fig. 12.11). Thus $\mathbf{S}'(t)$ is orthogonal to the plane P determined by $g_1'(x, y, z)$ and $g_2'(x, y, z)$. It is clear from our work above, however, that $F'(x, y, z)$ is also orthogonal to $\mathbf{S}'(t)$, hence that it must also be in the plane P. This means that at an extreme point, there are real numbers λ_1 and λ_2 so that

$$F'(x, y, z) = \lambda_1 g_1'(x, y, z) + \lambda_2 g_2'(x, y, z).$$

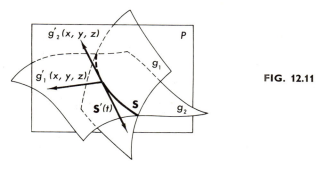

FIG. 12.11

In summary, we see that the extreme points of $F(x, y, z)$, subject to the side conditions $g_1(x, y, z) = 0$ and $g_2(x, y, z) = 0$ are obtained by solving simultaneously the equations

$$g_1(x, y, z) = 0$$
$$g_2(x, y, z) = 0,$$
$$F'(x, y, z) = \lambda_1 g_1'(x, y, z) + \lambda_2 g_2'(x, y, z).$$

Note that the last equation, when written out coordinate-wise, gives us three equations; hence, we have five equations in five unknowns λ_1, λ_2, x, y, z.

The procedure just outlined will fail if $g_1'(x, y, z)$ and $g_2'(x, y, z)$ are colinear, thereby not determining a plane P. It may also fail because of the difficulty of solving the simultaneous equations. On the other hand, the method does work often enough to have a name associated with it. The variables λ_1 and λ_2 introduced in solution are called *Lagrange multipliers*

EXAMPLE A

Find the point on the curve $y + z = 6$, $x^2 + (y - 3)^2 = 1$ that minimizes the distance to the origin (Fig. 12.10).

Since we may minimize the distance to the origin by minimizing its square, we shall take $F(x, y, z) = x^2 + y^2 + z^2$. The simultaneous equations are

$$y + z - 6 = 0,$$
$$x^2 + (y - 3)^2 - 1 = 0,$$
$$[2x \quad 2y \quad 2z] = \lambda_1[0 \quad 1 \quad 1] + \lambda_2[2x \quad 2(y - 3) \quad 0].$$

Writing out the coordinate equations from the last expression,

$$2x = \lambda_2(2x), \qquad\qquad \text{so} \quad \lambda_2 = 1,$$
$$2y = \lambda_1 + 2\lambda_2 y - 6\lambda_2, \qquad \text{so} \quad \lambda_1 = 6,$$
$$2z = \lambda_1, \qquad\qquad\qquad \text{so} \quad z = 3,$$

From the first equation it follows that $y = 3$, and finally we see that $x = \pm 1$. ∎

The discussion and example of this section should be taken as illustrative of a general procedure that can be used in similar situations. We emphasize this by using Lagrange multipliers to solve once again a problem first given as Example 12.3D.

EXAMPLE B
Find the point on the plane $x - y + 2z = 6$ that is closest to the origin.

As in the previous example, we shall minimize the square of the distance by choosing $F(x, y, z) = x^2 + y^2 + z^2$. Our equations are

$$x - y + 2z - 6 = 0,$$
$$[2x \quad 2y \quad 2z] = \lambda[1 \quad -1 \quad 2].$$

It follows from the vector equation that

$$2x = \lambda, \qquad x = \lambda/2,$$
$$2y = -\lambda, \qquad y = -\lambda/2,$$
$$2z = 2\lambda, \qquad z = \lambda.$$

Substitution in the first equation gives us $\lambda = 2$, hence a solution $(x, y, z) = (1, -1, 2)$. ∎

== 12·4 EXERCISES

In Exercises 1–6, find the largest and smallest values assumed by the given function when (x, y) is restricted to lie on the unit circle $x^2 + y^2 = 1$.

1. $F(x, y) = xy^2$.

2. $F(x, y) = 3x - 4y + 6$.

3. $F(x, y) = (x^2 + y^2)^{3/2}$.

4. $F(x, y) = \sqrt{x^2 + y^2}$.

5. $F(x, y) = e^{xy^2}$.

6. $F(x, y) = e^{3x - 4y + 6}$.

In Exercises 7–10, find the point in R^3 on the given curve that is closest to the origin.

7. $x + y + z = 6.$

8. $x - y + 2z = 4.$

9. $\dfrac{x - 3}{4} = \dfrac{y + 1}{2} = \dfrac{z - 1}{-2}.$

10. $\dfrac{x - 2}{5} = \dfrac{y - 1}{4} = \dfrac{z}{-2}.$

Find the shape of the rectangular box of greatest volume given the following constraints.

11 · The surface area is fixed.

12 · The total length of the edges is fixed.

13 · The diagonal has a fixed length.

14 · The area of one side is fixed.

CHAPTER

13

INTEGRATION OF FUNCTIONS OF SEVERAL VARIABLES

We have already defined the integral of a function $F: [a, b] \to R$. It is now our purpose to define the integral of a function $F: U \to R$ where $U \subset R^n$. We may proceed in a manner that parallels our work in the case of a function of one variable. There are new difficulties that beset us in the case of a function of several variables, however, and they are not easily set aside. To honestly face the questions we meet, and to provide adequate answers requires a small book. Several such books have been written.†

We will be honest about the questions, but it is beyond the scope of this course to see how they are answered. We restrict ourselves to gaining some familiarity with the way things work out after all the questions have been resolved.

13 · 1   The Integral on a Rectangle

New difficulties in the context of functions of several variables have been promised. A principal source of our problems may be identified quickly. Recall that the definition in the case of a single variable depended on forming an arbitrary partition of $[a, b]$. The resulting subintervals were used as bases of rectangles, the areas of which were added together to form Riemann sums. The thing to which we call attention here is that the partition always covered the original interval exactly.

Now consider the situation for a function of two variables defined in a region U of the plane. The natural generalization of a rectangle erected over a subinterval of $[a \quad b]$ is a "column" erected over a rectangular subset of U (Fig. 13.1). Here is the bind, however. An arbitrary region U cannot generally be partitioned into an exact finite number of rectangles.

We shall say more about these difficulties and the means of dealing with them in the next section. In the remainder of this section we avoid the problem by considering only those regions U that are rectangles. Then we can partition U into a finite number of subrectangles.

A rectangle in R^n is defined to be a set

$$I = \{(x_1, x_2, \ldots, x_n) : a_i \leq x_i \leq b_i\}.$$

---

† See W. W. Rogosinski, "Volume and Integral," Wiley (Interscience), 1962. This book is written from the point of view adopted in our text. It begins with intuitive ideas of volume and uses them to develop the Riemann integral. Deficiencies are then pointed out, and a variation of the original intuitive ideas is used to develop the Lebesque integral. Proofs for the assertions made in this chapter may be obtained from Rogosinski's book.

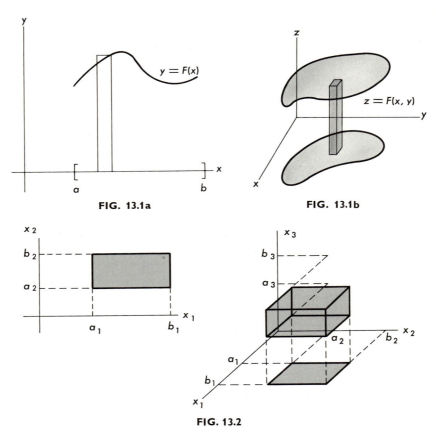

FIG. 13.1a   FIG. 13.1b

FIG. 13.2

When $n = 2$ or 3, the rectangle is just what we would expect (Fig. 13.2). All of the discussion of this section should in fact be checked against the image we have for $n = 2$ or 3. The volume of the rectangle I is defined to be

$$v(I) = (b_1 - a_1)(b_2 - a_2) \cdots (b_n - a_n).$$

The diameter of the rectangle is defined to be

$$d(I) = \sqrt{(b_1 - a_1)^2 + (b_2 - a_2)^2 + \cdots + (b_n - a_n)^2}.$$

Arbitrary hyperplanes (remember that in R^2, a line is the analog of a plane) passed parallel to the coordinate axes clearly partition a rectangle into non-overlapping rectangles, the union of which covers the original rectangle. These subrectangles constitute a partition of the original rectangle. The mesh of a partition P, denoted by $\delta(P)$, is the maximum of the diameters of the sub-rectangles.

We are now in a position to parallel our development of the integral in Chapter 6 in the case of a function F defined on a rectangle I in R^n. The next two definitions should be compared to Definitions 6.2A and 6.2B.

DEFINITION A   (Riemann Sums $S(F, P, \{p_k\})$)

Let P be a partition of I into m subrectangles $I_1, I_2, \ldots, I_m$. In each subrectangle I_k choose a point $\mathbf{p}_k$. Then form the Riemann sum

$$S(F, P, \{\mathbf{p}_k\}) = \sum_{k=1}^{m} F(\mathbf{p}_k) v(I_k).$$

DEFINITION B   (Riemann Integral of F on a Rectangle I)

Suppose we are given an arbitrary neighborhood N of some number r and that we are then able to guarantee that for any partition P with a small enough mesh (we specify how small, depending on the neighborhood of r), the number $S(F, P, \{\mathbf{p}_k\})$ will fall in the neighborhood N, no matter how the $\mathbf{p}_k$ are chosen in the I_k. Then the function F is said to be Riemann integrable on I, and the number r is called the *Riemann integral of F on I*. We write

$$r = \int_I F \, dA$$

More will be said later about the inclusion of the symbol dA in the expression at the right. For now, let it serve as a reminder that the sums were formed by using the function F and partitions of a rectangle I.

This puts us once again in the position of not being able, from the definition alone, to know whether a given function is integrable. (We shall understand the "Riemann" without repeating it each time.) This is because we could never try all possible partitions. Our deliverance is also parallel to a result in Section 6.2 (Theorem A).

THEOREM A

If F is continuous in the rectangle I, then it is integrable on I. See [6, page 90].

Now if we know that a function F is continuous on a rectangle I, any sequence of partitions with mesh tending to zero may be used to obtain $\int_I F$. The following example shows that the integral can be computed in this way. It also prohibits anyone from thinking that this is a good way.

EXAMPLE A

Let I be the rectangle in the plane described by

$$I = \{(x, y): \quad 1 \le x \le 2, \quad 1 \le y \le 2\}.$$

Let $z = F(x, y) = x + y$. Find $\int F$. (See Fig. 13.3)

We shall take our first partition of I to be the one that divides it into 4 equal squares; the second will divide it into 16 equal squares (each with side $1/2^2$); the nth will divide it into 4^n equal squares, each with side $1/2^n$. In each square, $\mathbf{p}_k$ will be the corner point nearest the origin.

$$S_1(F, P_1, \{\mathbf{p}_k\}) = \left\{ \left[F(1, 1) + F\left(1, \frac{2+1}{2}\right) \right] \right.$$

$$\vdots \quad + \left. \left[F\left(\frac{2+1}{2}, 1\right) + F\left(\frac{2+1}{2}, \frac{2+1}{2}\right) \right] \right\} \left(\frac{1}{2^1}\right)^2,$$

$$S_n(F, P_n, \{\mathbf{p}_k\}) = \sum_{j=0}^{2n-1} \sum_{i=0}^{2n-1} F\left(\frac{2^n + j}{2^n}, \frac{2^n + i}{2^n}\right) \left(\frac{1}{2^n}\right)^2,$$

$$= \sum_{j=0}^{2n-1} \sum_{i=0}^{2n-1} \frac{2^{n+1} + j + i}{2^n} \cdot \frac{1}{2^{2n}}. \tag{1}$$

We now use some of the things we learned about sigma notation in Section 6.1. Note that in working with the inner sigma sign first, j is treated as a constant.

$$S_n(F, P_n, \{\mathbf{p}_k\}) = \sum_{j=0}^{2n-1} \frac{1}{2^{3n}} \left\{ [2^{n+1} + j]2^n + \frac{2^n(2^n - 1)}{2} \right\},$$

$$= \frac{1}{2^{3n}} \left\{ \left[2^{2n+1} + \frac{2^n(2^n - 1)}{2} \right] 2^n + 2^n \frac{2^n(2^n - 1)}{2} \right\},$$

$$= 2 + 1 - \frac{1}{2^n} = 3 - \frac{1}{2^n}.$$

(As in Chapter 6, doubters are encouraged to compute $S_1(F, P_1, \{\mathbf{p}_k\})$ by using (1), and then by using $n = 1$ in the formula just derived.) Therefore

$$\lim_{n \to \infty} S_n(F, P_n, \{\mathbf{p}_k\}) = 3.$$

$$\int_I (x + y) \, dA = 3. \quad \blacksquare$$

In the case of a function of a single variable, we were rescued from such computations by the fundamental theorem of calculus. Here we are rescued

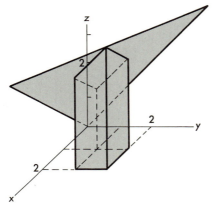

FIG. 13.3

by a theorem that, in a more sophisticated setting, is known as *Fubini's theorem*. Before we can state this extremely useful theorem, we need to explain what we mean by an iterated integral.

Let F be a function of n variables. If we "fix" all but the kth variable, then we may define

$$G(x_k) = F(x_1, \ldots, x_k, \ldots, x_n).$$

G may be integrable for $a_k \leq x_k \leq b_k$. Suppose it is. Then we write

$$\int_{a_k}^{b_k} G(x_k) \, dx_k = \int_{a_k}^{b_k} F(x_1, \ldots, x_k, \ldots, x_n) \, dx_k.$$

(Here, by the way, is another service performed by the dx_k inside the integral. It serves as a signpost, telling us which one of the variables is not being treated as a constant.)

EXAMPLE B

Let $F(x, y, z) = x^2 y + z \sin y - xz$.

Then

$$\int_0^{\pi} (x^2 y + z \sin y - xz) \, dy = \frac{x^2 y^2}{2} - z \cos y - xyz \bigg|_{y=0}^{y=\pi},$$

$$= \left(\frac{\pi^2 x^2}{2} + z - \pi xz \right) - (-z),$$

$$= \frac{\pi^2}{2} x^2 - \pi xz + 2z. \quad \blacksquare$$

As seen clearly from the example, the result of this "partial integration" is a function of the other $n - 1$ variables. We may in this function fix all the variables but one, say x_j, and then integrate again. This is indicated by writing

$$\int_{a_j}^{b_j} \left[\int_{a_k}^{b_k} F(x_1, x_2, \ldots, x_n) \, dx_k \right] dx_j. \tag{2}$$

The process of "partial integration" is repeated, first with respect to one variable, then another. This can be repeated n times. An expression of the form (2) is called an *iterated integral*. The next example illustrates the calculation of an iterated integral; and the result, when compared to Example A, hints at the content of Fubini's theorem.

EXAMPLE C

Evaluate the iterated integral $\int_1^2 \int_1^2 (x + y) \, dx \, dy$.

$$\int_1^2 \int_1^2 (x + y) \, dx \, dy = \int_1^2 \left(\frac{x^2}{2} + xy \right) \Big|_{x=1}^{x=2} dy,$$

$$= \int_1^2 \left[(2 + 2y) - \left(\frac{1}{2} + y \right) \right] dy = \int_1^2 \left(\frac{3}{2} + y \right) dy,$$

$$= \left(\frac{3}{2} y + \frac{y^2}{2} \right) \Big|_1^2 = 3 + 2 - \left(\frac{3}{2} + \frac{1}{2} \right) = 3. \; \blacksquare$$

THEOREM B

Let F be continuous on the rectangle $I = \{(x_1, \ldots, x_n) : a_i \le x_i \le b_i\}$. Then the integral of F over I, which we know exists, may be evaluated as an iterated integral.

$$\int_I F \, dA = \int_{a_n}^{b_n} \int_{a_{n-1}}^{b_{n-1}} \ldots \int_{a_1}^{b_1} F(x_1, \ldots, x_n) \, dx_1 \ldots dx_{n-1} \, dx_n.$$

Moreover, the order in which the integrations are performed is not important so long as the appropriate limits are used.

The last statement of the theorem means, for example, that

$$\int_1^2 \int_0^3 xy^2 \, dx \, dy = \int_0^3 \int_1^2 xy^2 \, dy \, dx.$$

This theorem is proved in exactly this form (for the case $n = 2$, but easily extended to arbitrary n) in [2, page 111, Theorem 8]. The theorem verifies that the computations of Example C constitute a valid procedure for evaluating the integral first encountered in Example A.

Figures 13.1 and 13.3 suggest a geometric interpretation for the integral $\int_I F(x, y)\, dA$. This number apparently may be taken as the volume of the solid having I as its floor and the surface $z = F(x, y)$ as its "roof." We will see that (as was the case in studying functions of one variable) an integral may have a variety of interpretations. Volume is one interpretation. We observe here that if $F(x, y) = 1$ for all $(x, y) \in I$, then $\int_I F\, dA$ gives the volume of a rectangle 1 unit high. This is, of course, numerically the same as the area of the rectangle I in square units and accounts to some extent for our choice of dA as a symbol.

13 · 1   EXERCISES

Verify the indicated equality by computing each of the iterated integrals in Exercises 1–4.

1. $\displaystyle \int_1^2 \int_0^1 xy\, dx\, dy = \int_0^1 \int_1^2 xy\, dy\, dx.$

2. $\displaystyle \int_0^2 \int_0^1 (x^2 y + 2x)\, dx\, dy = \int_0^1 \int_0^2 (x^2 y + 2x)\, dy\, dx.$

3. $\displaystyle \int_0^{\pi/3} \int_0^1 (e^x + x \sin y)\, dx\, dy = \int_0^1 \int_0^{\pi/3} (e^x + x \sin y)\, dy\, dx.$

4. $\displaystyle \int_0^2 \int_0^1 xye^x\, dx\, dy = \int_0^1 \int_0^2 xye^x\, dy\, dx.$

Compute the following integrals.

5. $\displaystyle \int_0^1 \int_1^2 xy^2 e^{xy}\, dy\, dx.$

6. $\displaystyle \int_0^1 \int_0^{\pi/3} 2xy \cos xy^2\, dx\, dy.$

7. $\displaystyle \int_1^2 \int_0^{\pi/3} \frac{x}{y^2} \cos \frac{x}{y}\, dx\, dy.$

8. $\displaystyle \int_0^1 \int_1^2 \frac{2y}{x + y^2}\, dy\, dx.$

13 · 2   The Integral on a Nice Set

We started Section 13.1 with the grand objective of defining the integral of a function $F: U \to R$ where U was a region of R^n. We quickly restricted ourselves, however, to the case in which U was a rectangle. We cited as the cause of difficulty the fact that an arbitrary region U cannot be partitioned into a finite number of rectangles that cover U.

There are a variety of ways of attacking the general region. All of them eventually come down to this. The success of attaching meaning to $\int_U F \, dA$ depends on whether U and its boundary are in some sense *nice*. To be specific about the meaning of nice would require a lot more work than we are inclined to embark on at this time. For this reason, we shall leave " nice set " intentionally vague.

We shall be primarily concerned with functions of either two or three variables. In this context the student will not go far wrong if he pictures nice regions as those bounded by segments of smooth curves in the plane and by portions of smooth surfaces in space. Beyond this general description, the reader may relax in the assurance that the examples, problems, and applications with which we concern ourselves here all meet the obscure requirement of being *nice*.

Suppose U is a region of the plane bounded by two curves described by $y = H_1(x)$ and $y = H_2(x)$ that intersect at $x = a$ and $x = b$ (Fig. 13.4). Let F be a real-valued function defined on U. If we let intuition be our guide, we might begin by leaning on a mnemonic device similar to the one used in the case of functions of one variable (Section 6.2). Thinking of $\int_U F \, dA$ as representing a volume, we look at the column in Fig. 13.4 and write down its volume as

$$F(x, y) \, dA \tag{1}$$

where dA is the area of the rectangular base; that is, dA is the product of dx and dy.

This column may be thought of as one of a series of columns forming a wall parallel to the x axis or, similarly, as contributing to a wall parallel to the y axis. Of the two choices, we know more about the wall parallel to the y axis; we know something about its bound on each end. The volume of this wall is the sum of the volumes of the constituent columns; that is, it is

$$\int_{H_1(x)}^{H_2(x)} F(x, y) \, dA.$$

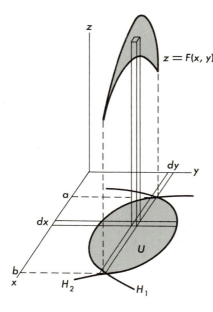

$$z = F(x, y)$$

FIG. 13.4

We have already said that dA is the product of dx and dy. We may use the order in which we write the factors to indicate that we have decided to sum the columns running parallel to the y axis first. To finish the problem, we then sum the "walls" perpendicular to the x axis. Thus

$$\int_U F(x, y)\, dA = \int_a^b \int_{H_1(x)}^{H_2(x)} F(x, y)\, dy\, dx.$$

Now we certainly have not proved this result, but we feel that (with the previously expressed reservations about U's being nice) this ought to be the case. The relevant theorem which can be proved [2, page 114], follows.

THEOREM A

Suppose U is bounded by the curves $y = H_1(x)$, $y = H_2(x)$, $x = a$, and $x = b$ where H_1 and H_2 are continuous functions on $[a, b]$ satisfying $H_1(x) < H_2(x)$ for $x \in (a, b)$. Then if F is continuous on U,

$$\int_U F\, dA = \int_a^b \int_{H_1(x)}^{H_2(x)} F(x, y)\, dy\, dx.$$

Consider Fig. 13.4 again. It is possible that in the situation pictured, the region U could also be described as the region enclosed by the two curves

$$x = G_1(y),$$

$$x = G_2(y),$$

intersecting at $y = c$ and $y = d$. We now see the same problem in the light of Fig. 13.5.

It now appears that we know more about the bounds of the pillars that form a wall parallel to the x axis, so we form our walls in this way. Then we sum these walls along the y axis. Our evaluation of the desired integral will then come from

$$\int_U F \, dA = \int_c^d \int_{G_1(y)}^{G_2(y)} F(x, y) \, dx \, dy.$$

Theorem A may of course be taken as descriptive of the conditions that must be satisfied by G_1 and G_2.

EXAMPLE A

U is the region enclosed by

$$y^2 - 4y + x = 0,$$

$$2y + x - 8 = 0.$$

Find $\int_U F \, dA$ where $F(x, y) = y$.

Solved for y, the curves bounding U are

$$y = 2 + \sqrt{4 - x},$$

$$y = -\tfrac{1}{2}x + 4.$$

Thus, the region U is the one drawn in Fig. 13.6. Knowing the upper and lower bounds of the region, we sum columns parallel to the vertical axis; then we sum "walls" along the x axis.

$$\int_U y \, dA = \int_0^4 \int_{-\frac{1}{2}x+4}^{2+\sqrt{4-x}} y \, dy \, dx = \int_0^4 [2\sqrt{4-x} - \tfrac{1}{8} x^2 + \tfrac{3}{2} x - 4] \, dx,$$

$$= (-\tfrac{4}{3}(4 - x)^{3/2} - \tfrac{1}{24} x^3 + \tfrac{3}{4} x^2 - 4x) \Big|_0^4 = 4. \tag{2}$$

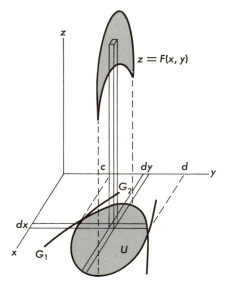

$z = F(x, y)$

FIG. 13.5

We also may solve this problem by first solving the equations that determine the boundaries for x. Then

$$x = 4y - y^2,$$

$$x = 8 - 2y.$$

Knowing the right and left bounds of the region, we may now sum our columns parallel to the horizontal axis; then sum the walls along the y axis.

$$\int_U y \, dA = \int_2^4 \int_{8-2y}^{4y-y^2} y \, dx \, dy = \int_2^4 y[(4y - y^2) - (8 - 2y)] \, dy,$$

$$= \int_2^4 (-y^3 + 6y^2 - 8y) \, dy = 4. \quad \blacksquare \tag{3}$$

We observe, by looking at (2) and (3), that the integration may be easier to perform in one order than in the other. This is a lesson we already learned in Exercise 13.1, of course.

Nothing was said in Example A about the meaning of the integral. We pause here to suggest several problems that give rise to the integral already evaluated.

(a) Find the moment of the region U about the x axis, assuming the density to be constantly ρ.

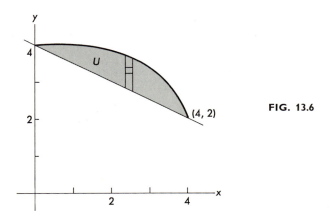

FIG. 13.6

The moment about the x axis of the element drawn in Fig. 13.6 is $\rho y \, dA$ where dA is the product of dy and dx.

$$M_x = \rho \int_U y \, dA.$$

(b) Find the mass of a thin sheet of material in the shape of the region U if the density of the material varies directly as its distance from the x axis.

The mass of our representative element is $yk \, dA$ where k is the constant of proportionality. Thus,

$$\text{Mass} = k \int_U y \, dA.$$

(c) Find the volume of the solid having U as its base and the plane $z = y$ as its top (Fig. 13.7).

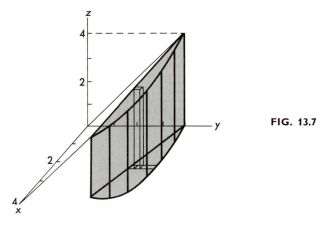

FIG. 13.7

The representative column has volume $z \, dA$. Summing these columns to form a wall parallel to the yz plane, and then summing the walls, we get

$$\text{Volume} = \int_0^4 \int_{-\frac{1}{2}x+4}^{2+\sqrt{4-x}} y \, dy \, dx.$$

The number of meanings we could assign to the integral seems to depend only on our imagination. Rather than tax ourselves any more in this direction, let us look at some similar problems that give rise to integrals defined on a region U in R^3.

EXAMPLE B

U is that region of R^3 enclosed by the positive coordinate planes, the surface $y = \sqrt[3]{1 - x^2}$, and the plane $z = y$. The density at a point of this region is directly proportional to the sum of its distances from the coordinate planes. Find the mass of the solid.

The basic building block is now a small cube. The cubes, when summed, build a column. The columns are then summed to form a wall, and finally the walls are summed to fill the region. We use $dz \, dy \, dx$ to represent the base unit, again relying on the order in which we write them down to indicate the order of our summing. Thus, $dz \, dy \, dx$ says the little cubes are taken parallel to the z axis, thus building a vertical column. The columns are then summed along the y axis, forming a wall parallel to the yz plane. Finally the walls are summed.

Now the mass of our representative element is, being proportional to the sum of its distances from the coordinate planes, given by

$$k(x + y + z) \, dz \, dy \, dx.$$

Therefore,

$$\text{Mass} = k \int_U (x + y + z) = k \int_0^1 \int_0^{(1-x^2)^{1/3}} \int_0^y (x + y + z) \, dz \, dy \, dx,$$

$$= k \int_0^1 \int_0^{(1-x^2)^{1/3}} \left(xy + \tfrac{3}{2} y^2 \right) dy \, dx,$$

$$= \tfrac{1}{2} k \int_0^1 [x(1 - x^2)^{2/3} + (1 - x^2)] \, dx,$$

$$= \tfrac{1}{2} k \tfrac{29}{30}. \; \blacksquare$$

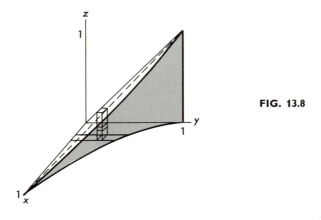

FIG. 13.8

EXAMPLE C

A solid of uniform density is bounded by the planes $z = 0$, $y = 0$, $y = x$, $x = 2$, and the surface $z = x^2 + y^2$. Find the center of gravity (the centroid).

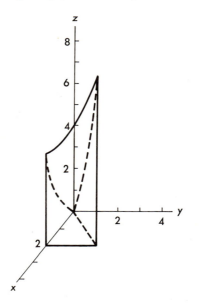

FIG. 13.9

We may assume the uniform density to be 1. We find the moments of our elements with respect to the coordinate planes. Thus,

$$M_{xy} = \int_0^2 \int_y^2 \int_0^{x^2+y^2} z \, dz \, dx \, dy = \tfrac{1}{2} \int_0^2 \int_y^2 (x^2 + y^2)^2 \, dx \, dy,$$

$$= \tfrac{1}{2} \int_0^2 \{\tfrac{32}{5} + \tfrac{16}{3} y^2 + 2y^4 - (\tfrac{1}{5} + \tfrac{2}{3} + 1)y^5\} \, dy = \tfrac{448}{45}.$$

$$M_{xz} = \int_0^2 \int_y^2 \int_0^{x^2+y^2} y \, dz \, dx \, dy = \int_0^2 \int_y^2 (x^2 y + y^3) \, dx \, dy,$$

$$= \int_0^2 \{(\tfrac{8}{3} y + 2y^3) - (\tfrac{1}{3} + 1)y^4\} \, dy = \tfrac{24}{5}.$$

$$M_{yz} = \int_0^2 \int_2^y \int_0^{x^2+y^2} x \, dz \, dx \, dy = \int_0^2 \int_y^2 (x^3 + xy^2) \, dx \, dy,$$

$$= \int_0^2 \{4 + 2y^2 - (\tfrac{1}{4} + \tfrac{1}{2})y^4\} \, dy = \tfrac{128}{15}.$$

$$\text{Mass} = \int_0^2 \int_2^y \int_0^{x^2+y^2} dz \, dx \, dy = \int_0^2 \int_y^2 (x^2 + y^2) \, dx \, dy,$$

$$= \int_0^2 \{(\tfrac{8}{3} + 2y^2) - (\tfrac{1}{3} + 1)y^3\} \, dy = \tfrac{16}{3},$$

$$\bar{z} = M_{xy}/\text{Mass} = \tfrac{448}{45} \cdot \tfrac{3}{16} = \tfrac{28}{15},$$

$$\bar{y} = M_{xz}/\text{Mass} = \tfrac{24}{5} \cdot \tfrac{3}{16} = \tfrac{9}{10},$$

$$\bar{x} = M_{yz}/\text{Mass} = \tfrac{128}{15} \cdot \tfrac{3}{16} = \tfrac{8}{5}. \ \blacksquare$$

In the preceding problem we made use of the notion of the moment of a body with respect to some fixed reference. Another useful notion in mechanics is the idea of the second moment, also called the *moment of inertia*, of a rotating mass about some fixed axis. If a particle of mass m rotates about a fixed axis at a distance r from the axis, then the moment of inertia I is defined to be the product

$$I = mr^2.$$

This definition is coupled with the idea of integration to define the moment of inertia of a solid body rotating about an axis.

EXAMPLE D
Find the moment of inertia of the solid described in Example C if it is rotated about the z axis.

We resort to the usual mnemonic device as a guide in setting up the integral. Our basic building block now looks like a block. If we assume its mass to be concentrated at the center, and if we continue to think of the density as 1, then the moment of inertia of our single element is

$$(x^2 + y^2)\, dz\, dx\, dy.$$

The desired moment of inertia is then given by

$$I = \int_0^2 \int_y^2 \int_0^{x^2+y^2} (x^2 + y^2)\, dz\, dx\, dy,$$

$$I = \int_0^2 (\tfrac{1}{5} x^5 + \tfrac{2}{3} x^3 y^2 + xy^4) \Big|_{x=y}^{x=2}\, dy,$$

$$= \int_0^2 (\tfrac{3\,2}{5} + \tfrac{1\,6}{3} y^2 + 2y^4 - \tfrac{2\,8}{1\,5} y^5)\, dy = 2^7 \cdot \tfrac{7}{4\,5}.$$

We found in Example C that the mass (volume with density assumed to be one) was $\tfrac{1\,6}{3}$. The distance at which a mass of $\tfrac{1\,6}{3}$ must be located from the z axis in order to have the same moment of inertia as our given solid is called the *radius of gyration* for the given solid.

$$\text{Radius of gyration} = \sqrt{\frac{2^7 \cdot 7}{45} \cdot \frac{3}{2^4}} = 2\sqrt{\frac{14}{15}}. \ \blacksquare$$

=== **13·2 EXERCISES**

In Exercises 1–6, express each of the iterated integrals as an iterated integral (or sum of iterated integrals) in which the integrations are performed in the reverse order.

1. $\int_0^4 \int_{y/2}^{\sqrt{y}} F(x, y)\, dx\, dy.$

2. $\int_0^1 \int_{2y}^{2\sqrt{y}} F(x, y)\, dx\, dy.$

3. $\int_3^4 \int_{4-y}^{\sqrt{4-y}} F(x, y)\, dx\, dy.$

4. $\int_0^2 \int_{(2-y)/2}^{\sqrt{4-y^2}} F(x, y)\, dx\, dy.$

5. $\int_0^2 \int_y^{\sqrt{8-y^2}} F(x, y)\, dx\, dy.$

6. $\int_0^1 \int_{-\sqrt{1-y^2}}^{1-y} F(x, y)\, dx\, dy.$

Evaluate integrals 7–14.

7. $\int_0^1 \int_y^1 ye^x\, dx\, dy.$

8. $\int_0^1 \int_y^1 (1 - y^2)\, dx\, dy.$

9 . $\int_0^1 \int_y^1 \sin x^2 \, dx \, dy.$

10 . $\int_0^1 \int_y^1 e^{x^2} \, dx \, dy.$

11 . $\int_0^1 \int_0^y \int_y^1 x \, dx \, dz \, dy.$

12 . $\int_0^1 \int_0^y \int_0^y z \, dx \, dz \, dy.$

13 . $\int_0^1 \int_0^y \int_0^y x(1 - y^2) \, dx \, dz \, dy.$

14 . $\int_0^1 \int_0^y \int_y^1 x^3 \cos xz \, dx \, dz \, dy.$

In Exercises 15–18, use triple integrals to find the volume of the indicated region of space.

15 . Bounded by the planes $z = y$, $z = 0$, $x = 1$ and the surface $y = x^2$.

16 . Bounded by the planes $y = 2x$, $y = 2$, $z = 0$, $x = 0$, and the surface $z = 5 - (x^2 + y^2)$.

17 . The region in the first octant bounded below by $x + y + z = 2$ and above by $x^2 + y^2 + z = 4$.

18 . Bounded by the planes $z = 0$, $x = 0$, $x + y + z = 6$ and the surface $y = x^2$.

In Exercises 19–22, find the centroid of the solid indicated. (You are to assume the density of the solid is uniform.)

19 . Exercise 15.

20 . Exercise 16.

21 . Exercise 17.

22 . Exercise 18.

In Exercises 23–26, find the moment of inertia about the z axis of the solid indicated.

23 . Exercise 15.

24 . Exercise 16.

25 . Exercise 17.

26 . Exercise 18.

The student will find the integrals to be quite formidable in Exercises 27–36. If the challenge is not such as to encourage him to try seeing the thing through, it should be carried far enough so that the difficulty is clearly seen. This will increase his appreciation for Section 13.3.

Each of the solids described in Exercises 27–30 is assumed to be regular and to have its altitude perpendicular to the base. In each exercise, we assume that a shaft of radius 1 unit has been drilled through the solid so that the altitude is the axis of the shaft. The problem, using triple integrals, is to find the volume of material remaining after the shaft is drilled out.

27 . A pyramid with height equal to 4 and a base that is a square having sides each 4 units long.

28 . A pyramid with height equal to 4 and a base that is an octagon having sides each 2 units long.

29 . A sphere of radius 4.

30· The paraboloid $z = 4 - (x^2 + y^2)$ cut off by the plane $z = 0$.

31–34· Find the moment of inertia of each of the "drilled out solids" of Exercises 27–30.

35· Find the moment of inertia of a sphere of radius a rotating about a diameter.

36· Find the moment of inertia of a sphere of radius a rotating about a line tangent to the sphere.

13 · 3 Change of Variables

In our work with the integral of a single variable, we sometimes found it helpful to make a substitution $u = G(x)$ in an integral

$$\int_a^b F(u)\, du. \tag{1}$$

The same thing happens when working with functions of several variables. For example, the change to polar coordinates in the case of R^2 has proved useful in other contexts, and it may be expected to help us here. In this situation, we are actually making a transformation that may be pictured as indicated in Fig. 13.10.

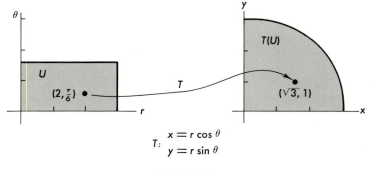

$$T: \begin{array}{l} x = r \cos \theta \\ y = r \sin \theta \end{array}$$

FIG. 13.10

Now suppose we want to make this substitution in an integral defined over a region in the XY plane. Such an integral takes the form

$$\iint_{T(U)} F(x, y)\, dy\, dx. \tag{2}$$

We recall that in the case of one variable, the du in (1) served as a reminder that substitution of $u = G(x)$) necessitated the introduction of the factor $du = DG(x)\, dx$. Our problem is to find the corresponding theorem for the case of two (or more) variables.

To get an intuitive idea of what to expect, we look at what happens to the basic building blocks (little rectangles) for double integrals. The general situation is this. We have a transformation T described by

$$T: \begin{array}{l} x = H(r, s), \\ y = G(r, s). \end{array}$$

Assuming T to be differentiable at (r, s), we then know that the image of the rectangle J (Fig. 13.11) is approximated by a parallelogram having the vertices

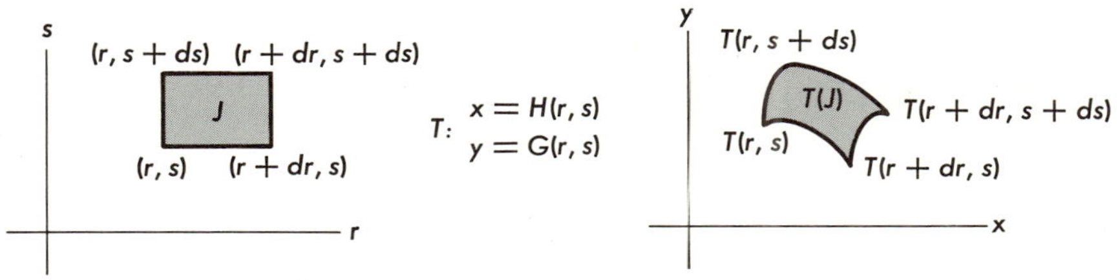

s

$(r, s + ds)$ $(r + dr, s + ds)$

J

(r, s) $(r + dr, s)$

r

$T: \begin{array}{l} x = H(r, s) \\ y = G(r, s) \end{array}$

y

$T(r, s + ds)$

$T(J)$

$T(r, s)$ $T(r + dr, s + ds)$

$T(r + dr, s)$

x

FIG. 13.11

$T(r, s)$,

$T(r, s) + T'(r, s)(dr, 0)$,

$T(r, s) + T'(r, s)(0, dr)$,

$T(r, s) + T'(r, s)(dr, ds)$.

$T'(r, s)$ is of course *linear*. Hence, this parallelogram, according to Theorem 10.4C, has an area given by

$$|\det[T'(r, s)]| \ dr \ ds. \tag{3}$$

This suggests replacing $dy\ dx$ in (2) by (3). (Note that this is exactly what we do in the case of one variable. In this simple situation, the matrix corresponding to $T'(x)$ is one by one. Hence, the determinant of the matrix is just $T'(x)$.)

With proper restrictions on a function $T: R^n \to R^n$ and on the set U, it may be proved [2, page 304] that if $F: T(U) \to R$ is continuous, then

$$\int_{T(U)} F(p) = \int_U F(T(p))|\det[T'(p)]| .$$

Because of its importance here and in other considerations, the number $|\det[T'(p)]|$ merits a special name. It is called the *Jacobian of the transformation*.

Return now to the change from polar to Cartesian coordinates.

$$T: \quad \begin{array}{l} x = r \cos \theta, \\ y = r \sin \theta, \end{array} \qquad \text{(Fig. 13.10)}.$$

The Jacobian at an arbitrary point (r, θ) is

$$\begin{vmatrix} \cos \theta & -r \sin \theta \\ \sin \theta & r \cos \theta \end{vmatrix} = r.$$

Then

$$\int_{T(U)} F(x, y) \, dy \, dx = \int_U F(r \cos \theta, r \sin \theta) r \, dr \, d\theta.$$

When $F(x, y) = 1$, this says that we find the area of $T(U)$ from either of the integrals

$$\int_{T(U)} dy \, dx = \int_U r \, dr \, d\theta.$$

This relationship is sometimes motivated (or remembered) by appeal to Fig. 13.12. The representative element is an (almost) rectangular-shaped

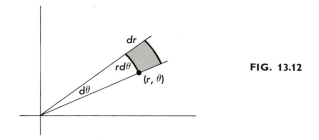

FIG. 13.12

element cut from a wedge corresponding to a change $d\theta$ in θ. One boundary of this "rectangle" is that portion of a circle of radius r cut off by a central angle $d\theta$; that is, it has length $r \, d\theta$. We thus use $r \, d\theta \, dr$ as the area of the element.

EXAMPLE A

Find the moment of inertia about the origin of a plane region of uniform density bounded by the lines $y = x$, $x = 0$, and the curve $y = \sqrt{9 - x^2}$.

The element of area in this case is the rectangle pictured in Fig. 13.13. The moment of inertia of this single element about the origin is

$$\rho(x^2 + y^2) \, dy \, dx.$$

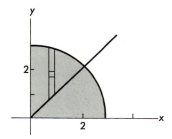

FIG. 13.13

(If the density was not constant, but rather varied with x and y, we would simply substitute the appropriate function of x and y for ρ in this expression.) The desired solution is therefore

$$I = \int_0^{3/\sqrt{2}} \int_x^{\sqrt{9-x^2}} \rho(x^2 + y^2)\, dy\, dx,$$

$$I = \rho \int_0^{3/\sqrt{2}} \{x^2[(9 - x^2)^{1/2} - x] + \tfrac{1}{3}[(9 - x^2)^{3/2} - x^3]\}\, dx. \qquad (4)$$

This can be integrated by methods we have studied. You are encouraged at this point, however, to be reluctant to begin such a task until you have thought over other ways to attempt the problem. Since circles and equations associated with circles play so prominent a place in this problem, you might think of using a coordinate system more amenable to such objects. Polar coordinates thus come to mind. In (4), the function $x^2 + y^2$ is replaced by r^2 and $dy\, dx$ is replaced by $r\, dr\, d\theta$. Limits of the transformed integral are most easily obtained directly from the figure.

$$I = \rho \int_{\pi/4}^{\pi/2} \int_0^3 r^2(r\, dr\, d\theta). \qquad (5)$$

Evaluation is now much more simple.

$$I = \rho\, \frac{1}{4} \int_{\pi/4}^{\pi/2} r^4 \Big|_{r=0}^{r=3}\, d\theta = \frac{81\rho}{4}\left(\frac{\pi}{2} - \frac{\pi}{4}\right) = \frac{81\rho\pi}{16}. \quad \blacksquare$$

The preceding example could have been worked from the beginning by means of polar coordinates of course. The picture we would have drawn in this case is indicated by Fig. 13.14. The moment of inertia of the indicated element is then written immediately as

$$\rho r^2 r\, dr\, d\theta.$$

This is also an appropriate place to mention that the integral of (5) may be taken as the volume of a certain solid in R^3. The equation $z = r^2$ describes, in

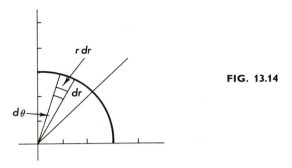

FIG. 13.14

cylindrical coordinates, the surface we more readily recognize as $z = x^2 + y^2$. If required to find the volume of the solid bounded by the planes $z = 0$, $y = x$, and the surfaces $z = x^2 + y^2$, $x^2 + y^2 = 9$, we would sketch Fig. 13.15. The elements to be summed have volume

$$zr\ dr\ d\theta = r^3\ dr\ d\theta.$$

This leads to the integral (5) where $\rho = 1$.

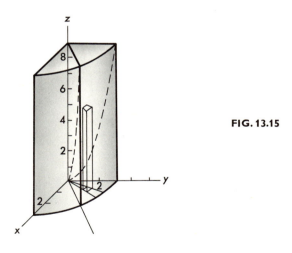

FIG. 13.15

PROBLEMS

A · A region of the plane is bounded by $\theta = \alpha$, $\theta = \beta$, and $r = G(\theta)$. Set up the double integral for the area of this region. Use it to derive line (1) in Section 9.5.

B · The change from cylindrical coordinates to Cartesian coordinates is described by

$$T: \quad \begin{aligned} x &= r \cos \theta, \\ y &= r \sin \theta, \\ z &= z. \end{aligned}$$

Find the Jacobian of this transformation and use it to show that

$$\int_{T(U)} F(x, y, z) \, dx \, dy \, dz = \int_T F(r \cos \theta, r \sin \theta, z) r \, dr \, d\theta \, dz.$$

C · Use the result of Problem B to solve Example 13.2D with cylindrical coordinates.

For the change from spherical coordinates to Cartesian coordinates, we need to appeal to the Jacobian of the transformation again.

$$T: \quad \begin{aligned} x &= \rho \sin \phi \cos \theta, \\ y &= \rho \sin \phi \sin \theta, \\ z &= \rho \cos \phi. \end{aligned}$$

$$\det[T'(p)] = \begin{vmatrix} \sin \phi \sin \theta & \rho \cos \phi \cos \theta & -\rho \sin \phi \sin \theta \\ \sin \phi \sin \theta & \rho \cos \phi \sin \theta & \rho \sin \phi \cos \theta \\ \cos \phi & -\rho \sin \phi & 0 \end{vmatrix}.$$

Evaluation gives $\rho^2 \sin \phi$. Hence,

$$\int_{T(U)} F(x, y, z) \, dx \, dy \, dz$$

$$= \int_U F(\rho \sin \phi \cos \theta, \rho \sin \phi \sin \theta, \rho \cos \phi) \, \rho^2 \sin \phi \, dp \, d\phi \, d\theta$$

When $F(x, y, z) = 1$, then the volume of $T(U)$ may be found from either of the integrals

$$\int_{T(U)} dx \, dy \, dz = \int_U \rho^2 \sin \phi \, d\rho \, d\phi \, d\theta$$

EXAMPLE B

Find the volume of the region bounded by the sphere $x^2 + y^2 + z^2 = 16$, the cone $z^2 = x^2 + y^2$, and the plane $z = 4$.

We wish to find 4 times the volume indicated in Fig. 13.16. The volume of our representative element is $\rho^2 \sin \phi \, dp \, d\phi \, d\theta$, with the order of $dp \, d\phi \, d\theta$ indicating, as usual, the order of summing. The element we picture is to be

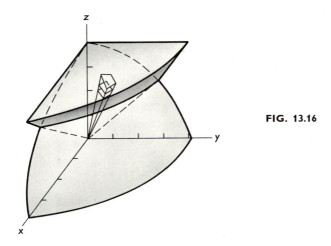

FIG. 13.16

summed within the "tilted column" from $\rho = 4$ (the sphere) up to $z = 4$. We need to observe that $z = \rho \cos \phi$, so in spherical coordinates, the plane is given by $\rho = 4 \sec \phi$. The "tilted columns" will then be summed from the vertical position ($\phi = 0$) to the cone ($\phi = \pi/4$).

$$\text{Volume} = 4 \int_0^{\pi/2} \int_0^{\pi/4} \int_4^{4 \sec \phi} \rho^2 \sin \phi \, d\rho d\phi \, d\theta.$$

$$V = 4 \int_0^{\pi/2} \int_0^{\pi/4} \frac{\rho^3}{3} \Big|_{\rho = 4}^{\rho = 4 \sec \phi} \sin \phi \, d\phi \, d\theta,$$

$$= \frac{4^4}{3} \int_0^{\pi/2} \int_0^{\pi/4} \left(\frac{\sin \phi}{\cos^3 \phi} - \sin \phi \right) d\phi \, d\theta,$$

$$= \frac{2^8}{3} \int_0^{\pi/2} (\tfrac{1}{2} \cos^{-2} \phi + \cos \phi) \Big|_{\phi = 0}^{\phi = \pi/4} d\theta.$$

$$= \frac{2^7}{3} (\sqrt{2} - 1) \frac{\pi}{2} = 2^6(\sqrt{2} - 1) \frac{\pi}{3}. \quad \blacksquare$$

13·3 EXERCISES

In Exercises 1–4, do the following.

(a) Compute $\int_A F(x, y) \, dA$ where A is the indicated region in the xy plane and $F(x, y)$ is the given function.

(b) Compute the Jacobian of the given transformation T. Also find the pre-image of A in the RS plane.

(c) Use the change of variables described by T to again evaluate the integral in part (a).

1. A is bounded by lines $x + y = 0$, $x + y = 3$, $x = 2y$, $x = 2y - 3$. $F(x, y) = 1$.

$$T: \quad \begin{aligned} x &= 2r - s, \\ y &= r + s. \end{aligned}$$

2. A is bounded by the lines $x + y = 0$, $x + y = 2$, $y = x$, $y = x + 2$. $F(x, y) = 1$.

$$T: \quad \begin{aligned} x &= r - s, \\ y &= r + s. \end{aligned}$$

3. A is bounded by the lines $y = 2x$, $y = 2x + 7$, $3x + 2y = 0$, $3x + 2y = 7$.

$$T: \quad \begin{aligned} x &= r - 2s, \\ y &= 2r + 3s. \end{aligned}$$

4. A is bounded by the lines $x + y = 0$, $x + y = 5$, $3x - 2y = 0$, $3x - 2y = 5$. $F(x, y) = xy - y$.

$$T: \quad \begin{aligned} x &= 2r + s, \\ y &= 3r - s. \end{aligned}$$

In Exercises 5–8, change the double integrals to an equivalent double integral in terms of polar coordinates and then evaluate.

5. $\displaystyle\int_{\sqrt{2}}^{2} \int_{0}^{\sqrt{4-x^2}} dy \, dx.$

6. $\displaystyle\int_{0}^{\sqrt{2}} \int_{0}^{\sqrt{4-x^2}} dy \, dx.$

7. $\displaystyle\int_{0}^{\infty} \int_{0}^{\infty} e^{-(x^2+y^2)} \, dx \, dy.$

8. $\displaystyle\int_{0}^{\infty} \int_{0}^{\infty} e^{-\sqrt{x^2+y^2}} \, dx \, dy.$

9. Use a triple integral to find the volume included between the plane $z = 0$ and the surfaces $x^2 + y^2 + z^2 = 15$ and $2z = x^2 + y^2$, using (a) rectangular coordinates; (b) cylindrical coordinates; (c) spherical coordinates.

10. Use a triple integral to find the volume of the region above the plane $z = 0$ and included between the surfaces $x^2 + y^2 + z^2 = 16$ and $z = 2 - \frac{1}{8}(x^2 + y^2)$, using (a) rectangular coordinates; (b) cylindrical coordinates; (c) spherical coordinates.

11–20. Use cylindrical coordinates to solve Exercises 27–36 from Section 13.2.

CHAPTER

14

INFINITE SERIES

Consider the functions

$$S(x) = \sin x, \qquad G(x) = \frac{1}{1 - x}.$$

Derivatives of all orders exist and are easily computed for both of these functions at $x = 0$. In Example 7.2D we found the Taylor polynomial of degree 3 corresponding to S. No difficulties are encountered if we extend the work done there to find the nth-degree polynomial. We get

$$S(x) = x - \frac{x^3}{3!} + \frac{x^5}{5!} - \cdots + R_n(0, x). \tag{1}$$

For the function G we have

$$DG(x) = (1 - x)^{-2} \qquad\qquad DG(0) = 1,$$
$$D^2G(x) = 2(1 - x)^{-3} \qquad\qquad D^2G(0) = 2,$$
$$\vdots \qquad\qquad\qquad\qquad \vdots$$
$$D^nG(x) = n!(1 - x)^{-(n+1)} \qquad D^nG(0) = n!$$

Again it is no trick to write down the corresponding Taylor polynomial of degree n. We get

$$G(x) = 1 + x + x^2 + \cdots + R_n(0, x). \tag{2}$$

Since the pattern is so clear and the terms can be computed for arbitrary n, someone is (was) sure to ask if we could not in some sense merely write

$$S(x) = \sin x = x - \frac{x^3}{3!} + \frac{x^5}{5!} - \cdots, \tag{3}$$

$$G(x) = \frac{1}{1 - x} = 1 + x + x^2 + x^3 + \cdots. \tag{4}$$

For these expressions to have meaning, it is clearly necessary for us to ask what is meant by a sum of an infinite collection of numbers; that is, what does it mean to write

$$S\left(\frac{1}{2}\right) = \frac{1}{2} - \frac{1}{2^3 \cdot 3!} + \frac{1}{2^5 \cdot 5!} - \cdots$$

We are encouraged in the whole project upon observing that we have used such "infinite sums" since our elementary school days. We are, for instance, accustomed to writing

$$G\left(\frac{1}{10}\right) = \frac{1}{1 - (1/10)} = 1 + \frac{1}{10} + \frac{1}{10^2} + \frac{1}{10^3} + \cdots \tag{5}$$

This is usually written a little differently:

$$\frac{10}{9} = 1.1111 \cdots.$$

We are not, of course, accustomed to having people bother us with questions about what

$$0.1111 \cdots$$

means, but we still feel comfortable about the idea. The reader may not feel so comfortable if someone else substitutes a 3 in (4), writing

$$\frac{1}{1-3} = 1 + 3 + 9 + 27 + \cdots. \tag{6}$$

We propose in this chapter to carefully define what is meant by an infinite sum. Once this is understood, we can talk about functions defined by an infinite series. We will then better understand the representation of S and G in the form (3) and (4). In particular, we will see why the representation of $G(\frac{1}{10})$ seems believable, whereas the representation of $G(3)$ appears to be nonsense.

14 · 1   Sequences and Series

Our goal is to understand an expression of the form

$$a_1 + a_2 + a_3 + \cdots + a_n + \cdots$$

where the a_i are real numbers. We shall get to the heart of the matter if we first learn something about sequences.

Sequences may be described informally as a string of numbers. The most interesting sequences for our purposes are those in which there is some pattern to this string of numbers. The reader's first encounter with a formal sequence may have been on an intelligence test, where he was asked to find the pattern in a sequence such as

$$1, 2, 5, 26, \ldots .$$

It is not hard to give a formal definition of a sequence of real numbers.

DEFINITION A   (Sequence)
A sequence A is a real-valued function that assigns to each non-negative integer some real number. (It may or may not be defined for zero.) The value of A at n is usually written A_n instead of $A(n)$, and the sequence itself is commonly represented by $\{A_n\}$.

As with any function, there may or may not be a formula giving the value of A at an arbitrary integer n. For instance, we may have

$$B_n = \frac{1}{n}, \quad \text{giving the sequence} \quad 1, \frac{1}{2}, \frac{1}{3}, \frac{1}{4}, \ldots .$$

On the other hand, we may have

$$P_n = n\text{th prime, giving the sequence}\quad 2, 3, 5, \ldots.$$

There is no known formula for the nth term of the second sequence.

Graphs of these two sequences are indicated in Figs. 14.1 and 14.2. Sometimes it is possible to find a fairly simple real-valued function F, defined on $(0, \infty)$, so that the graph of F passes through all the points that are the graph of the sequence $\{A_n\}$. This means, to state it another way, that we may be able to find an F defined on $(0, \infty)$ so that $F(n) = A_n$ for all n. For example, if we set $F(x) = 1/x$, then the graph of F passes through all the points on the graph of $\{B_n\}$ in Fig. 14.1.

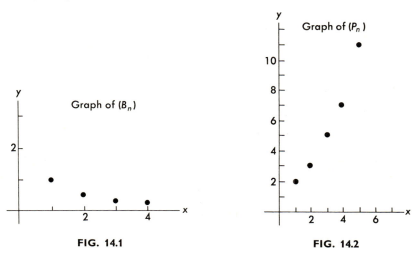

FIG. 14.1   FIG. 14.2

There are a number of situations in which it is useful to have a continuous function that agrees with the value of a sequence at each integer. We shall return to this idea. The reader may find it of immediate use in the solution of the following problem.

PROBLEM

A · Find $\lim_{n \to \infty} A_n$ for the following sequences.

(a) $A_n = \dfrac{1}{n}$.

(b) $A_n = \ln \dfrac{n}{n+1}$.

(c) $A_n = \dfrac{4(\frac{1}{2})^n}{1 - \frac{1}{2}}$.

(d) $A_n = n^{1/n}$.

DEFINITION B (The Infinite Series $\sum_{n=1}^{\infty} b_n$)

Let a sequence $\{b_n\}$ be given, and from it form a second sequence $\{B_n\}$ where

$$B_1 = b_1,$$

$$B_2 = b_1 + b_2,$$
$$\vdots$$
$$B_k = b_1 + b_2 + \cdots + b_k,$$
$$\vdots$$

$\{B_n\}$ is called the *sequence of partial sums corresponding to* $\{b_n\}$. If $\lim_{n \to \infty} B_n = r$, then we say the infinite series $\sum_{n=1}^{\infty} b_n$ converges to r. If the limit does not exist, we say the infinite series $\sum_{n=1}^{\infty} b_n$ diverges.

We warn the reader immediately that for a particular series, it is often not easy to tell whether it converges or diverges. The examples that follow are chosen to illustrate the definition, but they are not typical. We generally cannot find a formula for the numbers B_n so easily as we can in the examples presented.

EXAMPLE A
We show that

$$\sum_{n=1}^{\infty} \frac{1}{n(n+1)}$$

converges.

$$B_1 = \frac{1}{(1)(2)} = \frac{1}{2},$$

$$B_2 = \frac{1}{1(2)} + \frac{1}{2(3)} = \frac{2}{3},$$

$$B_3 = \frac{1}{1(2)} + \frac{1}{2(3)} + \frac{1}{3(4)} = \frac{3}{4},$$
$$\vdots$$
$$B_n = \frac{1}{1(2)} + \frac{1}{2(3)} + \frac{1}{3(4)} + \cdots + \frac{1}{n(n+1)} = ?$$

Our problem is to express B_n in a form that enables us to see what happens as n gets large. We have already warned the reader that this is difficult and that we chose our examples for a special reason. We can handle them. In the present case, all we need is a little inspiration in order to observe that

$$\frac{1}{k(k+1)} = \frac{1}{k} - \frac{1}{k+1}.$$

(Partial fractions can be used here if you want some explanation of how you might have come upon this yourself.) With this information, we can write B_n in the form

$$B_n = \left(\frac{1}{1} - \frac{1}{2}\right) + \left(\frac{1}{2} - \frac{1}{3}\right) + \left(\frac{1}{3} - \frac{1}{4}\right) + \cdots + \left(\frac{1}{n} - \frac{1}{n+1}\right).$$

It follows that $B_n = 1 - [1/(n+1)]$. (Check this for the values of B_1, B_2, B_3 already computed above.) Thus,

$$\lim_{n \to \infty} B_n = 1,$$

so by definition,

$$\sum_{n=1}^{\infty} \frac{1}{n(n+1)} = 1. \ \blacksquare$$

When asked to investigate the convergence of $\sum_{n=1}^{\infty} b_n$, we are really being asked to find $\lim_{n \to \infty} B_n$. This latter limit depends only on the behavior of the sequence $\{B_n\}$ for large values of n. The values of the first few terms are not crucial. Thus, the values of the first few terms of the sequence $\{b_n\}$ are not crucial either. For instance, the series

$$5 + 4 + 3 + 2 + 1 + \frac{1}{1 \cdot 2} + \frac{1}{2 \cdot 3} + \frac{1}{3 \cdot 4} + \cdots$$

converges because the series of Example A converges. The only difference is that this series will converge to

$$(5 + 4 + 3 + 2 + 1) + \left(\sum_{n=1}^{\infty} \frac{1}{n(n+1)}\right) = (15) + (1) = 16.$$

Our next example is an old high school problem dressed up in new clothes (the language of infinite series). This result is of great importance.

EXAMPLE B (The Geometric Series)
Let r be a real number in the interval $(0, 1)$. Then $\sum_{n=0}^{\infty} ar^n$ converges to $a/(1 - r)$.

To see this, note that the nth partial sum B_n is given by

$$B_n = a + ar + ar^2 + \cdots + ar^n. \tag{1}$$

As usual, when problems of this sort can be worked, it is by a trick. In this case, the trick is to multiply (1) by r and then subtract the result from (1),

$$rB_n = ar + ar^2 + \cdots + ar^n + ar^{n+1},$$

$$B_n - rB_n = a - ar^{n+1},$$

$$B_n = \frac{a - ar^{n+1}}{1 - r}. \tag{2}$$

Now since $r \in (0, 1)$, $\lim_{n \to \infty} ar^{n+1} = 0$.

$$\lim_{n \to \infty} B_n = \frac{a}{1 - r}.$$

This completes what we wanted to show. We can with little effort show even more. If $r = 1$, then (1) gives $B_n = (n + 1)a$. Assuming $a \neq 0$ (it is a very dull problem if $a = 0$),

$$\lim_{n \to \infty} B_n = \infty \qquad (\text{possibly} - \infty).$$

If $r > 1$, then $\lim_{n \to \infty} r^{n+1} = \infty$ and (2), which is valid for any $r \neq 1$, shows that again

$$\lim_{n \to \infty} B_n = \infty \qquad (\text{possibly} - \infty).$$

We have proved

$$\sum_{n=0}^{\infty} ar^n \text{ converges for } r \in (0, 1), \qquad \text{diverges for } r \geq 1. \blacksquare$$

In our next example we discuss the so-called harmonic series,

$$\sum_{n=1}^{\infty} \frac{1}{n} = 1 + \frac{1}{2} + \frac{1}{3} + \frac{1}{4} + \cdots.$$

This series will be shown to be divergent. The result is a useful one to know and has been proved in many ways. We have chosen a method that illustrates a technique that will be used again.

EXAMPLE C

Establish that the harmonic series diverges.

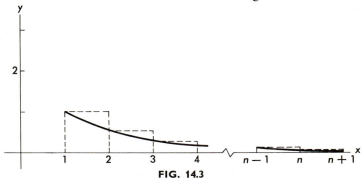

FIG. 14.3

It is clear from Fig. 14.3 that

$$B_n = 1 + \frac{1}{2} + \frac{1}{3} + \cdots + \frac{1}{n-1} + \frac{1}{n} > \int_1^{n+1} \frac{dt}{t}.$$

The integral on the right is simply $\ln(n + 1)$, so we have

$$B_n > \ln(n + 1).$$

We know (Problem 7.4F).

$$\lim_{n \to \infty} \ln(n + 1) = \infty.$$

Therefore,

$$\lim_{n \to \infty} B_n = \infty. \ \blacksquare$$

In each of the foregoing examples we were given a series $\sum_{n=1}^{\infty} b_n$. In each case we were to say something about the convergence or divergence of this series. In no case did we spend much time investigating the sequence $\{b_n\}$. Instead we formed $\{B_n\}$ and investigated it. That is as it must be, since convergence of $\sum_{n=1}^{\infty} b_n$ depends by definition on the behavior of $\{B_n\}$. There is a natural question to ask, however:

Question Is it possible, by applying some test to the sequence $\{b_n\}$, to predict whether or not $\lim_{n \to \infty} B_n$ will exist?

A multitude of answers have been given to this question. We shall now obtain an easy but important theorem that suggests one way of looking at $\{b_n\}$.

THEOREM A

If $\lim_{n \to \infty} b_n \neq 0$, then $\sum_{n=0}^{\infty} b_n$ diverges.

PROOF:

Suppose $\sum_{n=0}^{\infty} b_n$ converges, meaning that

$$\lim_{n \to \infty} B_n = r$$

for some finite real number r. Now note that $b_n = B_n - B_{n-1}$ so we have

$$\lim_{n \to \infty} b_n = \lim_{n \to \infty} (B_n - B_{n-1}),$$

$$= \lim_{n \to \infty} B_n - \lim_{n \to \infty} B_{n-1}.$$

Thus, $\lim_{n \to \infty} b_n = r - r = 0$.

We have now proved that if $\sum_{n=0}^{\infty} b_n$ converges, then $\lim_{n \to \infty} b_n = 0$, so if this limit is nonzero, the series cannot converge. ∎

This is a good place to pause for a short lesson in logic. Consider the promise Mrs. Jones makes to her children: "If grandma does not come tomorrow, we will go to the park." If grandma does not come the next day (and if Mrs. Jones is a woman of her word even to children), it is clear that they will go to the park. But suppose grandma does come tomorrow. We cannot logically draw any conclusions. The trip to the park may or may not be undertaken.

Return now to Theorem A. Things are clear enough if $\lim_{n \to \infty} b_n \neq 0$. The series $\sum_{n=1}^{\infty} b_n$ diverges. But suppose $\lim_{n \to \infty} b_n = 0$. We cannot logically draw any conclusions. The series may still diverge, as does the harmonic series in which

$$\lim_{n \to \infty} b_n = \lim_{n \to \infty} \frac{1}{n} = 0.$$

Although this is a simple idea, experience indicates that it must be made explicit, and must be called to mind from time to time.

EXAMPLE D

Investigate the convergence of

$$\sum_{n=1}^{\infty} \frac{n}{\sqrt{n^2 + 3n}} = \frac{1}{2} + \frac{2}{\sqrt{10}} + \frac{1}{\sqrt{2}} + \cdots.$$

We compute

$$\lim_{n \to \infty} \frac{n}{\sqrt{n^2 + 3n}} = \lim_{n \to \infty} \frac{n}{\sqrt{n^2[1 + (3/n)]}},$$

$$= \lim_{n \to \infty} \frac{1}{\sqrt{1 + (3/n)}} = 1.$$

Since the limit is nonzero, the series diverges. ∎

PROBLEM

B · Suppose that for a sequence $\{b_n\}$ of positive terms,

$$\lim_{n \to \infty} \frac{b_{n+1}}{b_n} = m > 1.$$

Can $\sum_{n=1}^{\infty} b_n$ converge?

14 · 1 EXERCISES

Determine whether or not each of the indicated series converges. Give a reason for your conclusion in each case. In the case of a convergent series, find the value to which it converges.

1. $\sum_{n=1}^{\infty} \frac{5}{2^n}.$

2. $\sum_{n=1}^{\infty} \frac{5}{3^n}.$

3. $\sum_{n=0}^{\infty} \frac{5}{2^{n+1}}.$

4. $\sum_{n=0}^{\infty} \frac{5}{3^{n+1}}.$

5. $\sum_{n=1}^{\infty} \frac{3}{(3n-2)(3n+1)}.$

6. $\sum_{n=1}^{\infty} \frac{1}{(2n-1)(2n+1)}.$

7. $\frac{1}{2000} + \frac{2}{2001} + \frac{3}{2002} + \frac{4}{2003} + \cdots.$

8. $\frac{1}{\sqrt{2}} - \frac{2}{\sqrt{5}} + \frac{3}{\sqrt{10}} - \frac{4}{\sqrt{17}}.$

9. $\sum_{n=1}^{\infty} \frac{5}{(5n-1)(5n+4)}.$

10. $\sum_{n=2}^{\infty} \frac{1}{(n-1)(n+1)}.$

11. $\sum_{n=0}^{\infty} \frac{3}{2^{n+2}}.$

12. $\sum_{n=0}^{\infty} \frac{2}{3^{n-4}}.$

13. $2(\sqrt{\pi} - 1) + 3(\sqrt[3]{\pi} - 1) + 4(\sqrt[4]{\pi} - 1) + \cdots.$

14. $\frac{e^{\sqrt{2}}}{\ln 2} + \frac{e^{\sqrt{3}}}{\ln 3} + \frac{e^{\sqrt{4}}}{\ln 4} + \cdots.$

15. $\sum_{n=1}^{\infty} \frac{4n+2}{(2n^2+1)(2n^2+4n+3)}.$

16. $\sum_{n=1}^{\infty} \frac{2n+1}{n^2(n+1)^2}.$

14 · 2 Positive Series

We have seen that the convergence or divergence of an infinite series $\sum_{n=1}^{\infty} b_n$ depends on the existence of

$$\lim_{n=\infty} B_n \qquad \text{where} \qquad B_n = b_1 + b_2 + \cdots + b_n. \tag{1}$$

We have also seen that it is generally not easy to express B_n in a form so that the desired limit can be computed. For this reason we were led in the last section to raise a question that we repeat here.

Question Is it possible, by applying some test to the sequence $\{b_n\}$, to predict whether or not $\lim_{n \to \infty} B_n$ will exist?

We have already learned that it is a good idea to check $\lim_{n \to \infty} b_n$, since if this limit is nonzero, the series will diverge. But suppose this limit is zero. Are there then further tests we can apply to $\{b_n\}$?

The answer is a resounding yes. There are, in fact, many more tests than we have time to consider. Our purpose here will be to introduce some very basic tests. These tests suffice for many of the series we commonly meet, and they are the basis on which more refined and specialized tests are often built.

In this section we shall study three tests that may be used on positive series; that is, on series $\sum_{n=1}^{\infty} b_n$ in which $b_n > 0$ for every n. The reader should note this restriction carefully. The tests of this section are stated for series all of whose terms are positive.

Like most basic results, the three tests we give here are simple to state, easy to believe, and hard to prove. Several of the proofs rest on a property of sequences which we may state as follows.

THEOREM A

Suppose the sequence $\{p_n\}$ has the following properties (meaning the graph looks like Fig. 14.4).

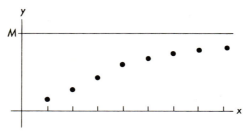

FIG. 14.4

(a)   It is a nondecreasing sequence of nonnegative terms; that is,

$$0 \le p_1 \le p_2 \le p_3 \le \cdots.$$

(b)   There is a number M so that for every n, $p_n < M$.
Then there must be an $L \le M$ so that $\lim_{n \to \infty} p_n = L$.

This theorem is an immediate consequence of a property of the real number system, a topic so far left for more advanced courses. We shall continue this plan, referring the interested student to [2, page 48] for a complete discussion of this matter.

Our first test is one suggested by the technique used in Example 14.1C. We suppose there is a function F defined on $[1, \infty)$ that is constantly decreasing and for which $F(n) = b_n$. It is then clear (see Fig. 14.5) that

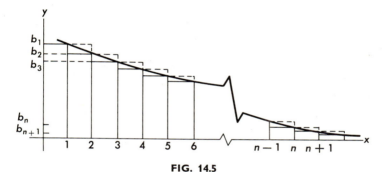

FIG. 14.5

$$\int_2^{n+1} F(x)\, dx < B_n - b_1 < \int_1^n F(x)\, dx < B_{n-1}.$$

Now if $\sum_{n=1}^{\infty} b_n$ converges, there must be a limit L to the sequence of partial sums $\{B_n\}$. Thus,

$$\lim_{n \to \infty} \int_1^n F(x)\, dx \le \lim_{n \to \infty} B_{n-1} = L,$$

so the integral converges. Conversely, if the integral converges, then

$$B_n - b_1 \le \lim_{n \to \infty} \int_1^n F(x)\, dx = K.$$

The sequence $\{B_n\}$ is thus bounded by $K + b_1 = M$. And since $\{B_n\}$ is increasing (all the b_n are positive), Theorem A implies that $\{B_n\}$ converges. We have established our first test which we state now as a theorem.

THEOREM B (The Integral Test)

Suppose F is a continuous decreasing function on $[1, \infty)$ for which $F(n) = b_n > 0$. Then

$$\sum_{n=1}^{\infty} b_n \quad \text{and} \quad \int_0^{\infty} F(x)\, dx$$

both converge or both diverge.

EXAMPLE A

$\sum_{n=1}^{\infty} 1/n^p$ diverges for $p \leq 1$, converges for $p > 1$.

For $p \neq 1$, this result follows from Problem 6.5B in which you proved

$$\int_1^{\infty} \frac{dx}{x^p} \text{ converges for } p > 1, \text{ diverges for } p < 1.$$

When $p = 1$, the series is simply the harmonic series, which we already know is divergent. We can, of course, use Theorem B to prove (once again) that the harmonic series is divergent.

$$\lim_{r \to \infty} \int_1^r \frac{dx}{x} = \lim_{r \to \infty} \ln r = \infty. \quad \blacksquare$$

PROBLEMS

A · Does $\sum_{n=1}^{\infty} e^{-n}$ converge?

B · Does $\sum_{n=2}^{\infty} 1/(n \ln n)$ converge?

Our next test is the basic test for convergence. Most other tests are simply variations of this one.

THEOREM C (Comparison Test)

Let $\sum_{n=1}^{\infty} a_n$ and $\sum_{n=1}^{\infty} b_n$ be two series of positive terms, and suppose that for all n, $a_n \leq b_n$. Then

(a) if $\sum_{n=1}^{\infty} a_n$ diverges, $\sum_{n=1}^{\infty} b_n$ diverges.
(b) if $\sum_{n=1}^{\infty} b_n$ converges, $\sum_{n=1}^{\infty} a_n$ converges.

PROOF:

Let the respective sequences of partial sums be $\{A_n\}$ and $\{B_n\}$. Both are increasing sequences of positive numbers, and we always have

$$A_n = a_1 + a_2 + \cdots + a_n \le b_1 + b_2 + \cdots + b_n = B_n.$$

Part (a)   If there is an M such that $A_n < M$ for each n, then Theorem A would tell us that $\lim_{n \to \infty} A_n$ exists. But this contradicts the assumption that $\sum_{n=1}^{\infty} a_n$ diverges; so no such M exists. This means

$$\lim_{n \to \infty} A_n = \infty.$$

Since $B_n \ge A_n$, it follows that $\lim_{n \to \infty} B_n = \infty$. Hence, $\sum_{n=1}^{\infty} b_1$ diverges.

Part (b)   If $\sum_{n=1}^{\infty} b_n$ converges, then by definition,

$$\lim_{n \to \infty} B_n = L$$

exists and is finite. This means that for each n, $B_n \le L$. (If there were a number k for which $B_k > L$, then for all $n > k$, $B_n \ge B_k > L$. This contradicts the fact that L is the limit.) Now $A_n \le B_n \le L$. We thus have the positive increasing sequence $\{A_n\}$ bounded above by L. Theorem A now assures us that

$$\lim_{n \to \infty} A_n$$

exists and is finite; $\sum_{n=1}^{\infty} a_n$ therefore converges. ∎

This theorem is not much help, of course, unless we already know of some series that converge or diverge. We do have some series in our storehouse that may be used as a basis for comparison.

$$\text{The } p \text{ series } \sum_{n=1}^{\infty} \frac{1}{n^p} \quad \begin{cases} \text{converges for } p > 1, \\ \text{diverges for } p \le 1. \end{cases}$$

$$\text{The geometric series } \sum_{n=1}^{\infty} ar^n \quad \begin{cases} \text{converges for } r \in [0, 1), \\ \text{diverges for } r \ge 1 \end{cases}$$

EXAMPLE B
Investigate the series

$$\text{(a)} \quad \sum_{n=1}^{\infty} \frac{\sin^2(\pi n/6)}{n^2}; \qquad \text{(b)} \quad \sum_{n=1}^{\infty} \left(\frac{3n+2}{2n}\right)^n.$$

Since $\dfrac{\sin^2(\pi n/6)}{n^2} \le \dfrac{1}{n^2}$ and $\displaystyle\sum_{n=1}^{\infty} \frac{1}{n^2}$ converges, the series (a) converges by comparison. (Explain why Theorem C could not be used in the same way to investigate the convergence of

$$\sum_{n=1}^{\infty} \frac{\sin(\pi n/6)}{n^2}.\Big)$$ (2)

To investigate series (b), note that

$$\left(\frac{3n+2}{2n}\right)^n > \left(\frac{3}{2}\right)^n \quad \text{and} \quad \sum_{n=1}^{\infty} \left(\frac{3}{2}\right)^n \text{ diverges}$$

The series (b) therefore diverges. We could also establish the divergence of series (b) by observing that the nth term does not approach 0. ∎

PROBLEMS

C · Does $\sum_{n=1}^{\infty} \sqrt{n}/(n+3)$ converge?

D · Suppose $\sum_{n=1}^{\infty} a_n$ and $\sum_{n=1}^{\infty} b_n$ are two positive series, both of which converge. Prove that $\sum_{n=1}^{\infty} a_n b_n$ converges. (Hint: $\lim_{n \to \infty} b_n = 0$. (Why?) Therefore there is an N for which $n > N$ implies $b_n < 1$.)

We get our final test by using the comparison test along with what we know about the geometric series. As usual, $\sum_{n=1}^{\infty} b_n$ is a positive series. Suppose that

$$\lim_{n \to \infty} \frac{b_{n+1}}{b_n} = k < 1.$$

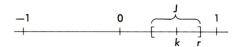

FIG. 14.6

Since $b_{n+1}/b_n > 0$, $k \in [0, 1)$. See Fig. 14.6. Choose a closed interval J about k that is contained in $(-1, 1)$. Let the right end point of this interval be r. We can choose an N large enough so that if $n \ge N$, then b_{n+1}/b_n is in J. In particular, this means

$$\frac{b_{N+1}}{b_N} < r \quad \text{so} \quad b_{N+1} < rb_N;$$

$$\frac{b_{N+2}}{b_{N+1}} < r \quad \text{so} \quad b_{N+2} < rb_{N+1} < r^2 b_N;$$

$$\frac{b_{N+3}}{b_{N+2}} < r \quad \text{so} \quad b_{N+3} < rb_{N+2} < r^3 b_N.$$

In the same way we get $b_{N+m} < r^m b_N$. Now consider the two infinite series

$$b_N + b_{N+1} + b_{N+2} + b_{N+3} + \cdots + b_{N+m} + \cdots,$$

$$b_N + rb_N + r^2 b_N + r^3 b_N + \cdots + r^m b_N + \cdots.$$

We recognize the second as the geometric series which, since $r < 1$, must converge. We have also seen that each term of the second is greater than the corresponding term of the first. Hence, by the comparison test, the first series must converge. Finally, the series that we set out to test was

$$\sum_{n=1}^{\infty} b_n = (b_1 + b_2 + \cdots + b_{N-1}) + (b_N + b_{N+1} + b_{N+2} + \cdots).$$

This is the sum of $N - 1$ numbers added to an infinite series known to converge; hence it converges.

We have proved that if $\lim_{n\to\infty} b_{n+1}/b_n = k < 1$, then $\sum_{n=1}^{\infty} b_n$ converges. This result, taken together with Problem 14.1B gives us the following test.

T H E O R E M D (Ratio Test)
Let $\sum_{n=1}^{\infty} b_n$ be a positive series for which

$$\lim_{n\to\infty} \frac{b_{n+1}}{b_n} = k$$

exists. Then the series converges if $k < 1$ and diverges if $k > 1$. If $k = 1$, no conclusion can be drawn.

EXAMPLE C
Investigate the series

$$\text{(a)} \quad \sum_{n=1}^{\infty} \frac{1}{n!} \qquad \text{and} \qquad \text{(b)} \quad \sum_{n=1}^{\infty} \frac{2^n}{n!}.$$

(a) $\dfrac{b_{n+1}}{b_n} = \dfrac{1}{(n+1)!} \cdot \dfrac{n!}{1} = \dfrac{1}{n+1} \to 0$ as $n \to \infty$.

Therefore series (a) converges.

(b) $\dfrac{b_{n+1}}{b_n} = \dfrac{2^{n+1}}{(n+1)!} \cdot \dfrac{n!}{2^n} = \dfrac{2}{n+1} \to 0$ as $n \to \infty$.

The second series also converges. ∎

PROBLEMS

E · Use the ratio test on $\sum_{n=1}^{\infty} 1/n^2$ (which we already know to be convergent) and on $\sum_{n=1}^{\infty} 1/n$ (which we already know to be divergent). What do these two problems illustrate?

F · Suppose $\sum_{n=1}^{\infty} b_n$ contains some negative terms. $\sum_{n=1}^{\infty} |b_n|$ is, of course, still a positive series. Suppose this latter series is tested by the ratio test and is found to be divergent. Does this mean that the original series diverges?

G · Suppose $\sum_{n=1}^{\infty} b_n$ contains some negative terms. $\sum_{n=1}^{\infty} |b_n|$ is a positive series. Suppose this latter series is tested by the integral test and found to be divergent. Does this mean that the original series diverges?

14·2 EXERCISES

Decide whether each of the infinite series below converges or diverges.

1 · $\sum_{n=1}^{\infty} \dfrac{n+5}{3^n}$.

2 · $\sum_{n=1}^{\infty} \dfrac{n^2}{2^n}$.

3 · $\sum_{n=1}^{\infty} \dfrac{1}{\sqrt{n^2+n}}$.

4 · $\sum_{n=1}^{\infty} \dfrac{e^n}{n^n}$.

5 · $\sum_{n=1}^{\infty} \dfrac{(n!)^2}{(2n)!}$.

6 · $\sum_{n=1}^{\infty} \dfrac{1}{\ln \pi n}$.

7 · $\sum_{n=1}^{\infty} \dfrac{n^n}{n!}$.

8 · $\sum_{n=1}^{\infty} \left(1+\dfrac{1}{n}\right)^n$.

9 · $\sum_{n=1}^{\infty} n(\pi^{1/n} - 1)$.

10 · $\sum_{n=1}^{\infty} \dfrac{n!}{n^n}$.

11 · $\sum_{n=1}^{\infty} \dfrac{3n^2}{4n^3+1}$.

12 · $\sum_{n=1}^{\infty} \dfrac{1}{\sqrt{n^3+n}}$.

13 · $\sum_{n=1}^{\infty} \dfrac{\sqrt{n+2}}{n^2(n+2)}$.

14 · $\sum_{n=1}^{\infty} \dfrac{n!}{e^n}$.

15 · $\sum_{n=1}^{\infty} \dfrac{1}{\ln \pi^n}$.

16 · $\sum_{n=1}^{\infty} \dfrac{n!}{(n+1)!}$.

17 · $\sum_{n=1}^{\infty} \dfrac{n}{2n^2+3}$.

18 · $\sum_{n=1}^{\infty} \dfrac{1}{ne^n}$.

19 · $\sum_{n=2}^{\infty} \dfrac{1}{n\sqrt{\ln n}}$.

20 · $\sum_{n=1}^{\infty} \dfrac{\ln n^3}{n^3}$.

21. $\displaystyle\sum_{n=1}^{\infty} \frac{\sqrt{n}}{n^2 + 1}$.

22. $\displaystyle\sum_{n=1}^{\infty} \frac{\arctan n}{1 + n^2}$.

23. $\displaystyle\sum_{n=1}^{\infty} \left(\frac{n+1}{n}\right)^n \frac{1}{n}$.

24. $\displaystyle\sum_{n=1}^{\infty} \left(\frac{n+1}{n}\right)^{2n} \frac{1}{\sqrt{n}}$.

25. $\displaystyle\sum_{n=1}^{\infty} \frac{3^{2n}}{3 \cdot 6 \cdot 9 \cdots (3n)}$.

26. $\displaystyle\sum_{n=1}^{\infty} \frac{1}{n^2} \left(\frac{3}{2}\right)^n$.

27. $\displaystyle\sum_{n=1}^{\infty} \frac{n3^n}{4^{n+1}}$.

28. $\displaystyle\sum_{n=1}^{\infty} \frac{n^2 2^n}{3^{n-2}}$.

14 · 3 Series with Some Negative Terms

None of the tests described in Section 14.2 apply to the series

$$1 - \frac{1}{4} + \frac{1}{9} - \frac{1}{16} + \cdots + (-1)^{n+1} \frac{1}{n^2} + \cdots. \tag{1}$$

Theorem 14.1A is not restricted to positive series, but since the nth term of (1) does go to zero, this theorem fails to give us any information.

It is natural, when looking at (1), to think of the series

$$1 + \frac{1}{4} + \frac{1}{9} + \frac{1}{16} + \cdots + \frac{1}{n^2} + \cdots. \tag{2}$$

This series is, of course, obtained from the first by taking the absolute value of each term. We know that (2) converges. Can information about this series help us with the first one? The point of the next theorem is that we can use what we know about (2) to settle the question of convergence for (1).

THEOREM A

If $\sum_{n=1}^{\infty} |b_n|$ converges, then $\sum_{n=1}^{\infty} b_n$ converges.

PROOF:

We introduce the following notation.

$$\sum_{n=1}^{\infty} b_n \text{ has } \{B_n\} \text{ as its sequence of partial sums,}$$

$$\sum_{n=1}^{\infty} |b_n| \text{ has } \{A_n\} \text{ as its sequence of partial sums}$$

$$c_n = b_n + |b_n|,$$

$$\sum_{n=1}^{\infty} c_n \text{ has } \{C_n\} \text{ as its sequence of partial sums} \tag{3}$$

Since $\sum_{n=1}^{\infty} |b_n|$ converges, $\lim_{n \to \infty} A_n = L_1$ exists for some finite L_1. Also, since $0 \le c_n \le 2|b_n|$, and since

$$\sum_{n=1}^{\infty} 2|b_n| = 2 \sum_{n=1}^{\infty} |b_n|$$

converges, it follows from the comparison test that $\sum_{n=1}^{\infty} c_n$ converges. Hence, $\lim_{n \to \infty} C_n = L_2$ for some finite L_2. Now from (3),

$$B_n = C_n - A_n,$$

so we may write, using Theorem 2.4B,

$$\lim_{n \to \infty} B_n = \lim_{n \to \infty} C_n - \lim_{n \to \infty} A_n = L_2 - L_1.$$

This means that $\sum_{n=1}^{\infty} b_n$ converges. ∎

On the basis of this theorem, (1) converges. It also settles the question of convergence for $\sum_{n=1}^{\infty} [\sin (\pi n/6)]/n^2$, first mentioned in line (2) of Example 14.2B.

As with all theorems, of course, we need to be careful in order not to use this theorem incorrectly. Thus, if $\sum_{n=1}^{\infty} |b_n|$ diverges, we may not conclude that $\sum_{n=1}^{\infty} b_n$ diverges. Theorem A says nothing, for instance, about the convergence or divergence of

$$1 - \frac{1}{2} + \frac{1}{3} - \frac{1}{4} + \cdots + (-1)^{n+1} \frac{1}{n} + \cdots. \tag{4}$$

So far as we now know, this series may converge even though its corresponding absolute series diverges. It is worth our time, in fact, to explore this possibility.

Let $C_k = B_{2k}$ be the sum of the first $2k$ terms. Thus,

$$C_k = \left(1 - \frac{1}{2}\right) + \left(\frac{1}{3} - \frac{1}{4}\right) + \cdots + \left(\frac{1}{2k - 1} - \frac{1}{2k}\right).$$

Each set of parentheses encloses a positive number. It is clear that the sequence $\{C_k\}$ is a positive, monotone increasing sequence. We may also write C_k in another form.

$$C_k = 1 - \left(\frac{1}{2} - \frac{1}{3}\right) - \left(\frac{1}{4} - \frac{1}{5}\right) - \cdots - \left(\frac{1}{2k - 2} - \frac{1}{2k - 1}\right) - \frac{1}{2k}.$$

Again each set of parentheses encloses a positive number, and each of these positive numbers, along with $1/2k$, is being subtracted from 1. This means $C_k < 1$. The positive, monotone increasing sequence $\{C_k\}$ is therefore bounded above by 1. According to Theorem 14.2A, this means there is a positive number L such that

$$\lim_{k \to \infty} C_k = L.$$

For the sum of the first $2k + 1$ terms, we have

$$B_{2k+1} = C_k + \frac{1}{2k + 1},$$

$$\lim_{k \to \infty} B_{2k+1} = \lim_{k \to \infty} C_k + \lim_{k \to \infty} \frac{1}{2k + 1} = L.$$

Since both the even- and odd-numbered partial sums converge to L, we must have

$$\lim_{n \to \infty} B_n = L \tag{5}$$

and so the series $\sum_{n=1}^{\infty} (-1)^{n+1}(1/n)$ converges to L. We now have an example of a convergent series for which the absolute series diverges. This suggests the introduction of some terminology.

DEFINITION A

If $\sum_{n=1}^{\infty} |b_n|$ converges, $\sum_{n=1}^{\infty} b_n$ is said to be absolutely convergent. If $\sum_{n=1}^{\infty} |b_n|$ diverges while $\sum_{n=1}^{\infty} b_n$ converges, then $\sum_{n=1}^{\infty} b_n$ is said to be conditionally convergent.

Theorem A simply says that an absolutely convergent series is convergent.

The method used to establish the convergence of (4) may be used to establish a test applying to similar problems.

THEOREM B (Alternating Series Test)

Let a series $\sum_{n=1}^{\infty} b_n$ satisfy the following.

(a) The series is strictly alternating (each term differs in sign from the preceding one).

(b) The terms are strictly decreasing in absolute value ($|b_{n+1}| < |b_n|$ for all n).

(c) $\lim_{n \to \infty} b_n = 0$.

Then the series converges.

PROBLEMS

A · Using the proof of the convergence of $\sum_{n=1}^{\infty} (-1)^{n-1}/n$ as a model, construct a proof for Theorem B.

B · We know that the number L to which $\sum_{n=1}^{\infty} (-1)^{n+1}/n$ converges is between 0 and 1. Show that L differs

(a) from $1 - \dfrac{1}{2} + \dfrac{1}{3} - \dfrac{1}{4}$ by less than $\dfrac{1}{5}$;

(b) from $1 - \dfrac{1}{2} + \dfrac{1}{3} - \dfrac{1}{4} + \dfrac{1}{5}$ by less than $\dfrac{1}{6}$.

Given a series $\sum_{n=1}^{\infty} b_n$, convergence depends on the existence of

$$\lim_{n \to \infty} B_n \qquad \text{where} \qquad B_n = b_1 + b_2 + \cdots + b_n. \tag{6}$$

When we can obtain an expression for B_n that enables us to actually compute the limit (6), then we can actually find the "sum of the series." In the last two sections we paid little attention to actually finding this sum, however. We have been content to merely ascertain whether or not the sum (that is, the limit (6)) actually exists or not.

When the sum $\sum_{n=1}^{\infty} b_n$ is known to exist (that is, when the series converges), then we may approximate the sum, if we so desire, by computing B_N for some N. The larger we take N to be, the closer we come to the desired sum. Our result is more useful, of course, if we have some measure of just how close we are to the desired sum. For example, in Problem B the reader was asked to show that

$$1 - \frac{1}{2} + \frac{1}{3} - \frac{1}{4} + \frac{1}{5} = 0.783$$

approximates the sum (4) to within $\frac{1}{6} = 0.167$. We can then be certain that the sum we desire is in the interval

$$(0.783 - 0.167, 0.783 + 0.167) = (0.616, 0.950).$$

A little more attention to the proof of Theorem B enables us to prove a further result for alternating series.

COROLLARY

Let $\sum_{n=1}^{\infty} b_n$ satisfy (a), (b), and (c) of Theorem B. Then it converges to some number L, and

$$|B_n - L| < |b_{n+1}|.$$

PROBLEM

C · Use your solution to Problem B as a guide in constructing a proof of the corollary.

14 · 3 EXERCISES

Decide whether each of the following infinite series converges or diverges; also tell whether it converges absolutely.

1 · $\displaystyle\sum_{n=1}^{\infty} \frac{(-1)^n}{n^2}$.

2 · $\displaystyle\sum_{n=1}^{\infty} \frac{\sin(\pi n/3)}{\sqrt{n^3}}$.

3 · $\displaystyle\sum_{n=1}^{\infty} \frac{(-1)^n}{n+\sqrt{n}}$.

4 · $\displaystyle\sum_{n=1}^{\infty} \frac{(-1)^n n^2}{3n^2+4}$.

5 · $\displaystyle\sum_{n=1}^{\infty} \frac{(-1)^n n}{1000+n}$.

6 · $\displaystyle\sum_{n=1}^{\infty} \frac{(-1)^n}{2n+3}$.

7 · $\displaystyle\sum_{n=1}^{\infty} \frac{n \sin(\pi n/2)}{2^n}$.

8 · $\displaystyle\sum_{n=1}^{\infty} \frac{(-1)^n n^2}{n!}$.

9 · $\displaystyle\sum_{n=1}^{\infty} \frac{(-1)^n e^n}{n!}$.

10 · $\displaystyle\sum_{n=1}^{\infty} \frac{(-1)^n}{n^2} \sin \frac{\pi}{n}$.

11 · $\displaystyle\sum_{n=1}^{\infty} \frac{(-1)^n}{\sqrt[3]{n}}$.

12 · $\displaystyle\sum_{n=1}^{\infty} (-1)^{n+1} \frac{\ln n}{n}$.

13 · $\displaystyle\sum_{n=1}^{\infty} \frac{(-1)^n}{2n} \left(\frac{2 \cdot 4 \cdots (2n)}{3 \cdot 5 \cdots (2n+1)} \right)$.

14 · $\displaystyle\sum_{n=1}^{\infty} (-1)^n \frac{(2n)^3(n-1)}{3e^n}$.

15 · $\displaystyle\sum_{n=1}^{\infty} \left(-\frac{3}{4} \right)^n$.

16 · $\displaystyle\sum_{n=1}^{\infty} (-1)^{n+1} \frac{n3^{1/n}}{2n+1}$.

17 · $\displaystyle\sum_{n=1}^{\infty} \frac{(-1)^{n+1}(n+3)}{2n+1}$.

18 · $\displaystyle\sum_{n=1}^{\infty} \frac{(-1)^n}{2^n} \left(\frac{3 \cdot 5 \cdots (2n+1)}{2 \cdot 4 \cdots (2n)} \right)$.

Show that each of the following series converges. Then find the sum accurate to within one unit in the third decimal place.

19 · $\displaystyle\sum_{n=1}^{\infty} \frac{(-1)^{n+1}}{n^3}$.

20 · $\displaystyle\sum_{n=0}^{\infty} \frac{(-1)^n}{n!}$.

21 · $\displaystyle\sum_{n=1}^{\infty} (-1)^{n+1} \frac{1}{2n!}$.

22 · $\displaystyle\sum_{n=0}^{\infty} (-1)^n \frac{1}{(2n+1)!}$.

14 · 4 Functions Defined by Infinite Series

It is possible to define a function by means of an infinite series. We know, for example, that $\sum_{n=1}^{\infty} 1/n^x$ converges for $x > 1$. If we choose $a > 1$, then $\sum_{n=1}^{\infty} 1/n^a$ is the corresponding number. Such a number is defined for each choice of $a > 1$. Hence, we have a function

$$H(x) = \sum_{n=1}^{\infty} \frac{1}{n^x}, \tag{1}$$

which has $(1, \infty)$ as its domain.

The value of H for a particular choice of x, say $x = 2$, is of course not easily determined. The function H, like the function $\sin x$, appears to be one of those functions for which we can only find numerical approximations to the actual value at a point x. For H, such approximations are obtained by using the first few terms of the series (1). Few is a relative term, depending on how the computing is being done. Computers make it possible to find $H(2)$, to continue our illustration, with great accuracy by adding the first "few" terms of (1).

The use of series in defining functions is very important. For this reason we will take the time to restate just how it is done, and to illustrate the procedure once again.

The General Procedure	Illustrated
Suppose that for each integer n we have a function $u_n(x)$.	Let $u_n(x) = x^n$.
Each function is assumed to be defined on a common interval (a, b).	Each $u_n(x)$ is defined on $(-\infty, \infty)$.
It is possible that for certain $x \in (a, b)$, the series $$\sum_{n=0}^{\infty} u_n(x)$$ may converge.	The series $\sum_{n=0}^{\infty} x^n$ converges for $x \in (-1, 1)$ (the geometric series).
If the series converges for $x \in (c, d)$, then we may define $$F(x) = \sum_{n=0}^{\infty} u_n(x).$$	We may define, for $x \in (-1, 1)$, $$G(x) = \sum_{n=0}^{\infty} x^n. \tag{2}$$

Once we have defined a function

$$F(x) = \sum_{n=0}^{\infty} u_n(x) \qquad \text{known to converge on } (c, d),$$

we may ask a variety of questions.

(A) Is the domain of F (that is, the x for which the series converges) necessarily one connected interval?

(B) Is F continuous?

(C) Will the series $\sum_{n=0}^{\infty} u_n'(x)$ converge on (c, d)? If so, will it converge to $F'(x)$?

(D) Will the series $\sum_{n=0}^{\infty} \int_c^d u_n(x)\, dx$ converge on (c, d)? If so, will it converge to $\int_c^d F(x)\, dx$?

Unfortunately, the answer to each question is no. Counterexamples have been given in each case, and are in fact not too hard to find. A thorough discussion of these questions is beyond the scope of our work. See [2, Section 4.2] and [7, Chapter 18].

We can restrict our attention to a certain kind of series for which the answer to all the questions above will be affirmative. This is the route we shall take.

DEFINITION A (Power Series about x_0)

A series of the form

$$\sum_{n=0}^{\infty} a_n(x - x_0)^n, \qquad a_n \text{ a constant for each } n,$$

is called a *power series about* x_0.

Thus, the series used to define F in (2) is a power series if and only if $u_n(x) = a_n(x - x_0)^n$ for each n. The function G used to illustrate (2) is defined by a power series about $x_0 = 0$. The function H of (1) is not defined by a power series.

As previously indicated, the questions posed above have very nice answers for functions defined by power series. The answers are given by the theorems below. All of these theorems may be proved in the form stated. See [2, Section 4.3] and [7, Chapter 19]. Proofs are better understood, however, in the context of functions of a complex variable. We shall simply state the results here and try to gain some facility in using them.

THEOREM A

The entire set of x for which the power series

$$\sum_{n=0}^{\infty} a_n(x - x_0)^n$$

converges may always be described in one of three ways

(1) The point x_0.
(2) An interval described by $|x - x_0| < r$ for some finite $r > 0$, together with the possible addition of one or both end points of this interval.
(3) All x.

The number r is called the *radius of convergence*. If we allow $r = 0$ and $r = \infty$, then all three cases may be described by some choice of r.

A most practical observation is that if a power series converges, it must converge absolutely except possibly at the end points of its interval of convergence.

EXAMPLE A

Find the domain of the function

$$F(x) = \sum_{n=1}^{\infty} \frac{2^n}{n}(x - 2)^n.$$

To find the interval of convergence, we investigate the absolute convergence of the series. For this purpose the ratio test is most convenient.

$$\lim_{n \to \infty} \frac{2^{n+1}|x - 2|^{n+1}}{n + 1} \cdot \frac{n}{2^n|x - 2|^n} = 2|x - 2|.$$

We have convergence if $2|x - 2| < 1$ and divergence if $2|x - 2| > 1$. The radius of convergence is therefore $\frac{1}{2}$, and we know the series converges on the interval $(\frac{3}{2}, \frac{5}{2})$. It remains to investigate the end points. For $x = \frac{3}{2}$

$$\sum_{n=1}^{\infty} \frac{2^n}{n}\left(-\frac{1}{2}\right)^n = \sum_{n=1}^{\infty} \frac{(-1)^n}{n},$$

which converges (alternating series). For $x = 5/2$,

$$\sum_{n=1}^{\infty} \frac{2^n}{n}\left(\frac{1}{2}\right)^n = \sum_{n=1}^{\infty} \frac{1}{n},$$

which diverges (harmonic series). The domain of F is therefore $[\frac{3}{2}, \frac{5}{2})$. ∎

PROBLEM

A · Determine the interval of convergence for each of the power series:

(a) $\displaystyle\sum_{n=1}^{\infty} n!(x - 2)^n$;

(b) $\displaystyle\sum_{n=1}^{\infty} \frac{3^n}{n^2}(x - 2)^n$;

(c) $\displaystyle\sum_{n=1}^{\infty} \frac{(x-2)^n}{n!}$.

Theorem A answered question A for power series. The next three theorems answer the corresponding questions. All of these theorems refer to a function F defined by

$$F(x) = \sum_{n=0}^{\infty} a_n (x - x_0)^n, \qquad \text{convergent for } |x - x_0| < r.$$

THEOREM B

If $r > 0$, then F is continuous on the open interval $|x - x_0| < r$.

THEOREM C

If $r > 0$, then F is differentiable at each x for which $|x - x_0| < r$, and

$$F'(x) = \sum_{n=0}^{\infty} na_n(x - x_0)^{n-1}.$$

The latter series also has r as its radius of convergence.

We note that the theorem can be applied to the series for $F'(x)$, enabling us to compute the second derivative. In fact, F must, by a repetition of this argument, have derivatives of all orders. They are obtained by repeated differentiation of the series defining F.

THEOREM D

If $r > 0$, and if c and d are points interior to the interval of convergence, then

$$\int_c^d F(x) = \sum_{n=0}^{\infty} \int_c^d a_n(x - x_0)^n.$$

PROBLEMS

The following problems refer to the two functions

$$C(x) = \sum_{n=0}^{\infty} (-1)^n \frac{x^{2n}}{(2n)!} = 1 - \frac{x^2}{2!} + \frac{x^4}{4!} - \cdots. \tag{3}$$

$$S(x) = \sum_{n=0}^{\infty} (-1)^n \frac{x^{2n+1}}{(2n+1)!} = x - \frac{x^3}{3!} + \frac{x^5}{5!} - \cdots. \tag{4}$$

B · Prove that $C'(x) = -S(x)$ and that $S'(x) = C(x)$.

C · Prove that $[S(x)]^2 + [C(x)]^2 = 1$. (Hint: Define $F(x) = [S(x)]^2 + [C(x)]^2$.)

The preceding problems are more than good exercises in the use of the theorems of this chapter. They also are designed to suggest a connection between the function S and the function $\sin x$ (and of course between C and cosine x). This same connection was suggested by other considerations in line (3) of the introduction to this chapter.

We shall now secure the connection. Let the Taylor polynomial $T_{S(0),n}$ corresponding to $\sin x$ be abbreviated to T_n. Then Taylor's theorem says

$$\sin x = T_n(x) + R_n(0, x), \tag{5}$$

which is what we wrote in (1) of Section 14.1. For fixed x, we now recognize that $T_n(x)$ is the nth partial sum of the series (4). This series, by definition, converges to $\lim_{n \to \infty} T_n(x)$. But from (5), $T_n(x) = \sin x - R_n(0, x)$, so

$$\lim_{n \to \infty} T_n(x) = \sin x - \lim_{n \to \infty} R_n(0, x).$$

Evidently the series (4) converges, for arbitrary x, to $\sin x$, providing that

$$\lim_{n \to \infty} R_n(0, x) = 0.$$

To establish the latter limit, we observe that

$$|R_n(0, x)| = \frac{|x|^{n+1}}{(n+1)!} |D^{n+1}S(\bar{x})| \leq \frac{|x|^{n+1}}{(n+1)!}.$$

For fixed x, it is certainly possible to choose N large enough so that

$$\frac{|x|}{N+1} < 1.$$

Then if n is greater than N, say $n = N + r$,

$$|R_n(0, x)| \leq \frac{|x|}{1} \cdot \frac{|x|}{2} \cdots \cdot \frac{|x|}{N} \cdot \frac{|x|}{N+1} \cdot \frac{|x|}{N+2} \cdots \cdot \frac{|x|}{N+r},$$

$$|R_n(0, x)| < \frac{|x|^N}{N!} \left(\frac{|x|}{N+1}\right)^r.$$

As n increases, the first factor remains constant while the second goes to zero.

The procedure we have followed for $\sin x$ may be followed for any function F that has derivatives of all orders at a point x_0. We form the Taylor polynomial $T_{F(x_0),n}$. Taylor's theorem gives

$$F(x) = T_n(x) + R_n(x_0, x). \tag{6}$$

Then if $\lim_{n \to \infty} R_n(x_0, x) = 0$ for all x in some interval J about x_0, the infinite series

$$\sum_{n=0}^{\infty} D^n F(x_0) \frac{(x - x_0)^n}{n!}$$

must converge to F for all x in J.

DEFINITION B (The Taylor Series)
Let a function F have derivatives of all orders at a point x_0. Then the Taylor series corresponding to F at x_0 is

$$\sum_{n=0}^{\infty} D^n F(x_0) \frac{(x - x_0)^n}{n!}.$$

By way of illustration, since $D^n e^x = e^x$, the corresponding Taylor series at $x_0 = 0$ is

$$e^x = 1 + x + \frac{x^2}{2} + \frac{x^3}{3!} + \cdots.$$

This particular series is, in fact, extremely important.

It is to be emphasized that the Taylor series may not converge; or it may converge, but not to the function F with which we started. In order to be certain of convergence, we must show that the remainder term $R_n(x_0, x)$ in (6) goes to zero. The typical pattern is for $R_n(x_0, x)$ to go to zero so long as x is restricted to an interval about x_0. On the other hand, some Taylor series, such as the one for e^x given above, converge to the desired function for all x.

When $x_0 = 0$, the Taylor series is more commonly referred to as the Maclaurin series.

PROBLEMS

D · The function $G(x) = 1/(1 - x)$ may be written in the form (2) of Section 14.1, as indicated in the introduction to this chapter. It happens that in this case, we can find an exact form for $R_n(0, x)$ instead of estimating it. Show that

$$R_n(0, x) = \frac{x^{n+1}}{1 - x}.$$

E · Explain why (5) gave a result we were willing to believe, while (6) did not.

It is now clear that many functions that are familiar to us may be represented by a power series. Moreover, we know how to find these series.

(a) We find an expression for the Taylor polynomial of arbitrary degree n about some point x_0.

(b) We find an expression for the remainder term $R_n(x_0, x)$.

(c) We show this remainder gets arbitrarily small as n increases. (This often requires us to restrict x to some neighborhood about x_0.)

P R O B L E M S †

F · Follow the pattern just given to represent the function $L(x) = \ln x$ as a power series about the point $x_0 = 1$.

G · Follow the pattern in an effort to represent the function $R(x) = x/(1 + x^2)$ in a Maclaurin series.

We have outlined a method for obtaining a power series expansion. We have also seen (Problem G) that this method can bog us down in a morass of computation. Sometimes there are other ways to obtain a power series representation. Consider again our function

$$G(x) = \frac{1}{1 - x}.$$

Long division gives

$$
\begin{array}{r}
1 + x + x^2 + \cdots + x^n \\
1 - x \overline{\smash{\big)}\ 1} \\
\underline{1 - x} \\
x \\
\underline{x - x^2} \\
x^2 \\
\underline{x^2 - x^3} \\
x^3 \\
\ddots \\
\underline{} \\
x^n \\
\underline{x^n - x^{n+1}} \\
x^{n+1}
\end{array}
$$

---

† These problems may prove difficult, but by trying them now the student will be better able to appreciate the ease with which they are handled later.

We therefore have

$$\frac{1}{1 - x} = 1 + x + x^2 + \cdots + x^n + \frac{x^{n+1}}{1 - x}.$$

It is clear that this series converges (that is, the remainder term goes to zero as n increases) if and only if $|x| < 1$.

We just used algebra to obtain a power series expansion of G about $x = 0$. The power series turned out to be the one we previously obtained by computing derivatives. This raises a question. Suppose we find by some trickery that a function F may be represented by a certain power series in a neighborhood of x_0. Will this power series always turn out to be the Taylor series? The next theorem says that the answer is yes.

THEOREM E

Suppose the function F is represented by a power series

$$F(x) = \sum_{n=0}^{\infty} a_n(x - x_0)^n,$$

which is known to converge in some interval about x_0. Then the series must be the Taylor series corresponding to F; that is

$$a_n = \frac{D^n F(x_0)}{n!}.$$

PROOF:

$$F(x) = a_0 + a_1(x - x_0) + a_2(x - x_0)^2 + \cdots + a_n(x - x_0)^n + \cdots;$$

$$DF(x) = \quad a_1 \quad + 2a_2(x - x_0) + \cdots + na_n(x - x_0)^{n-1} + \cdots;$$

$$\vdots$$

$$D^n F(x) = \quad\quad\quad\quad\quad\quad\quad n!a_n + (n + 1)!a_{n+1}(x - x_0) + \cdots.$$

Therefore

$$F(x_0) = a_0,$$

$$DF(x_0) = a_1,$$

$$\vdots$$

$$D^n F(x_0) = n!a_n. \quad\blacksquare$$

COROLLARY

If a function F is represented on (a, b) by two power series $\sum_{n=0}^{\infty} a_n(x - x_0)^n$ and $\sum_{n=0}^{\infty} b_n(x - x_0)^n$, then we must have $a_n = b_n$ for each n.

The corollary is immediate since, according to the theorem,

$$a_n = b_n = \frac{D^n F(x_0)}{n!}.$$

This opens the door to finding the Taylor series by any trick available.

EXAMPLE B

Solve Problems F and G by other methods.

Since $G(x) = 1/(1 - x) = 1 + x + x^2 + x^3 + \cdots$, $|x| < 1$, we may, upon replacing x by $-t$, get

$$\frac{1}{1 + t} = 1 - t + t^2 - t^3 + \cdots, \qquad |t| < 1.$$

We recall that

$$\ln x = \int_1^x \frac{1}{u}\, du.$$

Letting $u = 1 + t$, $du = dt$ and we have

$$\ln x = \int_0^{x-1} \frac{dt}{1 + t}.$$

Therefore

$$\ln x = \int_0^{x-1} \frac{dt}{1 + t} = \int_0^{x-1} [1 - t + t^2 - \cdots]\, dt.$$

Now according to Theorem D, if x is chosen so that $x - 1$ is interior to $(-1, 1)$, we may integrate term by term to obtain

$$\ln x = \left(t - \frac{t^2}{2} + \frac{t^3}{3} + \cdots \right) \Big|_{t=0}^{t=x-1}$$

$$\ln x = (x - 1) - \frac{(x - 1)^2}{2} + \frac{(x - 1)^3}{3} + \cdots \qquad \text{for } |x - 1| < 1.$$

Note how easily we have obtained the interval of convergence.

This expression is sometimes used to compute ln 2. Such a computation is not justified by Theorem D because 2 is not interior to the interval of convergence. But the computation can be justified on other grounds. (See [1, pages 97–98].)

The function R suggests that we begin with

$$G(-x^2) = \frac{1}{1+x^2} = 1 - x^2 + x^4 - x^6 + \cdots,$$

which converges for $|x| < 1$. Hence,

$$R(x) = \frac{x}{1+x^2} = x - x^3 + x^5 - x^7 + \cdots. \blacksquare$$

PROBLEMS

H · Represent $A(x) = $ Arctan x in a Maclaurin series.

I · Represent $G(x) = 1/(1-x)$ in a power series about $x_0 = \frac{1}{2}$.

Sometimes the reverse problem comes to our attention. We have a power series representation of a function. We seek to represent the function in terms of the familiar elementary functions.

EXAMPLE C
Can

$$F(x) = 2x - \frac{4x^3}{3!} + \frac{6x^5}{5!} - \frac{8x^7}{7!} + \cdots$$

be expressed in terms of elementary functions?

We observe immediately that the numerators appear to be the derivatives of more simple expressions

$$D^{-1}F(x) = x^2 - \frac{x^4}{3!} + \frac{x^6}{5!} - \frac{x^8}{7!} + \cdots.$$

The alternating terms, the factorials in the denominator, and the fact that every other power of x is missing all suggest the trigonometric functions.

$$D^{-1}F(x) = x\left[x - \frac{x^3}{3!} + \frac{x^5}{5!} - \frac{x^7}{7!} + \cdots\right],$$

$$D^{-1}F(x) = x \sin x,$$

$$F(x) = D(x \sin x) = x \cos x + \sin x.$$

This result may now be checked by multiplying (3) by x and adding it to (4). ■

We have now seen that it is very easy to work with power series representations of functions. Indeed, it is almost safe to summarize what we have learned by saying that one should proceed simply by doing the obvious thing whenever working with power series. This rule of thumb extends even to topics that have not yet been discussed. We can, for example, obtain the power series expansion of the sum of $F(x)$ and $G(x)$, given

$$F(x) = a_0 + a_1(x - x_0) + a_2(x - x_0)^2 + \cdots,$$
$$G(x) = b_0 + b_1(x - x_0) + b_2(x - x_0)^2 + \cdots,$$

by simply adding coefficients of corresponding terms. Thus

$$F(x) + G(x) = (a_0 + b_0) + (a_1 + b_1)(x - x_0) + (a_2 + b_2)(x - x_0)^2 + \cdots.$$

Similarly,

$$F(x)G(x) = a_0 b_0 + (a_0 b_1 + a_1 b_0)(x - x_0) \\ + (a_0 b_2 + a_1 b_1 + a_2 b_2)(x - x_0)^2 + \cdots.$$

Composition of functions and division of one function by another (nonzero) function may be obtained by direct computation. The student needs to be warned that this technique of proceeding with the obvious can lead to some serious problems if the series involved are not power series. This warning also applies to rearranging the terms of a series, an operation best confined to series known to be absolutely convergent.

14·4 EXERCISES

In Exercises 1–18, find the domain of the function F defined by the indicated series.

1. $\displaystyle\sum_{n=1}^{\infty} n(x + 2)^n.$

2. $\displaystyle\sum_{n=1}^{\infty} \frac{n^2(x - 1)^n}{2^n}.$

3. $\displaystyle\sum_{n=1}^{\infty} \frac{(-9)^n}{n+1} (x + 1)^{2n}.$

4. $\displaystyle\sum_{n=1}^{\infty} \frac{2n - 1}{n} x^n.$

5. $\displaystyle\sum_{n=1}^{\infty} \frac{n^2 x^{3n}}{8^n}.$

6. $\displaystyle\sum_{n=1}^{\infty} \left(\frac{\pi}{2n}\right)^n x^n.$

7. $\displaystyle\sum_{n=1}^{\infty} \frac{n! x^n}{n^{2n}}.$

8. $\displaystyle\sum_{n=1}^{\infty} \frac{(-1)^n x^{2n} n}{4^n}.$

9. $\displaystyle\sum_{n=1}^{\infty} (-1)^{n+1} \frac{(x - 3)^n}{n}.$

10. $\displaystyle\sum_{n=1}^{\infty} \frac{\pi^n x^n}{n^3}.$

11. $\displaystyle\sum_{n=1}^{\infty} (-1)^n \frac{(x-2)^n}{(n+3)3^n}.$

12. $\displaystyle\sum_{n=1}^{\infty} \frac{n}{x^n}.$

13. $\displaystyle\sum_{n=1}^{\infty} \left(\frac{2x}{3x+5}\right)^n.$

14. $\displaystyle\sum_{n=1}^{\infty} (-1)^n \frac{x^n}{n3^n}.$

15. $\displaystyle\sum_{n=1}^{\infty} n^n x^n.$

16. $\displaystyle\sum_{n=1}^{\infty} \frac{(-1)^n}{3n+1}\left(\frac{1-x}{1+x}\right)^n.$

17. $\displaystyle\sum_{n=1}^{\infty} \frac{n^2+2n}{n(x+1)^n}.$

18. $\displaystyle\sum_{n=1}^{\infty} \frac{(nx)^n}{n!}.$

In Exercises 19–24 we have listed some common functions for which you should know or be able to very quickly derive the Maclaurin expansion. You should also know the radius of convergence.

19. $\sin x.$

20. $\cos x.$

21. $e^x.$

22. $\dfrac{1}{1-x}.$

23. $\ln(x+1).$

24. Arctan $x.$

Use your knowledge of the Maclaurin series for the functions of Exercises 19–24 in order to find Maclaurin series for the following.

25. $\cos x^2.$

26. $\dfrac{\sin x}{x}.$

27. $\dfrac{1-x}{1+x^2}.$

28. $\dfrac{1}{e^{x^2}}.$

29. $\dfrac{\text{Arctan } x}{x}.$

30. $\ln(1-x).$

31. $\tan x.$

32. $\ln \dfrac{1-x}{1+x}.$

In each of the following exercises, you are given an infinite series that represents a function that can be expressed in terms of elementary functions. Do so.

33. $\displaystyle\sum_{n=1}^{\infty} nx^n.$

34. $\displaystyle\sum_{n=1}^{\infty} n^2 x^n.$

35. $\displaystyle\sum_{n=0}^{\infty} (n+1)(n+2)\frac{x^n}{2}.$

36. $\displaystyle\sum_{n=1}^{\infty} (n+1)x^n.$

37. $\displaystyle\sum_{n=0}^{\infty} \frac{(n+1)x^n}{(n+2)!}.$

38. $\displaystyle\sum_{n=1}^{\infty} \frac{(n+1)x^n}{n!}.$

39. $\displaystyle\sum_{n=0}^{\infty} \frac{(2n+2)x^{2n+1}}{(2n+3)!}(-1)^{n+1}.$

40. $\displaystyle\sum_{n=0}^{\infty} (-1)^n(2n+2)\frac{x^{2n+1}}{(2n)!}.$

41 · Find the Maclaurin series for $B(x) = (1 + x)^r$ where r is a rational number. Examine your solution for $r = 2, 3, 4$.

42 · Consider the series $\sum_{n=1}^{\infty} 39/(100)^n$. It converges to a number r. Express r as a rational number. Express r in decimal form.

43 · Find $\sin \frac{1}{2}$ accurate to within one unit in the third decimal place.

44 · Find $\cos \frac{1}{2}$ accurate to within one unit in the third decimal place.

45 · Find e^{-1} accurate to within one unit in the second decimal place. Use this answer to approximate e.

46 · Find $e^{1/2}$ accurate to within one unit in the second decimal place. Use this answer to approximate e.

47 · Find $\int_0^{1/2} \sqrt{1 + x^2}$ accurate to within one unit in the second decimal place.

48 · Find $\int_{1/2}^{1} \ln x$ accurate to within one unit in the second decimal place.

49 · Find $\int_0^{1/2} \cos x^2$ accurate to within one unit in the third decimal place.

50 · Find $\int_0^{1/2} \sin x^2$ accurate to within one unit in the third decimal place.

DIFFERENTIAL EQUATIONS

... AS FAR AS THE PROPOSITIONS OF MATHEMA-TICS REFER TO REALITY, THEY ARE NOT CERTAIN; AND SO FAR AS THEY ARE CERTAIN, THEY DO NOT REFER TO REALITY.

Albert Einstein
Geometry and Experience

At the end of Section 5.3, we studied the motion of a freely falling body. This study proceeded as follows.

I. We accepted at face value the statement of a physicist, a man who tries to explain what he observes in the world about him. This statement asserted that a freely falling body accelerates at a constant rate designated by g.

II. We recalled a mathematical model with which we had worked. In this model we had assumed that the distance s of a moving object from some fixed point at time t was given by

$$s = F(t).$$

We then defined the velocity at time t to be

$$v = \frac{ds}{dt} = F'(t)$$

and the acceleration to be

$$a = \frac{d^2 s}{dt^2} = F''(t).$$

III. Relating these two ideas, we expressed the statement of the physicist with a mathematical equation.

$$\frac{d^2 s}{dt^2} = g. \tag{1}$$

This statement, as far as the mathematician is concerned, is an assumption about the world in which we live (two assumptions really; one about the way in which bodies fall and one about the accuracy of the model). From this statement, the mathematician deduces the formula

$$S(t) = \tfrac{1}{2} g t^2 + v_0 t + s_0. \tag{2}$$

This formula may or may not accurately predict the position of a falling body in an actual experiment. If it does not, three things may be responsible for the failure:

(1) The physicist may have given us wrong information to start with. (Maybe the acceleration due to gravity is not constant.)

(2) The model we are using may not accurately reflect the world as it is.

(3) The measurements obtained in the experiment may be wrong.

Our point is that in no case is the mathematics at fault. Given (1), the work of the mathematician is to derive (2). The validity of his work is not tested by whether or not his result corroborates experience. It is tested by whether or not the function S described in (2), satisfies Eq. (1). We may not be certain that we have really described something in the world of experience, but we can be certain of our mathematics.

In this chapter, perhaps more than in any mathematics he has seen up to this point, the reader becomes aware that his work proceeds on the basis of assumptions. When this realization first dawns it can be unsettling.† The purpose of our discussion above was to emphasize that we know we are making assumptions, and we know our model only approximates what we observe. But having expressed our assumptions in terms of the model, we may proceed with assurance. The work we do will only be tested in the sense of verifying that our solution satisfies the equations derived from the model.

A part of the mathematician's work is to express the assumptions being made as mathematical statements about the model to be used (step III above). Such statements usually take the form of equations. If this equation involves a single unknown function of a single variable and one or more of its derivatives (as does (1) above), then it is called an *ordinary differential equation*. (If the unknown function involves several variables and some partial derivatives, then the equation is called a *partial differential equation*. We shall not be concerned with equations of this type.) Developing skill at writing down such equations is the task to which we address ourselves in Section 15.1.

We have accepted (2) as the solution to (1). Since v_0 and s_0 are arbitrary constants, it appears that the solution to a differential equation may be not a function, but rather a whole family of functions. It is a new experience for most students to solve problems in which the answer is not unique. We have some comments to make about this in Section 15.2.

We turn to the business of actually solving some differential equations in Sections 15.3–15.5. There is not *a* method for solving such equations. Rather, there are many methods to try. Facility is gained as one increases the number of methods he has at his disposal. Our aims in this direction are modest. We actually consider few methods. Our purpose is to understand what is meant by a solution, to see the role played by initial conditions, and to introduce the methods that, in our view, are most basic in elementary applications. This chapter may be viewed as an introduction to the fuller treatment presented in a book such as Rabenstein's "Introduction to Ordinary Differential Equations" [5] to which we often refer.

The basic ideas of calculus were introduced in this text by consideration of a problem in approximation first introduced in Section 2.7. We have returned to the problem several times. We return to it once again in the last section of the text, thus introducing the reader to approximate methods for solving differential equations.

---

† The author well remembers his own loss of confidence in what he was doing when he was encouraged to work with frictionless planes, air that offered no resistance, and electrical circuits in which inductive reactance between components did not exist.

15 · 1   Problems That Lead to Differential Equations

This section consists of six examples, most of which are classical problems. Our purpose here is to show how we translate assumptions into mathematical language. No attempt will be made in this section to solve the differential equations that we set up.

Our first example is a modification of the falling body problem reviewed in the introduction to this chapter. We modify the problem by considering a body falling in water so that it encounters resistance as it falls. Specifically, we proceed on the basis of two assumptions.

Assumptions

1.   (Newton's Law)   The force acting upon a moving body is equal to the mass of the body multiplied by its acceleration.
2.   The resistance encountered by a body falling through a medium is directly proportional to the velocity of the falling body.

EXAMPLE A

While leaning over the side of his boat to rinse his hand, a fisherman lets a lead sinker slip from his grasp. Find a formula giving the distance y of the sinker from the surface t seconds after it was released.

We shall designate the downward (negative) force due to the weight of the sinker as W. The resistance R is directed upward (positive). The total force acting upon the sinker is the sum of these (signed) forces;

$$F = W + R.$$

The mass of the sinker is $W/g = W/32$. Newton's law gives

$$W + R = \frac{W}{g} \frac{d^2 y}{dt^2}.$$

We are assuming that $R = -k \dfrac{dy}{dt}$ (where $k > 0$). Thus,

$$\frac{W}{g} \frac{d^2 y}{dt^2} = W - k \frac{dy}{dt}.$$

If we replace the arbitrary constant k by $(W/g)C$, C now being the arbitrary constant, the equation can be written in the more simple form

$$\frac{d^2y}{dt^2} + C\frac{dy}{dt} = g.$$

The statement of the problem implies that when $t = 0$, $dy/dt = y = 0$. ∎

PROBLEM

A . A weight W slides down an inclined plane (frictionless, naturally) that makes an angle of α with the horizontal. Let s represent the distance of the weight from the point where it was released at time $t = 0$. Set up a differential equation relating s to t.

Our second example depends upon an assumption about springs together with a second assumption due to Newton about forces acting upon a body.

Assumptions

3. (Hooke's Law) The force exerted by a stretched spring is proportional to the distance it has been stretched.
4. (Newton) If a particle is not moving, the forces acting upon it must be in equilibrium.

EXAMPLE B

A spring hangs from a support. Attached to the spring is a weight W which, when at rest, stretches the spring to what we will call the equilibrium position. Suppose the weight W is pulled down d in. from the equilibrium position and then released. Describe the motion of W; that is, give the position of the weight t sec after its release.

We shall take the lower end of the unstretched spring as the zero position. The equilibrium position is then at some point s_0 (negative), and the force of the spring is directed upward (positive). Hooke's law says that the force exerted by the spring when the weight W rests in equilibrium is

$$F_0 = -ks_0, \qquad k > 0.$$

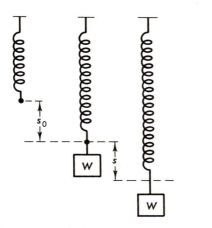

FIG. 15.1

Since W is at rest, this force must be balanced (Assumption 4) by the downward pull of gravity; that is, by the (negative) force mg where m is the mass of W and g is a negative constant. We have

$$mg = -ks_0. \tag{1}$$

At any time after its release, the distance of W from the equilibrium position is designated by s. The spring is then stretched $s + s_0$, and Hooke's law says the force exerted by the spring is

$$F_s = -k(s + s_0), \qquad k > 0. \tag{2}$$

(Note that if W is below the equilibrium position, s is negative, so F_s is larger than F_0. If W is above the zero position, then the positive number s is larger than $|s_0|$, so that $s + s_0 > 0$. Then the force given by (2) is negative, as we would expect.)

The weight W continues to exert a downward force $F_w = mg$. It is the imbalance between these two forces,

$$F = F_s - F_w,$$

that causes the weight to oscillate. According to assumption I at the beginning of this section, this force equals the mass of the weight multiplied by its acceleration.

$$F = m\frac{d^2s}{dt^2}.$$

Using our expressions above for F_s and F_w, and then using (1), we have

$$F = (-ks - ks_0) - (mg) = (-ks + mg) - mg = -ks.$$

Substitution for F gives us the desired differential equation.

$$m\frac{d^2s}{dt^2} = -ks, \qquad k > 0. \quad \blacksquare$$

PROBLEM

B · A particle moving on a straight line is said to be in simple harmonic motion if its acceleration is proportional to its distance from a fixed point on the line and is directed toward the point. Using the origin as the fixed point, set up a differential equation relating x to t if x gives the position at time t of a particle in simple harmonic motion along the X axis.

Our third example is based on an assumption about the way in which objects change temperature.

Assumption

5.   (Newton's Law of Heat Exchange).   The rate at which a body changes its temperature is directly proportional to the difference between the temperature of the body and the temperature of the surrounding medium.

EXAMPLE C
At noon a thermometer reading 75°F is taken outdoors, where the temperature is 15°F. Five minutes later the thermometer reading is 45°F. At 12:10 P.M. the thermometer is brought back inside, where the temperature is still 75°F. Find (a) the thermometer reading at 12:20 P.M.; (b) the time when the thermometer will show, to the nearest degree, the correct indoor temperature.

We let the thermometer reading at t minutes past noon be $y = F(t)$. We know that

$$F(0) = 75,$$

$$F(5) = 45.$$

Newton's law says the change in F is proportional to the difference between $F(t)$ and 15. This is expressed by writing

$$F'(t) = k(15 - y). \tag{3}$$

We have written $15 - y$ rather than $y - 15$ because we anticipate that $F'(t)$ will be negative. Thus, since we prefer to have k positive, we made the second factor negative (for y is surely greater than 15).

Equation (3) is valid so long as the surrounding medium is 15°; that is, valid for $t \in [0, 10]$. When the thermometer is moved back inside, we need a different equation. We will make a further assumption in writing this equation. We will assume that the constant of proportionality k depends only on the thermometer, thus remaining constant as the thermometer is moved. (This may not be the case, of course, if some other factor, such as a stiff breeze, contributed to the cooling outside.)

Our previous equation commits us to a $k > 0$. It is clear that $F'(t)$ will be positive after the thermometer has been brought inside again, and the temperature y will surely be less than 75°F. Hence, we have the equations

$$\frac{dy}{dx} = F'(t) = \begin{cases} k(15 - y), & t \in [0, 10], \\ k(75 - y), & t \geq 10. \end{cases}$$

PROBLEMS

C· A tank contains 10 gallons of brine in which there are 2 pounds of dissolved salt. Brine with 3/2 lb of salt per gallon enters the tank at 3 gal/min and the well-stirred mixture leaves the tank at 4 gal/min. Let s denote the amount of salt in the tank at time t. Set up a differential equation relating s to t for $t \in [0, 20]$.

D· In certain chemical reactions, the rate at which one substance is converted into another is proportional to the amount of the first substance that remains. Let x be the amount of the first substance at time t. Relate x and t by means of a differential equation.

EXAMPLE D
A right circular cylinder (a can) half full of water is rotated at a constant velocity about its axis. After a while the water and can will rotate together and the water will assume a shape similar to that pictured in Fig. 15.2. Describe the surface formed by the water.

We concentrate our attention on a particle P on the surface of the water. When stability has been achieved, the particle is stationary with respect to the can. This means (Assumption 4 above) that the forces acting upon it within the can must be in equilibrium. There are two forces acting on P (Fig. 15.3).

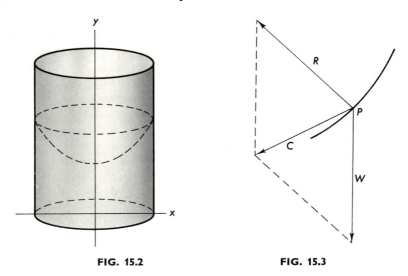

FIG. 15.2   **FIG. 15.3**

(a)   The weight $W = mg$ (its mass multiplied by the pull of gravity) acts downward.

(b)   The force R exerted by the other particles of water must act normal to the surface of the water at P. (Otherwise R would have a component tangent to the surface that would force P to move along the surface.)

Let the resultant of these two forces be the vector C. Impose a coordinate system as indicated in Fig. 15.4. Then P rotates about the y axis at a constant velocity that we shall designate by ω radians per second. We have shown (Example 9.4A) that in such a case, the acceleration vector for P is directed toward the center of rotation and has a magnitude equal to the radius multiplied by the square of the angular velocity.

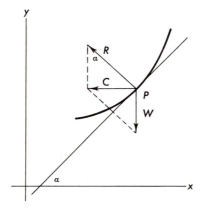

FIG. 15.4

For our example, this means that the acceleration vector at P is directed toward the y axis and has a magnitude of $x\omega^2$. According to assumption I above, force vector C must be the product of the mass m of P multiplied by the acceleration vector. That is, C must be directed toward the y axis (not tilted from the horizontal, as we tentatively suggested by our drawing in Fig. 15.3), and its length must be $mx\omega^2$.

We consider now the cross section (Fig. 15.4) of the surface made by the xy plane indicated in Fig. 15.2. Let α be the angle made with the positive x axis by the tangent to the curve at P. Then

$$\sin \alpha = \frac{m\omega^2 x}{|R|},$$

$$\cos \alpha = \frac{mg}{|R|}.$$

Hence,

$$\tan \alpha = \frac{m\omega^2 x}{mg}.$$

The tangent of this angle is, of course, the slope of the desired curve at the (arbitrary) point P.

$$\frac{dy}{dx} = \frac{\omega^2}{g}x. \quad\blacksquare$$

Our next example is based on some ideas about the way in which light is reflected.

Assumption

6. Light travels in rays, and when such a ray is reflected, the angle of incidence equals the angle of reflection.

EXAMPLE E

Design a reflector so that each ray of light from a small bulb will be reflected parallel to a fixed line.

We will let the bulb be at the origin and choose the x axis as the fixed line. The desired reflector will be generated by rotating a curve, the graph of $y = F(x)$, about the x axis. Our problem is to find the function F.

Choose any $P(x, y)$ on the curve. Draw the tangent line at P, and draw the ray from 0 to P. This ray is to be reflected parallel to the x axis, and we know that the ray and segment OP form equal angles α with the tangent line (Fig. 15.5). It is also clear, since PR is parallel to the x axis, that $\angle OAP$ is equal to α. Finally, while we are making random observations about our drawing, we note that $\beta = 2\alpha$. (The exterior angle of a triangle equals the sum of the two remote interior angles.)

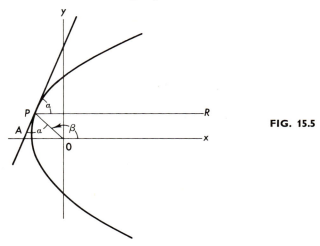

FIG. 15.5

Now what facts can we assemble? We note that since a is the angle formed by the tangent line with the positive x axis,

$$\frac{dy}{dx} = \tan \alpha.$$

It is also clear from the definition of the tangent function that

$$\tan \beta = \frac{y}{x}.$$

No relation appears involving 2α, but since $2\alpha = \beta$ we can write

$$\alpha = \beta - \alpha.$$

Then the information available suggests that we use the formula

$$\tan(\beta - \alpha) = \frac{\tan \beta - \tan \alpha}{1 + \tan \beta \tan \alpha}.$$

This gives

$$\tan \alpha = \frac{\tan \beta - \tan \alpha}{1 + \tan \alpha \tan \beta}.$$

Now using our expressions for tan α and tan β,

$$\frac{dy}{dx} = \frac{(y/x) - (dy/dx)}{1 + (y/x)\, dy/dx}.$$

Simplification gives

$$y = y\left(\frac{dy}{dx}\right)^2 + 2x\,\frac{dy}{dx}. \quad \blacksquare$$

PROBLEMS

E· Find a curve passing through (2, 4) that has the property that the line tangent to the curve at any point (x, y) cuts the x axis at $x/2$.

F· The equations $x^2 - y^2 = C$ determine a family F of curves. Find a second family of curves G so that whenever a curve from F intersects a curve from G, they intersect at right angles. (Either family is said to be an *orthogonal trajectory* for the other.)

Our last example† is again based on the idea of equilibrium of forces acting on a fixed point.

EXAMPLE F

A cable (wire, clothesline) is suspended between two points as indicated in Fig. 15.6. Find the shape that this cable assumes.

With an eye on symmetry, we choose the y axis to be midway between the points of support, intersecting the curve at A. We then look at a segment of the cable from A to an arbitrary point $P(x, y)$. The forces acting on this segement are (see Fig. 15.7):

(a)   its weight $|W| = \omega s$ where ω is a constant depending on the material from which the cable is made and s is the length of the arc AP;

(b)   the force H exerted by the pull of the cable at A;

(c)   the force T exerted by the pull of the cable at P.

Of course, H and T act along a line tangent to the cable at A and P, respectively. Let T make an angle of α with the positive x axis. Balancing the horizontal and vertical components of the forces acting upon the cable leads to the following equations.

$$|W| = |T|\,\sin \alpha$$
$$|H| = |T|\,\cos \alpha.$$

† Omit this example if Chapter 9 was skipped.

FIG. 15.6

Division gives

$$\frac{|W|}{|H|} = \tan \alpha.$$

Since $|W| = \omega s$ and $\tan \alpha = dy/dx$

$$\frac{dy}{dx} = \frac{\omega s}{|H|}. \tag{4}$$

Now this equation involves three variables; y, x, and s. But we recall that s is related to x and y by the differential equation

$$ds = \sqrt{1 + \left(\frac{dy}{dx}\right)^2}\, dx.$$

This suggests differentiating (4).

$$\frac{d^2y}{dx^2} = \frac{\omega}{|H|}\frac{ds}{dx} = \frac{\omega}{|H|}\sqrt{1 + \left(\frac{dy}{dx}\right)^2}. \; \blacksquare$$

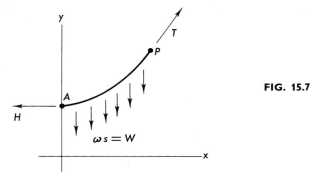

FIG. 15.7

If $y = F(x)$, then a common notational device in the study of differential equations is to use y' to designate $F'(x)$. This notation is extended in the obvious way to higher derivatives; hence $y'' = F''(x)$, etc. The six equations derived in this section are thus written

$$y'' + cy' = g, \tag{A}$$

$$ms'' + ks = 0, \tag{B}$$

$$y' + ky = 15k, \tag{C}$$

$$y' = \frac{\omega^2}{g} x, \tag{D}$$

$$y = y(y')^2 + 2xy', \tag{E}$$

$$y'' = \frac{\omega}{|H|} \sqrt{1 + (y')^2}. \tag{F}$$

We said in the introduction that an ordinary differential equation is an equation involving a single unknown function of a single variable and one or more of its derivatives. The *order* of a differential equation is the order of the highest order derivative of the unknown function that appears in the equation. Thus, the equations from Examples A, B, and F are second order; those from Examples, C, D, and E are first order.

A second-order differential equation of the form

$$a_2(x)y'' + a_1(x)y' + a_0(x)y = Q(x)$$

is said to be *linear*, and the extension of this definition to equations of order n is obvious. The linear equation is called *homogeneous* if $Q(x) = 0$. Equations A, B, C and D are linear, but only B is homogeneous.

For equations that are polynomials in the derivatives, the greatest exponent appearing on the derivative of highest order is called the *degree* of the equation. Equations D and E are first order, but of these two, only D is first degree. The notion of degree is not defined for the second order equation F.

===================================== 15·1 EXERCISES

Each of the following exercises leads to a differential equation. Write this equation.

1 · A pan of water is brought to a boil (212°F) and then allowed to stand in a room where the temperature is 72°F. Two minutes later the water is 196°F. Find the temperature after another 2 minutes have passed.

2 · An aluminum meat platter is heated in an oven to 400°F. It is then allowed to stand in a room where the temperature is 80°F. The temperature of the plate after 6 min is 240°F. How long will it be before the temperature of the platter reaches 100°F? (Assume the platter cools uniformly throughout, since the more realistic assumption of the edges cooling faster leads to a much more difficult problem.)

3. A solution containing 3 lb of salt per gallon flows into a tank at 6 gal/min. The tank originally had 180 lb of salt dissolved in 180 gal of liquid. If the well-mixed solution is drained off at the rate of 3 gal/min, how much salt will be in the tank at the end of 10 min?

4. A 500-gal tank is filled with water. A salt solution containing 2 lb of salt per gallon enters the tank at 5 gal/min. The well-mixed solution is being drained off at the same rate. When will the concentration of the salt in the solution being drained off reach 1 lb/gal?

5. The growth of population in a certain city is found to be directly proportional to the number of people in the city. It is also known that the city grew from 500,000 to 1 million from 1960 to 1970. Estimate the population in 1990.

6. Bacteria in a certain culture increase at a rate proportional to the number present. If the number of bacteria originally present doubles in 20 min, how will the number present in 1 hr compare with the original number?

7. Find the orthogonal trajectories of the family $x^2 = ay$. Sketch a graph showing the two curves that intersect at $(1, 2)$.

8. Find the orthogonal trajectories of the family $y = Cxe^x$. Sketch a graph showing the two curves that intersect at $(1, e)$.

9. An outside thermometer reading 100°F is brought into an air-conditioned room kept at 74°F. After 2 min it reads 86°F. Four minutes after being brought in, it is put in a refrigerator where the temperature is 36°F. Find the reading on the thermometer 4 min after it is put into the refrigerator.

10. A thermometer is brought from the outdoors into a room kept at 72°F. After 2 min the thermometer reads 48°F and after 4 min the thermometer reads 60°F. What was the temperature outside?

11. A mixture containing 2 lb of salt per gallon flows at the rate of 2 gal/min into a tank originally containing 100 gal of water. The well-mixed solution is drained off at the same rate. At noon the tank contains 100 lb of salt. What time was this process started?

12. Fresh water flows into a tank originally containing 200 lb of salt dissolved in 100 gal of water. The fresh water comes in at 2 gal/min and the well-mixed solution is drained off at 3 gal/min. How long after this process begins will the tank contain 100 lb of salt?

13. Suppose that the rate at which a cube of sugar dissolves is proportional to the surface area. If the edge of the cube is reduced from $\frac{1}{2}$ in. to $\frac{1}{4}$ in. in 20 sec, how long will it be before the edge is $\frac{1}{8}$ in.?

14. A radioactive substance decays at a rate directly proportional to the amount remaining. If the half-life (time it takes for half the original amount of decay) of such a substance is 1600 years, what percentage of the substance will be lost in 400 years?

15 · A curve $y = F(x)$ passing through $(4, 3)$ has the property that if P is on the curve, then the line from the origin to P and the line through P and perpendicular to the tangent at P will form an isosceles triangle having the x axis as its base. Find the curve.

16 · A curve $y = F(x)$ passing through $(1, 2)$ has the property that if a tangent is drawn at a point P on the curve, then that segment of the tangent between P and the x axis is bisected by the y axis. Find the curve.

17 · A chemical process transforms one substance into another at a rate proportional to the amount of the first substance remaining untransformed. If the process is begun with 100 gm and 50 gm remain at the end of 10 minutes, how much will remain at the end of of $\frac{1}{2}$ hr?

18 · A chemical process transforms chemical A into chemical B at a rate directly proportional to the amount of A that remains. If 40 gm of A remain after 5 min and 24 gm remain after 10 min, how much of chemical A was present at the start of the reaction?

15 · 2 Families of Solutions

What does it mean to say that we want to solve the differential equation

$$\frac{dy}{dx} = x + y? \tag{1}$$

It simply means that we wish to find a function F so that if we set $y = F(x)$, then

$$F'(x) = x + F(x).$$

For all we know, there may not be such a function. On the other hand, there may be many such functions. We shall never find out by simply staring at the problem, so we cautiously proceed by asking ourselves what would have to be true of any solution(s).

If we approach the problem geometrically, we observe that the graph of F would have an interesting property. The slope at any point (x, y) on the graph would equal the sum of the x and y coordinates. If there is a solution passing through $(1, 1)$, then the slope of the graph at that point would be 2. If there is a solution passing through $(1, 2)$, its slope would be 3.

A small line segment (called a *lineal element* in this context) drawn at an arbitrary point gives the direction of the graph of any solution passing through this point (Fig. 15.8).

Perhaps such lineal elements drawn at many points would help us understand our solutions. (We are reminded of the old experiment in which we learn something about a magnet by covering it with a piece of paper and sprinkling iron filings on the paper.)

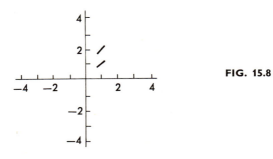

FIG. 15.8

Greater efficiency will result in our effort to plot lineal elements if we look for all places where the lineal elements have the same slope. For example, by setting $dy/dx = 2$ in (1), we see that all the lineal elements along the line $x + y = 2$ should be parallel to the one already drawn at $(1, 1)$. In like manner, all lineal elements along the line $x + y = k$ should have slope k.

A look (Fig. 15.9) at our resulting graph (called a *direction field* for the given differential equation) suggests two things:

(1)   There is something distinctive about the elements along the line $x + y = -1$. In fact, $y = F(x) = -x - 1$ has exactly the right slope at each point on its graph.

(2)   There appear to be many solutions for which the graphs have the general shape indicated by the sketch in Fig. 15.10.

From the first observation it is only a step to verify that $y = -x - 1$ is indeed a solution to (1). Finding other solutions is more difficult. Methods we shall develop later will enable us to recognize the other curves as graphs of the function $y = Ce^x - x - 1$.

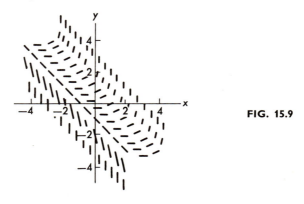

FIG. 15.9

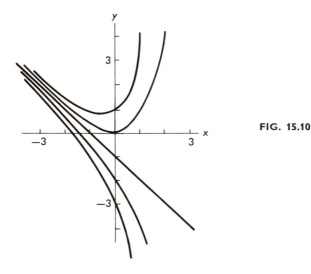

FIG. 15.10

PROBLEMS

A · Verify that for any constant C, $y = Ce^x - x - 1$ is a solution to (1).

B · Find a solution to (1) that passes through $(1, 1)$.

C · Sketch the direction field for the differential equation $x + y \, dy/dx = 0$. Then try to guess at the solutions.

We are not trying to suggest that graphing is the best way to solve a differential equation. We are trying to emphasize several other things.

(i) Solutions to differential equations are functions. These functions can be graphed, evaluated at a point, etc.

(ii) We may commonly expect a differential equation to have not one solution, but a whole family of solutions.

(iii) From among all the solutions, we can single out a special one if we specify a certain point through which its graph must pass.

The multiplicity of solutions is reflected in the formula of the solution functions by the appearance of arbitrary constants. Thus, solutions to the first-order differential equation

$$y' = x + y$$

are given by $y = Ce^x - x - 1$ involving an arbitrary constant C. We pointed out in the introduction to this chapter that solutions to the second-order differential equation

$$y'' = g$$

are given by $y = \frac{1}{2}gt^2 + v_0 t + s_0$ involving two arbitrary constants v_0 and s_0. It is generally the case that the solution of an nth order differential equation will involve n constants in such a way as to enable us to satisfy n special conditions. This rule of thumb does have exceptions, and it is sometimes difficult to tell whether our solution involves one or two constants. (For instance $Ce^{a+x} = Ce^a e^x = C_1 e^x$.) We shall have more to say about these matters, but we wish first to examine the important case of linear differential equations where the relation between the order of the equation and the number of arbitrary constants can be made very specific.

We begin with a review of some things learned in Chapter 10 about solving the linear equation

$$y_2 - 5y_1 + 6y_0 = 0. \tag{2}$$

I.   A solution (y_2, y_1, y_0) was properly viewed as a vector in the vector space R^3 of all ordered triples.

II.   It is possible to find two linearly independent solutions, such as $(5, 1, 0)$ and $(-6, 0, 1)$.

III.   All ordered triples of the form

$$(y_2, y_1, y_0) = r(5, 1, 0) + s(-6, 0, 1) \tag{3}$$

are also solutions. They form a subspace.

IV.   All solutions (y_2, y_1, y_0) of (2) may be represented in the form (3).

The subspace described in III is called the solution space for equation (2). The vectors $(5, 1, 0)$ and $(-6, 0, 1)$ form a basis for the solution space.

Guided by these reminders, we turn our attention to the linear differential equation

$$y'' - 5y' + 6y = 0. \tag{4}$$

I.   A solution to (4) is any function $F(x)$ such that $F''(x) - 5F'(x) + 6F(x) = 0$ for all x. Such a function may be viewed as a member of the set of functions continuous for all x. It is easy to verify from Definition 10.2A that this set is a vector space. (One needs to check the eight axioms, noting for example that if F and G are everywhere continuous, then so is $F + G$, etc.) Hence, a solution to (4) is properly viewed as a member of a vector space.

II.   It is possible to find two linearly independent solutions, such as $F(x) = e^{2x}$ and $G(x) = e^{3x}$. If the reader is willing to set aside for the moment questions of how we discovered these functions, they can both be readily verified by direct substitution in (4). Linear independence follows from the following consideration. Suppose that for all x,

$$re^{2x} + se^{3x} = 0.$$

Differentiating,

$$2re^{2x} + 3se^{3x} = 0.$$

Since these equations must hold for all x, including $x = 0$, we see that

$$r + s = 0,$$
$$2r + 3s = 0,$$

from which it follows that $r = s = 0$.

III.   All functions of the form

$$H(x) = rF(x) + sG(x) = re^{2x} + se^{3x} \tag{5}$$

are also solutions. (This is again easily verified by direct substitution.) They form a subspace.

IV.   All solutions of (4) may be represented in the form (5). (Although it is beyond the scope of our work, it can be shown that a second-order linear differential equation cannot have three linearly independent solutions. Hence, any solution to (4) must fall in the two dimensional subspace described by (5).)

The subspace described in III is called the solution subspace for Equation (4). The functions e^{2x} and e^{3x} form a basis for the solution space.

PROBLEMS

D · Show that the functions $F(x) = e^{ax}$ and $G(x) = e^{bx}$ are linearly independent for $a \neq b$.

E · Show that the functions $F(x) = e^{rx}$ and $G(x) = xe^{rx}$ are linearly independent.

F · Show that if $F(x)$ and $G(x)$ are two solutions to the differential equation $a_0 y'' + a_1 y' + a_2 y = 0$, then $H(x) = rf(x) + sG(x)$ is also a solution for any real values of r and s.

Although we shall not prove it, we should now be ready to understand and believe the following.

THEOREM A

The set of all solutions to the differential equation

$$a_2 y'' + a_1 y' + a_0 y = 0$$

forms a two-dimensional subspace of the space of all continuous functions.

One problem in solving a differential equation is to find a solution involving arbitrary constants. A second problem, often more difficult, is to be certain that any solution can be obtained by a proper choice of the constants.

EXAMPLE A

Consider the differential equation $y' + 2xy = 1 + 2x^2$.
(a) Verify that $y = Ce^{-x^2} + x$ is a solution for any choice of C.
(b) Verify that if $y = F(x)$ is a solution, then

$$F(x) = Ce^{-x^2} + x. \tag{6}$$

The first part is routine, and we leave it for the student to do himself. (The first part is instructive, and we urge the student to do it himself.) Our task in part (b) is to show that an arbitrary solution $F(x)$ must take the form (6). Given such an arbitrary solution $F(x)$, we have

$$F'(x) + 2xF(x) = 1 + 2x^2.$$

Now if we set $G(x) = e^{-x^2} + x$, it follows from part (a) that

$$G'(x) + 2xG(x) = 1 + 2x^2.$$

Subtraction gives

$$F'(x) - G'(x) + 2x[F(x) - G(x)] = 0.$$

The conclusion is obvious if $F(x) = G(x)$ for each x, so we suppose they are distinct. Then

$$\frac{F'(x) - G'(x)}{F(x) - G(x)} = -2x \qquad \text{wherever} \quad F(x) - G(x) \neq 0.$$

Therefore

$$\ln|F(x) - G(x)| = -x^2 + C,$$

$$|F(x) - G(x)| = e^{-x^2 + C} = e^{-x^2}e^C = C_1 e^{-x^2}. \tag{7}$$

$$F(x) - G(x) = \pm C_1 e^{-x^2}.$$

We see here that since F and G are distinct, $C_1 \neq 0$, so the right side is never zero. Hence, $F(x) - G(x) \neq 0$ for all x. This means that (7) is valid for all x, so (allowing C_1 to be either positive or negative)

$$F(x) = C_1 e^{-x^2} + G(x);$$

that is, using the definition of G,

$$F(x) = C_1 e^{-x^2} + e^{-x^2} + x,$$

$$F(x) = C_2 e^{-x^2} + x. \ \blacksquare$$

We now have an example of a first-order linear differential equation in which all the solutions may be obtained from a formula involving one constant. On the basis of Theorem A and its natural extension to nth order equations, we know that in the case of nth-order homogeneous linear differential equations with constant coefficients, all solutions may be obtained from a formula involving n constants. It is now time to run up a warning flag before the reader jumps to an all too obvious conclusion.

PROBLEM

G · Consider the differential equation $y = xy' + 2(y')^2 - y'$.

(a) Verify that for any C, $y = Cx + 2C^2 - C$ is a solution.
(b) Verify that $y = \frac{1}{8}(-x^2 + 2x - 1)$ is also a solution.

It is clear that no choice of C in part (a) will produce the solution of part (b) in the preceding problem. It is just because such things can happen that we have consistently hedged a bit in stating any rule about the number of constants appearing in the general solution. There is, in addition, the other problem to which we have alluded of deciding just how many constants are involved in the general solution. A painstaking, humorous, and sometimes impassioned discussion of this problem is to be found in R. P. Agnew, " Differential Equations," 2nd ed., Chap. 4 (McGraw-Hill, New York, 1960).

We shall continue to be guided by the intuitive notion that an nth order differential equation has a general solution involving n constants. We will bear in mind, however, that while such a solution often includes all other solutions (by choosing the constants), it can also happen that there are other solutions as well.

In this section we have raised two important questions concerning an arbitrary differential equation.

The existence question: Is there a function or family of functions that satisfies the equation?

The uniqueness question: If we succeed in finding a solution (or a family of solutions), can we ever be certain that we have found all the solutions?

Adequate answers to either of these questions are outside of the scope of this book. Indeed, it is not always possible to give answers when confronted with a specific example.

We will not go far wrong in the problems we consider if we simply assume that there is a solution. Fortunately, with respect to the second question, physical considerations giving rise to a differential equation suggest the kind of solution we expect. Hence, if we find such a solution, we commonly consider our problem solved without worrying about the possibility of other solutions. With the warning that things can go wrong, and with the encouragement to believe that they probably will not, we proceed.

15 · 3   Linear Differential Equations with Constant Coefficients

A second-order linear homogeneous equation having constant coefficients takes the form

$$a_2 y'' + a_1 y' + a_0 = 0. \tag{1}$$

Theorem 15.2A tells us that we are to expect solutions of the form

$$y = C_1 F(x) + C_2 G(x),$$

where F and G are linearly independent functions. As yet, however, we have no clue as to how the solutions $F(x)$ and $G(x)$ are to be determined. It is to this problem that we now address ourselves.

We will learn a great deal in a hurry if we tentatively set $y = e^{rx}$ where r is a constant not yet specified. Now $y' = re^{rx}$ and $y'' = r^2 e^{rx}$, so substitution in (1) gives

$$(a_2 r^2 + a_1 r + a_0)e^{rx} = 0.$$

Since e^{rx} is never zero, this product will be zero if and only if

$$a_2 r^2 + a_1 r + a_0 = 0. \tag{2}$$

This tells us how to choose the constant r. We can now shed light on how we obtained in II′ of Section 15.2 the solutions to $y'' - 5y' + 6y = 0$.

EXAMPLE A

Solve $y'' - 5y' + 6y = 0$.

Equation (2) becomes

$$r^2 - 5r + 6 = 0,$$

$$(r - 3)(r - 2) = 0.$$

Therefore $y = e^{2x}$ is a solution. A second solution is given by $y = e^{3x}$, and the general solution is $y = C_1 e^{2x} + C_2 e^{3x}$.

Equation (2) is called the *characteristic equation* for (1), and its roots are called the *characteristic roots*. We have in the past represented y' by Dy and y'' by D^2y. If we rewrite (1) using this notation, then we have

$$(a_2 D^2 + a_1 D + a_0)y = 0.$$

In this form the characteristic equation is immediately obvious; it is

$$a_2 D^2 + a_1 D + a_0 = 0.$$

Taking r_1 and r_2 to be the characteristic roots, it is natural to expect that the general solution will be

$$y = C_1 e^{r_1 x} + C_2 e^{r_2 x}.$$

This is almost the case, but there are some difficulties that we have not yet mentioned. An alert reader may have thought of them. Every reader will have thought of them after trying to solve the next two problems.

PROBLEMS

A · Solve $(D^2 + 2D + 1)y = 0$.

B · Solve $(D^2 + D + 1)y = 0$.

The first difficulty is that the characteristic roots may be equal, in which case our functions $e^{r_1 x}$ and $e^{r_2 x}$ are not linearly independent (Problem 15.2D). In this case, however, the differential equation takes the form

$$(D - r)^2 y = 0,$$

and it is easy to verify that in this case $y = e^{rx}$ and $y = xe^{rx}$ are both solutions. Since these two functions are linearly independent (Problem 15.2E), the general solution is

$$y = C_1 e^{rx} + C_2 x e^{rx}.$$

The second difficulty is that the roots of the characteristic equation may be complex. We have no definition for $e^{(a+bi)x}$. The obvious thing to do is to make a definition. The hard thing to do is to make one that seems sensible. As a first step, we observe that it would be nice to have $e^{(a+bi)x} = e^{ax}e^{bix}$. If we accept this much, we are only faced with having to define e^{bix}.

Our approach to this definition is via the series representation.

$$e^t = 1 + t + \frac{t^2}{2} + \frac{t^3}{3!} + \frac{t^4}{4!} + \frac{t^5}{5!} + \cdots.$$

Suppose we proceed formally. (We are not proving anything, after all; we are only trying to get an idea of how to make a reasonable definition.) Setting $t = bix$, we have

$$e^{bix} = 1 + bix + \frac{(bix)^2}{2} + \frac{(bix)^3}{3!} + \frac{(bix)^4}{4!} + \frac{(bix)^5}{5!} + \cdots.$$

Since $i^2 = -1$,

$$e^{bix} = \left[1 - \frac{(bx)^2}{2} + \frac{(bx)^4}{4!} - \cdots\right] + i\left[bx - \frac{(bx)^3}{3!} + \frac{(bx)^5}{5!} - \cdots\right].$$

We recognize the two series as $\cos bx$ and $\sin bx$. It therefore seems natural to define

$$e^{bix} = \cos bx + i \sin bx.$$

Now if the roots of a characteristic equation turn out to be $a \pm bi$, we write down as our solution

$$y = c_1 e^{(a+bi)x} + c_2 e^{(a-bi)x},$$

$$y = e^{ax}[c_1 e^{bix} + c_2 e^{-bix}],$$

$$y = e^{ax}[c_1(\cos bx + i \sin bx) + c_2(\cos(-bx) + i \sin(-bx))].$$

Finally, we combine constants to write

$$y = e^{ax}[C_1 \cos bx + C_2 \sin bx].$$

PROBLEMS

C • Show that $y = C_1 e^{-x} + C_2 xe^{-x}$ is a solution to the equation considered in Problem A.

D • Show that $y = e^{-\frac{1}{2}x}(C_1 \cos \sqrt{3}x/2 + C_2 \sin \sqrt{3}x/2)$ is a solution to the equation considered in Problem B.

E · Complete the solution to Example 15.1B.

F · Complete the solution to Problem 15.1B.

G · Complete the solution to Problem 15.1D.

We now summarize what we have learned. To solve

$$(a_2 D^2 + a_1 D + a_0)y = 0,$$

we find the roots r_1 and r_2 of the characteristic equation

$$a_2 D^2 + a_1 D + a_0 = 0.$$

(1)   If r_1 and r_2 are real and distinct,

$$y = C_1 e^{r_1 x} + C_2 e^{r_2 x}.$$

(2)   If $r_1 = r_2$, then (using r as the common value)

$$y = C_1 e^{rx} + C_2 x e^{rx}.$$

(3)   If $r_1 = a + bi$ and $r_2 = a - bi$, $b \neq 0$, then

$$y = e^{ax}(C_1 \cos bx + C_2 \sin bx).$$

This is another of those satisfying situations of which we spoke earlier. It can be proved that all possible solutions of our differential equation are included among those just listed.

Before taking leave of these considerations, we wish to comment on the equation

$$(a_2 D^2 + a_1 D + a_0)y = Q(x). \tag{3}$$

Solutions to this equation are closely related to the solution of the special case where $Q(x)$ is zero (the case we know how to solve). We can illustrate our point with several examples.

EXAMPLE B
Solve $(D^2 - 5D + 6)y = x + 1$.

Let us observe first of all that nothing of the form $y = e^{mx}$ will work, because no combination of derivatives of e^{mx} will give the polynomial on the right side. A little thought along this line will in fact convince most people that the solution will have to be of the form

$$y = P(x)$$

where P is a polynomial in x. Moreover, substitution of P into the equation gives

$$P''(x) - 5P'(x) + 6P(x) = x + 1. \tag{4}$$

The degree of the left side is the degree of P. Since this is an identity in x, and since the right side clearly has degree 1, we are forced to conclude that any solution will have to be of the form

$$y = P(x) = Ax + B.$$

Since $P'(x) = A$ and $P''(x) = 0$, substitution into (4) gives

$$-5A + 6(Ax + B) = x + 1.$$

Equating coefficients of like powers of x gives

$$6A = 1,$$

$$-5A + 6B = 1.$$

The solution is $y = \frac{1}{6}x + \frac{11}{36}$.

Now we capitalize on having learned that

$$(D^2 - 5D + 6)(C_1 e^{2x} + C_2 e^{3x}) = 0.$$

From this we conclude that

$$(D^2 - 5D + 6)(C_1 e^{2x} + C_2 e^{-x} + \tfrac{1}{6}x + \tfrac{11}{36}) = x + 1.$$

The general solution to our problem is

$$y = C_1 e^{2x} + C_2 e^{-x} + \tfrac{1}{6}x + \tfrac{11}{36}. \quad \blacksquare$$

We have illustrated the following principle. If trying to solve (3), try to find a particular solution by studying the form of $Q(x)$. (This is the hard part.) Then find the general solution to the auxiliary equation

$$(a_2 D^2 + a_1 D + a_0)y = 0.$$

The general solution to (3) is then obtained as the sum of the particular solution and the general solution to the auxiliary equation.

EXAMPLE C

Solve $(D^2 - D - 2)y = \sin 3x$.

No polynomial $y = P(x)$ will ever give the trigonometric function $\sin 3x$ on the right side. A little consideration, in fact, leads us to the conclusion that the solution must be some combination of $\sin 3x$ and $\cos 3x$. We try

$$y = A \sin 3x + B \cos 3x,$$

$$y' = 3A \cos 3x - 3B \sin 3x$$

$$y'' = -9A \sin 3x - 9B \cos 3x.$$

Substitution in the original equation gives

$$(-9A + 3B - 2A) \sin 3x + (-9B - 3A - 2B) \cos 3x = \sin 3x.$$

Since this is an identity,

$$-11A + 3B = 1,$$

$$-3A - 11B = 0.$$

Solution gives $A = -\frac{11}{30}$, $B = \frac{3}{130}$. The general solution is therefore

$$y = C_1 e^{2x} + C_2 e^{-x} - \tfrac{11}{130} \sin 3x + \tfrac{3}{130} \cos 3x. \quad \blacksquare$$

The method we have suggested for finding a particular solution is called the *method of undetermined coefficients*. It only enables us to find solutions when the right side of (3) is a very simple function. Other methods can be used, but once again we must stop somewhere. This seems like the place.

PROBLEMS

H · Complete the solution to Example 15.1A.

I · Complete the solution to Problem 15.1A.

J · In Problem 15.2A you were asked to verify that $y = Ce^x - x - 1$ solves the equation $y' = x + y$. Obtain this solution by the methods of this section.

K · Solve $(D^2 - 3D - 4)y = e^{2x}$.

━━━━━━━━━━━━━━━━━━━━━━━━━━━━━━━━━ **15 · 3   EXERCISES**

Solve the following:

1 · $(D^2 - 3D - 4)y = 0.$

2 · $(D^2 - 7D + 10)y = 0.$

3 · $(D^3 - 2D^2 - 3D)y = 0.$

4 · $(D^3 - D)y = 0.$

5 · $(D^3 - 1)y = 0.$

6 · $(D^3 + 8)y = 0.$

7 · $(D^3 - 2D^2 + D)y = 0.$

8 · $(D^3 + 4D^2 + 4D)y = 0.$

9 · $(D^2 + D + 1)(D + 3)^2 y = 0.$

10 · $(D^2 - D + 1)(D + 1)^2 y = 0.$

11 · $(D^2 - 3D - 4)y = e^x.$

12 · $(D^2 - 7D + 10)y = e^{-x}.$

13 · $(D^2 + 1)y = x^2 + x + 2.$

14 · $(D^2 + 4)y = 8x^2.$

15 · $(D^2 - 4)y = \sin x.$

16 · $(D^2 + D - 6)y = \cos x.$

15 · 4   First-Order Linear Differential Equations

The general homogeneous first-order linear differential equation takes the form

$$y' + P(x)y = 0. \tag{1}$$

We know that in the case where $P(x) = p$ is constant so that (1) takes the form $(D + p)y = 0$, the general solution is given by $y = Ce^{-px}$. If we write this solution in the form $y = Ce^{-\int p\, dx}$, then it is both natural to suspect and easy to verify that when $P(x)$ is not a constant, $y = Ce^{-\int P(x)\, dx}$ is a solution to (1).

The solution $y = Ce^{-\int P(x)dx}$ may be expressed in the form

$$ye^{\int P(x)\, dx} = C.$$

If we differentiate this equation implicitly with respect to x, remembering that $D\int P(x)\, dx = P(x)$, we have

$$\frac{dy}{dx}\, e^{\int P(x)\, dx} + yP(x)e^{\int P(x)\, dx} = \left[\frac{dy}{dx} + yP(x)\right]e^{\int P(x)\, dx} = 0.$$

This naturally leads us back to the differential equation just solved. It also provides a suggestion for solving the nonseparable equation

$$y' + P(x)y = Q(x). \tag{2}$$

If we multiply through by $e^{\int P(x)\, dx}$, then the left side of

$$\left[\frac{dy}{dx} + P(x)y\right]e^{\int P(x)\, dx} = Q(x)e^{\int P(x)\, dx} \tag{3}$$

is the derivative of $ye^{\int P(x)\, dx}$. The right side of (3) is the derivative of $\int Q(x)e^{\int P(x)\, dx}\, dx$. Therefore (3) is obtained as the derivative of

$$ye^{\int P(x)\, dx} = \int Q(x)e^{\int P(x)\, dx}\, dx + C.$$

Consequently, the solution to (2) is given by

$$y = Ce^{-\int P(x)\, dx} + e^{-\int P(x)\, dx}\int Q(x)e^{\int P(x)\, dx}\, dx. \tag{4}$$

Instead of memorizing this formula, the student should learn the method used to derive it. This method, multiplying through the original equation by $e^{\int P(x)\, dx}$, applied to a particular equation leads directly to the solution with no other memorization necessary. The following example illustrates the technique.

EXAMPLE A

Solve $y' + 2xy = 1 + 2x^2$.

This equation takes the form of (2) with $P(x) = 2x$:

$$e^{\int P(x)\,dx} = e^{\int 2x\,dx} = e^{x^2}.$$

Multiplication gives

$$y'e^{x^2} + 2xye^{x^2} = (1 + 2x^2)e^{x^2}.$$

Inspection of the left side (made easier by the fact that we know it is supposed to be something simple) reveals that it is the derivative with respect to x of $e^{x^2}y$. Therefore

$$e^{x^2}y = \int (1 + 2x^2)e^{x^2}\,dx + C. \tag{5}$$

The indicated antiderivative looks forbidding, but it is not too bad for readers who still remember how to use the method of parts:

$$\int (1 + 2x^2)e^{x^2}\,dx = \int e^{x^2}\,dx + \left[\int 2x^2 e^{x^2}\,dx \right].$$

In the bracketed term on the right, set

$$u = x, \qquad dv = 2xe^{x^2},$$
$$du = dx, \qquad v = e^{x^2},$$

$$\int (1 + 2x^2)e^{x^2}\,dx = \int e^{x^2}\,dx + \left[xe^{x^2} - \int e^{x^2}\,dx \right]$$

$$= xe^{x^2}.$$

Substitution in (5) gives

$$e^{x^2}y = xe^{x^2} + C.$$

Multiplication by e^{-x^2} gives

$$y = x + Ce^{-x^2}. \qquad ∎$$

The example just worked should be compared with Example 15.2A. Then Example 15.2A should be used as a guide to solving Problem A below.

PROBLEMS

A · Prove that if $y = F(x)$ is a solution to (2), then $F(x)$ is described by (4) for some choice of C. Hint: Let $G(x) = e^{-\int P(x)\,dx} \int Q(x)e^{\int P(x)\,dx}\,dx$. We know (set $C = 0$ in (4)) that $G(x)$ is a solution to (2). Show, following the work of Example 15.2A, that for some C,

$$F(x) = Ce^{\int p(x)\,dx} + G(x).$$

B · Is there a solution to $xy' + y = 2x$ that is continuous on $[-2, 2]$ that passes through $(1, -1)$?

C · Finish Problem 15.1C. Include a graph showing the value of s for $t \in [0, 20]$. When does the tank contain the maximum amount of salt?

D · Complete the solution to Problem 15.1E.

E · Complete the solution to Problem 15.1F.

F · A basic equation in electric circuit theory is

$$L\frac{di}{dt} + Ri = E(t),$$

where L is the inductance (constant) and R is the resistance (constant). Show that in the case where $E(t)$ is constant, say E_0, the current i tends to a limit as time goes by.

G · A function F satisfies the differential equation

$$y' = xy.$$

It is known that $F(\frac{1}{2}) = 1$. Find $F(\frac{1}{4})$ accurate to within one unit in the second decimal place.

At the end of Section 15.2 we pointed out that for a given differential equation, two questions may be asked; one of existence (is there a solution?) and one of uniqueness (have we found all the solutions?). For any equation of the form

$$y' + P(x)y = Q(x).$$

we know a solution exists because we have found it; in fact we have a formula for it (4). Moreover, as demonstrated in Problem A, the uniqueness question is also settled in this case.

A mathematician, like an artist, may leave certain pieces of work with a great sense of inner satisfaction. But unlike most artists or craftsmen who, we suspect, leave even their best work with a feeling that it might have been a little better, we are able to leave this equation with the feeling that the job is done and will not be improved upon. Novalis (1772–1801), the German romantic poet, said "Only mathematicians are happy men."

Find the general solution.

1 · $y' + (\cos x)y = e^{-\sin x}$.

2 · $y' + y - x = 0$.

3 · $xy' + 2y = x - 1$.

4 · $y' + (y - \sin x)\cos x = 0$.

5 · $y' + y = \sin x$.

6 · $xy' - 2y = x^3 e^{-x}$.

15 · 5   Other First-Order, First-Degree Differential Equations

A first-order, first-degree differential equation can always be written in the form

$$\frac{dy}{dx} + F(x, y) = 0. \tag{1}$$

The linear equation studied in Section 15.4 is a special case in which $F(x, y) = P(x)y - Q(x)$.

There is no standard way to solve the general equation of form (1). The would-be solver must equip himself with a variety of methods and learn which of them is most likely to work for various forms of the function $F(x, y)$. Our chances for solving the equation certainly increase as more methods and experience accumulate, but it should be understood that no collection of methods is complete in the sense of offering a way of solving any equation that comes along.

We shall introduce only two methods here. Together with our knowledge of linear equations, we are nevertheless equipped to solve a surprising variety of equations arising in applications. The first method hardly deserves the name. It is more in the nature of an observation, having to do with those equations in which the variables are said to be separable. This means we consider equations of the form (1) in which $F(x, y) = G(x)H(y)$. Setting $R(y) = 1/H(y)$, such an equation takes the form

$$R(y) \frac{dy}{dx} + G(x) = 0.$$

This is, however, the derivative with respect to x of

$$\int R(y)\, dy + \int G(x)\, dx = C.$$

EXAMPLE A

A can half full of water is rotated at a constant velocity of ω radians per second about its axis. Describe the surface formed by the water.

We saw in Example 15.1D that this problem leads us to the differential equation

$$\frac{dy}{dx} = \frac{\omega^2}{g} x,$$

$$dy = \frac{\omega^2}{g} x \, dx,$$

$$y = \frac{\omega^2}{2g} x^2 + C.$$

This means that water in a rotating cylinder assumes the shape of a parabola. The constant C gives the height of the water in the center of the container and obviously depends on the amount of water with which we started. ▮

EXAMPLE B

At noon a thermometer reading 75°F is taken outdoors, where the temperature is 15°F. Five minutes later the thermometer reading is 45°F. At 12:10 P.M. the thermometer is brought back inside, where the temperature is still 75°F. Find (a) the thermometer reading at 12:20 P.M.; (b) the time when the thermometer will show, to the nearest degree the correct indoor temperature.

We have already seen in Example 15.1C that this problem leads to the equations

$$\frac{dy}{dt} = \begin{cases} k(15 - y), & t \in [0, 10], \\ k(75 - y), & t \geq 10. \end{cases}$$

We begin with the solution for $t \in [0, 10]$.

$$\frac{dy}{dt} = k(15 - y),$$

$$\frac{dy}{15 - y} = k \, dt,$$

$$-\ln|15 - y| = kt + C.$$

Now when $t = 0$, $y = 75$, so $C = -\ln 60$. Substitution and algebraic simplification gives

$$\ln \frac{60}{|15 - y|} = kt,$$

$$\frac{60}{|15 - y|} = e^{kt},$$

$$|15 - y| = 60e^{-kt}.$$

When $t = 5$, $y = 45$, so $30 = 60e^{-5k} = 60[e^{-k}]^5$,

$$e^{-k} = \left(\frac{1}{2}\right)^{1/5}.$$

For $t \in [0, 10]$, the temperature y is certainly greater than 15, so (2) may be written

$$y - 15 = 60(e^{-k})^t.$$

Substitution for e^{-k} gives us

$$y = 15 + 60(\tfrac{1}{2})^{t/5}, \qquad t \in [0, 10].$$

It is easy (and wise) at this point to verify that when $t = 0$, $y = 75$ and when $t = 5$, $y = 45$. We also note with pleasure that if the thermometer were left outside, then our formula would give 15 as the limit of y as $t \to \infty$. Finally we note that when

$$t = 10, \qquad y = 30.$$

Now consider the solution for $t \geq 10$.

$$\frac{dy}{dt} = k(75 - y),$$

$$-\ln|75 - y| = kt + C_2.$$

When $t = 10$, $y = 30$; so $-\ln 45 = 10k + C_2$. Since $y \in [30, 75]$, this last equation takes the form

$$-\ln(75 - y) = kt - \ln 45 - 10k,$$

$$\ln \frac{75 - y}{45} = -k(t - 10),$$

$$\frac{75 - y}{45} = (e^{-k})^{t-10} = \left(\frac{1}{2}\right)^{(t-10)/5},$$

$$y = 75 - 45(\tfrac{1}{2})^{(t-10)/5}, \qquad t \geq 10.$$

It again builds up confidence to note that from this formula, we can see that when

$$t = 10, \qquad y = 30,$$

and as

$$t \rightarrow \infty, \qquad y = 75.$$

The complete solution, graphed in Fig. 15.11, is

$$y = \begin{cases} 15 + 60(\tfrac{1}{2})^{t/2}, & t \in [0, 10], \\ 75 - 45(\tfrac{1}{2})^{(t-10)/5}, & t \geq 10. \end{cases}$$

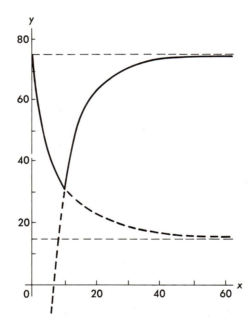

FIG. 15.11

The temperature reading when $t = 20$ is

$$75 - 45(\tfrac{1}{4}) = 63\tfrac{3}{4}.$$

Finally, we seek the time t for which $45(\tfrac{1}{2})^{(t-10)/5} < \tfrac{1}{2}$.

$$\ln 45 + \frac{t-10}{5} \ln \tfrac{1}{2} < \ln \tfrac{1}{2}.$$

Some algebraic simplification gives

$$t > 10 + 5 \frac{\ln 90}{\ln 2} \approx 42.$$

The thermometer gives the correct indoor reading at about 12:42.

PROBLEMS

A · Finish Problem 15.1D (again; you solved it in Problem 15.3G). If one quarter of the original substance is converted in the first 5 minutes, how long will it be before three quarters of the original substance has been converted ?

B · Finish Problem 15.1E. Compare with the solution obtained in Problem 15.4D.

C · Finish Problem 15.1F. Compare with the solution obtained in Problem 15.4E.

D · Solve $(1 + x^2)y' + 2xy = 0$.

The second method we wish to introduce is not restricted to first-order equations (as we shall see in Example D), but it can sometimes be useful in this setting. At any rate it is a good idea to introduce a new method in a rather simple situation. We shall go one step further and introduce our new method by using it to solve a problem we have already solved in another way (Example 15.4A).

EXAMPLE C

A function F satisfies the differential equation

$$y' + 2xy = 1 + 2x^2.$$

It is known that $F(0) = 2$. Find $F(\tfrac{1}{2})$ accurate to within one unit in the third decimal place.

Because of the special form of this particular example, we know a solution exists. Generally we do not, and we shall begin this solution as we would begin any. We merely assume a solution exists, and we also assume that the desired solution has a Maclaurin expansion; that is, we assume

$$
\begin{aligned}
y = F(x) &= a_0 + a_1 x + a_2 x^2 + \cdots + a_n x^n + \cdots . \\
y' = F'(x) &= a_1 + 2a_2 x + 3a_3 x^2 + \cdots + (n + 1)a_{n+1} x^n + \cdots . \\
2xy = 2xF(x) &= \quad\ + 2a_0 x + 2a_1 x^2 + \cdots + 2a_{n-1} x^n + \cdots .
\end{aligned}
$$

Adding the last two equations gives

$$y' + 2xy = a_1 + (2a_2 + 2a_0)x + (3a_3 + 2a_1)x^2 + \cdots$$
$$+ [(n + 1)a_{n+1} + 2a_{n-1}]x^n + \cdots.$$

Substitution of this series for the left side of our original equation gives

$$a_1 + (2a_2 + 2a_0)x + (3a_3 + 2a_1)x^2 + \cdots$$
$$+ [(n + 1)a_{n+1} + 2a_{n-1}]x^n + \cdots = 1 + 2x^2.$$

We may view this as the equality of two infinite series in which most coefficients of the series on the right are zero. According to the corollary to Theorem 14.4E, the coefficients of like powers must be equal. Therefore

$$1 = a_1,$$
$$0 = 2a_2 + 2a_0,$$
$$2 = 3a_3 + 2a_1,$$
$$\vdots$$
$$0 = (n + 1)a_{n+1} + 2a_{n-1}, \quad \text{for} \quad n \geq 3.$$

There seems to be no restriction on a_0. But once a_0 is chosen, all the coefficients are determined.

$$a_0 \text{ is arbitrary,}$$
$$a_1 = 1,$$
$$a_2 = -a_0,$$
$$a_3 = \frac{2 - 2a_1}{3} = 0,$$
$$\vdots$$
$$a_{n+1} = \frac{-2a_{n-1}}{(n + 1)}, \quad \text{for } n \geq 3.$$

The last expression enables us to compute as many coefficients as we please.

$$a_4 = \frac{-2a_2}{4} = \frac{2a_0}{4},$$

$$a_5 = \frac{-2a_3}{5} = 0,$$

$$a_6 = \frac{-2a_4}{6} = -\frac{2}{6}\frac{2a_0}{4} = -\frac{2^3 a_0}{2^3 3!}.$$

It is clear that all coefficients having an odd number greater than 1 as a subscript must be 0. The form in which we have written a_6 shows how the coefficients with even subscripts are determined.

$$a_{2k} = (-1)^k \frac{a_0}{k!}.$$

The Maclaurin expansion for F is

$$F(x) = \left[a_0 + x - a_0 x^2 + a_0 \frac{x^4}{2} - a_0 \frac{x^6}{3!} + \cdots + (-1)^k a_0 \frac{x^{2k}}{k!} + \cdots \right]. \qquad (3)$$

Since $F(0) = 2$, we see that $a_0 = 2$. We may then compute $F(\frac{1}{2})$ from the series. Since it is alternating, the error made in using the first few terms is no more than the first term ignored.

$$2 \frac{1}{4!} \left(\frac{1}{2} \right)^8 = \frac{1}{3072} < \frac{1}{3000} = 0.00033 \ldots .$$

Our answer will therefore be accurate to within one unit in the third decimal place if we use

$$F\left(\frac{1}{2} \right) \approx \left[2 + \frac{1}{2} - 2\frac{1}{4} + 2\frac{1}{32} - 2\frac{1}{384} \right];$$

that is,

$$F(\tfrac{1}{2}) \approx 2.057.$$

As a final observation, we point out that we could have proceeded differently after obtaining (3). We could have written this as

$$F(x) = x + a_0 \left[1 + (-x^2) + \frac{(-x^2)^2}{2!} + \frac{(-x^2)^2}{3!} + \cdots + \frac{(-x^2)^k}{k!} + \cdots \right].$$

Since

$$e^t = 1 + t + \frac{t^2}{2} + \frac{t^3}{3!} + \cdots + \frac{t^k}{k!} + \cdots,$$

the series represents e^{-x^2}.

$$F(x) = x + a_0 e^{-x^2}.$$

This is the solution we previously obtained (Example 15.4A). Some no doubt feel this latter answer is far better than (3). But that all depends on why you want it. It is easy enough to see that

$$F(0) = a_0 = 2$$

but the computation of $F(\frac{1}{2})$ now gives

$$F\left(\frac{1}{2}\right) = \frac{1}{2} + \frac{2}{e^{1/4}}.$$

To get three-place decimal accuracy from this, you are forced back to the series expansion for e^x or else to tables (which were computed using a series expansion for e^x). If you do use tables for e^x, you get

$$F\left(\frac{1}{2}\right) = \frac{1}{2} + \frac{2}{1.2840}.$$

It still leaves a divisional computation to get the desired answer. ∎

As indicated in remarks above, the method of assuming an infinite series solution is useful in some situations where the order of the equation is not one.

EXAMPLE D
Solve $y'' + c^2 y = 0$ where c is a constant.

Let

$$y = a_0 + a_1 x + \cdots + a_n x^n + \cdots.$$

Then

$$y' = a_1 + 2a_2 x + \cdots + (n + 1)a_{n+1} x^n + \cdots,$$
$$y'' = 2a_2 + 3 \cdot 2a_3 x + \cdots + (n + 2)(n + 1)a_{n+2} x^n + \cdots.$$

Using these expansions to substitute into our equation, we get

$$y'' + c^2 y = (2a_2 + c^2 a_0) + (3 \cdot 2a_3 + c^2 a_1)x + \cdots$$
$$+ [(n + 2)(n + 1)a_{n+2} + c^2 a_n]x^n + \cdots.$$

Since the left side is to be zero, each of the coefficients on the right side must be zero

$$2a_2 + c^2 a_0 = 0, \quad \text{so} \quad a_2 = \frac{-c^2 a_0}{2};$$

$$3 \cdot 2a_3 + c^2 a_1 = 0, \quad a_3 = \frac{-c^2 a_1}{3 \cdot 2},$$

$$\vdots \qquad\qquad \vdots$$

$$(n + 2)(n + 1)a_{n+2} + c^2 a_n = 0; \quad a_{n+2} = \frac{-c^2 a_n}{(n + 2)(n + 1)}.$$

The coefficients a_0 and a_1 are arbitrary, but once they are chosen, it is clear from the formula for a_{n+2} that all the rest are determined.

a_0 is arbitrary a_1 is arbitrary

$$a_2 = -\frac{c^2 a_0}{2},   \qquad a_3 = -\frac{c^2 a_1}{3 \cdot 2},$$

$$a_4 = -\frac{c^2 a_2}{4 \cdot 3} = \frac{c^4}{4!} a_0, \qquad a_5 = -\frac{c^2 a_3}{5 \cdot 4} = \frac{c^4}{5!} a_1,$$

$$\vdots \qquad\qquad\qquad \vdots$$

$$a_{2k} = (-1)^k \frac{c^{2k}}{(2k)!} a_0; \qquad a_{2k+1} = (-1)^{k+1} \frac{c^{2k}}{(2k+1)!} a_1.$$

The solution is therefore

$$y = a_0 \left[1 - \frac{c^2}{2} x^2 + \frac{c^4}{4!} x^4 + \cdots + (-1)^k \frac{c^{2k}}{(2k)!} x^{2k} + \cdots \right]$$

$$+ a_1 \left[x - \frac{c^2}{3 \cdot 2} x^3 + \cdots + (-1)^{k+1} \frac{c^{2k}}{(2k+1)!} x^{2k+1} + \cdots \right].$$

If we write this in the form

$$y = a_0 \left[1 - \frac{(cx)^2}{2} + \frac{(cx)^4}{4!} + \cdots + (-1)^k \frac{(cx)^{2k}}{(2k)!} + \cdots \right]$$

$$+ \frac{a_1}{c} \left[cx - \frac{(cx)^3}{3!} + \cdots + (-1)^{k+1} \frac{(cx)^{2k+1}}{(2k+1)!} + \cdots \right],$$

then the two series look familiar. After checking back to the expansions of various familiar functions, we see that by setting $a_0 = C_1$, $a_1/c = C_2$, the solution above may be written as

$$y = C_1 \cos cx + C_2 \sin cx.$$

We have been guilty of ignoring questions about rearranging series, the existence of a solution in the first place, and several other matters. At this point, however, we only need to substitute our solution in the original equation. It works. That puts a lot of criticism to rest. ∎

PROBLEMS

E· Solve Example D by the methods of Section 15.3.

F· Use the method of assuming a series solution to solve

$$(1 + x^2)y' + 2xy = 0.$$

Compare your answer to the one obtained for Example D above.

We do not have to look far to see that the use of infinite series has serious limitations. For example, if we let

$$y = x^{3/2},$$

then

$$y' = \tfrac{3}{2}x^{1/2} = \tfrac{3}{2}y/x.$$

This means that $2xy' = 3y$ is a differential equation having as its solution $y = x^{3/2}$. Bit the latter function will never be found by means of an assumed Maclaurin expansion because the Maclaurin expansion of $x^{3/2}$ does not exist.

There is a modification of the method of infinite series called the *method of Frobenius*. It enables us to solve the equation above and many others as well. We shall not describe this method here. See [5, Chapter 4].

We now have three methods to solve first-order, first-degree differential equations.

(1) Multiply $dy/dx + P(x)y = Q(x)$ by $e^{\int P(x)\,dx}$.
(2) Separate the variables.
(3) Assume a series expansion.

The last method has definite limitations, but we can always try it. It will sometimes work on equations of order higher than one.

Those desiring greater proficiency have already been referred to a text such as [5] where they will be introduced to many more methods. It is to be emphasized, however, that when all the standard methods have been learned (a standard method is one I know about), there is still a great deal to be said for a little luck and a lot of experience. In our last two examples we couple some of the things we have learned with some educated guesses in order to finish the two remaining examples of Section 15.1.

EXAMPLE E

Design a reflector so that each ray of light from a small bulb will be reflected parallel to a fixed line.

We have seen (Example 15.1E) that this problem leads us to the differential equation

$$y = y(y')^2 + 2xy'.$$

Suppose (perhaps after several tries at other things) we try treating this equation as a quadratic equation in y'. The formula then gives

$$y' = \frac{-2x \pm \sqrt{4x^2 + 4y^2}}{2y}.$$

Then after more fooling around, we hit on the idea of putting this in the form

$$2x + 2yy' = \pm 2\sqrt{x^2 + y^2}. \tag{4}$$

Here is where a little experience pays off. Anyone familiar with taking derivatives will observe that if we set $u = x^2 + y^2$, then

$$\frac{du}{dx} = 2x + 2y\frac{dy}{dx},$$

which is the left side of our equation. Thus, (4) becomes

$$\frac{du}{dx} = \pm 2u^{1/2}.$$

This is now separable.

$$\tfrac{1}{2}u^{-1/2}\,du = \pm dx,$$
$$u^{1/2} = \pm x + C.$$

Squaring both sides and replacing u by $x^2 + y^2$ again gives

$$(x^2 + y^2) = (\pm x + C)^2 = x^2 \pm 2Cx + C^2.$$

Allowing positive or negative choices for C, we may write

$$y^2 = 2Cx + C^2.$$

Graphs corresponding to the choices $C = \pm 1, \pm 2$ are indicated in Fig. 15.12. ∎

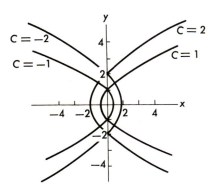

FIG. 15.12

EXAMPLE F

A cable is suspended between two points (which we take to be symmetric with respect to the y axis) as indicated in Fig. 15.13. Find the shape that the cable assumes.

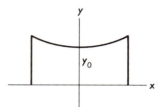

FIG. 15.13

We saw in Example 15.1F that this problem led to the differential equation

$$\frac{d^2y}{dx^2} = \frac{\omega}{|H|} \sqrt{1 + \left(\frac{dy}{dx}\right)^2}.$$

This is a second-order equation, but since there is no y term, an immediate simplification suggests itself. Set

$$p = \frac{dy}{dx} \qquad \text{so} \qquad \frac{dp}{dx} = \frac{d^2y}{dx^2}.$$

Our equation becomes

$$\frac{dp}{dx} = \frac{\omega}{|H|} \sqrt{1 + p^2},$$

which is separable.

$$\frac{dp}{\sqrt{1 + p^2}} = \frac{\omega}{|H|} \, dx.$$

The antiderivative of the left side may be obtained by a trigonometric substitution that leads to a logarithmic expression. An equivalent and neater antiderivative (see Problem 7.5C) is obtained by recalling that

$$D \, \text{Argsinh} \, u = \frac{1}{\sqrt{1 + u^2}} \frac{du}{dx}.$$

Therefore,

$$\text{Argsinh} \, p = \frac{\omega}{|H|} x + C.$$

When $x = 0$, $p = dy/dx = 0$. This determines $C = 0$.

$$p = \frac{dy}{dx} = \sinh \frac{|H|}{\omega} x,$$

$$y = \frac{|H|}{\omega} \cosh \frac{\omega}{|H|} x + C_1.$$

We have not yet specified the value of y_0 at which the curve crosses the y axis. When $x = 0$,

$$y_0 = \frac{|H|}{\omega} \cosh 0 + C_1 = \frac{|H|}{\omega} + C_1.$$

If we choose $y_0 = |H|/\omega$, then $C_1 = 0$ and the equation becomes

$$y = \frac{|H|}{\omega} \cosh \frac{\omega}{|H|} x.$$

If we set $|H|/\omega = a$, then

$$y = a \cosh \frac{x}{a}.$$

That's the way the cable hangs. See Fig. 7.24 in Section 7.5. ∎

───

15 · 5 EXERCISES

Equations 1–6 are separable. Solve them.

1. $y' = \dfrac{2x^2 + 2x + 1}{2xy + 2y}$.

2. $y' = \dfrac{x \sin x}{2y} e^{-y^2}$.

3. $y' - yy' = x(1 + yy' - y')$.

4. $y' = 2x + 2xy^2$.

5. $y' = y \ln x$.

6. $xy - y' = y(y' + \ln x)$.

Equations 7–12 are to be solved by assuming a series expansion for the solution. This solution is to satisfy the initial condition given in each case. (Give at least three nonzero terms of the Maclaurin expansion of the solution in each case.)

7. $y' = (x + y)e^x$
   When $x = 0$, $y = 1$.

8. $y' = (x + y)e^{-x}$
   When $x = 0$, $y = 1$.

9. $xy' = (x + 2)y$
   When $x = 1$, $y = e$

10. $x(y + y') = y$
   When $x = 1$, $y = 1/e$

11. $y + y'' = 2e^{-x}$
   When $x = 0$, $y = 1$

12. $y + y'' = 2e^x$
   When $x = 0$, $y = 2$.

In each exercise that follows, find a solution subject to the initial conditions.

13. $y' = 2x\sqrt{1 - y^2}$
When $x = 0$, $y = -1$

14. $y' + y \tan x = \cos x$
When $x = 0$, $y = 2$

15. $y' = \dfrac{1 + y^2}{\sqrt{1 - x^2}}$.
When $x = 0$, $y = 0$

16. $y' = \dfrac{\sqrt{1 - y^2}}{1 + x^2}$.
When $x = 0$, $y = 0$

17. $y' + 2xy = x$
When $x = 0$, $y = 2$

18. $y' = x^2 y' + 2y$
When $x = 0$, $y = 3$

19. $y' + (2x \cot x^2)y = 2x \cos x^2$
When $x = \dfrac{\sqrt{\pi}}{2}$, $y = \sqrt{2}$

20. $(x^2 - 1)y' - x(y + x^2 - 2) = 0$
When $x = 2$, $y = 1$

21. $y'' - y = 0$
When $x = 0$, $y = 2$ and $y' = -2$

22. $y'' - 4y = 0$
When $x = 0$, $y = 3$ and $y' = 6$

23. $y'' = (4x^2 - 2)y$
When $x = 0$, $y = 1$ and $y' = 0$

24. $xy'' = (x + 2)y$
When $x = 0$, $y = 0$ and $y' = 1$

25–42. Complete the solutions to the exercises following Section 15.1.

15 · 6 Approximating Solutions

In Example 15.5C we considered the differential equation

$$y' + 2xy = 1 + 2x^2.$$

We wanted a solution F for which $F(0) = 2$. We were then supposed to find $F(\tfrac{1}{2})$. This problem takes the form of an old friend encountered earlier (Section 2.7).

The Approximation Problem

Suppose F is defined and well behaved in some interval containing x_0, and suppose $y_0 = F(x_0)$ is known. Can we devise a method for easily approximating $F(x_0 + h)$ in the event that h is small and that the exact value of $F(x_0 + h)$ is either very difficult or impossible to find?

This is exactly what we were doing in Example 15.5C. The problem is made more difficult than it seemed back in Section 2.6 because we no longer know what the function F is. All we know is that it satisfies a certain differential equation.

We have already seen that the problem may be solved by finding the Taylor series representing the function F in a neighborhood of the point x_0

where we have some information. (In the example already cited, we had information at $x_0 = 0$, so we used the Maclaurin series.) We now want to illustrate a very nice way to obtain this series in the situation where initial conditions are given. (You will recall that in the previous method, we merely assumed that a series representation worked even where no initial conditions were given. See Example 15.5C.) We use the problem already cited for purposes of illustration.

EXAMPLE A

A function F satisfies $y' + 2xy = 1 + 2x^2$. Given that $F(0) = 2$, find $F(\frac{1}{2})$.

We know that

$$F(0) = 2.$$

Since $y = F(x)$ is a solution of the equation, we may substitute, getting

$$F'(x) = -2xF(x) + 1 + 2x^2. \tag{1}$$

Thus,

$$F'(0) = 0 + 1 + 0 = 1.$$

Differentiating (1) twice gives

$$F''(x) = -2xF'(x) - 2F(x) + 4x,$$
$$F'''(x) = -2xF''(x) - 2F'(x) - 2F'(x) + 4.$$

Substituting $x = 0$ in these last expressions gives

$$F''(0) = -2F(0) = -4,$$
$$F'''(0) = -4F'(0) + 4 = 0.$$

We could clearly continue the process as long as we cared to. With the information we have so far, we can write the Taylor polynomial of degree 3 corresponding to F at $x = 0$.

$$F(x) \approx F(0) + F'(0)x + \frac{F''(0)}{2}x^2 + \frac{F'''(0)}{3!}x^3$$

$$\approx 2 + x - 2x^2 + 0x^3.$$

We have obtained by a new method the first few terms of the series (3) of Section 15.5. We could obviously complete our problem here the way we did it before by substituting $x = \frac{1}{2}$. ∎

PROBLEMS

A · Continue the example above to obtain the Taylor polynomial of degree 6. Compare your result with (3) of Section 15.5.

B · Using the method of Example A, solve Problem 15.4G. Note that in this example, you will get a Taylor series about the point $x_0 = \frac{1}{2}$.

We wish to introduce one final method for solving the approximation problem when it relates to a differential equation of the form $y' = G(x, y)$. Suppose we seek a solution $y = F(x)$ for which $y_0 = F(x_0)$. Such a solution must satisfy

$$F'(x) = G(x, F(x)). \tag{2}$$

Recall Theorem 6.4A, which says that whenever $F'(x)$ is continuous,

$$F(x) = \int_a^x F'(t)\, dt + C$$

where a and C may be any constants whatever. Therefore, by (2),

$$F(x) = \int_{x_0}^x G(t, F(t))\, dt + y_0. \tag{3}$$

The choice of our constants is motivated by the requirement that $F(x_0) = y_0$.

If we knew what F was, it would have to satisfy (3). The trouble is that we do not know F. Suppose we had a function that approximated the function F in a neighborhood of x_0. We might guess that such a function would almost satisfy (3). Now the only thing we have to go on is that the constant function

$$F_0(x) = y_0 \qquad \text{(for all } x\text{)}$$

matches our desired function at x_0. If we substitute $F_0(x)$ into (3), we would not expect equality. With a little luck, though, substitution of $F_0(x)$ on the right side of (3) might define a function $F_1(x)$ that would at least come closer to approximating F than does the function F_0. We thus define

$$F_1(x) = y_0 + \int_{x_0}^x G(t, F_0(t))\, dt.$$

We have at least preserved the fact that our new function matches the desired function at x_0; that is $F_1(x_0) = y_0$. Now if $F_1(x)$ comes closer to satisfying (3) than does $F_0(x)$, the same reasoning leads us to use it in defining

$$F_2(x) = y_0 + \int_{x_0}^x G(t, F_1(t))\, dt.$$

Once we have the idea, there is no limit to how long we may play this game.

$$F_3(x) = y_0 + \int_{x_0}^{x} G(t, F_2(t)) \, dt$$
$$\vdots$$

It is obvious that all these functions pass through (x_0, y_0). It can be proved [L. R. Ford, Differential Equations, 2nd ed., Chapter 5, McGraw-Hill, New York, 1955] that under very general conditions, the sequence of functions $F_0, F_1, F_2, F_3, \cdots$ approach the function F that we desire. This method of solution is of great theoretical as well as practical importance. It is known as *Picard's method*.

EXAMPLE B
Solve Example A by Picard's method.

In this example, $x_0 = 0$, $y_0 = F_0(x_0) = 2$, and

$$G(x, y) = -2xy + 1 + 2x^2.$$

Substitution in the formula for F_1 gives

$$F_1(x) = 2 + \int_0^x (-4t + 1 + 2t^2) \, dt,$$

$$F_1(x) = 2 - 2x^2 + x + \tfrac{2}{3}x^3.$$

Now

$$G(t, F_1(t)) = -2t[2 + t - 2t^2 + \tfrac{2}{3}t^3] + 1 + 2t^2,$$
$$= 1 - 4t + 4t^3 - \tfrac{4}{3}t^4.$$

Substitution in the formula for F_2 gives

$$F_2(x) = 2 + \int_0^x \left(1 - 4t + 4t^3 - \frac{4}{3}t^4\right) dt,$$

$$F_2(x) = 2 + x - 2x^2 + x^4 - \frac{4}{15}x^5.$$

When $x = \tfrac{1}{2}$ we have

$$F_1\left(\frac{1}{2}\right) = 2 - \frac{2}{4} + \frac{1}{2} + \frac{2}{3}\frac{1}{8} = 2.083,$$

$$F_2\left(\frac{1}{2}\right) = 2 + \frac{1}{2} + 2\frac{1}{4} + \frac{1}{16} - \frac{4}{15}\frac{1}{43} = 2.054.$$

Since we have proved (Example 15.5C) that $F(\frac{1}{2}) = 2.057$ is correct to within one unit in the third decimal place, it appears that our approximations are getting closer. ∎

PROBLEM

C· Use Picard's method to solve Problem 15.4G. Compare this with the solution you obtained for Problem B.

15·6 EXERCISES

For each of the following differential equations, find, in a neighborhood of the indicated x_0, the fourth approximation $F_3(x)$ as given by the method of Picard. Find $F_3(x_1)$. Finally, use other methods to find the solution $F(x)$ in terms of elementary functions, and then compute $F(x_1)$.

1· $y' + y = x$; $x_0 = 0$ and $y_0 = F(x_0) = 1$; $x_1 = \frac{1}{2}$.

2· $y' - 4y = 0$; $x_0 = 0$, $y_0 = 1$, $x_1 = \frac{1}{2}$.

3· $y' - y^2 = 1$; $x_0 = 0$, $y_0 = 0$, $x_1 = \frac{1}{2}$.

4· $y' - ye^x = 0$; $x_0 = 0$, $y_0 = 2$, $x_1 = \frac{1}{2}$.

5· $y' + x^2 + y^2$; $x_0 = 0$, $y_0 = 1$, $x_1 = \frac{1}{4}$.

6· $y' = x + y^2$; $x_0 = 0$, $y_0 = 0$, $x_1 = \frac{1}{4}$.

WHAT NEXT?

You have now been introduced to some of the main ideas of calculus. The introduction was not formal. At times, as noted in the presentation, we have proceeded as if nothing could go wrong. You were led to guess at how a thing would turn out. A theorem would be quoted (generally not proved) to assure you that most of the time things do turn out as you would expect. In this way, it has been possible to get you familiar with a lot of mathematics in a reasonably short time.

We have said that you are familiar with a lot of mathematics. Some people object to getting familiar until a proper introduction has been made. They argue that without time in which basic understandings can be built up and allowed to mature, the acquaintanceship is destined to be superficial. And they are right in pointing out this real danger. You will do well to remember that you have reached your present level by skipping over some very important ideas.

Let us briefly review some things to which we have given short shrift. In Chapter 2, we discussed functions, and we listed in Section 2.6 a lot of properties of continuous functions. These properties were not proved for a very good reason. Their proof depends on some fundamental ideas about the real number system. These same properties will be of aid in your understanding of limits, the theory of integration, and in your understanding of sequences and series. Note, for example, Theorem 14.2A.

Before we could obtain the computational rules for finding derivatives, we had need for a lot of theorems about limits which we had thoughtfully listed (but not proved) in Section 2.4. Example 2.4A was given to warn you that the theory of limits is tricky business. Needless to say, these matters deserve further attention from anyone who expects to forge ahead in mathematics.

Succeeding chapters were similarly punctuated with crucial theorems that we merely called upon when they were most needed. These needs were probably most obvious in the chapters about integration and in the chapter on infinite series where we spoke somewhat vaguely of the theory of a complex variable. They were also lurking in the background of Chapter 9 where, you may recall, we gave some references for those who are attracted to dark corners.

In short, if you are serious about learning mathematics, you will surely need to go back and fill in the gaps we have left. Our hope is that you will be more interested in these rather difficult studies because you know how badly we need the results that they make possible. For those who wish to set about this business, the proper courses are generally given under such titles as advanced calculus, theory of a real variable, or foundations of analysis.

It is also possible to continue forward instead of returning to a close examination of fundamentals. One can, for example, pursue the ideas of matrices as introduced in Chapter 10. We have already seen their importance to our subject, and have in fact (as was our custom) called upon this body of knowledge for some results we needed. But in addition, this study carries one into a host of new ideas and applications of mathematics. These matters are taken up in a course generally called linear algebra. You should be prepared to venture into such a course.

There are other options for those wishing to move forward. A full blown study of differential equations was suggested in Chapter 15. It is just as reasonable to suggest a course in mathematical statistics. Courses in applied mathematics or mathematics for engineering are still other possibilities.

Finally, we shall make passing reference to several areas in which, at first encounter, the point of view seems to be entirely different. We refer to those courses that typically begin with a few undefined terms (a concept needing some explanation) like points and lines, and a few postulates. The spirit is similar to that of a high-school geometry course. Beginning with the postulates, the idea is to see what things they imply by logical deduction. Although the applications of such mathematics are numerous and important, the beginning student is not likely to see much connection with the world of experience. Courses taught from this point of view are abstract algebra, noneuclidian geometry, and point set topology. It is also true that if you seek a deeper understanding of the material in this book, you ultimately will come once again to the business of undefined terms, postulates, and proofs of theorems. This is the warp and woof of mathematics.

In short, there are a number of topics in mathematics that you are now prepared to investigate. Each one contributes in its own way to your mathematical development as well as to your general intellectual development. The choice is yours.

BIBLIOGRAPHY

Reference has been made in the text to the books below. There is no attempt here to even suggest the wealth of material that may be used to supplement and extend the content of this text. These references are given because they are, for the most part, the ones that have shaped the author's ideas. It is natural to except that our point of view is consistent with these texts. The reader may therefore find them the natural ones to consult.

[1] Boas, Ralph P., *A Primer of Real Functions*. Carus Mathematical Monographs, 1961.

[2] Buck, R. Creighton, *Advanced Calculus*, 2nd Ed. New York: McGraw-Hill, 1965.

[3] Dieudonne, J., *Foundations of Modern Analysis*. New York: Academic Press.

[4] Nevanlinna, F. and R., *Absolute Analysis*, Berlin, Germany: Springer, 1959.

[5] Rabenstein, Albert L., *Introduction to Ordinary Differential Equations* New York: Academic Press, 1966.

[6] Rogosinski, W. W., *Volume and Integral*. New York: Wiley (Interscience), 1962.

[7] Taylor, Angus E., *Advanced Calculus*. Boston: Ginn (Blaisdell), 1955.

TABLE OF ANTIDERIVATIVES

1. $\displaystyle\int (ax+b)^n x\, dx = \frac{1}{a^2(n+2)}(ax+b)^{n+2} - \frac{b}{a^2(n+1)}(ax+b)^{n+1},$
$$a \neq 0, \quad n \neq -1, \quad -2$$

2. $\displaystyle\int \frac{x\, dx}{ax+b} = \frac{x}{a} - \frac{b}{a^2}\ln(ax+b), \qquad a \neq 0$

3. $\displaystyle\int \frac{x^2\, dx}{ax+b} = \frac{1}{a^3}[\tfrac{1}{2}(ax+b)^2 - 2b(ax+b) + b^2\ln(ax+b)], \qquad a \neq 0$

4. $\displaystyle\int \frac{dx}{x(ax+b)} = \frac{1}{b}\ln\frac{x}{ax+b}, \qquad b \neq 0$

5. $\displaystyle\int x^2\sqrt{ax+b}\, dx = \frac{(30a^2x^2 - 24abx + 16b^2)(ax+b)^{\frac{3}{2}}}{105a^3}, \qquad a \neq 0$

6. $\displaystyle\int \frac{x\, dx}{\sqrt{ax+b}} = \frac{2ax-4b}{3a^2}\sqrt{ax+b}, \qquad a \neq 0$

7. $\displaystyle\int \frac{dx}{x\sqrt{ax+b}} = \frac{1}{\sqrt{b}}\ln\frac{\sqrt{ax+b}-\sqrt{b}}{\sqrt{ax+b}+\sqrt{b}}, \qquad b > 0$

8. $\displaystyle\int \frac{dx}{(ax+b)(cx+d)} = \frac{1}{bc-ad}\ln\frac{cx+d}{ax+b}, \qquad bc-ad \neq 0$

9. $\displaystyle\int \frac{x\, dx}{(ax+b)(cx+d)} = \frac{1}{bc-ad}\left[\frac{b}{a}\ln(ax+b) - \frac{d}{c}\ln(cx+d)\right], \qquad bc-ad \neq 0$

10. $\displaystyle\int \frac{dx}{a^2-x^2} = \frac{1}{2a}\ln\frac{a+x}{a-x}, \qquad a \neq 0$

11. $\displaystyle\int \frac{dx}{ax^2+b} = \frac{1}{\sqrt{ab}}\arctan\left(x\sqrt{\frac{a}{b}}\right), \qquad a > 0, \quad b > 0$

12. $\displaystyle\int \frac{dx}{ax^2+b} = \frac{1}{2\sqrt{-ab}}\ln\frac{x\sqrt{a}-\sqrt{-b}}{x\sqrt{a}+\sqrt{-b}}, \qquad a > 0, \quad b < 0$

13. $\displaystyle\int \frac{x\, dx}{ax^2+b} = \frac{1}{2a}\ln(ax^2+b), \qquad a \neq 0$

14. $\displaystyle\int \sqrt{a^2-x^2}\, dx = \tfrac{1}{2}[x\sqrt{a^2-x^2} + a^2\arcsin(x/a)], \qquad a \neq 0$

15. $\displaystyle\int \frac{dx}{\sqrt{x^2 \pm a^2}} = \ln(x + \sqrt{x^2 \pm a^2})$

16. $\displaystyle\int \sqrt{ax^2+b}\, dx = \frac{x}{2}\sqrt{ax^2+b} + \frac{b}{2\sqrt{a}}\ln(x\sqrt{a} + \sqrt{ax^2+b}), \qquad a > 0,$

$$= \frac{x}{2}\sqrt{ax^2+b} + \frac{b}{2\sqrt{-a}}\arcsin\left(x\sqrt{\frac{-a}{b}}\right), \qquad a < 0.$$

17. $\displaystyle\int \frac{dx}{\sqrt{ax^2 + b}} = \frac{1}{\sqrt{a}}\,\ln(x\sqrt{a} + \sqrt{ax^2 + b}),\qquad a > 0$

$\displaystyle\qquad\qquad = \frac{1}{\sqrt{-a}}\,\arcsin\left(x\sqrt{\frac{-a}{b}}\right),\qquad a < 0,\quad b > 0$

18. $\displaystyle\int x\sqrt{ax^2 + b}\,dx = \frac{1}{3a}(ax^2 + b)^{\frac{3}{2}},\qquad a \neq 0$

19. $\displaystyle\int x^2\sqrt{ax^2 + b}\,dx = \frac{x}{4a}(ax^2 + b)^{\frac{3}{2}} - \frac{bx}{8a}\sqrt{ax^2 + b}$

$\displaystyle\qquad\qquad - \frac{b^2}{8\sqrt{a^3}}\,\ln(x\sqrt{a} + \sqrt{ax^2 + b}),\qquad a > 0$

$\displaystyle\qquad\qquad = \frac{x}{4a}(ax^2 + b)^{\frac{3}{2}} - \frac{bx}{8a}\sqrt{ax^2 + b}$

$\displaystyle\qquad\qquad - \frac{b^2}{8a\sqrt{-a}}\,\arcsin\left(x\sqrt{\frac{-a}{b}}\right),\qquad a < 0,\qquad b >$

20. $\displaystyle\int \frac{\sqrt{ax^2 + b}}{x}\,dx = \sqrt{ax^2 + b} + \sqrt{b}\,\ln\frac{\sqrt{ax^2 + b} - \sqrt{b}}{x},\qquad b > 0$

$\displaystyle\qquad\qquad = \sqrt{ax^2 + b} - \sqrt{-b}\,\arctan\frac{\sqrt{ax^2 + b}}{\sqrt{-b}},\qquad b < 0$

21. $\displaystyle\int \frac{dx}{x\sqrt{a^2 \pm x^2}} = -\frac{1}{a}\,\ln\left(\frac{a + \sqrt{a^2 \pm x^2}}{x}\right),\qquad a \neq 0$

22. $\displaystyle\int \frac{dx}{x\sqrt{x^2 - a^2}} = \frac{1}{a}\,\arccos\left(\frac{a}{x}\right),\qquad a \neq 0$

23. $\displaystyle\int \frac{dx}{x\sqrt{ax^2 + b}} = \frac{1}{\sqrt{b}}\,\ln\frac{\sqrt{ax^2 + b} - \sqrt{b}}{x},\qquad b > 0$

$\displaystyle\qquad\qquad = \frac{1}{\sqrt{-b}}\,\operatorname{arcsec}\left(x\sqrt{-\frac{a}{b}}\right),\qquad b < 0,\quad a > 0$

24. $\displaystyle\int (ax^2 + b)^{3/2}\,dx = \frac{x}{8}(2ax^2 + 5b)\sqrt{ax^2 + b} + \frac{3b^2}{8\sqrt{a}}\,\ln(x\sqrt{a} + \sqrt{ax^2 + b}),\ a > 0$

$\displaystyle\qquad\qquad = \frac{x}{8}(2ax^2 + 5b)\sqrt{ax^2 + b} + \frac{3b^2}{8\sqrt{-a}}\,\arcsin\left(x\sqrt{\frac{-a}{b}}\right),\ a < 0,\quad b > 0$

25. $\displaystyle \int \frac{dx}{(ax^2 + b)^{\frac{3}{2}}} = \frac{x}{b\sqrt{ax^2 + b}}, \qquad b \neq 0$

26. $\displaystyle \int \frac{dx}{x(ax^n + b)} = \frac{1}{bn} \ln \frac{x^n}{ax^n + b}, \qquad b \neq 0$

27. $\displaystyle \int \frac{dx}{x\sqrt{ax^n + b}} = \frac{1}{n\sqrt{b}} \ln \frac{\sqrt{ax^n + b} - \sqrt{b}}{\sqrt{ax^n + b} + \sqrt{b}}, \qquad b > 0$

$\displaystyle \qquad\qquad = \frac{2}{n\sqrt{-b}} \operatorname{arcsec} \sqrt{\frac{-ax^n}{b}}, \qquad b < 0$

28. $\displaystyle \int \frac{dx}{ax^2 + bx + c} = \frac{1}{\sqrt{b^2 - 4ac}} \ln \frac{2ax + b - \sqrt{b^2 - 4ac}}{2ax + b + \sqrt{b^2 - 4ac}}, \qquad b^2 > 4ac$

$\displaystyle \qquad\qquad = \frac{2}{\sqrt{4ac - b^2}} \arctan \frac{2ax + b}{\sqrt{4ac - b^2}}, \qquad b^2 < 4ac$

$\displaystyle \qquad\qquad = -\frac{2}{2ax + b}, \qquad b^2 = 4ac$

29. $\displaystyle \int \sqrt{2ax - x^2}\, dx = \frac{1}{2}\left[(x - a)\sqrt{2ax - x^2} + a^2 \arcsin \frac{x - a}{a}\right], \qquad a \neq 0$

30. $\displaystyle \int \frac{dx}{\sqrt{2ax - x^2}} = \arccos\left(\frac{a - x}{a}\right), \qquad a \neq 0$

31. $\displaystyle \int \sqrt{\frac{x + a}{x + b}}\, dx = \sqrt{x + b}\sqrt{x + a} + (a - b)\ln(\sqrt{x + b} + \sqrt{x + a})$

32. $\displaystyle \int \sqrt{\frac{1 + x}{1 - x}}\, dx = \arcsin x - \sqrt{1 - x^2}$

33. $\displaystyle \int \sin^4 ax\, dx = \frac{3x}{8} - \frac{3 \sin 2ax}{16a} - \frac{\sin^3 ax \cos ax}{4a}, \qquad a \neq 0$

34. $\displaystyle \int \sin^n ax\, dx = -\frac{\sin^{n-1} ax \cos ax}{na} + \frac{n - 1}{n} \int \sin^{n-2} ax\, dx, \qquad n \text{ positive integer}$

35. $\displaystyle \int \frac{dx}{\sin^n ax} = -\frac{1}{a(n - 1)} \frac{\cos ax}{\sin^{n-1} ax} + \frac{n - 2}{n - 1} \int \frac{dx}{\sin^{n-2} ax}, \qquad n \text{ integer} > 1,\ a \neq 0$

36. $$\int \frac{dx}{b + c \sin ax} = \frac{-2}{a\sqrt{b^2 - c^2}} \arctan\left[\sqrt{\frac{b - c}{b + c}} \tan\left(\frac{\pi}{4} - \frac{ax}{2}\right)\right], \qquad b^2 > c^2$$

$$= \frac{-1}{a\sqrt{c^2 - b^2}} \ln \frac{c + b \sin ax + \sqrt{c^2 - b^2} \cos ax}{b + c \sin ax}, \qquad c^2 > b^2$$

37. $$\int \sin ax \sin bx \, dx = \frac{\sin (a - b)x}{2(a - b)} - \frac{\sin (a + b)x}{2(a + b)}, \qquad a^2 \neq b^2$$

38. $$\int \cos^4 ax \, dx = \frac{3x}{8} + \frac{3 \sin 2ax}{16a} + \frac{\cos^3 ax \sin ax}{4a}, \qquad a \neq 0$$

39. $$\int \cos^n ax \, dx = \frac{\cos^{n-1} ax \sin ax}{na} + \frac{n - 1}{n}\int \cos^{n-2} ax \, dx, \qquad n \text{ positive integer}$$

40. $$\int \sec^n ax \, dx = \frac{1}{a(n - 1)} \frac{\sin ax}{\cos^{n-1} ax} + \frac{n - 2}{n - 1}\int \sec^{n-2} ax \, dx, \qquad n \text{ integer} > 1$$

41. $$\int \frac{dx}{b + c \cos ax} = \frac{1}{a\sqrt{b^2 - c^2}} \arctan\left(\frac{\sqrt{b^2 - c^2} \sin ax}{c + b \cos ax}\right), \qquad b^2 > c^2$$

$$= \frac{1}{a\sqrt{c^2 - b^2}} \operatorname{arctanh}\left(\frac{\sqrt{c^2 - b^2} \sin ax}{c + b \cos ax}\right), \qquad c^2 > b^2$$

42. $$\int \frac{dx}{b \sin ax + c \cos ax} = \frac{1}{a\sqrt{b^2 + c^2}} \ln\left[\tan \frac{1}{2}\left(ax + \arctan \frac{c}{b}\right)\right], \qquad ab \neq 0$$

43. $$\int \tan^n ax \, dx = \frac{1}{a(n - 1)} \tan^{n-1} ax - \int \tan^{n-2} ax \, dx, \qquad n \text{ integer} > 1$$

44. $$\int \cot^n ax \, dx = -\frac{1}{a(n - 1)} \cot^{n-1} ax - \int \cot^{n-2} ax \, dx, \qquad n \text{ integer} > 1$$

45. $$\int \sec^3 ax \, dx = \frac{1}{2a}\left[\tan ax \sec ax + \ln \tan\left(\frac{ax}{2} + \frac{\pi}{4}\right)\right], \qquad a \neq 0$$

46. $$\int \sec^n ax \, dx = \frac{1}{a(n - 1)} \frac{\sin ax}{\cos^{n-1} ax} + \frac{n - 2}{n - 1}\int \sec^{n-2} ax \, dx, \qquad n \text{ integer} > 1, \, a \neq 0$$

47. $$\int \csc^3 ax \, dx = \frac{1}{2a}\left[-\cot ax \csc ax + \ln \tan \frac{ax}{2}\right], \qquad a \neq 0$$

48. $\displaystyle\int \csc^n ax\, dx = -\frac{1}{a(n-1)} \frac{\cos ax}{\sin^{n-1}ax} + \frac{n-2}{n-1}\int \csc^{n-2}ax\, dx,$ n integer >1

49. $\displaystyle\int x^2 \sin ax\, dx = \frac{2x}{a^2} \sin ax + \frac{2}{a^3} \cos ax - \frac{x^2}{a} \cos ax,$ $a \neq 0$

50. $\displaystyle\int x^2 \cos ax\, dx = \frac{2x}{a^2} \cos ax - \frac{2}{a^3} \sin ax + \frac{x^2}{a} \sin ax$

51. $\displaystyle\int x^n e^{ax}\, dx = \frac{1}{a} x^n e^{ax} - \frac{n}{a}\int x^{n-1}e^{ax}\, dx,$ n positive

52. $\displaystyle\int \frac{dx}{b + ce^{ax}} = \frac{1}{ab}[ax - \ln(b + ce^{ax})],$ $ab \neq 0$

53. $\displaystyle\int \frac{dx}{be^{ax} + ce^{-ax}} = \frac{1}{a\sqrt{bc}} \arctan\left(e^{ax}\sqrt{\frac{b}{c}}\right),$ b and c positive.

54. $\displaystyle\int e^{ax} \sin bx\, dx = \frac{e^{ax}}{a^2 + b^2}(a \sin bx - b \cos bx),$ $a^2 + b^2 \neq 0$

55. $\displaystyle\int e^{ax} \cos bx\, dx = \frac{e^{ax}}{a^2 + b^2}(a \cos bx + b \sin bx),$ $a^2 + b^2 \neq 0$

56. $\displaystyle\int e^{ax} \cos^n bx\, dx = \frac{e^{ax} \cos^{n-1} bx\, (a \cos bx + nb \sin bx)}{a^2 + n^2 b^2}$

$$+ \frac{n(n-1)b^2}{a^2 + n^2 b^2}\int e^{ax} \cos^{n-2} bx\, dx,\qquad a^2 + n^2 b^2 \neq 0$$

57. $\displaystyle\int e^{ax} \sin^n bx\, dx = \frac{e^{ax} \sin^{n-1} bx\, (a \sin bx - nb \cos bx)}{a^2 + n^2 b^2}$

$$+ \frac{n(n-1)b^2}{a^2 + n^2 b^2}\int e^{ax} \sin^{n-2} bx\, dx\qquad a^2 + n^2 b^2 \neq 0$$

58. $\displaystyle\int x^n (\ln ax)^m\, dx = \frac{x^{n+1}}{n+1}(\ln ax)^m - \frac{m}{n+1}\int x^n(\ln ax)^{m-1}\, dx,$ $n \neq -1$

59. $\displaystyle\int (\arcsin ax)^2\,dx = x(\arcsin ax)^2 - 2x + \frac{2}{a}\sqrt{1 - a^2x^2}\,\arcsin ax\,, \qquad a \neq 0$

60. $\displaystyle\int x^n \arcsin ax\,dx = \frac{x^{n+1}}{n+1}\arcsin ax - \frac{a}{n+1}\int \frac{x^{n+1}\,dx}{\sqrt{1 - a^2x^2}}\,, \qquad n \neq -1$

SOLUTIONS TO
PROBLEMS

Section 1 · 1

A · $(x - y)(x + y) = 0$. The equation is satisfied by points where $x = y$ or where $x = -y$ (Fig. 1).

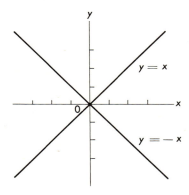

FIG. 1

B · See Fig. 2. The distance of the arbitrary point (x, y) from the y axis is x; its distance from $(2, 0)$ is allowing for directed distances, either $\pm \sqrt{(x - 2)^2 + y^2}$. The required equation is $x = \pm \sqrt{(x - 2)^2 + y^2}$ or $y^2 - 4x + 4 = 0$.

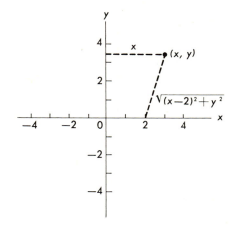

FIG. 2

Section 1 · 2

A · $\dfrac{1}{|\mathbf{v}|} \mathbf{v} = \dfrac{1}{\sqrt{9 + 25}} (3\mathbf{i} + 5\mathbf{j}) = \dfrac{3}{\sqrt{34}} \mathbf{i} + \dfrac{5}{\sqrt{34}} \mathbf{j}$.

B · For $A(1, -9)$, $B(3, -4)$, and $C(7, 6)$, the vectors $\mathbf{AB} = 2\mathbf{i} + 5\mathbf{j}$ and $\mathbf{BC} = 4\mathbf{i} + 10\mathbf{j}$ are parallel since $\mathbf{BC} = 2\mathbf{AB}$.

C · $\mathbf{BA} = -7\mathbf{i} + \mathbf{j}$ and $\mathbf{BC} = -4\mathbf{i} - 3\mathbf{j}$, so $\cos \angle ABC = \dfrac{28 - 3}{\sqrt{50}\sqrt{25}} = \dfrac{1}{\sqrt{2}}$. $\angle ABC = 45°$.

D · $\mathbf{v} \cdot \mathbf{v} = a^2 + b^2 = |\mathbf{v}|^2$.

Section 1 · 3

A · Substitution in $y - y_1 = m(x - x_1)$ gives $y - 1 = \frac{1}{2}(x + 4)$. From the graph (Fig. 3), starting at $(-4, 1)$ and going over four and up two (achieving a slope of

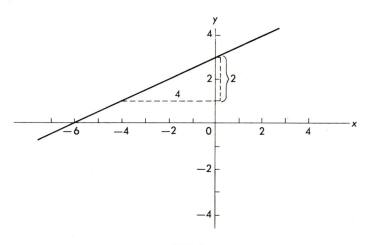

FIG. 3

$2/4 = 1/2$), the y intercept is 3. Then substitution in $y = mx + b$ gives $y = \frac{1}{2}x + 3$. Finally note that

$$y - 1 = \tfrac{1}{2}(x + 4) = \tfrac{1}{2}x + 2$$

which simplifies to $y = \frac{1}{2}x + 3$.

B · $d = \dfrac{3(0) - 4(0) - 10}{-\sqrt{9+16}} = 2$ meaning that the origin lies above the line. The nor-

mal form, with $\rho = -2$, also implies that the origin is above the line.

Section 2 · 2

A · The function H is defined whenever $4 - x^2 \geq 0$; that is, $x^2 \leq 4$. Hence the domain is $[-2, 2]$. Since $0 \leq \sqrt{4 - x^2} \leq 2$, the range of H is $[3, 5]$.

B · The points $(1, 6)$ and $(1, -2)$ are both in S. There is no function G for which we
can have both $G(1) = 6$ and $G(1) = -2$, so the set S cannot be the graph of any
function G. Solution of the given equation for y gives

$$y = 2 \pm 2\sqrt{x+3}$$

Set $G_1(x) = 2 + 2\sqrt{x+3}$; $G_2(x) = 2 - 2\sqrt{x+3}$. The graphs of these two functions
now include those points and only those points which are members of S.

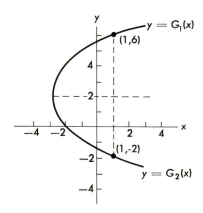

FIG. 4

C · The only place at which R is undefined is at $x = 3$ where the denominator is
zero. Hence, the domain is the set of real numbers excluding 3. It is very difficult at
this point to say much about the range. It is clear, since both the numerator and
denominator of the formula for $R(x)$ remain positive for all x, that the range of R
contains no negative numbers.

Section 2 · 3

A · If a vertical line $(x = c)$ cuts a subset of the plane in two or more points, the subset cannot be the graph of a single function F, for $F(c)$ would be ambiguous (Fig. 5).

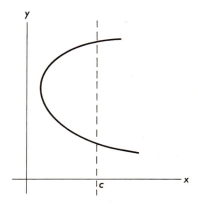

FIG. 5

B · Since we may divide by $x + 1$ when $x \neq -1$, we have (Fig. 6)

$$y = \begin{cases} x - 1 & \text{for } x \neq -1 \\ \text{undefined} & \text{for } x = -1 \end{cases}$$

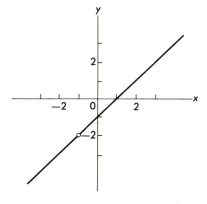

FIG. 6

C · $y = \dfrac{(x+3)(x-2)}{(x-1)(x+1)}$ (see Fig. 7).

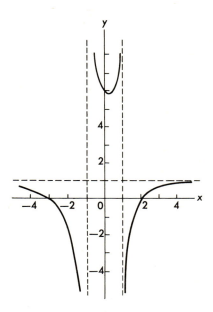

FIG. 7

D · Fig. 8.

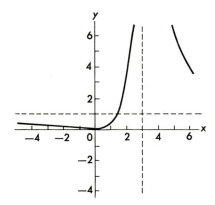

FIG. 8

Section 2 · 4

A · (a) If $x = a$, equality is obvious. If $x > a$, then by definition

$$|x - a| = x - a$$
$$|a - x| = x - a.$$

Similarly for $x < a$.

(b) $|x| = |x - 0| = \begin{cases} x - 0 = x & \text{if } x \geq 0 \\ 0 - x = -x & \text{if } x \leq 0 \end{cases}$.

Compare with (1) of Section 2.1.
For any number k,

$$|t + k|^2 = (t + k)^2 = t^2 + 2tk + k^2 \leq |t|^2 + 2|t||k| + |k|^2$$

Therefore, $|t + k|^2 \leq (|t| + |k|)^2$ and since both sides are positive, $|t + k| \leq |t| + |k|$.

(c) Take $k = b$ above.
(d) Take $k = -b$ above.
(e) The corner is at $x = 0$. Fig. 4.2 in Chapter 4.

B · (a) $\lim\limits_{x \to 0^-} F(x) = \lim\limits_{x \to 0^+} F(x) = 3$

(b) $\lim\limits_{x \to 2^-} F(x) = \lim\limits_{x \to 2^+} F(x) = 1$

(c) $\lim\limits_{x \to 4^-} F(x) = -2$

$\lim\limits_{x \to 4^+} F(x) = 0$

(d) $\lim\limits_{x \to 0} F(x) = 3$

$\lim\limits_{x \to 2} F(x) = 1$

$\lim\limits_{x \to 4} F(x)$ does not exist
because the limits from the
two sides are not equal.

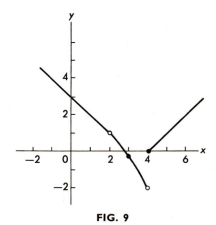

FIG. 9

C · $\dfrac{3x^2 + 6x}{2x^3 - 7x^2 + 11} = \dfrac{\dfrac{3}{x} + \dfrac{6}{x^2}}{2 - \dfrac{7}{x} + \dfrac{11}{x^3}}$. As x gets numerically large (positively or

negatively), the fraction approaches $0/2 = 0$.

D· $F(x) = \dfrac{x^2 + 4x - 3}{x^2 - 4x + 3} = \dfrac{x^2 - 4x + 3}{(x-3)(x-1)}$

$= \dfrac{1 + \dfrac{4}{x} - \dfrac{3}{x^2}}{1 - \dfrac{4}{x} + \dfrac{3}{x^2}}$. From the second

expression we see that the graph has vertical asymptotes at $x = 1$ and $x = 3$; it is zero at $x = -2 \pm \sqrt{7}$. From the third expression, we see that $\lim\limits_{x \to \infty} F(x) = 1$ (Fig. 10).

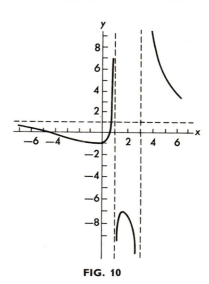

FIG. 10

E· When $n > m$, multiply the numerator and denominator by $1/x^m$. Then

$$\lim_{x \to \infty} R(x) = \lim_{x \to \infty} \frac{a_n x^{n-m} + \cdots + (a_0/x^m)}{b_m + \cdots + (b_0/x^m)} = \infty$$

When $n = m$, multiply the numerator and denominator by $1/x^m = 1/x^n$. Then show $\lim\limits_{x \to \infty} R(x) = a_n/b_m$. Finally, use a similar technique to show that $\lim\limits_{x \to \infty} R(x) = 0$ when $n < m$.

F· $G(x) = \begin{cases} x^2 + 3x & \text{if } x \neq 1 \\ \text{undefined} & \text{at } x = 1 \end{cases}$

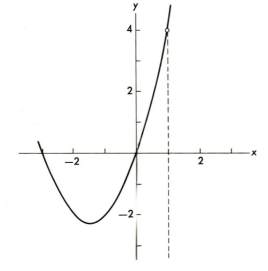

FIG. 11

G • The numerator is not a polynomial. The shortcuts you have been learning do not apply. Reread the third paragraph after Problem 2.3A in the text. Then do the best you can with this problem. After doing your best, if you can't stand the suspense, you may peek ahead to the solution of Problem 5.1B.

H • As it stands, nothing can be determined since both the numerator and the denominator go to zero as h goes to zero. However, if we rationalize the numerator we have

$$\lim_{h \to 0} \frac{\sqrt{4+h} - \sqrt{4}}{h} \frac{\sqrt{4+h} + \sqrt{4}}{\sqrt{4+h} + \sqrt{4}} = \lim_{h \to 0} \frac{(4+h) - 4}{h[\sqrt{4+h} + \sqrt{4}]}$$

$$= \lim_{h \to 0} \frac{1}{\sqrt{4+h} + \sqrt{4}} = \frac{1}{4}$$

Section 2 · 5

A • $y = G(x) = \dfrac{2x+1}{x-3}$. Solve for x, getting $x = G^{-1}(y) = \dfrac{3y+1}{y-2}$. Note that $G(5)$
$= \frac{11}{2}$ and (happily) $G^{-1}(\frac{11}{2}) = 5$.

B • $y = H(x) = \dfrac{x}{x^2 + 1}$, so $yx^2 - x + y = 0$. Treating this as a quadratic in x, we get from the formula,

$$x = \frac{1 \pm \sqrt{1 - 4y^2}}{2y}.$$

This defines two functions

$$x = G_1(y) = \frac{1 - \sqrt{1 - 4y^2}}{2y}$$

$$x = G_2(y) = \frac{1 + \sqrt{1 - 4y^2}}{2y}.$$

$H(2) = \frac{2}{5}$. Since $G_1(\frac{2}{5}) = \frac{1}{2}$ and $G_2(\frac{2}{5}) = 2$, it is clear that G_2 is the inverse we seek.

Section 2 · 6

A • Let $\varepsilon > 0$ be given. For any two numbers x_1 and x_2 in (a, b), $|F(x_1) - F(x_2)| = |c - c| = 0 < \varepsilon$. We don't even have to get x_1 and x_2 close in this case.

B • Let $\varepsilon > 0$ be given. Choose $\delta = \varepsilon$. Then $|L(x) - L(x_0)| = |x - x_0| < \delta = \varepsilon$.

C · $F(x) = a_n$ is continuous by A. $L(x) = x$ is continuous by B. $F(x)L(x) = a_n x$ is therefore continuous by Theorem B. Similarly for each term of the polynomial. Then $P(x)$ is continuous by Theorem A.

D · $R(x)$ is continuous whenever $Q(x)$ is nonzero.

E · Let $\varepsilon > 0$ be given. We want to choose δ so that if $|x - 2| < \delta$, then $|G(x) - G(2)| < \varepsilon$. Observe $|G(x) - G(2)| = ||2x - 4| - 0| = |2x - 4| = 2|x - 2|$. If we choose $\delta = \varepsilon/2$, then $|G(x) - G(2)| < 2(\varepsilon/2) = \varepsilon$.

F · No. $F(x) = 1/x$ does not attain a maximum value on $(0, 1)$.

G · $M(\tfrac{1}{2}) = 8$; $M(\tfrac{3}{2}) = 16$. Pick $c = 9$ (or any number such that $8 < c < 16$). There is no $\bar{x}$ such that $M(\bar{x}) = 9$.

H · (a)   Continuity at all $x \neq -\tfrac{1}{2}$ follows from Problem D above.
(b)   The numerator, when graphed, is a parabola. Its high point is 18. The numerator is thus never more than 18, and on $[1, 6]$, the minimum of $2x + 1$ is 3. The quotient cannot possibly be as big as $18/3$.

(c) $R(3) = \dfrac{10}{7} < \sqrt{10} < \dfrac{10}{3} = R(1)$. Theorem E applies. There is an $\bar{x}$ so that $R(\bar{x}) = \sqrt{10}$.

Section 3·1

A · Suppose R is everywhere equal to c. Then

$$\lim_{h \to 0} \frac{R(x + h) - R(x)}{h} = \lim_{h \to 0} \frac{c - c}{h} = 0.$$

B · When F is differentiable, then

$$\lim_{h \to 0} \frac{F(x_0 + h) - F(x_0)}{h} = F'(x_0).$$

Consider

$$\lim_{h \to 0^-} \frac{F(x_0 + h) - F(x_0)}{h}.$$

The numerator remains negative because $F(x_0) > F(x_0 + h)$. The denominator, since $h < 0$, is negative. The quotient is therefore positive, and we have $F'(x_0) \geq 0$. Now consider

$$\lim_{h \to 0^+} \frac{F(x_0 + h) - F(x_0)}{h}.$$

Study of signs now shows $F'(x_0) \leq 0$. It follows that $F'(x_0) = 0$.

C · Argue as in B; this time the numerator remains positive.

Section 3 · 2

A · $F'(2) = \lim\limits_{h \to 0} \dfrac{1}{h}[F(2 + h) - F(2)]$

$F(2 + h) - F(2) = [6(2 + h) - (2 + h)^2 - 5] - [3]$

$F'(2) = \lim\limits_{h \to 0} \dfrac{1}{h}[2h - h^2] = \lim\limits_{h \to 0} [2 - h] = 2$

$\Delta y = F(2 + \tfrac{1}{2}) - F(2) = \dfrac{15}{4} - 3 = \tfrac{3}{4}$

$dy = [F'(2)]dx = 2(\tfrac{1}{2}) = 1$

Section 3 · 5

A · We are to find a *linear* function L so that

(*) $\qquad\qquad\qquad F(x_0 + h) - F(x_0) = L(h) + |h|r(x_0, h)$

where $\lim\limits_{h \to 0} r(x_0, h) = 0$. Since $F(x_0 + h) = F(x_0) + F(h)$, this says we want

$$F(h) = L(h) + |h|r(x_0, h).$$

If we set $L = F$ (which we may do since F is *linear*), then we have $|h|r(x_0, h) = 0$ and it is clear that $r(x_0, h) = 0$ for any h.

B · If F is constant, the left side of (*) in the solution to 1.1 is 0. If we choose $L = 0$, we again have $r(x_0, h) = 0$ for any h.

C · $[F'(x)] = [3x^2]$ so $[F'(2)](3) = [12](3) = 36$

Section 4 · 1

A · $\dfrac{2 - 1}{3} = \dfrac{1}{2} x^{-1/2}; \; x = \dfrac{9}{4}$

B · For any $x \in (a, b]$, there exists x_0 so that

$$\frac{F(x) - F(a)}{x - a} = F'(x_0).$$

But $F'(x_0) = 0$, so $F(x) = F(a)$.

C · If $F'(x_0) = c$, then

$$\frac{F(x) - F(a)}{x - a} = c.$$

D · Direct substitution gives $G(a) = 0$; $G(b) = 0$. Therefore (Rolle's theorem), there exists an $x_0 \in (a, b)$ so that $G'(x_0) = 0$. But

$$G'(x_0) = F'(x_0) - \frac{F(b) - F(a)}{b - a}.$$

Substituting $G'(x_0) = 0$ gives the desired result.

Section 4 · 2

A · See Fig. 5.5 of Section 5.1.

B · Answers to these questions are discussed in Section 5.1. The important thing here is for you to think about the questions.

C · No. In fact,

$$D^n P(x) = \left(\frac{1}{2}\right)\left(-\frac{1}{2}\right) \cdots \left(-\frac{2n-3}{2}\right) x^{-(2n-1)/2}$$

Section 4 · 3

A · $F(2) = 14$; $F'(2) = 7$; $F''(2) = 6$; $F'''(2) = 6$.

$$F(x) = 14 + 7(x - 2) + \frac{6}{2}(x - 2)^2 + \frac{6}{3!}(x - 2)^3.$$

B · (See Fig. 12).
  (a) 1
  (b) $1 + x$
  (c) $1 + x + x^2$
  (d) $1 + x + x^2 + x^3$

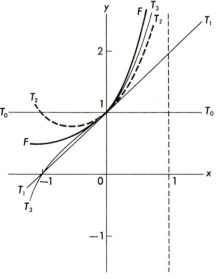

FIG. 12

C . $c = -\dfrac{4}{3}$. $D^3G(x) = \dfrac{6}{(1-x)^4} - 8$. Setting this expression equal to 0, $\bar{x} = 1 - (\frac{3}{4})^{1/4}$.

D . $R(0, \frac{1}{4}) = \dfrac{1}{4^3(1-\bar{x})^4}$, $\bar{x} \in [0, \frac{1}{4}]$.

Because of the restriction on x, the denominator is always greater than or equal to $4^3(\frac{3}{4})^4$. Hence, $R(0, \frac{1}{4}) < \dfrac{1}{3^4/4} = \dfrac{4}{3^4}$. $F\left(\frac{1}{4}\right) - T_2\left(\frac{1}{4}\right) = \dfrac{4}{3} - \dfrac{21}{16} = \dfrac{1}{48} \sim 0.021$ is the actual error. To get $R_n(0, \frac{1}{4}) < 0.005$, we must choose n so that

$$\frac{1}{4^{n+1}(\frac{3}{4})^{n+2}} = \frac{4}{3^{n+2}} < \frac{5}{1000}$$

Trial and error shows that $n = 5$ will do.

Section 5 · 1

A . Set $R'(x) = 0$; then $x = 0$ or $x = \frac{3}{2}$. From the sketch drawn previously, $(0, R(0)) = (0, -1)$ is seen to be a relative minimum; $(\frac{3}{2}, -7)$ is a relative maximum.

B . (See Fig. 13.) $S'(x) = \dfrac{16(3-x)}{3x^3(x-2)^{1/3}}$

$\lim\limits_{x \to 2^-} S'(x) = -\infty$

$\lim\limits_{x \to 2^+} S'(x) = \infty$

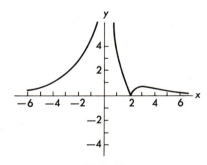

FIG. 13

C . $6y\dfrac{dy}{dx} - 12\dfrac{dy}{dx} - 4x - 12 = 0$ so $\dfrac{dy}{dx} = \dfrac{2(x+3)}{3(y-2)}$

When $x = -2$, $y = 2 \pm \dfrac{\sqrt{6}}{3}$ so $\dfrac{dy}{dx} = \pm\dfrac{2}{\sqrt{6}}$

When $x = -1$, $y = 2 \pm \dfrac{2\sqrt{6}}{3}$ so $\dfrac{dy}{dx} = \dfrac{4}{\pm 2\sqrt{6}} = \dfrac{2}{\pm\sqrt{6}}$

We get the same value of dy/dx at $x = 0, 1, 2$. This is explained by noting that the graph of the given equation consists of two intersecting lines having slopes of $2/\sqrt{6}$ and $2/(-\sqrt{6})$.

D· (See Fig. 14.) $R(x) = \dfrac{-2(2x-1)(2x-7)}{2x+1}$.

$$R'(x) = 0 \text{ when } x = \tfrac{3}{2},\ -\tfrac{5}{2}.$$

$$R(\tfrac{3}{2}) = 4.$$

The maximum occurs at $x_1 = \tfrac{3}{2}$.
The minimum occurs at $x_2 = 6$.

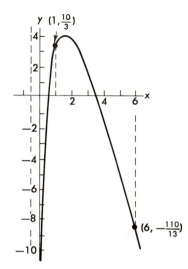

FIG. 14

E· A speculator may hold a certain stock for its growth potential. After a period of rapidly increasing profits, he notices that the profits, while increasing, are not increasing as fast. He may consider this to be the time to sell; that is, he may wish to sell when the rate of increase of profits (second derivative) falls even though the change in profits (first derivative) continues to rise.

Section 5·2

A· The area to be minimized now is $A(x) = 8x^2 + 2\pi xh = 8x^2 + 128/x$. Setting $\dfrac{dA}{dx} = 16x - \dfrac{128}{x^2} = 0$, we find $x = 2$.

B· The time on land is now $\dfrac{1-x}{5}$. Consequently,

$$T = \tfrac{1}{3}(x^2 + \tfrac{9}{16})^{1/2} + \tfrac{1}{5} - \tfrac{1}{5}x.$$

The derivative is as in Example 5.2B. We are interested in $x \in [0, 1]$. The solution (obviously a minimum from Fig. 5.9) is $x = \tfrac{9}{16}$.

C· Maximum is at $(\tfrac{9}{2}, 13\tfrac{3}{8})$, an end point. Minimum is $(3, 1)$. The relative maximum at $(\tfrac{1}{3}, \tfrac{283}{27})$ is not the maximum on $[0, \tfrac{9}{2}]$.

Section 5 · 3

A · $x^2 + y^2 = 18^2$

$$2x \frac{dx}{dt} + 2y \frac{dy}{dt} = 0$$

It is given that $dx/dt = 3$. When $x = 6$, $y = 12\sqrt{2}$.
Therefore $dy/dt = -3/(2\sqrt{2})$ ft/sec.

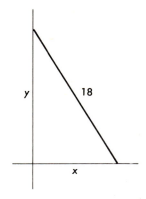

FIG. 15

B · $G'(x) = 0$ so $G(x) = F_1(x) - F_2(x) = C$

C · (a) $\frac{3}{2}x^{2/3} + \frac{3}{4}x^{4/3} + C$ (b) $-x^{-1} + C$

    (c) You will not get this one until you have completed Section 7.4.

    (d) $\frac{16}{45}x^{9/4} + C$ (e) $\frac{1}{3}(x^2 + 1)^{3/2} + C$

    (f) Chances are that you did not get this one, but you may verify that $\dfrac{x}{\sqrt{x^2 + 1}}$

works. Methods enabling you to get this antiderivative yourself are given in Chapter 8.

D · If $p \neq -1$, then $D^{-1}x^p = \dfrac{1}{p+1} x^{p+1}$. For the case $p = -1$, you must read Section 7.4.

E · When thrown from the ground with initial velocity v_0, the height at time t is $s = -16t^2 + v_0 t$. The maximum height occurs when $ds/dt = 0$; that is, at $t = v_0/32$. This height must be $-16(v_0/32)^2 + v_0(v_0/32) = 324$. Solution gives $v_0 = 144$. (See Exercise 6 of this section.)

F · The initial vertical velocity is 200 ft/sec. The height at time t is $s = -16t^2 + 200t$. The projectile will therefore return to earth when $t = 25/2$. During that time the projectile will travel: (horizontal component of initial velocity) × (time in flight) = $(200\sqrt{3})(25/2) = 2500\sqrt{3}$ ft.

Section 6 · 1

A · $\displaystyle\sum_{k=4}^{7} k^2 = 16 + 25 + 36 + 49 = 126$

B · $\displaystyle\sum_{k=0}^{n}(ca_k + d) = (ca_0 + d) + (ca_1 + d) + \cdots + (ca_n + d)$

$$= c[a_0 + a_1 + \cdots + a_n] + (n+1)d$$

C · Suppose the formula is true for $n - 1$; that is, suppose

$$1^2 + 2^2 + \cdots + (n-1)^2 = \frac{(n-1)n(2n-1)}{6}.$$

Then

$$1^2 + 2^2 + \cdots + (n-1)^2 + n^2 = \frac{(n-1)n(2n-1) + 6n^2}{6}.$$

The numerator simplifies to the desired expression.

D · $\displaystyle S_n = \frac{2}{n}\left[G\left(0 \cdot \frac{2}{n}\right) + G\left(1 \cdot \frac{2}{n}\right) + \cdots + G\left((n-1) \cdot \frac{2}{n}\right)\right]$

$$= \frac{2}{n}\sum_{k=0}^{n-1} G\left(k \cdot \frac{2}{n}\right) = \frac{2}{n}\sum_{k=0}^{n-1}\left[\left(\frac{2k}{n}\right)^2 + 2\right]$$

$$= \frac{2}{n}\left\{\frac{4}{n^2}\sum_{k=0}^{n-1} k^2 + n(2)\right\}$$

$$= \frac{8}{n^3}\frac{(n-1)n(2n-1)}{6} + 4 = \frac{20}{3} - \frac{4}{n} + \frac{4}{3n^2}.$$

$$\lim_{n\to\infty} S_n = \lim_{n\to\infty}\left(\frac{20}{3} - \frac{4}{n} + \frac{4}{3n^2}\right) = \frac{20}{3}.$$

Section 6 · 2

A · [0, 1] is partitioned into four subintervals. The first four terms of the Riemann sum are

$$8(\tfrac{1}{4}) + 8(\tfrac{1}{4}) + 8(\tfrac{1}{4}) + 8(\tfrac{1}{4}) = 8.$$

Similarly, the four terms corresponding to [1, 2] contribute 16, and those from [2, 3] contribute 24 to the total. The answer is 48.

B · Consider the difference

$$\sum_{k=1}^{n} F(x_k)(x_k - x_{k-1}) - \sum_{k=1}^{n} F(x_{k-1})(x_k - x_{k-1})$$

$$= \sum_{k=1}^{n}[F(x_k) - F(x_{k-1})](x_k - x_{k-1}) \le \delta(P)\sum_{k=1}^{n}[F(x_k) - F(x_{k-1})]$$

where $\delta(P)$ is the mesh of the partition. Now the sum

$$\sum_{k=1}^{n} [F(x_k) - F(x_{k-1})] = F(x_n) - F(x_0) = F(b) - F(a)$$

which is constant. Thus, the difference goes to 0 as $\delta(P) \to 0$. (In accord with our policy of warning the reader when something has been done without full justification, we remark that we have here assumed that if two sequences $\{a_n\}$ and $\{b_n\}$ have the property that $|a_n - b_n| \to 0$, then they both converge to the same number. Proof involves things we have not discussed.)

C . Given an arbitrary $\varepsilon > 0$, choose $\delta = \varepsilon/Mr$. Corresponding to the partition $P = \{a = x_0, \ldots, x_n = b\}$ suggested in the hint, we have

$$|\sum_{k=1}^{n} F(t_k)(x_k - x_{k-1})| \leq \sum' |F(t_k)| \, |x_k - x_{k-1}| + |\sum'' F(t_k)(x_k - x_{k-1})|$$

where $\sum'$ extends over the r intervals including a c_k and $\sum''$ extends over the rest of the intervals. Since $|F(t_k)| \, |x_k - x_{k-1}| < M\delta$ and $M\delta r = \varepsilon$,

$$|\sum_{k=1}^{n} F(t_k)(x_k - x_{k-1})| \leq \varepsilon + |\sum'' F(t_k)(x_k - x_{k-1})|.$$

Refinements within each of the intervals included in $\sum''$ will, since F is here continuous, lead to within ε of a finite limit L. Then

$$\lim_{\delta(P) \to 0} |\sum_{k=1}^{n} F(t_k)(x_k - x_{k-1})| < 2\varepsilon + L$$

and the integral exists.

D . (a) Area $= \dfrac{25\pi}{4}$ (b) Area $= \dfrac{18}{2} + 3 = 12$ (c) $\displaystyle\int_0^2 (4 - x^2) = \int_0^2 [6 - (x^2 + 2)]$

$$= 12 - \frac{20}{3} = \frac{16}{3}$$

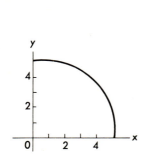 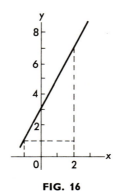 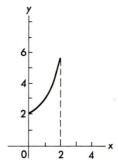

FIG. 16

E · (a) Find the area bounded by the positive coordinate axes, the line $x = 2$, and the curve $y = \pi x^2$.

(b) The region bounded by the lines $x = 0$, $y = x$, and $y = 2$ is rotated about the y axis. Find the volume.

(c) Find the moment about the y axis of the region bounded by the lines $y = 0$, $x = 2$, $y = \pi x$.

F · Geometrically, look at Fig. 17. Analytically, argue as we did for the function M on [0, 1] in Problem A above.

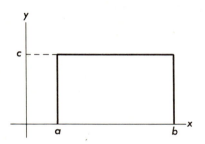

FIG. 17

Section 6 · 3

A · $\int_0^2 (x^2 + 2)\, dx = \int_0^2 x^2\, dx + \int_0^2 2\, dx$;   or $\dfrac{20}{3} = \int_0^2 x^2\, dx + 4$.

Thus, $\int_0^2 x^2\, dx = \dfrac{8}{3}$.

B · $\int_{-1}^2 x\, dx = 2 - \tfrac{1}{2} = \tfrac{3}{2}$.   See Fig. 18.

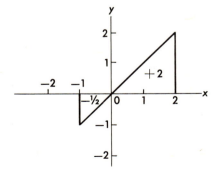

FIG. 18

$$\int_{-1}^{2} (2x + 3)\, dx = \int_{-1}^{2} 2x\, dx + \int_{-1}^{2} 3\, dx = 2\int_{-1}^{2} x\, dx + 3[2 - (-1)] = 3 + 9 = 12.$$

C. $\int_{0}^{3} (x^2 + 2)\, dx = \int_{0}^{2} (x^2 + 2)\, dx + \int_{2}^{3} (x^2 + 2)\, dx.$ $15 = \dfrac{20}{3} = + \int_{2}^{3} (x^2 + 2)\, dx.$

D. $(x - \sqrt{2})^2\, dx \geq 0$ so $\int_{0}^{2} (x - \sqrt{2})^2\, dx \geq 0.$

E. (a) $\frac{2}{3}(\frac{1}{2}) + \frac{1}{2}(\frac{1}{2}) < L(2) < 1(\frac{1}{2}) + \frac{2}{3}(\frac{1}{2}).$ See Fig. 19

(b) Between 2 and 3. (If you guessed 2.7, you are either inspired or else cheating.)

(c) and (d) You will get a chance to check these later.

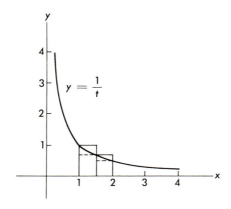

FIG. 19

Section 6 · 4

A. $\int_{0}^{2} x^2\, dx = \dfrac{x^3}{3}\bigg|_{1}^{2} = \dfrac{8}{3}.$

B. $\int_{-1}^{2} (2x + 3)\, dx = (x^2 + 3x)\bigg|_{-1}^{2} = (4 + 6) - (1 - 3) = 12.$

C. $\int_{0}^{3} (x^2 + 2)\, dx = \left(\dfrac{x^3}{3} + 2x\right)\bigg|_{0}^{3} = 9 + 6 = 15.$

D. $\int_{0}^{2} (x^2 - 2\sqrt{2}x + 2)\, dx = \left(\dfrac{x^3}{3} - \sqrt{2}x^2 + 2x\right)\bigg|_{0}^{2}.$

$= \frac{8}{3} - 4\sqrt{2} + 4 \approx 6.67 - 5.6 > 0.$

E. $\int_{0}^{2} t\, dt = \dfrac{t^2}{2}\bigg|_{0}^{2} = 2.$

F · $\int_0^2 2x\sqrt{x^2 + 9}\, dx = \int_0^{25} u^{1/2}\, du$ since $u = x^2 + 9,\ du = 2x\, dx$.

$$= \tfrac{2}{3} u^{3/2} \Big|_9^{25} = \tfrac{2}{3}[125 - 27] = \tfrac{2}{3}(98).$$

G · $u = bt$ so $du = b\, dt$.

$$\int_1^{ab} \frac{du}{u} = \int_{1/b}^a \frac{b\, dt}{bt} = \int_{1/b}^a \frac{dt}{t}.$$

Section 6 · 5

A · $\displaystyle\int_0^\infty \frac{dx}{x\sqrt{x^2 + x}} = \int_0^1 \frac{dx}{x\sqrt{x^2 + x}} + \int_1^\infty \frac{dx}{x\sqrt{x^2 + x}}.$

On $[0, 1]$, $x^4 \le x^3$. Therefore,

$$\int_0^1 \frac{dx}{\sqrt{x^4 + x^3}} \ge \int_0^1 \frac{dx}{\sqrt{2x^3}} = \lim_{r \to 0+} \frac{1}{\sqrt{2}} \int_r^1 x^{-3/2}\, dx$$

$$= \lim_{r = 0+} \frac{-2}{\sqrt{2}} x^{-1/2} \Big|_r^1 = \infty$$

The integral diverges since the first integral on the right diverges.

B · When $p \ne 1$, $\displaystyle\int_1^b x^{-p}\, dx = \frac{1}{-p + 1} x^{-p+1} \Big|_1^b$. If $p > 1$, then $\displaystyle\lim_{b \to \infty} \frac{1}{1 - p} \left[\frac{1}{b^{p-1}} - 1 \right]$

$= \dfrac{1}{p - 1}$. If $p < 1$, then $\displaystyle\lim_{b \to \infty} \frac{1}{1 - p} [b^{1-p} - 1] = \infty$. Thus, we have convergence to

$\dfrac{1}{p - 1}$ if $p > 1$; divergence for $p < 1$. We again defer the case $p = 1$ until Section 7.4.

Section 7 · 1

A · $2 \sin^2 \beta/2 = 1 - \cos \beta = 1 - (-\tfrac{4}{5}) = \tfrac{9}{5}$.
$\sin^2 \beta/2 = \tfrac{9}{10}$. Since β is in the third quadrant, $\beta/2$ must be in the second quadrant. Thus, $\sin \beta/2 = 3/\sqrt{10}$.

B · α and β are in the first quadrant. If $(\alpha + \beta)$ is in the second quadrant, $\cos(\alpha + \beta)$ will be negative, thus telling us we are in the second quadrant. $\sin(\alpha + \beta)$ will remain positive in either quadrant.

C · Use the method of Problem B.

Section 7 · 2

A · Use the fact from trigonometry that $\sin(-t) = -\sin t$. Then

$$\lim_{t \to 0+} \frac{-t}{-\sin t} = \lim_{t \to 0+} \frac{t}{\sin t} = 1.$$

B · $D \tan x = D \dfrac{\sin x}{\cos x}$ which we recognize as a quotient.

$$D \tan x = \frac{(\cos x) \cos x - \sin x(-\sin x)}{\cos^2 x} = \frac{1}{\cos^2 x} = \sec^2 x.$$

The others are similar.

C · $D \tan(x^2 + 1)^{1/2} = x(x^2 + 1)^{-1/2} \sec^2(x^2 + 1)^{1/2}.$

D · $\displaystyle\int (\cos x - \cos x \sin^2 x) \, dx = \sin x - \tfrac{1}{3} \sin^3 x + C.$

E · $\displaystyle\int \cos^2 x \, dx = \int \frac{1 + \cos 2x}{2} \, dx = \tfrac{1}{2}x + \tfrac{1}{4} \sin 2x + C.$

Section 7 · 3

A · If $y = \text{Arccos } x$, then $x = \cos y$. Implicit differentiation gives $1 = (-\sin y) \dfrac{dy}{dx}$ from which

$$\frac{dy}{dx} = \frac{-1}{\sin y} = \frac{-1}{\sqrt{1 - \cos^2 y}} = \frac{-1}{\sqrt{1 - x^2}}$$

where we choose $\sin y = +\sqrt{1 - \cos^2 y}$ because $y \in [0, \pi]$.

For $y = \text{Arctan } x$, $x = \tan y$, $1 = (\sec^2 y) \dfrac{dy}{dx}$, and

$$\frac{dy}{dx} = \frac{1}{\sec^2 y} = \frac{1}{1 + \tan^2 y} = \frac{1}{1 + x^2}.$$

The computation for $y = \text{Arccot } x$ is similar.

B · For $y = \text{Arccsc } x$, $x = \csc y$ and implicit differentiation gives $1 = (-\csc y)(\cot y) \dfrac{dy}{dx}$. Now

$$\cot y = \begin{cases} -\sqrt{\csc^2 y - 1} & \text{if } y \in \left[-\dfrac{\pi}{2}, 0\right) \\ +\sqrt{\csc^2 y - 1} & \text{if } y \in \left(0, \dfrac{\pi}{2}\right] \end{cases}$$

Thus

$$\frac{dy}{dx} = \begin{cases} \dfrac{1}{-x(-\sqrt{x^2-1})} & \text{if } y\in\left[-\dfrac{\pi}{2},0\right) \text{ which means } x\leq -1. \\[4mm] \dfrac{1}{-x(+\sqrt{x^2-1})} & \text{if } y\in\left(0,\dfrac{\pi}{2}\right] \text{ which means } x\geq 1. \end{cases}$$

Since $x = -|x|$ when $x \leq -1$ and $-x = -|x|$ when $x \geq 1$,

$$\frac{dy}{dx} = \frac{1}{-|x|\sqrt{x^2-1}}.$$

C · (a) $\dfrac{-2x(x^2+1)^{-2}}{\sqrt{1-(x^2+1)^{-2}}} = \dfrac{-2x}{(x^2+1)\sqrt{x^4+2x^2}}$

(b) $\dfrac{\cos x}{1+\sin^2 x}$

Section 7 · 4

A · $L(ab) = \displaystyle\int_1^{ab}\frac{dt}{t} = \int_{1/b}^{a}\frac{dt}{t}$ by Problem 6.4G.

$$= \int_{1/b}^{1}\frac{dt}{t} + \int_1^{a}\frac{dt}{t}$$

$$= \int_1^{a}\frac{dt}{t} - \int_1^{1/b}\frac{dt}{t} = L(a) - L\left(\frac{1}{b}\right).$$

B · (a) Set $a=1$ in Problem A, remembering that $L(1)=0$.
(b) Use part (a) in the result of Problem A.

(c) $\dfrac{a}{b} = a\left(\dfrac{1}{b}\right)$. $L\left(\dfrac{a}{b}\right) = L(a) + L\left(\dfrac{1}{b}\right) = L(a) - L(b)$.

C · $DL(x^p) = \dfrac{px^{p-1}}{x^p} = p\,\dfrac{1}{x} = D(pL(x))$.

$\therefore L(x^p) = pL(x) + C$. Set $x=1$ to see that $C=0$.

D · (a) $D\ln[x+(x^2-1)^{1/2}] = \dfrac{1+x(x^2-1)^{-1/2}}{x+(x^2-1)^{1/2}} \cdot \dfrac{(x^2-1)^{1/2}}{(x^2-1)^{1/2}}$.

E · (b) $DS(t) = \dfrac{e^t-(-e^{-t})}{2} = C(t)$.

F · Let $y = \ln x$ so $x = e^y < 3^y$. As $x \to \infty$, it is clear that y must also approach ∞.

Section 7 · 5

A · $2 \sinh t \cosh t = 2 \dfrac{e^t - e^{-t}}{2} \cdot \dfrac{e^t + e^{-t}}{2} = \frac{1}{2}[e^{2t} - e^{-2t}].$

B · $D \tanh t = D \dfrac{\sinh t}{\cosh t} = \dfrac{\cosh t(\cosh t) - \sinh t(\sinh t)}{\cosh^2 t}.$

The identity $\cosh^2 t - \sinh^2 t = 1$ is now used to complete the proof.

C · The solutions parallel examples in the text.

D · The solutions parallel examples in the text.

Section 8 · 1

A · $\displaystyle\int x \sin^2 x^2 \, dx = \int x \frac{1 - \cos 2x^2}{2} \, dx = \frac{x^2}{4} - \frac{\sin 2x^2}{8} + C$

B · $\displaystyle\int \frac{\csc u(\cot u - \csc u)}{\cot u - \csc u} \, du = \ln|\cot u - \csc u| + C$

Section 8 · 2

A · Let $x = \tan \theta;\ dx = \sec^2 \theta \, d\theta$

$$\int \frac{dx}{(x^2 + 1)^2} = \int \frac{\sec^2 \theta \, d\theta}{\sec^4 \theta} = \int \cos^2 \theta$$

$$= \int \frac{1 + \cos 2\theta}{2} = \frac{\theta}{2} + \frac{\sin 2\theta}{4} + C$$

$$= \frac{\theta}{2} + \frac{\sin \theta \cos \theta}{2}$$

$$= \frac{1}{2} \operatorname{Arctan} x + \frac{1}{2} \cdot \frac{x}{1 + x^2} + C.$$

Section 8 · 3

A · According to the theorem, with $u = 1/t,\ v = t,$

$$\int_1^2 \frac{dt}{t} = \frac{1}{t} \cdot t \, \bigg|_{t=1}^{t=2} - \int_1^2 -\frac{dt}{t}.$$

Now the function $\dfrac{1}{t} \cdot t = 1$ is a constant function C so that $C|_{t=1}^{t=2} = C - C = 0$, not 1 as in our "proof."

Section 8 · 4

A · Set $\dfrac{x + 10}{(x + 2)(x^2 + 4)} = \dfrac{A}{x + 2} + \dfrac{Bx + C}{x^2 + 4}$

$x + 10 = A(x^2 + 4) + (Bx + C)(x + 2)$

$x = -2;\ 8 = 8A$ so $A = 1$.

$x + 10 = (B + 1)x^2 + (2B + C)x + (4 + 2C)$.

Therefore, $B = -1;\ C = 3$.

B · $\displaystyle\int \frac{3x^3 + x^2 + 10}{(x^2 + 1)^2(x - 3)}\,dx = -\int \frac{x}{x^2 + 1} - 3\int \frac{1}{(x^2 + 1)^2} + \int \frac{1}{x - 3}$

$$= -\frac{1}{2}\ln|x^2 + 1| - 3\left[\frac{1}{2}\,\text{Arctan } x + \frac{1}{2}\frac{x}{1 + x^2}\right]$$

$$+ \ln|x - 3| + C$$

$$= \frac{1}{2}\ln\frac{(x - 3)^2}{x^2 + 1} - \frac{3}{2}\,\text{Arctan } x - \frac{3x}{2(1 + x^2)} + C.$$

C · (a) $\displaystyle\int \frac{dx}{(x + 1)^2 + 4} = \frac{1}{2}\,\text{Arctan }\frac{x + 1}{2} + C$

(b) $\dfrac{3}{2}\displaystyle\int\left[\frac{2x + 2}{x^2 + 2x + 5} + \frac{4}{3}\frac{1}{x^2 + 2x + 5}\right]dx$

$$= \frac{3}{2}\ln|x^2 + 2x + 5| + \left(\frac{3}{2}\right)\left(\frac{4}{3}\right)\left(\frac{1}{2}\right)\text{Arctan }\frac{x + 1}{2} + C.$$

D · $\displaystyle\int \frac{x^3 - 2x}{(x^2 + 1)(x^2 + x + 1)} = -3\int \frac{1}{2x + 1} + \frac{1}{2}\int \frac{2x + 1}{x^2 + x + 1} + \frac{5}{2}\int \frac{1}{x^2 + x + 1}$

$$= -3\,\text{Arctan } x + \frac{1}{2}\ln|x^2 + x + 1| + \frac{5}{2}\int \frac{1}{(x + \frac{1}{2})^2 + \frac{3}{4}}$$

$$= -3\,\text{Arctan } x + \frac{1}{2}\ln|x^2 + x + 1| + \frac{5}{2}\frac{2}{\sqrt{3}}\,\text{Arctan }\frac{2x + 1}{\sqrt{3}}.$$

Section 9 · 2

A · A few of the many possibilities:

(a) $x = \cos t, y = \sin t, t \in [0, \pi]$

(b) $x = -\cos t, y = \sin t, t \in [0, \pi]$

(c) $x = t, y = \sqrt{1 - t^2}, t \in [-1, 1]$

(d) $x = \ln t, y = \sqrt{1 - \ln^2 t}, t \in [1/e, e]$.

Section 9 · 3

A · Since both F and G are differentiable at t_0, we can for small h write

$$\mathbf{R}(t_0 + h) = F(t_0 + h)\mathbf{i} + G(t_0 + h)\mathbf{j}$$

$$= [F(t_0) + F'(t_0)h + |h|r_1(t_0, h)]\mathbf{i} + [G(t_0) + G'(t_0)h + |h|r_2(t_0, h)]\mathbf{j}$$

where both $r_1(t_0, h)$ and $r_2(t_0, h)$ go to 0 with h. Hence,

$$\mathbf{R}(t_0 + h) - \mathbf{R}(t_0) = [F'(t_0)\mathbf{i} + G'(t_0)\mathbf{j}]h + |h|[r_1(t_0, h)\mathbf{i} + r_2(t_0, h)\mathbf{j}]$$

$$= \mathbf{R}'(t_0)h + |h|\mathbf{r}(t_0, h)$$

where $\mathbf{r}(t_0, h) = r_1(t_0, h)\mathbf{i} + r_2(t_0, h)\mathbf{j}$ is a vector-valued function that goes to $\mathbf{0}$ as h approaches 0.

B · $\mathbf{R}(t) \cdot \mathbf{S}(t) = F(t)H(t) + G(t)K(t)$. Differentiate, using the product rule on the right side.

C · $\mathbf{R}'(t) = t \cos t\mathbf{i} + t \sin t\mathbf{j}$ (see Fig. 20).

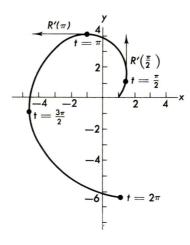

FIG. 20

D · Parametric equations for the path are $x = \dfrac{5t}{1+t}$, $y = \dfrac{1-5t}{1+t}$. Elimination of t gives $6x + 5y = 5$. For the straight line,

$$S(t) = \sqrt{x^2 + (y-1)^2} = \left[\frac{25t^2}{(1+t)^2} + \frac{36t^2}{(1+t)^2}\right]^{1/2} = \frac{t}{t+1}\sqrt{61}$$

$$|\mathbf{R}'(t)| = \frac{1}{(1+t)^2}\sqrt{61} = S'(t).$$

E · Corresponding to the answers given for Problem 9.2A,

(a) $L = \displaystyle\int_0^{\pi} \sqrt{(-\sin t)^2 + \cos^2 t}\; dt = \pi$

(b) $L = \displaystyle\int_0^{\pi} \sqrt{(\sin^2 t) + \cos^2 t}\; dt = \pi$

(c) $L = \displaystyle\int_{-1}^{1} \sqrt{1 + [t(1-t^2)^{-1/2}]^2}\; dt$

$$= \int_{-1}^{1} \frac{dt}{\sqrt{1-t^2}} = \text{Arcsin } t\Big|_{-1}^{1} = \frac{\pi}{2} - \left(-\frac{\pi}{2}\right) = \pi$$

(d) $L = \displaystyle\int_{1/e}^{e} \sqrt{\left(\frac{1}{t}\right)^2 + \left[\frac{(1-\ln^2 t)^{-1/2}\ln t}{-t}\right]^2}\; dt$

$$= \int_{1/e}^{e} \frac{dt}{t\sqrt{1-\ln^2 t}} = \text{Arcsin } \ln t\Big|_{1/e}^{e} = \frac{\pi}{2} - \left(-\frac{\pi}{2}\right) = \pi$$

Section 9 · 4

A · $\mathbf{R}''\left(\dfrac{\pi}{2}\right) = -\dfrac{\pi}{2}\mathbf{i} + \mathbf{j}$; $\mathbf{R}''(\pi) = -\mathbf{i} - \pi\mathbf{j}$

B · $\mathbf{T}(t) = \cos t\,\mathbf{i} + \sin t\,\mathbf{j}$

$$\kappa(t) = \left|(-\sin t\,\mathbf{i} + \cos t\,\mathbf{j})\frac{dt}{ds}\right|$$

Now $\dfrac{ds}{dt} = |\mathbf{R}'(t)| = t$, so

$$\kappa(t) = \frac{1}{t}; \quad \kappa\left(\frac{\pi}{2}\right) = \frac{2}{\pi}; \quad \kappa(\pi) = \frac{1}{\pi}$$

C · $\dfrac{d}{dt}\mathbf{T}(t) = -\sin t\,\mathbf{i} + \cos t\,\mathbf{j} = \mathbf{N}(t)$

$$\mathbf{N}\left(\frac{\pi}{2}\right) = -\mathbf{i}, \quad \mathbf{N}(\pi) = -\mathbf{j}$$

D · $a_T(t) = \dfrac{d}{dt}\left(\dfrac{ds}{dt}\right) = 1$

$a_N(t) = \left(\dfrac{ds}{dt}\right)^2 \kappa(t) = t^2\left(\dfrac{1}{t}\right) = t$

$\mathbf{R}''\left(\dfrac{\pi}{2}\right) = 1[\mathbf{j}] + \dfrac{\pi}{2}\,[-\mathbf{i}]$

$\mathbf{R}''(\pi) = 1[-\mathbf{i}] + \pi[-\mathbf{j}]$

E · $(a_T\mathbf{T} + a_N\mathbf{N}) \cdot (a_T\mathbf{T} + a_N\mathbf{N}) = a_T{}^2\mathbf{T} \cdot \mathbf{T} + 2a_Ta_N\mathbf{T} \cdot \mathbf{N} + a_N{}^2\mathbf{N} \cdot \mathbf{N}$.
Now $\mathbf{T} \cdot \mathbf{N} = 0$, and $\mathbf{T} \cdot \mathbf{T} + \mathbf{N} \cdot \mathbf{N} = 1$, so we have our result.

Section 9 · 5

A ·

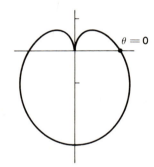

FIG. 21

B ·

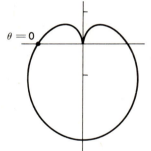

FIG. 22

C · $\dfrac{A}{\beta - \alpha} = \dfrac{\pi a^2}{2\pi}$

$A = \tfrac{1}{2}a^2(\beta - \alpha)$

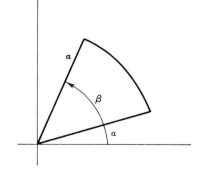

FIG. 23

Section 10 · 2

A · We seek a_1 and a_2 such that $-4\mathbf{i} - 8\mathbf{j} = a_1(4\mathbf{i} + 2\mathbf{j}) + a_2(-2\mathbf{i} + \mathbf{j})$. Equating coefficients of the linearly independent vectors $\mathbf{i}$ and $\mathbf{j}$,

$$4a_1 - 2a_2 = -4$$
$$2a_1 + a_2 = -8.$$

This leads to $a_1 = -\frac{5}{2}$, $a_2 = -3$.

B · The same reasoning as in Problem A leads to the equations

$$4a_1 - 2a_2 = r$$
$$2a_1 + a_2 = s$$

and consequently to $a_1 = (r + 2s)/8$, $a_2 = (r - 2s)/(-4)$.

C · Vectors $\mathbf{u}_1 = a_1\mathbf{i} + b_1\mathbf{j}$, $\mathbf{u}_2 = a_2\mathbf{i} + b_2\mathbf{j}$, $\mathbf{u}_3 = a_3\mathbf{i} + b_3\mathbf{j}$ are linearly dependent if we can find three real numbers x, y, z not all zero, such that $x\mathbf{u}_1 + y\mathbf{u}_2 + z\mathbf{u}_3 = \mathbf{0}$. This leads (following the procedure of Problem A) to

$$a_1x + a_2y + a_3z = 0$$
$$b_1x + b_2y + b_3z = 0.$$

Since any set of two homogeneous equations in three unknowns must have a non-trivial solution, the vectors must be linearly dependent.

D · One way to proceed is to note that $\mathbf{i} = -\mathbf{u}_1$, $\mathbf{j} = \frac{1}{2}(\mathbf{u}_2 + \mathbf{u}_3)$, $\mathbf{k} = \frac{1}{2}(\mathbf{u}_2 - \mathbf{u}_3)$. Thus, for arbitrary $\mathbf{v}$,

$$\mathbf{v} = a\mathbf{i} + b\mathbf{j} + c\mathbf{k} = a(-\mathbf{u}_1) + (b/2)(\mathbf{u}_2 + \mathbf{u}_3) + (c/2)(\mathbf{u}_2 - \mathbf{u}_3)$$
$$= (-a)\mathbf{u}_1 + [(b/2) + (c/2)]\mathbf{u}_2 + [(b/2) - (c/2)]\mathbf{u}_3.$$

This proves that $\mathbf{u}_1, \mathbf{u}_2, \mathbf{u}_3$ span R^3. Now if for three real numbers x, y, z we have $x\mathbf{u}_1 + y\mathbf{u}_2 + z\mathbf{u}_3 = 0$, then $x(-\mathbf{i}) + y(\mathbf{j} + \mathbf{k}) + z(\mathbf{j} - \mathbf{k}) = \mathbf{0}$.
This leads to the system

$$-x \qquad = 0$$
$$y + z = 0$$
$$y - z = 0$$

for which the trivial solution is the only solution; that is, $x = y = z = 0$ and so $\mathbf{u}_1, \mathbf{u}_2, \mathbf{u}_3$ are linearly independent.

E · Using the work in Problem D where $a = -4$, $b = -1$, $c = -5$, we write $\mathbf{v} = 4\mathbf{u}_1 - 3\mathbf{u}_2 + 2\mathbf{u}_3$. This is verified by writing $\mathbf{v} = 4(-\mathbf{i}) - 3(\mathbf{j} + \mathbf{k}) + 2(\mathbf{j} - \mathbf{k}) = -4\mathbf{i} - \mathbf{j} - 5\mathbf{k}$.

F · Solving $\mathbf{v}_1 = x\mathbf{w}_1 + y\mathbf{w}_2$ leads to the system

$$x + 2y = 5$$
$$2x - y = 5$$
$$-x \qquad = -3$$

for which a solution is given by $x = 3$, $y = 1$. In a similar way, one is able to solve other systems, finally coming to

$$\mathbf{v}_1 = 3\mathbf{w}_1 + \mathbf{w}_2$$
$$\mathbf{v}_2 = \mathbf{w}_1 - \mathbf{w}_2$$
$$\mathbf{v}_3 = \mathbf{w}_1 + 2\mathbf{w}_2$$

Solving $\mathbf{w}_1 = x\mathbf{v}_1 + y\mathbf{v}_2 + z\mathbf{v}_3$ leads to the system

$$5x - y + 5z = 1$$
$$5x + 3y = 2$$
$$-3x - y - z = -1$$

for which a solution (one of many) is $x = 1, y = -1, z = -1$. Solving a similar system corresponding to $\mathbf{w}_2$, we can write

$$\mathbf{w}_1 = \mathbf{v}_1 - \mathbf{v}_2 - \mathbf{v}_3$$
$$\mathbf{w}_2 = \mathbf{v}_1 - 2\mathbf{v}_2 - \mathbf{v}_3.$$

G · All three properties follow from the obvious computation. For example, (iii) is established by

$$\mathbf{u} \cdot r\mathbf{v} = (a_1, \ldots, a_n) \cdot (rb_1, \ldots, rb_n) = a_1 rb_1 + \cdots + a_n rb_n$$
$$= r[a_1 b_1 + \cdots + a_n b_n] = r[\mathbf{u} \cdot \mathbf{v}].$$

H · We seek a unit vector $\mathbf{u}_3 = (x, y, z)$ for which $\mathbf{u}_1 \cdot \mathbf{u}_3 = 0$ and $\mathbf{u}_2 \cdot \mathbf{u}_3 = 0$. The latter two equations lead to

$$x - y = 0$$
$$2x + 2y + z = 0$$

for which the solution space is spanned by $(1, 1, -4)$. The desired unit vector is either of $\mathbf{u}_3 = \pm (1/3\sqrt{2}, 1/3\sqrt{2}, -4/3\sqrt{2})$.

I · Since $\mathbf{v} \cdot \mathbf{u}_1 = r\mathbf{u}_1 \cdot \mathbf{u}_1 + s\mathbf{u}_2 \cdot \mathbf{u}_1 = r$ (because $\mathbf{u}_1 \cdot \mathbf{u}_1 = 1$, $\mathbf{u}_2 \cdot \mathbf{u}_1 = 0$), and similarly $\mathbf{v} \cdot \mathbf{u}_2 = s$, we have immediately $\mathbf{v} = -\frac{26}{5}\mathbf{u}_1 + \frac{7}{5}\mathbf{u}_2$. The same technique for $\mathbf{w} = r\mathbf{u}_1 + s\mathbf{u}_2$ gives $r = \mathbf{w} \cdot \mathbf{u}_1$, $s = \mathbf{w} \cdot \mathbf{u}_2$.

J · $(\mathbf{p} + t\mathbf{q}) \cdot (\mathbf{p} + t\mathbf{q}) = |\mathbf{p}|^2 + 2t\mathbf{p} \cdot \mathbf{q} + t^2|\mathbf{q}|^2 \geq 0$. Set $t = |\mathbf{p}|/|\mathbf{q}|$. Algebraic simplification leads to $\mathbf{p} \cdot \mathbf{q} \geq - |\mathbf{p}||\mathbf{q}|$. The same calculations, beginning with $(\mathbf{p} - t\mathbf{q}) \cdot (\mathbf{p} - t\mathbf{q})$ lead to $\mathbf{p} \cdot \mathbf{q} \leq |\mathbf{p}||\mathbf{q}|$. Together, $-|\mathbf{p}||\mathbf{q}| \leq \mathbf{p} \cdot \mathbf{q} \leq |\mathbf{p}||\mathbf{q}|$, from which the desired inequality follows.

Section 10 · 3

A · For any vector $\mathbf{p} \neq \mathbf{0}$, $\mathbf{p}/|\mathbf{p}|$ is a unit vector and we have through use of linearity properties of T, $T(\mathbf{p}) = |\mathbf{p}| T(\mathbf{p}/|\mathbf{p}|) = |\mathbf{p}|(0) = 0$.

B · $[T]_s\mathbf{u}_1 = \begin{bmatrix} -2 & 6 \\ 4 & -2 \end{bmatrix}\begin{bmatrix} 1 \\ 1 \end{bmatrix} = \begin{bmatrix} 4 \\ 2 \end{bmatrix};$ $[T]_s\mathbf{u}_2 = \begin{bmatrix} -2 & 6 \\ 4 & -2 \end{bmatrix}\begin{bmatrix} 1 \\ -1 \end{bmatrix} = \begin{bmatrix} -8 \\ 6 \end{bmatrix}.$

C · From Problem B, $T(\mathbf{u}_1) = 4\varepsilon_1 + 2\varepsilon_2 = (2\mathbf{u}_1 + 2\mathbf{u}_2) + (\mathbf{u}_1 - \mathbf{u}_2) = 3\mathbf{u}_1 + \mathbf{u}_2$.
$T(\mathbf{u}_2) = -8\varepsilon_1 + 6\varepsilon_2 = (-4\mathbf{u}_1 - 4\mathbf{u}_2) + (3\mathbf{u}_1 - 3\mathbf{u}_2) = -\mathbf{u}_1 - 7\mathbf{u}_2$. Therefore,

$$[T]_B = \begin{bmatrix} 3 & -1 \\ 1 & -7 \end{bmatrix}.$$

D · $\mathbf{p} = 3\mathbf{u}_1 - 2\mathbf{u}_2 = 3(\varepsilon_1 + \varepsilon_2) - 2(\varepsilon_1 - \varepsilon_2) = \varepsilon_1 + 5\varepsilon_2$.

$$[T]_B \begin{bmatrix} 3 \\ -2 \end{bmatrix} = \begin{bmatrix} 3 & -1 \\ 1 & -7 \end{bmatrix} \begin{bmatrix} 3 \\ -2 \end{bmatrix} = \begin{bmatrix} 11 \\ 17 \end{bmatrix};$$

$$[T]_S \begin{bmatrix} 1 \\ 5 \end{bmatrix} = \begin{bmatrix} -2 & 6 \\ 4 & -2 \end{bmatrix} \begin{bmatrix} 1 \\ 5 \end{bmatrix} = \begin{bmatrix} 28 \\ -6 \end{bmatrix}.$$

To verify that our results are compatible, we note that the image with respect to (wrt) basis B is $11\mathbf{u}_1 + 17\mathbf{u}_2 = 11(\varepsilon_1 + \varepsilon_2) + 17(\varepsilon_1 - \varepsilon_2) = 28\varepsilon_1 - 6\varepsilon_2$, the image wrt basis S.

E · For points $\mathbf{p}$ and $\mathbf{q}$ and real number t, $H(\mathbf{p} + t\mathbf{q}) = G[F(\mathbf{p} + t\mathbf{q})] = G[F(\mathbf{p}) + tF(\mathbf{q})]$ because F is linear; and then because G is linear, $H(\mathbf{p} + t\mathbf{q}) = G[F(\mathbf{p})] + tG[F(\mathbf{q})] = H(\mathbf{p}) + tH(\mathbf{q})$.

F · $(AB)C = \begin{bmatrix} 7 & -6 \\ 2 & -2 \end{bmatrix} \begin{bmatrix} 5 & 7 \\ 2 & 3 \end{bmatrix} = \begin{bmatrix} 23 & 31 \\ 6 & 8 \end{bmatrix} = \begin{bmatrix} 1 & 3 \\ 0 & 2 \end{bmatrix} \begin{bmatrix} 14 & 19 \\ 3 & 4 \end{bmatrix} = A(BC).$

G · $AB = \begin{bmatrix} 7 & -6 \\ 2 & -2 \end{bmatrix}; \quad BA = \begin{bmatrix} 4 & 6 \\ 1 & 1 \end{bmatrix}.$

H · $AI = IA = A.$

I · $A(B + C) = AB + AC; \quad (B + C)A = BA + CA.$

J · $AR = RA = I; \quad BS = SB = I.$

K · $C^{-1} = \begin{bmatrix} 3 & -7 \\ -2 & 5 \end{bmatrix}$

L · There is no inverse. Fancy that!

M · The inverse of $N = \begin{bmatrix} a & b \\ c & d \end{bmatrix}$ exists if and only if $D = ad - bc \neq 0$, and in this case, $N^{-1} = \begin{bmatrix} -d/D & -b/D \\ -c/D & a/D \end{bmatrix}$. Thus, $R^{-1} = \begin{bmatrix} \frac{2}{5} & -1 \\ -\frac{3}{5} & 2 \end{bmatrix}$ and $S^{-1} = \begin{bmatrix} -4 & 7 \\ 3 & -5 \end{bmatrix}$

N · $(AB)^t = B^t A^t; \quad (BA)^t = A^t B^t.$

O · $A^{-1} = \begin{bmatrix} -\frac{3}{11} & \frac{8}{11} \\ \frac{4}{11} & \frac{7}{11} \end{bmatrix}$

P · $\begin{bmatrix} -\frac{3}{11} & \frac{8}{11} \\ \frac{4}{11} & -\frac{7}{11} \end{bmatrix} \begin{bmatrix} 7 & 8 \\ 4 & 3 \end{bmatrix} \begin{bmatrix} x \\ y \end{bmatrix} = \begin{bmatrix} -\frac{3}{11} & \frac{8}{11} \\ \frac{4}{11} & -\frac{7}{11} \end{bmatrix} \begin{bmatrix} 10 \\ 1 \end{bmatrix}$

$$\begin{bmatrix} x \\ y \end{bmatrix} = \begin{bmatrix} -2 \\ 3 \end{bmatrix}$$

Section 10 · 4

A · $\det(A)\det(B) = (29)(-46) = -1334 = \det\begin{bmatrix} 0 & -5 & 14 \\ 7 & 6 & 17 \\ 11 & -5 & 29 \end{bmatrix} = \det(AB).$

B · $\det(BA) = -1334.$

C · $B = BI = B(AC) = (BA)C = IC = C.$

D · If $AB = I$, then $BA = I$, so $C = IC = (BA)C = B(AC) = BI = B.$

E · $(AB)(B^{-1}A^{-1})\,A(BB^{-1})A^{-1} = I$, and since $(AB)^{-1}$ is unique, $(AB)^{-1} = B^{-1}A^{-1}.$

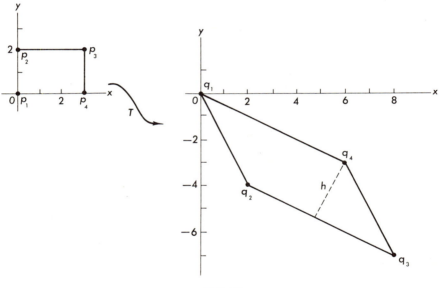

FIG. 24

F · $q_1 = T(p_1) = \begin{bmatrix} 0 \\ 0 \end{bmatrix}$; $q_2 = T(p_2) = \begin{bmatrix} 2 \\ -4 \end{bmatrix}$; $q_3 = T(p_3) = \begin{bmatrix} 8 \\ -7 \end{bmatrix}$;

$q_4 = T(p_4) = \begin{bmatrix} 6 \\ -3 \end{bmatrix}.$

The area of $\square\, q_1q_2q_3q_4$ is (see Fig. 24) $|q_3 - q_2|h$ where h is the height of the parallelogram with $q_3 - q_2$ as a base. The line through the *points* q_2 and q_3 is $y + 4 = -\frac{3}{6}(x - 2)$, or in normal form $(1/\sqrt{5})x + (2/\sqrt{5})y = -6/\sqrt{5}$ from which we see that $h = 6/\sqrt{5}$ and $|q_3 - q_2|h = |6\mathbf{i} - 3\mathbf{j}|(6/\sqrt{5}) = \sqrt{45}(6/\sqrt{5}) = 18$. Now $\det[T] = -3$ and the area of $\square\, p_1p_2p_3p_4$ is 6. The product is -18 which, except for the sign (due to the negative determinant), is the area of the image parallelogram.

Section 10 · 6

A · $B(\mathbf{v}, \mathbf{u}) = x_1 x_2 a_{11} + x_2 y_1 a_{12} + x_1 y_2 a_{21} + y_1 y_2 a_{22}$.

B · The expression in Problem A equals (3) in Section 10.6 for all $\mathbf{u}$ and $\mathbf{v}$ if and only if $a_{12} = a_{21}$.

C · $B_1\mathbf{u} = \begin{bmatrix} \frac{3}{5} & -\frac{4}{5} \\ \frac{4}{5} & \frac{3}{5} \end{bmatrix}\begin{bmatrix} 2 \\ 1 \end{bmatrix} = \begin{bmatrix} \frac{2}{5} \\ \frac{11}{5} \end{bmatrix}$; $B_1\mathbf{v} = \begin{bmatrix} -\frac{9}{5} \\ \frac{13}{5} \end{bmatrix}$

$B_2\mathbf{u} \begin{bmatrix} -\frac{3}{5} & \frac{4}{5} \\ \frac{4}{5} & \frac{3}{5} \end{bmatrix}\begin{bmatrix} 2 \\ 1 \end{bmatrix} = \begin{bmatrix} -\frac{2}{5} \\ \frac{11}{5} \end{bmatrix}$; $B_2\mathbf{v} = \begin{bmatrix} \frac{9}{5} \\ \frac{13}{5} \end{bmatrix}$

$B_1\mathbf{u}$ and $B_1\mathbf{v}$ have been rotated through $\theta = \text{Arcsin } \frac{4}{5}$. $B_2\mathbf{u}$ and $B_2\mathbf{v}$ have been rotated through the same angle and then reflected in the y axis (Fig. 25).

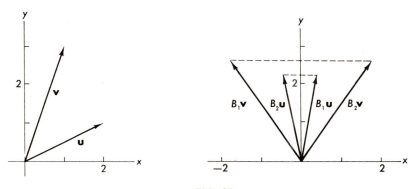

FIG. 25

D · $BB^t = I$, so $\det(B)\det(B^t) = \det(I) = 1$. Since $\det(B^t) = \det(B)$, we have $(\det B)^2 = 1$, so $\det(B) = \pm 1$.

E · $B^tB = \begin{bmatrix} \mathbf{v}_1{}^t \\ \vdots \\ \mathbf{v}_n{}^t \end{bmatrix}[\mathbf{v}_1, \ldots, \mathbf{v}_n] = [\mathbf{v}_i{}^t \cdot \mathbf{v}_j] = I$. Therefore $\mathbf{v}_i{}^t \cdot \mathbf{v}_j = \begin{cases} 1 & \text{if } i = j \\ 0 & \text{if } i \neq j \end{cases}$

which says that vectors $\mathbf{v}_1, \ldots, \mathbf{v}_n$ form an orthonormal basis.

F · $|T\mathbf{v}|^2 = [T\mathbf{v}]^t[T\mathbf{v}] = \mathbf{v}^t T^t T\mathbf{v} = \mathbf{v}^t I\mathbf{v} = |\mathbf{v}|^2$.

G · The angle between $T\mathbf{u}$ and $T\mathbf{v}$ has a cosine given by

$$\frac{T\mathbf{u} \cdot T\mathbf{v}}{|T\mathbf{u}|\,|T\mathbf{v}|} = \frac{[T\mathbf{u}]^t[T\mathbf{v}]}{|\mathbf{u}|\,|\mathbf{v}|} = \frac{\mathbf{u}^t T^t T\mathbf{v}}{|\mathbf{u}|\,|\mathbf{v}|} = \frac{\mathbf{u}^t\mathbf{v}}{|\mathbf{u}|\,|\mathbf{v}|} = \frac{\mathbf{u} \cdot \mathbf{v}}{|\mathbf{u}|\,|\mathbf{v}|}$$

which is the cosine of the angle formed by $\mathbf{u}$ and $\mathbf{v}$.

Section 10 · 7

A · If S is positive so that $P^tSP = D$ is a diagonal matrix with entries $r_1, \cdots, r_n$ all nonnegative, then for arbitrary $\mathbf{v}$, define $\mathbf{u} = P^{-1}\mathbf{v}$. Then $\mathbf{v}^tS\mathbf{v} = \mathbf{u}^tP^tSP\mathbf{u} = \mathbf{u}^tD\mathbf{u} = r_1u_1{}^2 + \cdots + r_nu_n{}^2$ where $\mathbf{u}^t = [u_1, \cdots, u_n]$. Since all $r_i \geq 0$, it follows that $\mathbf{v}^tS\mathbf{v} \geq 0$.

On the other hand, suppose that S is symmetric with the property that for any $\mathbf{v}$, $\mathbf{v}^tS\mathbf{v} \geq 0$. There is a nonsingular matrix P such that $P^tSP = D$, the entries $r_1, \cdots, r_n$ in the diagonal matrix D being eigenvalues of S. Choose $\mathbf{v}_i = P\epsilon_i$ where ϵ_i is the usual standard basis vector. Then $\mathbf{v}_i{}^tS\mathbf{v}_i = \epsilon_i{}^tPSP^t\epsilon_i = \epsilon_i{}^tD\epsilon_i = r_i$ which by the stated property of S must be ≥ 0. This being the case for arbitrary i, S is positive.

The argument for positive definite matrices is similar.

Section 12 · 2

A · $\nabla z = [-2x \quad -2y]$
From $(2, 1)$, one should move in the direction of $[-4, -2]$ to obtain maximum growth of z.

$$w = 9 - x^2 - y^2 - z$$
$$\nabla w = [-2x \quad -2y \quad -1]$$

At $(2, 1, 4)$, $\nabla w = [-4 \quad -2 \quad -1]$ which is perpendicular to the graph at that point. From (2) of Section 12.1, since $F'(1, 2) = [-4 \quad -2]$,

$$z = T(x, y) = F(1, 2) + [-4 \quad -2]\begin{bmatrix} x - 2 \\ y - 1 \end{bmatrix}$$
$$z = 4 - 4(x - 1) - 2(y - 2).$$

From Theorem 11.2A, we have as the equation of this plane

$$-4(x - 1) - 2(y - 2) - 1(z - 4) = 0.$$

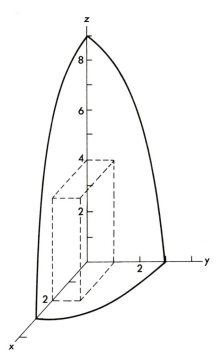

FIG. 26

B · $F(p_2) - F(p_1) = 6 - 14 = F'(p_1 + t(p_2 - p_1))(p_2 - p_1)$.
Now $p_2 - p_1 = (-1, 2)$ and $F'(x, y) = [y + 6x \quad x]$.
$p_1 + t(p_2 - p_1) = (2, 1) + t(-1, 2) = (2 - t, 1 + 2t)$.
$F'(2 - t, 1 + 2t) = [13 - 4t \quad 2 - t]$.

Substitution in our first equation above gives

$$-8 = [13 - 4t \quad 2 - t]\begin{bmatrix} -1 \\ 2 \end{bmatrix}$$

$$-8 = (-1)(13 - 4t) + 2(2 - t)$$

$$t = \tfrac{1}{2}.$$

Section 12 · 3

A. $[3 \quad 0 \quad 4]\begin{bmatrix} 2 & 0 & -1 \\ 3 & 1 & 2 \\ -2 & 1 & -1 \end{bmatrix}\begin{bmatrix} 1 \\ -1 \\ 2 \end{bmatrix} = [3 \quad 0 \quad 4]\begin{bmatrix} 0 \\ 6 \\ -5 \end{bmatrix} = -20$

$[1 \quad -1 \quad 2]\begin{bmatrix} 2 & 0 & -1 \\ 3 & 1 & 2 \\ -2 & 1 & -1 \end{bmatrix}\begin{bmatrix} 3 \\ 0 \\ 4 \end{bmatrix} = [1 \quad -1 \quad 2]\begin{bmatrix} 2 \\ 17 \\ -10 \end{bmatrix} = -35$

Thus, $B(h, k) = -20$ and $B(k, h) = -35$.

B. Using (3.6), we have

$$F''(x, y, z) \leftrightarrow \begin{bmatrix} 2y & 2x + 2yz & y^2 \\ 2x + 2yz & 2xz & 2xy \\ y^2 & 2xy & 0 \end{bmatrix}$$

$$F''(1, -1, 2) \leftrightarrow \begin{bmatrix} -2 & -2 & 1 \\ -2 & 4 & -2 \\ 1 & -2 & 0 \end{bmatrix}$$

C. $T_1(x, y, z) = F(1, -1, 2) + [F'(1, -1, 2)]\begin{bmatrix} x - 1 \\ y + 1 \\ z - 2 \end{bmatrix}$

$[F'(x, y, z)] = [2xy + y^2z \quad x^2 + 2xyz \quad xy^2]$

$T_1(x, y, z) = 1 + [0 \quad -3 \quad 1]\begin{bmatrix} x - 1 \\ y + 1 \\ z - 2 \end{bmatrix}$

$T_1(x, y, z) = 1 - 3(y + 1) + (z - 2)$

$T_2(x, y, z) = T_1(x, y, z) + \tfrac{1}{2}[x - 1 \quad y + 1 \quad z - 2][F''(1, -1, 2)]\begin{bmatrix} x - 1 \\ y + 1 \\ z - 2 \end{bmatrix}$

Using the answer from Problem B, the last term is

$$\tfrac{1}{2}[x-1 \quad y+1 \quad z-2]\begin{bmatrix} -2 & -2 & 1 \\ -2 & 4 & -2 \\ 1 & -2 & 0 \end{bmatrix}\begin{bmatrix} x-1 \\ y+1 \\ z-2 \end{bmatrix}$$

$$= \tfrac{1}{2}\{-2(x-1)^2 - 4(x-1)(y+1) + 2(x+1)(z-2)$$
$$+ 4(y+1)^2 - 4(y+1)(z-2)\}$$
$$T_2(x, y, z) = 1 - 3(y+1) + (z-2) - (x-1)^2 - 2(x-1)(y+1)$$
$$+ (x-1)(z-2) + 2(y+1)^2 - 2(y+1)(z-2)$$

It remaihs to substitute values, getting

$$T_1(1.13, -0.92, 1.88) = 0.64$$
$$T_2(1.13, -0.92, 1.88) = 0.6187$$
$$F(1.13, -0.92, 1.88) = 0.6233.$$

D $\cdot \displaystyle\lim_{t \to 0-} \dfrac{F(p_0 + th) - F(p_0)}{t} = F'(p)h \geq 0$

(The inequality is because, for t close to zero and negative, the fraction is necessarily positive.) Similarly,

$$\lim_{t \to 0+} \dfrac{F(p_0 + th) - F(p_0)}{t} = F'(p)h \leq 0.$$

Hence, for any unit vector h, $F'(p)h = 0$. Then $F'(p) = 0$.

E $\cdot$ The inequalities of Problem D are simply reversed.

F $\cdot$ $G''(y, z) = \begin{bmatrix} 4 & -4 \\ -4 & 10 \end{bmatrix}$. The characteristic values, being solutions to

$$\begin{vmatrix} x-4 & 4 \\ 4 & x-4 \end{vmatrix} = 0, \text{ are } x = 12 \text{ and } x = 2.$$

Thus, $G(-1, 2)$ is positive definite and $(-1, 2)$ is a minimum.

G $\cdot$ See Fig. (27).

$y = 3x - 10$

Point	y_i	$\bar{y}_i$	d_i
(2, 7)	7	-4	11
(7, 12)	12	11	1
(12, 14)	14	26	-12

$$M = 0$$

$y = x + 5$

Point	y_i	$\bar{y}_i$	d_i
(2, 7)	7	7	0
(7, 12)	12	12	0
(12, 14)	14	17	-3

$$M = -3$$

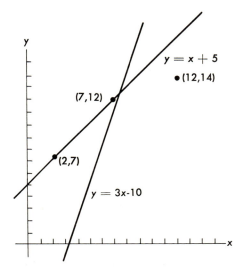

FIG. 27

Observe that although $M = 0$ for the line $y = 3x - 10$, we have the feeling from looking at the picture that $y = x + 5$ " fits " the points better.

Section 13 · 3

A · $\displaystyle\int_{\alpha}^{\beta}\int_{0}^{G(\theta)} r\, dr\, d\theta = \frac{1}{2}\int_{\alpha}^{\beta} r^{2}\,\Big|_{r=0}^{G(\theta)} d\theta.$

B · $\begin{bmatrix} \cos\theta & -r\sin\theta & 0 \\ \sin\theta & r\cos\theta & 0 \\ 0 & 0 & 1 \end{bmatrix} = r.$

The conclusion then follows from the theorem.

C · The line $x = 2$ is also described by $r = 2\sec\theta.$

$$I = \int_{0}^{\pi/4}\int_{0}^{2\sec\theta}\int_{0}^{r^{2}} (r^{2})r\, dz\, dr\, d\theta$$

$$= \int_{0}^{\pi/4}\int_{0}^{2\sec\theta} r^{5}\, dr\, d\theta = \frac{1}{6}\int_{0}^{\pi/4} 2^{6}\sec^{2}\theta(\tan^{2}\theta + 1)^{2}\, d\theta$$

$$= \frac{32}{3}\left[\frac{1}{5}\tan^{5}\theta + \frac{2}{3}\tan^{3}\theta + \tan\theta\right] = \frac{2^{7}\cdot 7}{45}$$

Section 14 · 1

A · (a) 0   (b) 0   (c) 0

(d) Set $F(x) = x^{1/x}$ so $A_n = F(n)$. Set $y = x^{1/x}$ so $\ln y = \dfrac{\ln x}{x}$. We may now use L'Hôpital's rule to find $\lim\limits_{x \to \infty} \ln y = 0$. Thus, $\lim\limits_{x \to \infty} x^{1/x} = 1$.

B · Let $r \in (1, m)$. Since $\lim\limits_{n \to \infty} \dfrac{b_{n+1}}{b_n} = m$, there must be some N so that if $i \geq N$, then $\dfrac{b_{i+1}}{b_i} > r$. In particular,

$$\frac{b_{N+1}}{b_N} > r \qquad \text{so } b_{N+1} > rb_N$$

$$\frac{b_{N+2}}{b_{N+1}} > r \qquad \text{so } b_{N+2} > rb_{N+1} > r^2 b_N.$$

Similarly, $b_{N+k} > r^k b_N$, so $\lim\limits_{k \to \infty} b_{N+k} = \infty$. Then $\lim\limits_{n \to \infty} b_n \neq 0$, so $\sum_{n=1}^{\infty} b_n$ diverges.

Section 14 · 2

A · $\displaystyle\int_1^{\infty} e^{-x} = \lim_{b \to \infty} -e^{-x} \Big|_1^b = \frac{1}{e}.$

The series $\sum_{n=1}^{\infty} e^{-n}$ converges.

B · $\displaystyle\int_2^{\infty} \frac{1}{x \ln x} = \lim_{b \to \infty} \ln \ln x \Big|_2^b = \infty.$

Note that we use 2 as a lower limit to avoid the problems encountered by using 1. The point is that the lower limit is of no importance insofar as this test is concerned, just as the first few terms of an infinite series do not determine the convergence of the series.

C · $\dfrac{\sqrt{n}}{n+3} \geq \dfrac{\sqrt{n}}{2n}$ for $n \geq 3$. We know

$$\sum_{n=3}^{\infty} \frac{\sqrt{n}}{2n} = \frac{1}{2} \sum_{n=3}^{\infty} \frac{1}{n^{1/2}}$$

diverges, since this is a p-series with $p = \frac{1}{2}$. Hence,

$$\sum_{n=1}^{\infty} \frac{\sqrt{n}}{n+3} = \frac{1}{4} + \frac{\sqrt{2}}{5} + \sum_{n=3}^{\infty} \frac{\sqrt{n}}{n+3}$$

diverges. (We again see that if a test applies for all n except a few, we may use the test anyhow.)

D • For $n > N$, $a_n b_n < a_n 1$ and $\sum_{n=1}^{\infty} a_n$ is known to converge. Therefore, $\sum_{n=1}^{\infty} a_n b_n$ converges.

E • $\lim\limits_{n \to \infty} \dfrac{1}{(n+1)^2} \cdot \dfrac{n^2}{1} = 1$; $\lim\limits_{n \to \infty} \dfrac{1}{n+1} \cdot \dfrac{n}{1} = 1$.

Do not draw any conclusion if the ratio test produces a 1.

F • We showed in Problem 14.1B that if $\lim\limits_{n \to \infty} |b_{n+1}|/|b_n| = m > 1$, then $\lim\limits_{n \to \infty} |b_n| \neq 0$. But then certainly $\lim\limits_{n \to \infty} b_y \neq 0$ either. Hence, $\sum_{n=1}^{\infty} b_n$ diverges (since Theorem 14.1A applies to any series, not just positive ones).

G • There is no reason to think that the divergence of $\sum_{n=1}^{\infty} |b_n|$ will imply the divergence of $\sum_{n=1}^{\infty} b_n$ (except in the special case where divergence of $\sum_{n=1}^{\infty} |b_n|$ is established by using the ratio test).

Section 14 · 3

A • The proof for the alternating series $\sum_{n=1}^{\infty} b_n$ is obtained by substituting $b_n = (-1)^{n+1}/n$ in the proof that $\sum_{n=1}^{\infty} (-1)^{n+1}/n$ converges.

B • Using the notation introduced in the proof that $\sum_{n=1}^{\infty} (-1)^{n+1}/n$ converges, we recall that $\{C_k\}$ is monotone increasing to L. Now look at two terms of the sequence of partial sums having odd subscripts.

$$B_{2k+3} = C_{2k} + \frac{1}{2k+1} + \frac{-1}{2k+2} + \frac{1}{2k+3}$$

$$B_{2k+1} = C_{2k} + \frac{1}{2k+1}$$

Then

$$B_{2k+1} - B_{2k+3} = \frac{1}{2k+2} - \frac{1}{2k+3}.$$

Now $-1/(2k+2)$ is negative so $-(-1/(2k+2))$ is positive. From this we subtract the positive term $1/(2k+3)$ which is smaller than $|-1/(2k+2)|$. Therefore, the sequence $\{B_{2k+1}\}$ is monotone decreasing; and we know it converges to L. We thus have the situation pictured in Fig. 28. It is clear that the difference between two successive partial sums must be greater than the difference between either one of them and L; and the difference between two successive partial sums is always the absolute value of the first term not used.

$$B_{2k-2} \quad B_{2k} \qquad L \qquad B_{2k-1} \quad B_{2k-3}$$

FIG. 28

C • Set $b_k = (-1)^{k+1}/k$ in the solution to Problem B.

Section 14 · 4

A · (a) $\lim\limits_{n \to \infty} \left| \dfrac{(n+1)!(x-2)^{n+1}}{n!(x-2)^n} \right| = \lim\limits_{n \to \infty} (n+1)|x-2| = \infty$ unless $x=2$. The radius of convergence is 0.

(b) $\lim\limits_{n \to \infty} \left| \dfrac{3^{n+1}(x-2)^{n+1}(n+1)^2}{3^n(x-2)^n n^2} \right| = 3|x-2| < 1$ if and only if $|x-2| < \frac{1}{3}$.

When $x = 7/3$, we have $\sum\limits_{n=1}^{\infty} \dfrac{1}{n^2}$ which converges, and for $x = 5/3$, $\sum\limits_{n=1}^{\infty} \dfrac{(-1)^n}{n^2}$ also converges. The interval of convergence is $\left[\dfrac{5}{3}, \dfrac{7}{3} \right]$.

(c) $\lim\limits_{n \to \infty} \left| \dfrac{(x-2)^{n+1}}{(n+2)!} \cdot \dfrac{n!}{(x-2)^n} \right| = 0$ for all x. The radius of convergence is ∞.

B · $C'(x) = -x + \dfrac{x^3}{3} - \cdots = -1 \left[x - \dfrac{x^3}{3!} + \cdots \right] = -S(x)$.

Similarly, $S'(x) = C(x)$.

C · Using the chain rule, $F'(x) = 2S(x)S'(x) + 2C(x)C'(x)$; that is, $F'(x) = 2S(x)C(x) + 2C(x)[-S(x)] = 0$. Thus, F is constant. But it is obvious that $F(0) = 1$.

D · Simply multiply

$$\frac{1}{1-x} = 1 + x + \cdots + x^n + \frac{x^{n+1}}{1-x}$$

through by $(1-x)$ to get the desired identity. (Note that this holds for all $x \neq 1$.)

E · $\lim\limits_{n \to \infty} R_n(0, x) = \begin{cases} 0 & \text{if } |x| < 1 \\ \infty & \text{if } |x| > 1 \end{cases}$

We therefore have convergence only for $x \in (-1, 1)$.

F · $D^n L(x) = (-1)^{n+1}(n-1)! \, x^{-n}$ so $D^n L(1) = (-1)^{n+1}(n-1)!$ for $n = 1, 2, \ldots$. Then for arbitrary n

$$T_n(x) = (x-1) - \frac{(x-1)^2}{2} + \frac{(x-1)^3}{3} + \cdots + (-1)^{n+1} \frac{(x-1)^n}{n}.$$

$$|R_n(1, x)| = \left| \frac{n!}{(\tilde{x})^{n+1}} \frac{(x-1)^{n+1}}{(n+1)!} \right|. \quad \lim\limits_{n \to \infty} R_n(1, x) = \lim\limits_{n \to \infty} \frac{|x-1|}{n+1} \left| \frac{x-1}{\tilde{x}} \right| \left| \frac{1}{\tilde{x}} \right|^n.$$

Since L is not defined at 0, we might expect that the interval of convergence about 1 will not extend beyond 0. The correct interval is $(0, 2)$, but it is hard to show that the expression above goes to zero if x (hence $\tilde{x}$) is restricted to $(0, 2)$. The reader can easily show that the limit is 0 for $x \in [\frac{1}{2}, \frac{3}{2}]$, however.

G · The difficulty here is that one does not easily find an expression for $D^n R(x)$. It is, of course, easy enough to find the first few terms, and to tentatively write (ignoring questions of convergence, etc.),

$$R(x) \approx x - x^3 + x^5 - x^7 + \cdots.$$

H · Two key observations are that Arctan $x = \int_0^x \dfrac{dt}{1+t^2}$ and

$$\frac{1}{1+t^2} = 1 - t^2 + t^4 - t^6 + \cdots.$$

Putting them together with appropriate theorems,

$$\text{Arctan } x = \int_0^x (1 - t^2 + t^4 - t^6 + \cdots)$$

$$= x - \frac{x^3}{3} + \frac{x^5}{5} - \frac{x^7}{7} + \cdots \qquad |x| < 1.$$

I · Concentrate on the fact that you want powers of $x - \frac{1}{2}$. This leads you to write

$$G(x) = \frac{1}{\frac{1}{2} - (x - \frac{1}{2})}.$$

If only the factor $(x - \frac{1}{2})$ were subtracted from 1 instead of $\frac{1}{2}$, we would know what to do. That can be arranged.

$$G(x) = \frac{2}{1 - [2(x - \frac{1}{2})]} = 2\{1 + [(2x - \frac{1}{2})] + [2(x - \frac{1}{2})]^2 + \cdots\}$$

$$G(x) = 2 + 4(x - \frac{1}{2}) + 8(x - \frac{1}{2})^2 + \cdots, \qquad 2|x - \frac{1}{2}| < 1.$$

Section 15 · 1

A · The force F tending to push the weight down the plane is

$$F = W \sin \alpha.$$

The mass of the weight is W/g. By Newton's law,

$$W \sin \alpha = \frac{W}{g} \frac{d^2 s}{dt^2}$$

$$\frac{d^2 s}{dt^2} = 32 \sin \alpha.$$

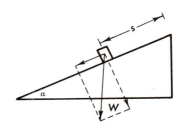

FIG. 29

B · Let the fixed point on the line be the origin. When the moving particle is to the right of **O**, then

$$\frac{d^2x}{dt^2} = -kx.$$

(The negative sign is to keep $k > 0$, since we know $d^2x/dt^2 < 0$ and $x > 0$.) The reader should convince himself that the signs are correct for $x < 0$.

C · Salt comes in at $3(\frac{3}{2}) = \frac{9}{2}$ lb/min. Salt leaves at $4s/(10 - t)$ for $t < 10$. (After $t = 10$, the new solution passes right through the tank.)

$$\frac{ds}{dt} = \begin{cases} \dfrac{9}{2} - \dfrac{4s}{10 - t} & t < 10 \\ 0 & t \geq 10 \end{cases}$$

D · $\dfrac{dx}{dt} = kx.$

E · The slope of the curve is dy/dx. It is also given by

$$\frac{y - 0}{x - \dfrac{x}{2}} = \frac{2y}{x}.$$

$$\frac{dy}{dx} = \frac{2y}{x}$$

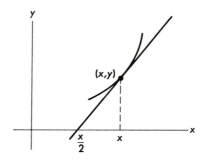

FIG. 30

F · At any point (x, y) on $x^2 - y^2 = C$, $dy/dx = x/y$. The slope of a curve in the orthogonal trajectory must have $-y/x$ as a slope; that is, the differential equation of the orthogonal trajectory is

$$\frac{dy}{dx} = -\frac{y}{x}.$$

Section 15 · 2

A · Since $y = Ce^x - x - 1$, $dy/dx = Ce^x - 1 = x + y$.

B · Substituting $x = 1$, $y = 1$ in the solution gives $1 = Ce - 2$ so the required solution is $y = (3e^{-1})e^x - x - 1 = 3e^{x-1} - x - 1$.

C · (see Fig. 30)

$\dfrac{dy}{dx}$	Points
-2	$2y = x$
-1	$y = x$
0	$x = 0$
1	$y = -x$
2	$2y = -x$

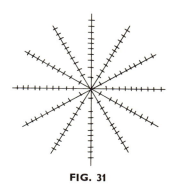

FIG. 31

Even timid souls might boldly guess that this directional field results from families of the form

$$x^2 + y^2 = C^2.$$

As before, verification should be carried out by substitution.

D · If $re^{ax} + se^{bx} = 0$ for all x, then differentiation shows that we also have $rae^{ax} + sbe^{bx} = 0$ for all x. Setting $x = 0$ in these two equations gives us the system

$$r + s = 0$$

$$ar + bs = 0.$$

The coefficient matrix has determinant $b - a \neq 0$. Therefore the trivial solution $r = s = 0$ is the only solution which means that e^{ax} and e^{bx} are linearly independent.

E · Proceeding as in Problem D, we have $re^{ax} + sxe^{ax} = 0$ and $rae^{ax} + s[xae^{ax} + e^{ax}] = 0$. Setting $x = 0$, it is immediately clear that $r = s = 0$, hence that the functions are linearly independent.

F · $a_0 H''(x) + a_1 H'(x) + a_2 H(x)$

$$= a_0[rF(''x) + sG''(x)] + a_1[rF'(x) + sG'(x)] + a_2[rF(x) + sG(x)]$$

$$= r[a_0 F''(x) + a_1 F'(x) + a_2 F(x)] + s[a_0 G''(x) + a_1 G'(x) + a_2 G(x)]$$

$$= r[0] + s[0] = 0.$$

G · Direct substitution in each case.

Section 15 · 3

A · $(D + 1)^2 = 0$ is the characteristic equation. $y = C_1 e^{-x} + C_2 e^{-x} = C_3 e^{-x}$ where $C_3 = C_1 + C_2$.
We seem to have violated our rule of thumb about having two constants.

B · The characteristic roots are complex. Anyone know what $\exp\left(-\dfrac{1}{2} + \dfrac{\sqrt{3}}{2}i\right)$ means? Try an "easier" question. What does e^t mean?

C · Substitute. **D ·** Substitute.

E · We are to solve $(mD^2 + k)s = 0, k > 0$. The characteristic equation $mD^2 + k = 0$ has solutions $D = \pm i\sqrt{k/m}$. The solution is

$$s = C_1 \cos\sqrt{\frac{k}{m}}\, t + C_2 \sin\sqrt{\frac{k}{m}}\, t.$$

Substituting $t = 0$ gives $s = C_1$, so $C_1 = s_0$. We also know that $\dfrac{ds}{dt} = 0$ when $t = 0$.

$$\frac{ds}{dt} = s_0\left(-\sqrt{\frac{k}{m}}\right)\sin\sqrt{\frac{k}{m}}\, t + C_2\sqrt{\frac{k}{m}}\cos\sqrt{\frac{k}{m}}\, t.$$

Substituting $t = 0$ gives

$$\frac{ds}{dt} = C_2\sqrt{\frac{k}{m}} \quad \text{and so } C_2 = 0.$$

F · $x = C_1 \cos\sqrt{k}x + C_2 \sin\sqrt{k}x$.

G · We are to solve $(D - k)x = 0$. The characteristic equation $D - k = 0$ has the root $D = k$ leading to the solution $x = Ce^{kt}$.

H · We are to solve $(D^2 + CD)y = g$ subject to the conditions that when $t = 0$, $y' = y = 0$. The general solution to the auxiliary equation, having characteristic equation $D(D + C) = 0$, is $y = C_1 + C_2e^{-Ct}$. To obtain a particular solution $P(t)$ satisfying $P''(t) + CP'(t) = g$, we try $P(t) = At + B$. Substitution gives $CA = g$, so $A = g/C$. Combining constant terms, $y = C_3 + C_2e^{-Ct} + (g/C)t$. The initial conditions (when $t = 0$) tell us that $0 = C_3$ and $0 = -CC_2 + g/C$. We conclude that $y = (g/C^2)e^{-Ct} + (g/C)t$.

I · We are to solve $D^2s = 32\sin\alpha$ where α is constant and we know that when $t = 0$, $s = 0$ and $ds/dt = 0$. The general solution of the auxiliary equation $D^2s = 0$ is $s = C_1t + C_2$. It is easy to see from $s = D^{-2}\, 32\sin\alpha$ that $s = 16t^2\sin\alpha$ is a particular solution (in fact this technique could be used from the beginning in this case) leading us to write for the general solution $s = 16t^2\sin\alpha + C_1t + C_2$. The initial conditions show that $C_1 = C_2 = 0$.

J · The equation $(D - 1)y = x$ has auxiliary solution $y = Ce^x$ and finding $P(x)$ so that $P'(x) - P(x) = x$ suggests trying $P(x) = Ax + B$. Substitution gives $A - (Ax + B) = x$, so $A = -1$ and $B = -1$. We have the general solution $y = Ce^x - x - 1$.

K · The auxiliary solution is $y = C_1 e^{4x} + C_2 e^{-x}$. For a particular solution we try $y = A e^{2x}$. Substituting,

$$(4 - 6 - 4)A e^{2x} = e^{2x}.$$

Hence, $A = -\frac{1}{6}$ and we have

$$y = C_1 e^{4x} + C_2 e^{-x} - \tfrac{1}{6} e^{2x}.$$

Section 15 · 4

A · We know that $F(x)$ is a solution to (2), so

$$F'(x) + P(x)F(x) = Q(x).$$

Since $G(x)$ as defined is also a solution,

$$G'(x) + P(x)G(x) = Q(x).$$

Subtraction gives

$$F'(x) - G'(x) + P(x)[F(x) - G(x)] = 0.$$

The conclusion is obvious if $F(x) = G(x)$ for each x, so we suppose the functions are distinct. Then

$$\frac{F'(x) - G'(x)}{F(x) - G(x)} = -P(x) \qquad \text{wherever} \qquad F(x) - G(x) \neq 0.$$

Therefore

$$\ln|F(x) - G(x)| = - \int P(x) \, dx + C$$

and as in Example 15.2A, we have for all x

$$F(x) - G(x) = C e^{-\int P(x) \, dx}.$$

B · Solve in the usual way to obtain the solution

$$y = x + \frac{c}{x}.$$

If we set $C = -2$, the curve $y = x + 2/x$ passes through $(1, -1)$ but is not continuous on $[-2, 2]$. According to the previous problem, there are no other solutions through $(1, -1)$, so we cannot obtain the desired continuous curve through $(1, -1)$.

C · $\dfrac{ds}{dt} + \dfrac{4}{10 - t} s = \dfrac{9}{2}$

$$(10 - t)^{-4} \frac{ds}{dt} + 4(10 - t)^{-5} s = \frac{9}{2} (10 - t)^{-4}$$

$$(10 - t)^{-4} s = \frac{3}{2} (10 - t)^{-3} + C$$

When $t = 0$, $s = 2$ so $C = -\dfrac{13}{10^4}$

$$s = \frac{3}{2}(10 - t) - 13\left(\frac{10 - t}{10}\right)^4.$$

The maximum amount of salt is in the tank when

$$\frac{ds}{dt} = 0; \qquad \text{that is, when } t = 1 - \sqrt[3]{\frac{15}{52}}.$$

D • $\dfrac{dy}{dx} - \dfrac{2}{x}y = 0$. Multiply by $e^{-\int(2/x)\,dx} = e^{\ln x^{-2}} = 1/x^2$. $x^{-2}y' - x^{-3}y = 0$ so $x^{-2}y = C$. When $x = 2$, $y = 4$ so $C = 1$ and $y = x^2$.

E • The procedure of Problem D leads to multiplication by x and then to the solution $xy = C$.

F • $\dfrac{di}{dt} + \dfrac{R}{L}i = \dfrac{E_0}{L}$

$$ie^{(R/L)t} = \frac{E_0}{R}e^{(R/L)} + C$$

$$i = \frac{E_0}{R} + Ce^{-(R/L)t}$$

As t increases, i approaches E_0/R.

G • One easily obtains the solution

$$y = C\exp[\tfrac{1}{2}x^2]$$

and $1 = Ce^{1/8}$ implies that $C = e^{-1/8}$ so

$$y = F(x) = \exp[\tfrac{1}{2}x^2 - \tfrac{1}{8}]$$

$$F(\tfrac{1}{4}) = e^{-3/32}.$$

To get accuracy within one unit in the second decimal place, one can either use tabulated values of e^{-x} (from which interpolation will probably be necessary) or one can use

$$e^{-3/32} \approx 1 - \frac{3}{32} \approx 0.91$$

since

$$\left(\frac{3}{32}\right)^2\frac{1}{2!} < \left(\frac{3}{30}\right)^2\frac{1}{2} = \frac{1}{200} = 0.005.$$

Section 15 · 5

A . $\dfrac{dx}{x} = k\,dt.$

In $x = kt + C$ so $x = e^{kt+C} = e^{C}e^{kt} = C_1 e^{kt}$. Let $x_0 =$ amount at $t = 0$. Then x, the amount of the first substance at time t, is $\tfrac{3}{4}x_0$ when $t = 5$ (for $\tfrac{1}{4}x_0$ has been converted).

$$x_0 = C_1 e^0 \qquad\qquad \text{so } C_1 = x_0$$

$$\tfrac{3}{4}x_0 = x_0\, e^{5k} \qquad \text{so } e^{5k} = \tfrac{3}{4}$$

When $x = \tfrac{1}{4}x_0$, then $\tfrac{1}{4} = e^{kt}$. We can either solve $e^{5k} = \tfrac{3}{4}$ for k, or we can write

$$\tfrac{1}{4} = (e^{5k})^{t/5} = (\tfrac{3}{4})^{t/5}$$

$$\tfrac{1}{5}t\,\ln\tfrac{3}{4} = \ln\tfrac{1}{4}$$

$$t \approx 22.3 \text{ min}$$

B . $\dfrac{dy}{y} = \dfrac{2dx}{x}$

$\ln y = 2\ln x + C$

$y = C_1 x^2$ and $4 = C_1(2)^2$ so $C_1 = 1$

$y = x^2$

C . $\dfrac{dy}{y} = -\dfrac{dx}{x}$

$\ln y = -\ln x + C$

$\ln xy = C$ so $xy = C_1$

D . $\dfrac{dy}{y} + \dfrac{2x\,dx}{1+x^2} = 0$

$\ln y + \ln(1 + x^2) = C$

$y = \dfrac{C_1}{1+x^2}$

E . $(D^2 + c^2)y = 0$ has characteristic equation $D^2 + c^2 = 0$ with roots $D = \pm ic$ leading to the general solution $y = C_1 \cos cx + C_2 \sin cx$.

F . $y = \displaystyle\sum_{n=0}^{\infty} a_n x^n$ so $y' = \displaystyle\sum_{n=0}^{\infty} n a_n x^{n-1}$.

The equation is now

$$\sum_{n=0}^{\infty} n a_n x^{n-1} + \sum_{n=0}^{\infty} n a_n x^{n+1} + 2\sum_{n=0}^{\infty} a_n x^{n+1} = 0.$$

The first summation may be written

$$a_1 + \sum_{n=0}^{\infty} (n+2)a_{n+2}x^{n+1}.$$

We then have

$$a_1 + \sum_{n=0}^{\infty} [(n+2)a_{n+2} + (n+2)a_n]x^{n+1} = 0.$$

Consequently, $a_1 = 0$ and $a_{n+2} = -a_n$. Apparently, a_0 is arbitrary, but all odd subscripts give a zero coefficient and we have

$$y = a_0[1 - x^2 + x^4 - x^6 + \cdots] = \frac{a_0}{1+x^2}.$$

Compare this with the solution to Problem D.

Section 15 · 6

A · $T_6(x) = 2 + x - 2x^2 + x^4 - \frac{1}{3}x^6$.

B · We are given that F is to satisfy $F'(x) = xF(x)$ and that $F(\frac{1}{2}) = 1$. Then $F'(\frac{1}{2}) = \frac{1}{2}$, and

$$F''(x) = xF'(x) + F(x) \qquad F''(\frac{1}{2}) = \frac{1}{4} + 1 = \frac{5}{4}$$

$$F'''(x) = xF''(x) + 2F'(x) \qquad F'''(\frac{1}{2}) = \frac{1}{2}(\frac{5}{4}) + 2(\frac{1}{2}) = \frac{13}{8}$$

$$F(x) = 1 + \frac{1}{2}(x - \frac{1}{2}) + \frac{5}{8}(x - \frac{1}{2})^2 + \frac{13}{48}(x - \frac{1}{2})^3 + \cdots$$

$$F(\frac{1}{4}) \approx 1 - \frac{1}{8} + \frac{5}{128} = \frac{117}{128} = 0.9141$$

The series is alternating, so if we stop as indicated, our error will be less than $13/48$ $(64) \approx 0.00423 < 0.005$ so we are within one unit in the second decimal by using 0.91.

C · Begin with $F_0(x) = 1$ and $G(x, F(x)) = xF(x)$. Set

$$F_1(x) = 1 + \int_{1/2}^{x} t \, dt = 1 + \frac{1}{2}x^2 - \frac{1}{8} = \frac{1}{2}x^2 + \frac{7}{8}$$

$$F_2(x) = 1 + \int_{1/2}^{x} (\frac{1}{2}t^3 + \frac{7}{8}t) \, dt = \frac{1}{8}x^4 + \frac{7}{16}x^2 + \frac{113}{128}$$

From these we get as the first three estimates:

$$F_0(\frac{1}{4}) = 1$$

$$F_1(\frac{1}{4}) = \frac{29}{32}$$

$$F_2(\frac{1}{4}) = \frac{1865}{2048} \approx 0.9106 \approx 0.91$$

ANSWERS TO
ODD EXERCISES

Exercise 1 · 2

1 · $\mathbf{i} - 2\mathbf{j}$

5 · $(-8, 4)$

9 · $(2, 0)$

13 · $(1/\sqrt{5})(2\mathbf{i} + \mathbf{j})$

3 · $(1/\sqrt{5})(-\mathbf{i} + 2\mathbf{j})$

7 · $(-\frac{7}{3}, \frac{8}{3})$

11 · $\pi/4$ or $45°$

Exercise 1 · 3

1a · $y = (\frac{4}{3})x + 4$

  b · $(x/-3) + (y/4) = 1$

  c · $-\frac{4}{5}x + \frac{3}{5}y = 12/5$

5a · $y = -\frac{3}{2}x + 2$

  b · $x/\frac{4}{3} + y/2 = 1$

  c · $(3/\sqrt{13})x + (2/\sqrt{13})y = 4/\sqrt{13}$

9 · $\mathbf{v} = 3\mathbf{i} + 4\mathbf{j}; \mathbf{w} = -4\mathbf{i} + 3\mathbf{j}$

13 · $y - 2 = -\frac{1}{10}(x - 4)$

17 · $y = \frac{1}{3}x + 2$

21 · $y - 3 = \frac{4}{3}(x - 5)$

23 · $(10 + 4\sqrt{5})x + (5 - 3\sqrt{5})y = 70 - 12\sqrt{5}$

25 · $x + y = 5$ is one answer.

29 · $5x + 3y = 15$

3a · $y = -2x + 14$

  b · $(x/-7) + (y/14) = 1$

  c · $(2/\sqrt{5})x + (1\sqrt{5})y = 14/\sqrt{5}$

7a · $y = \frac{1}{2}x - \frac{5}{4}$

  b · $x/\frac{5}{2} + y/-\frac{5}{4} = 1$

  c · $(-1/\sqrt{5})x + (2/\sqrt{5})y = -5/2\sqrt{5}$

11 · $\text{Arccos}(1/\sqrt{5})$

15 · $y - 3 = 3(x - 1)$

19 · $y = \frac{1}{3}x + 3$

27 · $x + 2y = 8$

Exercise 1 · 4A

1 · $(x + 2)^2 + y^2 = 4$

5 · $(x + 1)^2 + (y - 3)^2 = 0$

9 · $(x - 6)^2 + (y \pm 4\sqrt{2})^2 = 36$

11 · $[x - \frac{2}{3}]^2 + [y - \frac{2}{3}]^2 = \frac{4}{9}$ and $(x + 2)^2 + (y - 2)^2 = 4$

13 · $(x - 7)^2 + (y - 3)^2 = 50$

15 · $x^2 + (y + 4)^2 = 5$ and $(x + 2)^2 + y^2 = 5$

3 · $x^2 + (y + 1)^2 = \frac{4}{3}$

7 · $(x + 1)^2 + [y - \frac{5}{4}]^2 = \frac{25}{16}$

Exercise 1 · 4B

1 · foci $(0, \pm \sqrt{6}/2$; directricies $y = \pm 3\sqrt{6}/4$

3 · foci $(1 \pm \sqrt{5}, 3)$; directricies $x = 1 \pm (9\sqrt{5}/5)$

5 · foci $(-1, 2 \pm \sqrt{3})$; directricies $y = 2 \pm (4\sqrt{3}/3)$

7 · foci $\left(\dfrac{3}{2}, \dfrac{-1 \pm \sqrt{2}}{2}\right)$; directricies $y = -\dfrac{1}{2} \pm \dfrac{3\sqrt{2}}{4}$

9 · $x^2 + 3y^2 = 6$ **11 ·** $5x^2 + 9y^2 = 45$

13 · $16x^2 + 64x + 7y^2 - 28y + 29 = 0$

Exercise 1 · 4C

1 · foci $(1 \pm \sqrt{2}, -2)$; directricies $x = \pm \sqrt{2}/2$

3 · foci $(-2, \pm \sqrt{5}/2)$; directricies $y = \pm \sqrt{5}/10$

5 · foci $(-2/3, (1/2) \pm (\sqrt{13}/6))$; directricies $y = (1/2) \pm (3\sqrt{13}/26)$

7 · lines intersecting at $(1, -1)$

9 · $4x^2 + 8x - y^2 + 6y - 9 = 0$ **11 ·** $y^2 - 6y - 4x^2 - 8x + 1 = 0$

13 · $3x^2 - y^2 = 192$ **15 ·** $8y^2 - 104y - x^2 + 4x + 316 = 0$

Exercise 1 · 4D

1 · focus $(0, \frac{3}{8})$ **3 ·** focus $(\frac{1}{2}, \frac{1}{2})$

5 · focus $(-1, -\frac{9}{4})$ **7 ·** focus $(\frac{11}{8}, \frac{1}{2})$

9 · $x^2 - 8x + 8y = 16$ **11 ·** $4y^2 - 8y + 12x = 41$

13 · hyperbola; center $(1, -2)$; $a = b = \sqrt{3}$

15 · ellipse; center $(-\frac{1}{2}, 2)$; $a = \sqrt{5}$; $b = \sqrt{5}/2$

17 · hyperbola; center $(\frac{1}{3}, -\frac{1}{2})$; $a = \frac{1}{3}$; $b = \frac{1}{2}$

19 · parabola; vertex $(\frac{1}{2}, \frac{1}{2})$

21 · circle; center $(\frac{1}{3}, -\frac{2}{3})$; $r = \sqrt{5}/3$

23 · parabola; vertex $(3, -\frac{9}{4})$

Exercise 1 · 4E

1 · hyperbola (rotate through $\theta = \pi/4$ or $45°$)

3 · hyperbola (rotate through $\theta = \pi/6$ or $30°$)

5 · ellipse (rotate through $\theta = \text{Arccos } 1\sqrt{10}$)

Exercise 2 · 1

1 · $-(x+1)$

3 · $\dfrac{1}{\sqrt{1+3x}}$ if $x>0$; $\dfrac{-1}{\sqrt{1+3x}}$ if $x<0$

5 · x^2+1

7 · -3

9 · 3

11 · 32

13 · -2

19 · (a) 6   (b) 2
      (c) 1   (d) 3
      (e) $\dfrac{\pi}{2}$ (f) 3

Exercise 2 · 2

1 · domain $[-3, 3]$
range $[0, 3]$

3 · domain $[-5, 1]$
range $[0, 3]$

5 · domain $[-\frac{3}{2}, 0]$
range $[0, 3]$

7 · all x not -2 or -3

9 · all x

11 · $[-\sqrt{3}, 0)$ and $(0, \sqrt{3}]$

13 · domain $[-5, -1]$
range $[0, 2]$

15 · domain $[-2, 1]$
range $[0, 3]$

Exercise 2 · 4

1 · $\frac{3}{2}$

3 · $-\frac{1}{3}$

5 · 0

7 · e^2

9 · $-\infty$

11 · 0

13 · 1

15 · e^2

Exercise 2 · 5

1 · $x=\dfrac{y-3}{2}$

3 · $x=\dfrac{1-2y}{y}$

5 · $x=\frac{1}{2}(y-4)^2$

7 · $x=\dfrac{y+1}{2-y}$

9 . $x = \dfrac{3 + \sqrt{9 + 4y}}{2}$ **11 .** $x = \dfrac{\sqrt{y}}{2 - \sqrt{y}}$

13 . $y = \dfrac{3x + 1}{6x + 4}$ **15 .** $y = \dfrac{2x + 4}{5 + x}$

Exercise 2 · 7

1 . $T(x) = 2 + m(x - 4)$ **3 .** $T(x) = 19 + m(x - 2)$

5 . $T(x) = 1 + m(x - 3)$ **7 .** $T(x) = \frac{2}{5} + m(x - 4)$

9 . $T(x) = 6 + m(x - 8)$

Exercise 3 · 1

1 . 4 **3 .** 11

5 . $-\frac{1}{25}$ **7 .** $\frac{1}{6}$

9 . $\frac{4}{25}$ **11 .** $4x + 3$

13 . $-\dfrac{2}{x^3}$ **15 .** $\dfrac{1}{2\sqrt{x}}$

17 . $y - 4 = 11(x - 1)$ **19 .** $y - 3 = \frac{1}{6}(x - 9)$

Exercise 3 · 2

1 . $\Delta y = 0.84$ **3 .** $\Delta y \approx 3.422$
     $dy = 0.80$ $dy = 2.750$

5 . $\Delta y \approx 0.00408$ **7 .** $\Delta y \approx -0.0504$
     $dy = 0.00400$ $dy = -0.0500$

9 . $\Delta y \approx 0.0308$ **11 .** $\Delta y = -1.48$
     $dy = 0.0320$ $dy = -1.50$

13 . $\Delta y \approx -0.0193$ **15 .** $\Delta y \approx 0.0373$
     $dy = -0.0222$ $dy = 0.0375$

Exercise 3 · 4

1 . $12x^3 - 5$ **3 .** $15x^2 - 20x$

5 . $5x^4 + 4x^3 - 18x^2 + 10x + 15$ **7 .** $5x^4 - 28x^3 + 15x^2 - 84x + 49$

9 · $\dfrac{1 - x^2}{(x^2 + 1)^2}$

11 · $\dfrac{-5}{(x - 3)^2}$

13 · $-\dfrac{x^4 + 3}{(x^3 + 3x)^2}$

15 · $\frac{3}{2}x^{-1/2} + \frac{1}{3}x^{-2/3}$

17 · $(x + \frac{3}{2})(x^2 + 3x)^{-1/2}$

19 · $7(4x^3 + 3x - 1)^6(12x^2 + 3)$

21 · $dy = (5x^4 + 28x^3 + 24x^2 - 38x - 21)\, dx$

23 · $dy = \dfrac{3}{(x + 2)^2}\, dx$

25 · $dy = \dfrac{7x^2 - 1}{3x^{2/3}}\, dx$

27 · $dy = -\dfrac{(3x - 1)\, dx}{2\sqrt{x}(3x + 1)^2}$

29 · $dy = 30x(3x^2 + 4)^4\, dx$

31 · $\dfrac{-1}{\sqrt{x}(\sqrt{x} - 1)^2}$

33 · $\dfrac{3x^2 - 1}{2x\sqrt{x}}$

35 · $2x + 1 - \dfrac{1}{x^2}$

37 · $\dfrac{-1}{x^2(x^2 + 1)^{1/2}}$

39 · $-\dfrac{3x^2 + 8x}{(x^3 + 4x^2)^2}$

41 · $1 + t^{-1/2}$

43 · $4t(3x^5 + 12x^3 + 9x)$ where $x = t^2 + 1$

45 · $\dfrac{3t + 1}{2(t + 1)^2\sqrt{2t^2 + t}}$

47 · -2

49 · $\frac{1}{4}$

51 · does not exist

53 · $\dfrac{-5}{8(17)^2}$

55 · -1

57 · $-\frac{1}{4}$

59 · 1

61 · does not exist

63 · $\Delta y = 0.01$
    $dy = 0$

65 · $\Delta y \approx 0.0494$
    $dy = 0.0500$

67 · $\Delta y = 0.375$
    $dy = 0.300$

69 · $\Delta y = 0.400$
    dy is not defined

71 · $y = 1$

73 · $y - 2 = \frac{1}{4}(x - 3)$

75 · $y - 3 = -3(x - 1)$

77 · $x = 0$

Exercise 3 · 5A

1 · $y - 7 = 2(x - 3)$

3 · $y + 2 = 4(x - 1)$

5 · $y + 1 = 0$

7 · $y = \frac{3}{2}x$

9 · $y = -\frac{2}{3}x$

11 · $y = \frac{3}{4}x$

13 . $\dfrac{45}{7}$

15 . 0

17 . $-\dfrac{14}{3}$

19 . $\dfrac{22}{3}$

Exercise 4 · 1

1 . 1

3 . $-1 + \dfrac{\sqrt{3}}{2}$

5 . $\pm \frac{1}{3}\sqrt{3}$

Exercise 4 · 2

1 . $3x^{-1/2}$

3 . $\dfrac{-1}{(x+1)^3}$

5 . $(x^2 + 1)^{-3/2}$

7 . $\dfrac{2x^3 - 6x}{(x^2 + 1)^3}$

9 . $\dfrac{20x^3 - 10x^4}{(x+1)^7}$

11 . $-2(a - b)(x - b)^{-3}$

13 . $\frac{8}{27}x^{-7/3}$

15 . $-6(x + 1)^{-4}$

17 . $-3x(x^2 + 1)^{-5/2}$

19 . $(-1)^n(2)(n!)(x - 1)^{-(n+1)}$ where $n! = n(n - 1) \cdots (2)(1)$

21 . $\dfrac{(-1)^{n-1}}{2^n} (3)(5) \cdots (2n - 3)(x + 1)^{-(2n-1)/2}$

Exercise 4 · 3

1 . $-1 + 3(x - 2) + 8(x - 2)^2$

3 . $16 + 38(x - 3) + 29(x - 3)^2$

5 . $1 - x^2$

7 . $(x + 2) + (x + 2)^2$

9 . $1 - 2x + 2x^2$

11 . $|R_2(2, \frac{9}{4})| = (4\bar{x} - 3)(\frac{1}{4})^3 < 6(\frac{1}{4})^3 = \frac{3}{32}$
$|R_2(3, \frac{9}{4})| = (4\bar{x} - 3)(\frac{3}{4})^3 < 9(\frac{3}{4})^3 = \frac{243}{64}$

13 . For Exercise 1, $|F(\frac{9}{4}) - T_2(\frac{9}{4})| = \frac{21}{256} < \frac{24}{256} = \frac{3}{32}$
For Exercise 3, $|F(\frac{9}{4}) - T_2(\frac{9}{4})| = \frac{891}{256} < \frac{972}{256} = \frac{243}{64}$

15 . $T_2(\frac{1}{4}) = \frac{5}{8}; |R_2(0, \frac{1}{4})| < \frac{1}{32}$

Exercise 5 · 1A

1 · $x = \frac{1}{3}$ (max)
  $x = 3$ (min)

3 · $x = -2$ (min)
  $x = 1$ (inf)

5 · $x = -2$ (min)
  $x = \frac{1}{2}$ (inf)

7 · $x = -2$ (max)
  $x = 0$ (inf)
  $x = \frac{3}{5}$ (min)

9 · $x = 2$ (max)

11 · $x = 0$ (min)
  $x = \frac{8}{3}$ (max)

13 · $x = 0$ (max; end point)
  $x = \frac{12}{5}$ (min)

15 · $x = 0$ (max)
  $x = \frac{8}{5}$ (min)

17 · $x = \pm\frac{2}{5}\sqrt{10}$ (min)
  $x = 0$ (max)

19 · $\dfrac{4x - 4}{3y^2 + 3}$

21 · $\dfrac{-2(y + 6x^2)}{x}$

23 · $\dfrac{3x^2 - y^2}{2xy + 4}$

25 · $\dfrac{y^3 + y}{x - xy^2}$

27 · $\dfrac{y^2 - 3x^2y^3}{xy + 6}$

Exercise 5 · 1B

1 · $x = 0$ (min)
  $x = 1$ (inf)
  $x = -2/5$ (max)

3 · $x = \frac{3}{2}$ (min)

5 · $x = 1$ (min)
  $x = -1$ (max)

7 · Case I $(a < b)$ Case II $(a > b)$
  $x = a$ (max) $x = a$ (min)
  $x = b$ (inf) $x = b$ (inf)
  $x = \dfrac{3a + 2b}{5}$ (min) $x = \dfrac{3a + 2b}{5}$ (max)

13 · local maximum $M = \frac{17}{2}$ occurs at $x = \frac{1}{2}$; local minimum $m = -\frac{15}{2}$ occurs at $x = \frac{1}{2}$

15 · end point maximum $M = 5$ occurs at $x = 0$; local minimum $m = -15$ occurs at $x = 2$.

Exercise 5 · 2

1 · (a) $C = 2kx^2 + 4(2k)xy + kx^2$ (b) $x^2y = 48$ (c) $x = 4$ ft; $y = 3$ ft
3 · (a) $V = \pi x^2 y$ (b) $2x + y = 12$ (c) $V = 64\pi$

5 • (a) $s = 8\left[\dfrac{x}{y} - \dfrac{x}{8}\right]$　(b) $x^2 + 8 = y^2$　(c) $y = 4$

7 • 6 　　　　　　　　　　　**9 •** 375×750

13 • $\dfrac{2a}{\sqrt{3}}$ 　　　　　　　**15 •** height equals three times the radius

17 • 22

19 • Use $\dfrac{400}{4 + \pi}$ for the square to obtain minimum area. Maximum area is only obtained if we use all the wire for the circle.

21 • One-third of the way from sixth to seventh

23 • underwater direct to factory

25 • 4 　　　　　　　　　　　**27 •** $(-2, -1)$

Exercise 5 · 3A

1 • 10 ft/sec 　　　　　　　　**3 •** (a) about 12:22　(b) ≈ 73 mph

5 • $\frac{1}{4}(7)^{-2/3}$ in./sec 　　　　**7 •** $-\frac{1}{18}$ in./sec

9 • $\frac{1}{2}(9\pi)^{-1/3}$ in./sec

Exercise 5 · 3B

1 • $\frac{81}{4}$ ft 　　　　　　　　　**3 •** $\frac{5}{2}$ sec

7 • yes; 100 ft/sec > 68 mph 　　**9 •** (a) 242 ft　(b) 60.5 ft

11 • $\frac{2}{3}x^3 - \frac{3}{2}x^2 + 4x + C$ 　　**13 •** $\frac{4}{9}x^{3/2} + \frac{3}{8}x^{2/3} + C$

15 • $\frac{2}{9}(4 + x^3)^{3/2} + C$ 　　　**17 •** Try this one after reading Chapter 7

19 • $3(1 + x^2)^{1/2} + C$

Exercise 6 · 1

1 • (a) 196　(b) 224　(c) 20　(d) (25)(13)(17)

Exercise 6 · 2

1 • $\displaystyle\int_0^1 (2 - x^2)\, dx$ 　　　　**3 •** $\displaystyle\int_2^4 (x^2 - 4x + 7)\, dx$

5 • $\int_0^1 \pi(2 - x^2)^2\, dx$ **7 •** $\int_2^4 \pi(x^2 - 4x + 7)^2\, dx$

9 • $\int_0^1 2\pi x(2 - x^2)\, dx$ **11 •** $\int_2^4 2\pi x(x^2 - 4x + 7)\, dx$

13 • $\int_0^1 \frac{1}{2}(2 - x)^2\, dx$ **15 •** $\int_2^4 \frac{1}{2}(x^2 - 4x + 7)^2\, dx$

17 • $\int_0^1 (2x - x^3)\, dx$ **19 •** $\int_2^4 (x^3 - 4x^2 + 7x)\, dx$

Exercise 6 · 4A

1 • $\dfrac{5}{3}$ **3 •** $\dfrac{26}{3}$ **5 •** $\dfrac{43\pi}{15}$ **7 •** $\dfrac{202\pi}{5}$

9 • $\dfrac{3\pi}{2}$ **11 •** $\dfrac{164\pi}{3}$ **13 •** $\dfrac{43}{30}$ **15 •** $\dfrac{101}{5}$

17 • $\dfrac{3}{4}$ **19 •** $\dfrac{82}{3}$ **21 •** $\dfrac{3}{2}$ **23 •** $\dfrac{1}{2}$

25 • $\dfrac{2}{3}$ **27 •** $\dfrac{7}{72}$ **29 •** $149\dfrac{5}{12}$

Exercise 6 · 4B

1 • $\dfrac{3}{10}$ **3 •** $-2\sqrt{3}$ **5 •** $2\sqrt{3} - \dfrac{2}{3}$ **7 •** $\dfrac{15}{4}$

9 • $2\sqrt{2} - 2$ **11 •** $\dfrac{26}{3}$ **13 •** $\dfrac{4}{3}$ **15 •** $8\frac{1}{4}$

17 • 6π **19 •** $\dfrac{632\pi}{15}$ **21 •** $\left(0, \dfrac{158}{65}\right)$ **23 •** $\dfrac{8\pi}{3}$

25 • $\dfrac{64\pi}{15}$ **27 •** $\left(1, \dfrac{8}{5}\right)$

Exercise 6 · 5

1 • 2   **3 •** $\frac{1}{2}$

5 • $\frac{3}{2}$ **7 •** 1

9 • diverges   **11 •** converges to $\frac{4}{3}$

13 • converges   **15 •** diverges

Exercise 7 · 1A

1. (a) $\dfrac{7\pi}{4}$

(b) $\dfrac{3\pi}{4}$

(c) $\dfrac{7\pi}{6}$

(d) $-\dfrac{2\pi}{3}$

(e) $-\dfrac{\pi}{4}$

(f) $\dfrac{7\pi}{180}$

3. (a) $\dfrac{1}{2}, -\dfrac{\sqrt{3}}{2}, -\dfrac{1}{\sqrt{3}}$

(b) $-\dfrac{1}{2}, -\dfrac{\sqrt{3}}{2}, \dfrac{1}{\sqrt{3}}$

(c) $-\dfrac{1}{\sqrt{2}}, -\dfrac{1}{\sqrt{2}}, 1$

(d) $-\dfrac{\sqrt{3}}{2}, \dfrac{1}{2}, -\sqrt{3}$

(e) $-1, 0, -$

(f) $1, 0, -$

5. (a) $-\dfrac{4}{5}, -\dfrac{3}{4}$

(b) $-\dfrac{2}{\sqrt{5}}, -\dfrac{1}{2}$

(c) $-\dfrac{5}{13}, \dfrac{12}{-5}$

(d) $-\dfrac{2\sqrt{2}}{3}, \dfrac{-1}{2\sqrt{2}}$

Exercise 7 · 1B

1. (a) $-\dfrac{\sqrt{3}}{2}, \dfrac{1}{2}, -\sqrt{3}$

(b) $-\dfrac{1}{2}, \dfrac{\sqrt{3}}{2}, -\dfrac{1}{\sqrt{3}}$

(c) $\dfrac{1}{2}, \dfrac{-\sqrt{3}}{2}, -\dfrac{1}{\sqrt{3}}$

(d) $0, -1, 0$

3. (a) $\dfrac{2}{5\sqrt{5}}$

(b) $\dfrac{1}{\sqrt{5}}$

(c) $-\dfrac{4}{5}$

(d) $\dfrac{7}{25}$

(e) $-\sqrt{\dfrac{5 - 2\sqrt{5}}{10}}$

(f) $\dfrac{3}{\sqrt{10}}$

5. $\dfrac{\sqrt{7} - 3}{4\sqrt{2}}$

7. $\sin 2\alpha = \dfrac{24}{25}, \cos 2\alpha = \dfrac{-7}{25}$

9. $2\pi n, \dfrac{2\pi}{3} + 2\pi n, \dfrac{4\pi}{3} + 2\pi n$

Exercise 7 · 2A

1 · (a) 1   (b) 1
(c) 0   (d) $-\infty$

3 · (a) $-x(x^2+1)^{-3/2}\cos(x^2+1)^{-1/2}$ (b) $x\sec^2 x+\tan x$

(c) $\dfrac{(\tan x+1)\cos x-\sin x\sec^2 x}{(\tan x+1)^2}$ (d) $-2x^3\sin x^2+2x\cos x^2$

5 · (a) $3\sin^2 x\cos x$ (b) $3x^2\cos x^3$ (c) $3\sec^3 x\tan x$
(d) $\sec x(1+2\tan^2 x)$

7 · $\dfrac{1}{1+\cos x}$ **9 ·** $16\sin^3 x\cos x$

11 · $\dfrac{(x^2+1)\cos x-2x\sin x}{(x^2+1)^2}$

Exercise 7 · 2B

1 · $2y-\sqrt{3}+1=(1+\sqrt{3})\left(\dfrac{2\pi}{3}-x\right)$

3 · $\sqrt{13}$ **5 ·** 80π miles/min

7 · 27; if the pole is $<9\sqrt{5}$, fasten wire at the top.

11 · $T_1(x)=x$; $T_2(x)=x$; $T_3(x)=x-\dfrac{x^3}{6}$

13 · If $n=2k-1$ or $2k$ $(k=1,2,\ldots)$, then

$$T_n(x)=x-\frac{x^3}{3!}+\cdots+(-1)^{k+1}\frac{x^{2k-1}}{(2k-1)!}$$

15 · $-\frac{1}{3}\cos 3x+C$ **17 ·** $\dfrac{x}{2}-\dfrac{1}{12}\sin 6x+C$

19 · $\frac{1}{2}\tan^2 x+C$ **21 ·** $\dfrac{1}{8}x-\dfrac{1}{32}\sin 4x+C$

23 · $-\cot x-x+C$ **25 ·** $2-\dfrac{\pi}{4}$

27 · $1-\dfrac{\pi}{4}$ **29 ·** $\dfrac{15}{4}$

31 · $\dfrac{2\pi+3\sqrt{3}}{24}$ **33 ·** 1

35 · $\dfrac{\pi}{192}$ with an error $< \displaystyle\int_0^{1/4} \dfrac{\pi^3 x^4}{3!} < 0.001$

37 · $\dfrac{\pi}{32}\left[1 + \dfrac{\pi^2}{96}\right]$ with an error < 0.002

Exercise 7 · 3

1 · $\dfrac{1}{2\sqrt{x - x^2}}$

3 · $\dfrac{-1}{\sqrt{1 - x^2}}$ if $x > 0$

5 · $\dfrac{6x}{(x^2 + 2)\sqrt{x^4 + 4x^2 - 5}}$

7 · $\dfrac{8}{1 + 16x^2}$ Arctan $4x$

9 · $\dfrac{-1}{2\sqrt{x}(x + 1)}$

11 · $\dfrac{-2(2x^4 - 8x^2 + 3)}{\sqrt{4 - x^2}}$

15 · Arctan $x^2 + C$

17 · $-$Arcsin $\dfrac{3x}{2} + C$

19 · $\dfrac{1}{2}$ Arcsin $\dfrac{x^2}{3} + C$

21 · Arctan $(x^2 + 2x) + C$

23 · $\dfrac{\pi}{6}$

25 · $\dfrac{\pi}{12\sqrt{3}}$

27 · $\dfrac{1}{8\sqrt{2}}$ rad/sec

29 · 50 ft

31 · Underwater cable subtends an arc with central angle $2\pi/3$.

33 · $T_5(x) = x + \dfrac{x^3}{3!} + \dfrac{9x^5}{5!}$

$\pi/6 = B(\tfrac{1}{2}) \approx T_5(\tfrac{1}{2}) = 0.5216$; $\pi \approx 3.1296$

Exercise 7 · 4

1 · $\dfrac{2x + 1}{x^2 + x}$

3 · cot x

5 · tan x

7 · $\dfrac{5}{\sqrt{x^2 + 1}}$

9 · $1 + \ln x$

11 · $(2x + 1)\exp[x^2 + x]$

13 · $\dfrac{1}{2\sqrt{x}(\sqrt{1 - x})}$

15 · $-2 \cos x \sin x \exp[\cos^2 x]$

17 · $\dfrac{e^x}{x} + e^x \ln x$

19 · $\exp[x^2][2x^{-1} + 4x \ln x]$

21 · no elementary antiderivative

23 · $-\ln \cos x + C$

25 · $\frac{1}{2} \exp[x^2] + C$

27 · $\text{Arctan } e^x + C$

29 · $\frac{1}{2} \ln(1 + e^{2x}) + C$

31 · no critical points

33 · $x = 2$ (max); $x = 0$ (min)

35 · $x = 1$

37 · no critical points

39 · $x^{\sqrt{x}-1/2}[1 + \frac{1}{2} \ln x]$

41 · $(\sin x)^x[x \cot x + \ln \sin x]$

43 · $(2x)^x[1 + \ln 2x]$

45 · $1 - x + \dfrac{x^2}{2!} - \dfrac{x^3}{3!} + \dfrac{x^4}{4!} - \dfrac{x^5}{5!}$

47 · $T_6(x) = 2x + 2x^3/3 + 2x^5/5$

$\ln 2 = F(\frac{1}{3}) \approx T_6(\frac{1}{3}) = 0.676$

Exercise 7 · 5

9 · $2t \text{ sech } t^2$

11 · $4t \text{ csch } 2t^2$

13 · $|\sec t|$

15 · $2te^{t^2} \sinh 2e^{t^2}$

Exercise 7 · 6

1 · $\frac{3}{5}$

3 · ∞

5 · -1

7 · 0

9 · $-\infty$

11 · ∞

13 · 2

15 · $-\sqrt{3}$

17 · 1

19 · $1 - \ln 2$

21 · $\frac{1}{6}$

23 · 1

25 · $-\infty$

27 · 0

29 · $\dfrac{1}{\sqrt{e}}$

Exercise 8 · 1

1 · $-\ln |\cos x| + C$

3 · $-\dfrac{5}{18}(1 - x^3)^{6/5} + C$

5 · $\frac{1}{3} \sin(3x + 1) + C$

7 · $-\dfrac{1}{2} \cot x^2 + C$

9 · $\frac{1}{2} \tan^2 x + C$

11 · $\frac{1}{3} e^{x^3} + C$

13 · $\frac{1}{2} \ln(x^2 + 1) + C$

15 · $\frac{1}{2} \text{Arctan } x^2 + C$

17 · $x + C$

19 · $-\frac{1}{2} \cos e^{x^2} + C$

21 · $\frac{1}{4} \sin^4 x + C$

23 · $-\frac{2}{3} \cos^3 x + C$

25 • $-\cos x + \frac{1}{3}\cos^3 x + C$ **27 •** $\tan x - x + C$

29 • $\frac{3}{8}x - \frac{1}{4}\sin 2x + \frac{1}{32}\sin 4x + C$ **31 •** $-\frac{1}{6}\cot^3 2x + C$

33 • $\frac{3}{4}(x^2+1)^{2/3} + C$ **35 •** $\frac{1}{2}[\sec 2x + \cos 2x] + C$

37 • $\frac{1}{2}\ln(x^2 + 4x + 5) + C$ **39 •** $e^{x \sin x} + C$

Exercise 8 · 2

1 • $-(a^2 - x^2)^{1/2} + C$ **3 •** $\frac{a^2}{2}\,\mathrm{Arcsin}\,\frac{x}{a} - \frac{x}{2}(a^2 - x^2)^{1/2} + C$

5 • $\frac{1}{2}\ln(4 + x^2) + C$

7 • $(x^2 - a^2)^{1/2} - a\,\mathrm{Arctan}\,\dfrac{(x^2 - a^2)^{1/2}}{a} + C$

9 • $\ln\left[\dfrac{4 + 2\sqrt{3}}{3 + \sqrt{5}}\right]$ **11 •** π

13 • $\sqrt{3} - \dfrac{\pi}{3}$

Exercise 8 · 3

1 • $x \ln x - x + C$ **3 •** $\dfrac{3x^2}{8}(1 + x^2)^{4/3} - \dfrac{9}{56}(1 + x^2)^{7/3} + C$

5 • $\tan x + C$ **7 •** $s\,\mathrm{Arcsin}\,x + (1 - x^2)^{1/2} + C$

9 • $\frac{1}{2}[\tan x \sec x + \ln\,|\sec x + \tan x|] + C$

11 • $\frac{1}{5}e^x(\cos 2x + 2 \sin 2x) + C$ **13 •** $2[(\sin 1) - 1] + \cos 1$

15 • $\dfrac{8}{3}\ln 2 - \dfrac{7}{9}$

Exercise 8 · 4

1 • $\dfrac{x^3}{3} + \dfrac{1}{2}\ln(x^2 + 2) + C$ **3 •** $\dfrac{3}{2}\ln|x - 1| - \dfrac{5}{3}\ln|x + 2| + C$

5 • $\ln|x| + \dfrac{4}{\sqrt{3}}\,\mathrm{Arctan}\,\dfrac{2x + 1}{\sqrt{3}} + C$ **7 •** $\dfrac{3}{2}\,\mathrm{Arctan}\,\dfrac{x}{2} + \dfrac{1}{2}\ln(x^2 + 1) + C$

9 • $-\dfrac{2}{x} + 4\ln|x - 1| + \dfrac{5}{2}\ln|x + 1| + C$

11 · $\dfrac{x^3 + x^2 - 6x + 2}{2(x-2)} + C$

13 · $7 \ln|x+2| + \dfrac{3}{16} \text{Arctan} \dfrac{x}{2} + \dfrac{3x}{8(x^2+4)} + C$

15 · $3 \ln|x-1| + \ln|x^2 - x + 1| - \dfrac{5}{x-1} - 2\sqrt{3}\ \text{Arctan} \dfrac{2x-1}{\sqrt{3}} + C$

Exercise 8 · 5

1 · $\frac{1}{2} \ln|\sec x^2| + C$ **3 ·** $2(1 + \tan x)^{1/2} + C$

5 · $\dfrac{x^3}{3} \text{Arctan } x - \dfrac{1}{6} x^2 + \dfrac{1}{6} \ln(x^2+1) + C$

7 · $\frac{1}{2}(\ln x)^2 + C$ **9 ·** $\text{Arctan } e^x + C$

11 · $\ln|x + \frac{1}{2} + \sqrt{1 + x + x^2}| + C$ **13 ·** $\text{Arcsin } \dfrac{2x+1}{\sqrt{5}} + C$

15 · $\dfrac{1}{6} \ln \dfrac{(x-1)^2}{|x^2+x+1|} + \dfrac{1}{\sqrt{3}} \text{Arctan} \dfrac{2x+1}{\sqrt{3}} + C$

17 · $-\text{Arctan}(\cos x) + C$ **19 ·** $-\dfrac{1}{x} \text{Arctan } x + \dfrac{1}{2} \ln \dfrac{x^2}{x^2+1} + C$

21 · $\ln(e^x + 1) + C$ **23 ·** $\frac{1}{3} \sec^2 x \tan x + \frac{2}{3} \tan x + C$

25 · $\text{Arcsin } \dfrac{x-3}{3} + C$

27 · $\dfrac{2}{5}(x+2)^{5/2} - \dfrac{8}{3}(x+2)^{3/2} + 8(x+2)^{1/2} + C$

29 · $\dfrac{1}{6} \text{Arctan} \dfrac{\sqrt{x^2-9}}{3} - \dfrac{\sqrt{x^2-9}}{2x^2} + C$

31 · $\dfrac{x^2}{2} + 5x + \dfrac{21}{2} \ln|x^2 - 5x + 6| + \dfrac{45}{2} \ln\left|\dfrac{x-3}{x-2}\right| + C$

33 · no elementary antiderivative

35 · $2x^2 \sin x + 4x \cos x - 4 \sin x + C$

37 · $\dfrac{1}{2} \ln \dfrac{(x+1)^2}{|x^2+x+1|} - \dfrac{1}{x+1} - \dfrac{1}{\sqrt{3}} \text{Arctan} \dfrac{2x+1}{\sqrt{3}} + C$

39 · $\dfrac{x^2-8}{3} \sqrt{x^2+4}$ **41 ·** $\dfrac{1}{10} e^{3x}(3 \sin x - \cos x) + C$

43 · $\dfrac{3}{10} \ln \dfrac{|x-3|(x+2)^4}{|x+3|^5} + C$ **45 ·** $\dfrac{2}{45}(4 + 9x^2)^{2/3}(3x^2 - 2) + C$

47 · 0.5

49 · 0.05

51 · 0.4

53 · 1.6

55 · 0.15

57 · diverges

59 · 0.6

61 · diverges

63 · 0.5

65 · 0.1

67 · (0.52, 1.80)

69 · πab

71 · $\dfrac{4}{3} \rho b a^2$

Exercise 9 · 1

1 · $Dy = -\cot t$
$D^2 y = -\frac{1}{2} \csc^3 t$
$x^2 + y^2 = 4$

3 · $Dy = -\tan t$
$D^2 y = -\frac{1}{2} \sec^3 t$
$x^2 + y^2 = 4$

5 · $Dy = 2te^{t^2 - t}$
$D^2 y = (4t^2 - 2t + 2)e^{t^2 - 2t}$
$y = x^{\ln x}$

7 · $Dy = \sin t$
$D^2 y = \cos^3 t$
$x^2 + 1 = y^2$

9 · $Dy = \frac{1}{2}$
$D^2 y = 0$
$2y = x - 1$

11 · $Dy = \frac{1}{2}$
$D^2 y = 0$
$2y = x - 1$

13 · $x = 5 \cos t - \cos 5t$
$y = 5 \sin t - \sin 5t$

15 · $x = 4 \cos t + 4t \sin t$
$y = 4 \sin t - 4t \cos t$

Exercise 9 · 2

1 · $A: x^2 + 4y = 4$
$B: 3x = 4y$
intersect at $(1, \frac{3}{4})$
no collision

3 · $A: y = x^3 - 3x + 1$
$B: y = x + 1$
intersect and collide at
$(0, 1)$ when $t = 0$
$(2, 3)$ when $t = 1$

5 · $A: xy = 1$
$B: xy = 1$
Paths coincide from $(1, 1)$ to $(2^{2\pi}, 2^{-2\pi})$.
collision when $t = \dfrac{2\pi \ln 2}{1 + \ln 2}$

7 · $A: y = (x + 1)^{\ln(x + 1)}$
$B: x + y^2 = 1$
no intersection

9 · $A: xy = 1$
$B: x^2 = y$
intersect and collide at
$(1, 1)$ when $t = 1 + 1/\sqrt{2}$

Exercise 9 · 3

1 · $\sqrt{17} + \frac{1}{4}\ln(4 + \sqrt{17})$

3 · $2 - \sqrt{2} + \ln\dfrac{1 + \sqrt{2}}{\sqrt{3}}$

5 · 12

7 · $\sqrt{2}(e^{2\pi} - 1)$

9 · $\frac{1}{2}(e^{4\pi} + \pi - 1)$

11 · 10

13 · $\dfrac{9}{2}\pi^2$

Exercise 9 · 5

11 · $x + y = 3$

13 · $4x^2 + 3y^2 + 16y = 64$

15 · $9x^2 + 16y^2 = 144$

17 · $\dfrac{3\pi}{2}$

19 · $\dfrac{\pi}{2}$

21 · 2

23 · 11π

Exercise 10 · 1

1 · $(0, 0, 0)$

3 · $z(-\frac{3}{2}, -\frac{1}{2}, 1)$

5 · $y(-3, 1, 5)$

7 · $z(0, 2, 0, 1)$

9 · $(-y - 3z, 5y - 5z, 4y, 4z)$

11 · $\left(\dfrac{18 - 4z}{7}, \dfrac{26 - 5z}{7}, z\right)$

13 · no solutions

15 · $(2y - 3, 5 - 3y, y, 5 - 2y)$

17 · $(0, 3, 1, 0)$

Exercise 10 · 2

1 · no

3 · yes

5 · Arccos $4/\sqrt{30}$

7 · $\left(\dfrac{8}{3} + \dfrac{1}{\sqrt{2}}\right)\mathbf{u}_1 + \left(\dfrac{8}{3} + \dfrac{1}{\sqrt{2}}\right)\mathbf{u}_2 + \left(\dfrac{4}{3} - 2\sqrt{2}\right)\mathbf{u}_3$

9 · $\dfrac{2}{3}\mathbf{u}_1 + \dfrac{2}{3}\mathbf{u}_2 + \dfrac{1}{3}\mathbf{u}_3$

11 · $\mathbf{v}_1 = (1/\sqrt{2}, 0, 1/\sqrt{2})$

$\mathbf{v}_2 = (-1/3\sqrt{2}, 2\sqrt{2}/3, 1/3\sqrt{2})$

$\mathbf{v}_3 = (-2/3, -1/3, 2/3)$

13 · $v_1 = (1/\sqrt{5},\, 0,\, 2/\sqrt{5})$

$v_2 = (2/\sqrt{6},\, 1/\sqrt{6},\, -1/\sqrt{6})$

$v_3 = (2/\sqrt{30},\, -5/\sqrt{30},\, -1/\sqrt{30})$

15 · $v_1 = (1/\sqrt{2},\, 0,\, 0,\, 1/\sqrt{2})$

$v_2 = (2/3\sqrt{2},\, -1/\sqrt{2},\, 1/3\sqrt{2},\, -2/3\sqrt{2})$

17 · The solution space has dimension $n - r$.

Exercise 10 · 3A

1 · $\begin{bmatrix} 16 \\ 1 \end{bmatrix},\ \begin{bmatrix} 6 \\ 10 \end{bmatrix},\ \begin{bmatrix} 5 \\ -2 \end{bmatrix}$

3 · $\begin{bmatrix} 1 & 0 & 2 \\ -1 & 1 & 1 \\ 4 & 3 & -1 \\ 3 & 4 & 3 \end{bmatrix}$

5 · $[L] = \begin{bmatrix} 2 & 3 \\ 1 & -2 \end{bmatrix}$

$L(3,\, 2) = (12,\, -1)$

Exercise 10 · 3B

1 · $[T] = \begin{bmatrix} 3 & 1 \\ -9 & 7 \end{bmatrix}$

3 · $\begin{bmatrix} -50 & -17 & -59 \\ -24 & -7 & -18 \\ -4 & -3 & -19 \end{bmatrix}$

$T(3,\, 4) = (13,\, 1)$

5 · $BC = \begin{bmatrix} 2 & 0 & -4 \\ 8 & 3 & 11 \\ -2 & -2 & -11 \end{bmatrix}$; $CB = \begin{bmatrix} 1 & 3 & 4 \\ 5 & -1 & -7 \\ 3 & -1 & -6 \end{bmatrix}$

13 · $x = -5,\ y = -2,\ z = 3$

15 · no

Exercise 10 · 4

1 · 26

3 · 0

9 · 384

15 · Yes, since the determinant will be zero and the volume of the plane figure is zero.

17 · 9

19 · $\begin{bmatrix} \frac{4}{3} & \frac{1}{3} & -\frac{1}{3} & -\frac{2}{3} \\ -\frac{1}{3} & 0 & 0 & -\frac{1}{3} \\ -1 & 0 & 1 & 1 \\ -\frac{4}{3} & 0 & 0 & -\frac{1}{3} \end{bmatrix}$

Exercise 10 · 5

1 · $[T]_C = \begin{bmatrix} -1 & -9 \\ 2 & 8 \end{bmatrix}$

3 · $[T]_B = \begin{bmatrix} 11 & \frac{5}{2} \\ -30 & -7 \end{bmatrix}$

5 · $[T]_B = \begin{bmatrix} 1 & \frac{19}{6} & \frac{17}{6} \\ -3 & \frac{31}{6} & -\frac{7}{6} \\ -1 & -\frac{11}{6} & -\frac{13}{6} \end{bmatrix}$

11 · $[T]_B = \begin{bmatrix} 8 & 3 \\ 2 & 1 \\ -6 & -2 \end{bmatrix}$

Exercise 10 · 6

1 · $[B] = \begin{bmatrix} 12 & 8 \\ 9 & 5 \end{bmatrix}$

3 · $[B] = \begin{bmatrix} 10 & 7 \\ 7 & 4 \end{bmatrix}$

Exercise 10 · 7

1 · $3u^2 - 2v^2 + 4w^2 = 0$

3 · $9v^2 - 9w^2 = 6$

5 · $x^2 + 2y^2 + 2z^2 + 2xy + 2xz = 18$

Exercise 11 · 1

1 · no

3 · no

5 · no

7 · $2x + 3y - z = -6$

9 · $\sin \theta = \dfrac{\sqrt{6}}{6}$

11 · $3\sqrt{5}$

13 · $(2, 3, 6)$

Exercise 11 · 2

1 · $\dfrac{x-1}{0} = \dfrac{y+1}{1} = \dfrac{z-2}{1}$

3 · $\left(\dfrac{41}{14}, \dfrac{-6}{7}, \dfrac{-9}{14} \right)$

5 · $\text{Arccos } \dfrac{2}{\sqrt{42}}$

7 · $\left(\dfrac{17}{7}, \dfrac{1}{7}, \dfrac{16}{7} \right)$

9 · $y + 4z = 7$

11 · $(-1, -5, -3)$

13 · $2x - z = 1$

Exercise 11 · 3

1 · $x - y + 2z = 6$ 3 · hyperboloid of one sheet

5 · elliptic paraboloid   7 · ellipsoid

9 · hyperboloid of two sheets   11 · hyperbolic paraboloid

13 · elliptic cone   15 · elliptic cylinder

23 · (a) $x^2 + y^2 = 2y$ (b) $\rho \sin \phi = 2 \sin \theta$

25 · (a) $r^2 = z$ (b) $\rho \sin^2 \phi = \cos \phi$

27 · (a) $x^2 + y^2 + z^2 = z$ (b) $r^2 + z^2 = z$

Exercise 12 · 1

1 · $\begin{bmatrix} y & x+z & y \\ yz & xz & xy \\ 2x & 2y & 2z \end{bmatrix}$ 3 · $\begin{bmatrix} e^{yz} & xze^{yz} & xye^{yz} \\ 0 & (yz+1)e^{yz} & y^2e^{yz} \\ ze^x + e & xe^y & e^x \end{bmatrix}$

5 · $\begin{bmatrix} \sin yz & xz \cos yz & xy \cos yz \\ 0 & -\dfrac{y}{z} \sin \dfrac{y}{z} + \cos \dfrac{y}{z} & \dfrac{y^2}{z^2} \sin \dfrac{y}{z} \\ z \sec^2 x + y \sec x \tan x & \sec x & \tan x \end{bmatrix}$

7 · $F(1.1, 1.85, -1.15) = (-0.0925, -2.3402, 5.9550)$
$T(1.1, 1.85, -1.15) = (-0.1, -2.35, 5.9)$

9 · $z = T(x, y) = 2 - \frac{1}{2}(x - 1) + \frac{1}{2}(y + 1)$

11 · $\dfrac{\partial r}{\partial x} = (v + w)ye^{xy} + (u + w)\dfrac{1}{x + y} + (u + v)y \cos xy$

$\dfrac{\partial r}{\partial y} = (v + w)xe^{xy} + (u + w)\dfrac{1}{x + y} + (u + v)x \cos xy$

$\dfrac{\partial s}{\partial x} = 2uye^{xy} + 2v\dfrac{1}{x + y} + 2wy \cos xy$

$\dfrac{\partial s}{\partial y} = 2uxe^{xy} + 2u\dfrac{1}{x + y} + 2wx \cos xy$

13 · $(\pi^2/2)e^{\pi^2/4}$ 15 · $\dfrac{372}{7}$

17 · $\dfrac{5^7 \cdot 9 \cdot 43}{(29)^3 2^5}$ 19 · decreasing at 57π cu units/sec

Exercise 12 · 2

1. $\dfrac{-4(x+y)}{\sqrt{6z^3}}$

3. $\dfrac{xy-3}{\sqrt{10}}\,e^{yz}$

5. $\dfrac{2x+y}{\sqrt{5}}\cos xy$

7. $\dfrac{x^2-14xy-y^2}{5(x^2+y^2)^2}$

9. $\dfrac{-\sqrt{5}(x+y)}{(z+w)^2}$

11. $\dfrac{x-2}{3}=\dfrac{y-3}{2}=\dfrac{z-6}{-1}$

13. $(-1, 2)$ is neither a high nor a low point.

15. $(2, 1, 12)$

17. $600:200:7:1$

Exercise 12 · 3

1. $-6-8(x-3)-3(y+2)+2(x-3)(y+2)-2(x-3)^2+3(y+2)^2$

3. $-28-14(x-2)-6(y-5)+2(z-1)-3(x-2)(y-5)+(x-2)(z-1)$

5. $(x-\pi)+\pi(y-1)+(x-\pi)(y-1)-(x-\pi)(z+1)+\pi(y-1)(z+1)$

7. $\dfrac{3}{17}(x+4)+\dfrac{4}{17}(y-3)-\dfrac{8}{289}(x+4)(y-3)-\dfrac{24}{289}(x+4)^2$

9. 3.014

11. $(-3, 2)$; min

13. $(0, 0)$; neither max nor min

15. $(0, 0)$; neither max nor min

17. $(0, 0)$; neither max nor min $(4, 4)$; min

19. critical points at $(1, -\frac{1}{4}, -\frac{1}{4})$ and $(-1, \frac{1}{4}, -\frac{1}{4})$

21. max of $\frac{49}{4}$ at $(\frac{3}{2}, 3)$

23. max of 4 at $(0, 0)$

25. $(2^{-1/4}, \pm 2^{1/4}, 2^{-1/4})$

27. $y=\dfrac{24}{49}x+\dfrac{17}{49}$

29. $y=\dfrac{12}{35}x+\dfrac{180}{105}$

Exercise 12 · 4

1. $M=2/3\sqrt{3};\ m=-2/3\sqrt{3}$

3. F is constant on $x^2+y^2=1$

5. $M=e^{2/9};\ \mathrm{m}=e^{-2/9}$

7. $(1, 2, -1)$

9. $(\frac{5}{3}, -\frac{5}{3}, \frac{5}{3})$

11. cubical

13. cubical

Exercise 13 · 1

5. $e + 1$

7. $\sqrt{3} - \dfrac{3}{2}$

Exercise 13 · 2

1. $\displaystyle\int_0^2 \int_{x^2}^{2x} F(x, y)\, dy\, dx$

3. $\displaystyle\int_0^1 \int_{4-x}^{4-x^2} F(x, y)\, dy\, dx$

5. $\displaystyle\int_2^{2\sqrt{2}} \int_0^{\sqrt{8-x^2}} F(x\ y)\, dy\, dx + \int_0^2 \int_0^x F(x, y)\, dy\, dx$

7. $\left(\dfrac{e}{2} - 1\right)$

9. $\dfrac{1}{2} - \dfrac{1}{2}\cos 1$

11. $\dfrac{1}{8}$

13. $\dfrac{1}{24}$

15. $\dfrac{1}{10}$

17. 4

19. $\left(\dfrac{5}{6}, \dfrac{10}{21}, \dfrac{5}{21}\right)$

21. $\left(\dfrac{19}{30}, \dfrac{19}{30}, \dfrac{161}{90}\right)$

23. $\dfrac{25}{252}$

25. $\dfrac{208}{45}$

27. $\dfrac{128 - 24\pi + 16\sqrt{2}}{6}$

29. $20\pi\sqrt{15}$

31. $\dfrac{512}{15} - 2\pi + \dfrac{16}{5\sqrt{2}}$

33. $140\pi\sqrt{15}$

35. $\dfrac{2}{5} Ma^2$ where $M = \dfrac{4}{3}\pi a^3$

Exercise 13 · 3

1. $\displaystyle\int_0^1 \int_0^1 |3|\, dr\, ds = 3$

3. $\displaystyle\int_0^1 \int_0^1 |7|\,[(r - 2s)(2r + 3s) - (r - 2s)]\, dr\, ds = -\dfrac{91}{12}$

5. $\displaystyle\int_0^{\pi/4} \int_{\sqrt{2}\sec\theta}^2 r\, dr\, d\theta = \dfrac{\pi}{2} - 1$

7. $\displaystyle\int_0^{\pi/2} \int_0^\infty re^{-r^2}\, dr\, d\theta = \dfrac{\pi}{4}$

9. 27π

Exercise 14 · 1

1 · 5   **3 ·** 5   **5 ·** 1   **7 ·** diverges

9 · $\frac{1}{4}$ **11 ·** $\frac{3}{2}$ **13 ·** diverges   **15 ·** $\frac{1}{3}$

Exercise 14 · 2

1 · converges   **3 ·** diverges   **5 ·** converges   **7 ·** diverges

9 · diverges   **11 ·** diverges   **13 ·** converges   **15 ·** diverges

17 · diverges   **19 ·** diverges   **21 ·** converges   **23 ·** diverges

25 · converges   **27 ·** converges

Exercise 14 · 3

1 · absolutely convergent   **3 ·** conditionally convergent

5 · divergent   **7 ·** absolutely convergent

9 · absolutely convergent   **11 ·** conditionally convergent

13 · absolutely convergent   **15 ·** absolutely convergent

17 · divergent   **19 ·** 0.902

21 · 0.316

Exercise 14 · 4

1 · $(-3, -1)$ **3 ·** $[-\frac{4}{3} \quad -\frac{2}{3}]$

5 · $(-2, 2)$ **7 ·** all x

9 · $(2, 4]$ **11 ·** $(-1, 5]$

13 · $(-\infty, -5)$ and $(-1, \infty)$ **15 ·** $x = 0$

17 · $(-\infty, -2)$ and $(0, \infty)$ **19 ·** $\sin x = x - \dfrac{x^3}{3!} + \dfrac{x^5}{5!} - \dfrac{x^7}{7!} + \cdots$

21 · $e^x = 1 + x + \dfrac{x^2}{2!} + \dfrac{x^3}{3!} + \cdots$ **23 ·** $\ln(x + 1) = x - \dfrac{x^2}{2} + \dfrac{x^3}{3} - \dfrac{x^4}{4} + \cdots$

25 · $\cos x^2 = 1 - \dfrac{x^4}{2!} + \dfrac{x^8}{4!} - \dfrac{x^{12}}{6!} + \cdots$

27 · $\dfrac{1 - x}{1 + x^2} = 1 - x - x^2 + x^3 + x^4 - x^5 - x^6 + \cdots$

29 · $\dfrac{\text{Arctan } x}{x} = 1 - \dfrac{x^2}{3} + \dfrac{x^4}{5} - \dfrac{x^6}{7} + \cdots$

31 · $\tan x = x + \dfrac{x^3}{3} + \dfrac{2x^5}{15} + \dfrac{17x^7}{315} + \cdots$

33 · $\dfrac{x}{(1-x)^2}$

35 · $\dfrac{1}{(1-x)^3}$

37 · $\dfrac{xe^x - e^x + 1}{x^2}$

39 · $\dfrac{x \cos x - \sin x}{x^2}$

41 · $B(x) = 1 + rx + \dfrac{r(r-1)}{2} x^2 + \dfrac{r(r-1)(r-2)}{3 \cdot 2} x^3 + \cdots$

This is the binomial theorem!

43 · 0.479

45 · $0.37;\ e \approx 2.7$

47 · 0.54

49 · 0.497

Exercise 15 · 1

1 · $\dfrac{du}{dt} = k(u - 72)$

3 · $\dfrac{ds}{dt} = 18 - \dfrac{s}{60 + t}$

5 · $\dfrac{dp}{dt} = kp$

7 · $\dfrac{dy}{dx} = -\dfrac{x}{2y}$

9 · $\dfrac{du}{dt} = \begin{cases} -k(u - 74) & t \in [0,\ 4] \\ -k(u - 36) & t > 4 \end{cases}$

11 · $\dfrac{ds}{dt} = 4 - \dfrac{2s}{100}$

13 · $\dfrac{ds}{dt} = -6ks^2$

15 · $\dfrac{dy}{dx} = \dfrac{x}{y}$

17 · $\dfrac{dx}{dt} = -kx$

Exercise 15 · 3

1 · $y = C_1 e^{4x} + C_2 e^{-x}$

3 · $y = C_1 + C_2 e^{-x} + C_3 e^{3x}$

5 · $y = C_1 e^x + e^{-x/2} \left[C_2 \cos \dfrac{\sqrt{3}}{2} x + C_3 \sin \dfrac{\sqrt{3}}{2} x \right]$

7 · $y = C_1 + C_2 e^x + C_3 x e^x$

9 · $y = e^{-x/2} \left[C_1 \cos \dfrac{\sqrt{3}}{2} x + C_2 \sin \dfrac{\sqrt{3}}{2} x \right] + C_3 e^{-3x} + C_4 x e^{-3x}$

11 · $y = C_1 e^{4x} + C_2 e^{-x} - \frac{1}{6} e^x$

13 · $y = C_1 \cos x + C_2 \sin x + x^2 + x$

15 · $y = C_1 e^{2x} + C_2 e^{-2x} - \frac{1}{5} \sin x$

Exercise 15 · 4

1 · $y = (x + C)e^{-\sin x}$

3 · $y = \frac{1}{3}x - \frac{1}{2} + \dfrac{C}{x^2}$

5 · $y = C_1 e^{-x} + \frac{1}{2}\sin x - \frac{1}{2}\cos x$

Exercise 15 · 5

1 · $y^2 - x^2 = \ln|x + 1| + C$

3 · $x + \dfrac{y^2}{2} = y + \ln|x + 1| + C$

5 · $y = C\left(\dfrac{x}{e}\right)^x$

7 · $1 + x + \dfrac{3}{2}x^2 + \dfrac{4}{3}x^3 + \cdots$

9 · $y = \begin{cases} x^2\left(1 + x + \dfrac{x^2}{2} + \cdots\right) \\ x^2 e^x \end{cases}$

11 · $y = \begin{cases} \left(1 - x + \dfrac{x^2}{2} - \cdots\right) + \left(x - \dfrac{x^3}{3!} + \cdots\right) \\ e^{-x} + \sin x \end{cases}$

13 · $y = -\cos x^2$

15 · $y = x(1 - x^2)^{-1/2}$

17 · $y = \dfrac{1}{2} + \dfrac{3}{2}e^{-x^2}$

19 · $y \sin x^2 + \dfrac{1}{4}\cos 2x^2 = 1$

21 · $y = 2e^{-x}$

23 · $y = e^{-x^2}$

25 · $182°$

27 · 321 lb

29 · 4 million

31 · $2y^2 + x^2 = c$

33 · $47\frac{1}{8}$

35 · ≈ 34.7 min before noon

37 · 60 sec

39 · $x^2 - y^2 = 7$

41 · 12.5 grams

Exercise 15 · 6

1 · $F_3(x) = \dfrac{1}{24}x^4 - \dfrac{x^3}{3} + x^2 - x + 1$

$F_3(\frac{1}{2}) = 0.7129$
$F(x) = 2e^{-x} + x - 1$
$F(\frac{1}{2}) = 0.7131$

3 · $F_3(x) = \dfrac{1}{63}x^7 + \dfrac{2}{15}x^5 + \dfrac{1}{3}x^3 + x$

$F_3(\frac{1}{2}) = 0.5459$
$F(x) = \tan x$
$F(\frac{1}{2}) = 0.5452$

5 · $F_3(x) = 1 + x + x^2 + \dfrac{4}{3}x^3 + \dfrac{7}{6}x^4$

$F_3(\frac{1}{4}) = 1.3378$

INDEX

Numbers in boldface refer to Section numbers.